Edited by
Rainer Klages, Wolfram Just, and Christopher Jarzynski

Nonequilibrium Statistical Physics of Small Systems

Reviews of Nonlinear Dynamics and Complexity

Schuster, H. G. (ed.)

Reviews of Nonlinear Dynamics and Complexity
Volume 1

2008

ISBN: 978-3-527-40729-3

Schuster, H. G. (ed.)

Reviews of Nonlinear Dynamics and Complexity
Volume 2

2009

ISBN: 978-3-527-40850-4

Schuster, H. G. (ed.)

Reviews of Nonlinear Dynamics and Complexity
Volume 3

2010

ISBN: 978-3-527-40945-7

Grigoriev, R. and Schuster, H. G. (eds.)

Transport and Mixing in Laminar Flows
From Microfluidics to Oceanic Currents

2011

ISBN: 978-3-527-41011-8

Lüdge, K. (ed.)

Nonlinear Laser Dynamics
From Quantum Dots to Cryptography

2011

ISBN: 978-3-527-41100-9

Klages, R., Just, W., Jarzynski, C. (eds.)

Nonequilibrium Statistical Physics of Small Systems
Fluctuation Relations and Beyond

2013

ISBN: 978-3-527-41094-1

Pesenson, M. M. (ed.)

Multiscale Analysis and Nonlinear Dynamics
From Genes to the Brain

2013

ISBN: 978-3-527-41198-6

Niebur, E., Plenz, D., Schuster, H. G. (eds.)

Criticality in Neural Systems

2014

ISBN: 978-3-527-41104-7
ISBN: 978-0-471-66658-5

Edited by Rainer Klages, Wolfram Just, and Christopher Jarzynski

Nonequilibrium Statistical Physics of Small Systems

Fluctuation Relations and Beyond

WILEY-VCH

WILEY-VCH Verlag GmbH & Co. KGaA

The Editors

Dr. Rainer Klages
Queen Mary University of London
School of Mathematical Sciences
London, UK
r.klages@qmul.ac.uk

Dr. Wolfram Just
Queen Mary University London
School of Mathematical Sciences
London, UK

Prof. Christopher Jarzynski
University of Maryland
Dept. of Chemistry & Biochemistry
Institute for Physical Science and Technology
College Park, USA

The Series Editor

Prof. Dr. Heinz Georg Schuster
Saarbrücken, Germany

Cover picture

Cover, upper picture by Prof. Dr. Sergio Ciliberto et al., Universite de Lyon, ENSL and CNRS
Distributions of the classical work of a Brownian particle trapped in a nonlinear double-well potential produced by two laser beams, and Fluctuation Theorem formula.
Cover, lower picture by Prof. Dr. Felix Ritort et al., Universitat de Barcelona
Experimental setup for the mechanical folding/unfolding of a DNA hairpin, and Jarzynski work relation.

Library of Congress Card No.: applied for

British Library Cataloguing-in-Publication Data
A catalogue record for this book is available from the British Library.

Bibliographic information published by the Deutsche Nationalbibliothek
The Deutsche Nationalbibliothek lists this publication in the Deutsche Nationalbibliografie; detailed bibliographic data are available on the Internet at http://dnb.d-nb.de.

Print ISBN: 978-3-527-41094-1
ePDF ISBN: 978-3-527-65873-2
ePub ISBN: 978-3-527-65872-5
mobi ISBN: 978-3-527-65871-8
oBook ISBN: 978-3-527-65870-1

Cover Design Spieszdesign, Neu-Ulm

Typesetting Thomson Digital, Noida, India

Printing and Binding Markono Print Media Pte Ltd, Singapore

Printed in Singapore

Contents

Preface

The term *small systems* denotes objects composed of a limited, small number of particles, as is typical for matter on meso- and nanoscales. The interest of the scientific community in small systems has been boosted by the recent advent of micromanipulation techniques and nanotechnologies. These provide scientific instruments capable of measuring tiny energies in physical systems under *non-equilibrium* conditions, that is, when these systems are exposed to external forces generated by gradients or fields. Prominent examples of small systems exhibiting nonequilibrium dynamics are biopolymers stretched by optical tweezers (as shown in the lower picture on the book cover), colloidal particles dragged through a fluid by optical traps, and single molecules diffusing through meso- and nanopores.

Understanding the *statistical physics* of such systems is particularly challenging, because their small size does not allow one to apply standard methods of statistical mechanics and thermodynamics, which presuppose large numbers of particles. Small systems often display an intricate interplay between microscopic nonlinear dynamical properties and macroscopic statistical behavior leading to highly non-trivial fluctuations of physical observables (cf. the upper picture on the book cover). They can thus serve as a laboratory for understanding the emergence of complexity and irreversibility, in the sense that for a system consisting of many entities the dynamics of the whole is more than the sum of its single parts.

Studying the behavior of small systems on different spatiotemporal scales becomes particularly interesting in view of nonequilibrium transport phenomena such as diffusion, heat conduction, and electronic transport. Understanding these phenomena in small systems requires novel theoretical concepts that blend ideas and techniques from nonequilibrium statistical physics, thermodynamics, stochastic theory, and dynamical systems theory. More recently, it has become clear that a central role in this field is played by *fluctuation relations*, which generalize fundamental thermodynamic relations to small systems in nonequilibrium situations.

The aim of this book is to provide an introduction for both theorists and experimentalists to small systems physics, fluctuation relations, and the associated research topics listed in the word cloud diagram shown below. The book should also be useful for graduate-level students who want to explore this new field of research. The single chapters have been written by internationally recognized experts in small systems physics and provide in-depth introductions to the directions of their research. This approach of a multi-author reference book appeared to be particularly useful in view of the vast amount of literature available on different forms of fluctuation relations. While there exist excellent reviews highlighting single facets of fluctuation relations, we feel that the field lacks a reference that brings together the most important contributions to this topic in a comprehensive manner. This book is an attempt to fill the gap. In a way, it may act itself as a complex system, in the sense that the book as a whole ideally yields a new picture on small systems physics and fluctuation relations emerging from a synergy of the individual chapters. Along these lines, our intention was to embed research on fluctuation relations into a wider context of small systems research by pointing out cross-links to other theories and experiments. We thus hope that this book may serve as a catalyst both to fuse existing theories on fluctuation relations and to open up new directions of inquiry in the rapidly growing area of small systems research.

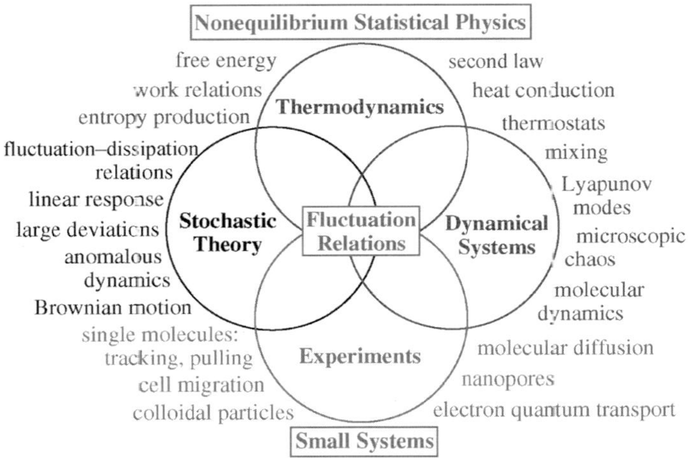

Accordingly, the book is organized into two parts. Part I introduces both the theoretical and experimental foundations of *fluctuation relations*. It starts with a threefold opening on basic theoretical ideas. The first chapter features a pedagogical introduction to fluctuation relations based on an approach that was coined "stochastic thermodynamics." The second chapter outlines a fully deterministic theory of fluctuation relations by working it out both analytically and numerically for a particle in an optical trap. The third chapter generalizes these deterministic ideas by also establishing cross-links to the Gallavotti–Cohen fluctuation theorem, which historically was the first to be established, with mathematical rigor, for nonequilibrium steady states. After this theoretical opening, the following two chapters summarize groundbreaking experimental work on two fundamental types of fluctuation relations. Along the lines of Gallavotti and Cohen, the first subset of them is often referred to as "fluctuation theorems" generalizing the second law of thermodynamics to small systems (see the first formula on the book cover). This type of fluctuation formulas is tested experimentally in systems where particles are confined by optical traps under nonequilibrium conditions. "Work relations," on the other hand, generalize an equilibrium relation between work and free energy to nonequilibrium (see the second formula on the book cover). The result is tested in experiments where single DNA and RNA chains are unzipped by optical tweezers. The remaining three chapters of Part I elaborate on aspects of fluctuation relations that moved into the focus of small systems research more recently. The first one introduces the nonequilibrium thermodynamics of information processing by using feedback control. The second one reviews quantum mechanical generalizations of fluctuation relations applied to electron transport in mesoscopic circuits. The third one discusses generalizations of fluctuation relations for stochastic anomalous dynamics with cross-links to experiments on biological cell migration.

Part II goes *beyond fluctuation relations* by reviewing topics that, while centered around nonequilibrium fluctuations in small systems, do not elaborate in particular on fluctuation relations. It starts with a discussion of fluctuation–dissipation relations, which are intimately related to, but may not be confused with, fluctuation relations. A cross-link to the foregoing chapter is provided in terms of partially studying anomalous dynamics, a topic that becomes particularly important for heat conduction in nanostructures, as is demonstrated from both an experimental and a theoretical point of view in the subsequent chapter. Fluctuation relations bear an important relation to large deviation theory, as is outlined in the next chapter, with applications to interacting particle systems. The book concludes with a summary about Lyapunov modes, which provide important information about the phase space dynamics in deterministically chaotic interacting many-particle systems, and experiments about diffusion in meso- and nanopores by performing single-molecule spectroscopy.

We finally remark that the various points of view expressed in the single chapters may not always be in full agreement with each other. This became clear in lively discussions between different groups of authors when the book was in preparation. As editors, we do not necessarily aim to achieve a complete consensus among all authors, as differences in opinions are typical for a very active field of research such as the one presented in this book.

We are most grateful to Heinz-Georg Schuster, the editor of the series *Reviews of Nonlinear Dynamics and Complexity*, in which this book is published as a Special Issue, for his invitation to edit this book, and for his help in getting the project started. We also thank Vera Palmer and Ulrike Werner from Wiley-VCH Publishers for their kind and efficient assistance in editing this book. C.J. gratefully acknowledges financial support from the National Science Foundation (USA) under grant DMR-0906601. W.J. is grateful for support from the British EPSRC by grant EP/H04812X/1. We finally thank all book chapter authors for sharing their expertise in this multi-author monograph. Their strong efforts and enthusiasm for this project were indispensable for bringing it to success.

Summer 2012

London — *Rainer Klages*
London — *Wolfram Just*
College Park, MD — *Christopher Jarzynski*

List of Contributors

Anna Alemany
Universitat de Barcelona
Departament de Física Fonamental
Small Biosystems Lab
Avda. Diagonal 647
08028 Barcelona
Spain

and

Instituto de Salud Carlos III
CIBER-BBN de Bioingeniería
Biomateriales y Nanomedicina
C/ Sinesio Delgado 4
28029 Madrid
Spain

L. Bellon
Université de Lyon
Ecole Normale Supérieure de Lyon
Laboratoire de Physique (CNRS
UMR 5672)
46 Allée d'Italie
69364 Lyon Cedex 07
France

Christoph Bräuchle
Ludwig-Maximilians-Universität
München
Department Chemie
Lehrstuhl für Physikalische
Chemie I
Butenandtstr. 11
81377 Munich
Germany

Aleksei V. Chechkin
National Science Center "Kharkov
Institute of Physics and Technology"
(NSC KIPT)
Institute for Theoretical Physics
Akademicheskaya Street 1
Kharkov 61108
Ukraine

Sergio Ciliberto
Université de Lyon
Ecole Normale Supérieure de Lyon
Laboratoire de Physique (CNRS
UMR 5672)
46 Allée d'Italie
69364 Lyon Cedex 07
France

Peter Dieterich
Technische Universität Dresden
Medizinische Fakultaet "Carl Gustav Carus"
Institut für Physiologie
Fetscherstrasse 74
01307 Dresden
Germany

Denis J. Evans
Australian National University
Research School of Chemistry
Building 35 Science Rd
Canberra, ACT 0200
Australia

Ian Ford
University College London
Department of Physics and Astronomy and London Centre for Nanotechnology
Gower Street
London WC1E 6BT
UK

Pierre Gaspard
Université Libre de Bruxelles
Center for Nonlinear Phenomena and Complex Systems and Department of Physics
Campus Plaine
Code Postal 231
Boulevard du Triomphe
1050 Brussels
Belgium

J.R. Gomez-Solano
Université de Lyon
Ecole Normale Supérieure de Lyon
Laboratoire de Physique (CNRS UMR 5672)
46 Allée d'Italie
69364 Lyon Cedex 07
France

G. Gradenigo
Universita degli Studi di Roma "La Sapienza"
Dipartimento di Fisica
Piazzale A. Moro 2
00185 Rome
Italy

Rosemary J. Harris
Queen Mary University of London
School of Mathematical Sciences
Mile End Road
London E1 4NS
UK

O.G. Jepps
Griffith University
School of Biomolecular and Physical Sciences
Queensland Micro- and Nanotechnology Centre
170 Kessels Road
Brisbane, Qld 4111
Australia

Rainer Klages
Queen Mary University of London
School of Mathematical Sciences
Mile End Road
London E1 4NS
UK

Baowen Li
National University of Singapore
Department of Physics and Centre for Computational Science and Engineering
Science Drive 2
Singapore 117542
Singapore

and

NUS Graduate School for Integrative Sciences and Engineering
28 Medical Drive
Singapore 117456
Singapore

Sha Liu
National University of Singapore
Department of Physics and Centre
for Computational Science and
Engineering
Science Drive 2
Singapore 117542
Singapore

and

NUS Graduate School for Integrative
Sciences and Engineering
28 Medical Drive
Singapore 117456
Singapore

Stephan Mackowiak
Ludwig-Maximilians-Universität
München
Department Chemie
Lehrstuhl für Physikalische
Chemie I
Butenandtstr. 11
81377 Munich
Germany

A. Petrosyan
Université de Lyon
Ecole Normale Supérieure de Lyon
Laboratoire de Physique (CNRS
UMR 5672)
46 Allée d'Italie
69364 Lyon Cedex 07
France

A. Puglisi
Universita degli Studi di Roma
"La Sapienza"
Dipartimento di Fisica
Piazzale A. Moro 2
00185 Rome
Italy

Günter Radons
Chemnitz University of Technology
Institute of Mechatronics and
Institute of Physics
09107 Chemnitz
Germany

James C. Reid
The University of Queensland
Australian Institute for
Bioengineering and Nanotechnology
AIBN Building (75)
Corner Cooper & College Roads
Brisbane, Qld 4072
Australia

Marco Ribezzi-Crivellari
Universitat de Barcelona
Departament de Fisica Fonamental
Small Biosystems Lab
Barcelona
Spain

and

Instituto de Salud Carlos III
CIBER-BBN de Bioingenieria
Biomateriales y Nanomedicina
Madrid
Spain

Felix Ritort
Universitat de Barcelona
Departament de Fisica Fonamental
Small Biosystems Lab
Barcelona
Spain

and

Instituto de Salud Carlos III
CIBER-BBN de Bioingenieria
Biomateriales y Nanomedicina
Madrid
Spain

Lamberto Rondoni
Dipartimento di Scienze
Matematiche
Politecnico di Torino
Corso Duca degli Abruzzi 24
10129 Torino
INFN, Sezione di Torino
Via P. Giuria 1
10125 Torino, Italy

Takahiro Sagawa
Kyoto University
The Hakubi Center for Advanced
Research
iCeMS Complex 1 West Wing
Yoshida-Ushinomiya-cho, Sakyo-ku
Kyoto 606-8302
Japan

and

Kyoto University
Yukawa Institute of Theoretical
Physics
Kitashirakawa Oiwake-Cho, Sakyo-ku
Kyoto 606-8502
Japan

A. Sarracino
Universita degli Studi di Roma
"La Sapienza"
Dipartimento di Fisica
Piazzale A. Moro 2
00185 Rome
Italy

Debra J. Searles
The University of Queensland
Australian Institute for
Bioengineering and Nanotechnology
and School of Chemistry and
Molecular Biosciences
AIBN Building (75)
Corner Cooper & College Roads
Brisbane, Qld 4072
Australia

Richard Spinney
University College London
Department of Physics and
Astronomy and London Centre
for Nanotechnology
Gower Street
London WC1E 6BT
UK

Hugo Touchette
Queen Mary University of London
School of Mathematical Sciences
Mile End Road
London E1 4NS
UK

Masahito Ueda
The University of Tokyo
Department of Physics
7-3-1 Hongo, Bunkyo-ku
Tokyo 113-0033
Japan

D. Villamaina
Universita degli Studi di Roma
"La Sapienza"
Dipartimento di Fisica
Piazzale A. Moro 2
00185 Rome
Italy

A. Vulpiani
Universita degli Studi di Roma
"La Sapienza"
Dipartimento di Fisica
Piazzale A. Moro 2
00185 Rome
Italy

Stephen R. Williams
Australian National University
Research School of Chemistry
Building 35 Science Rd
Canberra, ACT 0200
Australia

Hong-Liu Yang
Chemnitz University of Technology
Institute of Mechatronics and
Institute of Physics
09107 Chemnitz
Germany

Gang Zhang
Peking University
Key Laboratory for the Physics and
Chemistry of Nanodevices and
Department of Electronics
Yiheyuan Road 5
Beijing 100871
China

and

National University of Singapore
Department of Physics and Centre
for Computational Science and
Engineering
Science Drive 2
Singapore 117542
Singapore

Color Plates

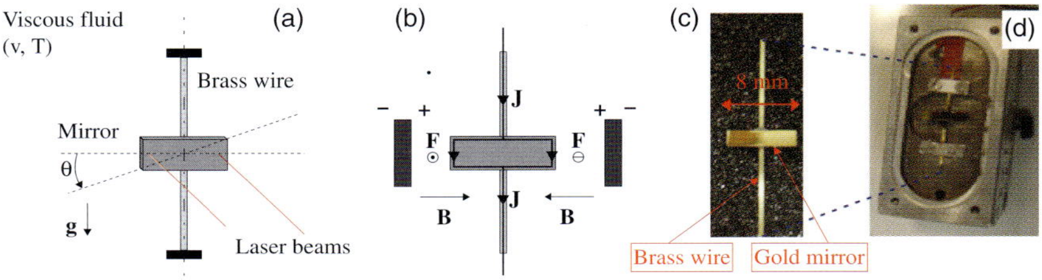

Figure 4.1 (a) The torsion pendulum. (b) The magnetostatic forcing. (c) Picture of the pendulum. (d) Cell where the pendulum is installed.

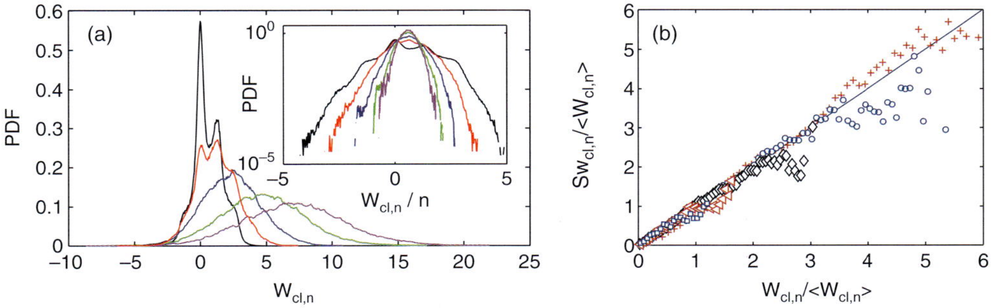

Figure 4.12 (a) Distribution of classical work W_{cl} for different numbers of period $n = 1, 2, 4, 8$, and 12 ($f = 0.25$ Hz). *Inset*: Same data in lin-log. (b) Normalized symmetry function as a function of the normalized work for $n = 1$ (+), 2 (○), 4 (◇), 8 (Δ), and 12 (□).

Nonequilibrium Statistical Physics of Small Systems: Fluctuation Relations and Beyond, First Edition.
Edited by Rainer Klages, Wolfram Just, and Christopher Jarzynski.

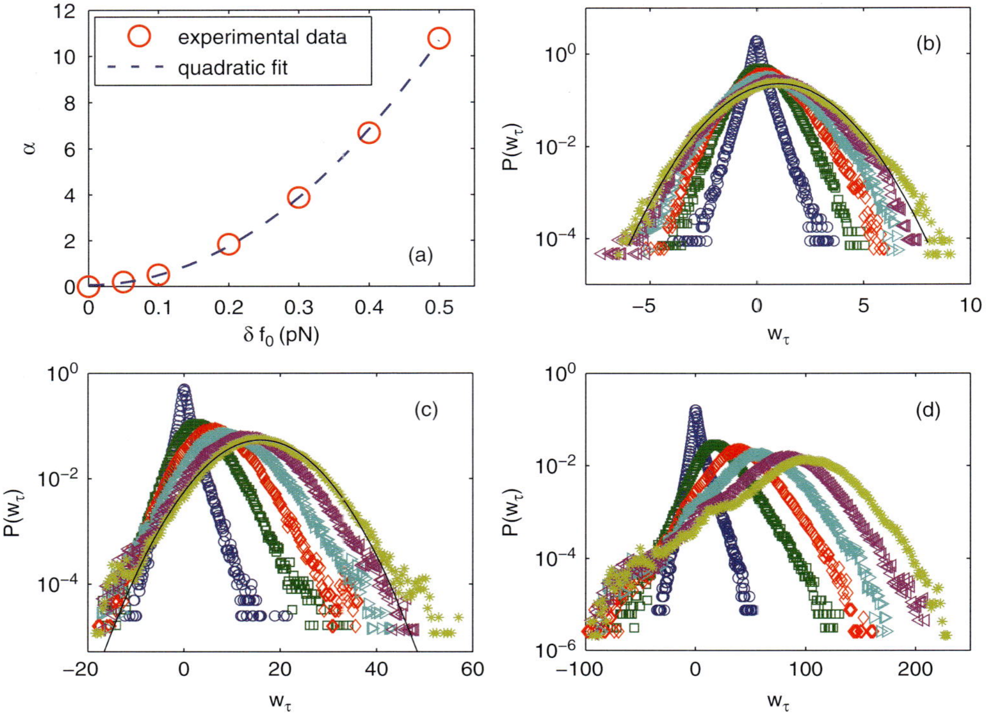

Figure 4.14 (a) Dependence of the parameter α on the standard deviation of the Gaussian exponentially correlated external force f_0 acting on the colloidal particle. Probability density functions of the work w_τ for (b) $\alpha = 0.20$, (c) $\alpha = 3.89$, and (d) $\alpha = 10.77$. The symbols correspond to integration times $\tau = 5$ ms (○), 55 ms (□), 105 ms (◇), 155 ms (◁), 205 ms (▷), and 255 ms (∗). The solid black lines in (b) and (c) are Gaussian fits.

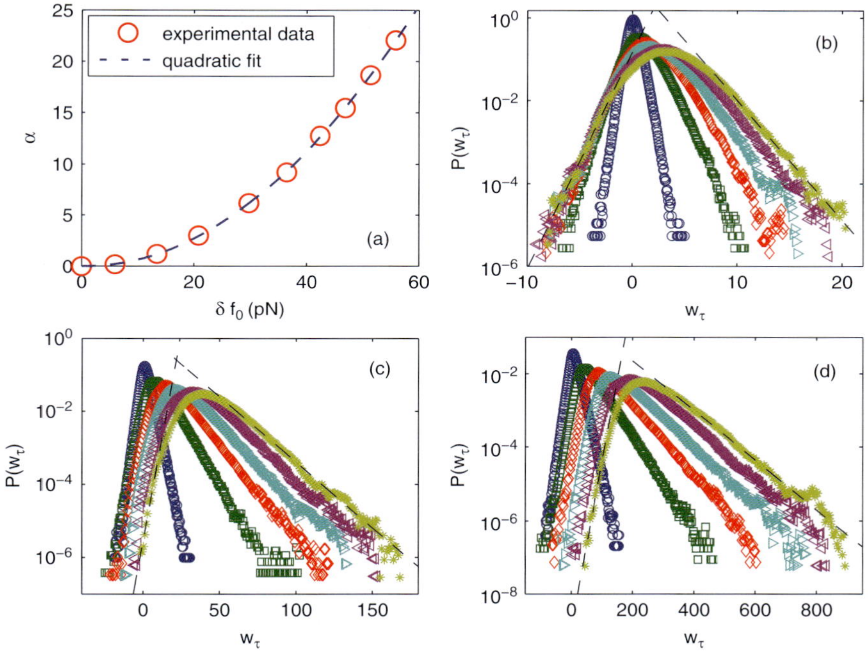

Figure 4.16 (a) Dependence of the parameter α on the standard deviation of the Gaussian white external force f_0 acting on the cantilever. Probability density functions of the work w_τ for (b) $\alpha = 0.19$, (c) $\alpha = 3.03$, and (d) $\alpha = 18.66$. The symbols correspond to integration times $\tau = 97\ \mu s$ ($\circ$), $1.074\ \mu s$ ($\square$), 2.051 ms ($\diamond$), 3.027 ms ($\triangleleft$), 4.004 ms ($\triangleright$), and 4.981 ms ($*$). The black dashed lines in (b)–(d) represent the exponential fits of the corresponding tails.

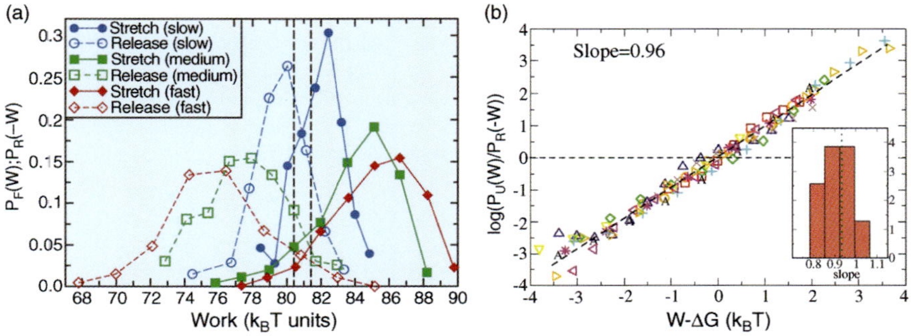

Figure 5.3 The Crooks fluctuation relation. (a) Work distributions for the hairpin shown in Figure 5.1 measured at three different pulling speeds: 50 nm s^{-1} (blue), 100 nm s^{-1} (green), and 300 nm s^{-1} (red). Unfolding or forward (continuous lines) and refolding or reverse work distributions (dashed lines) cross each other at a value of $81.0 \pm 0.2\ k_BT$ independent of the pulling speed. (b) Experimental test of the CFR for 10 different molecules pulled at different speeds. The log of the ratio between the unfolding and refolding work distributions is equal to $(W - \Delta G)$ in k_BT units. The inset shows the distribution of slopes for the different molecules that are clustered around an average value of 0.96. (Figure taken from Ref. [19].)

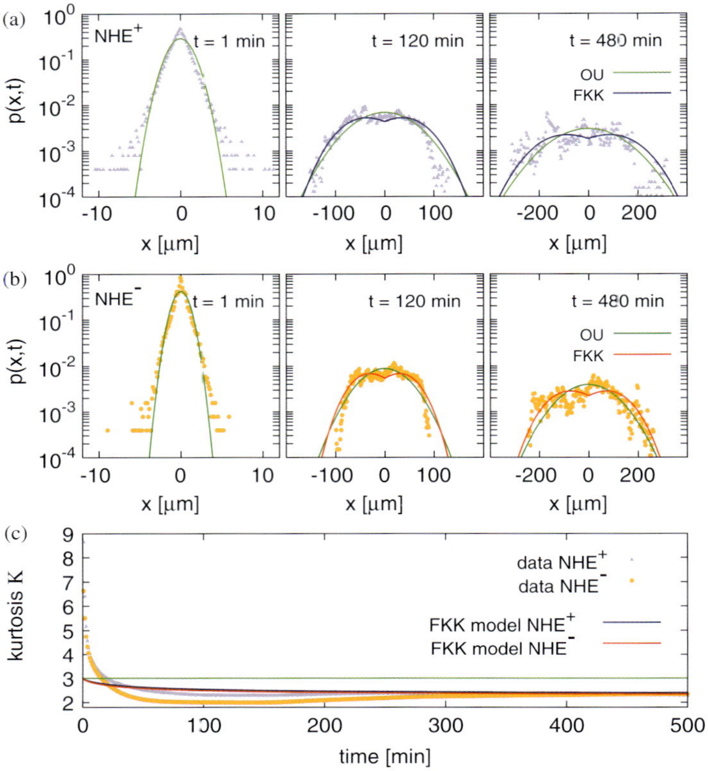

Figure 8.8 Spatiotemporal probability distributions $P(x, t)$. (a and b) Experimental data for both cell types at different times in semilogarithmic representation. The dark lines, labeled FKK, show the long-time asymptotic solutions of our model (Eq. (8.31)) with the same parameter set used for the MSD fit. The light lines, labeled OU, depict fits by the Gaussian distributions (Eq. (8.11)) representing Brownian motion. For $t = 1$ min, both $P(x, t)$ show a peaked structure clearly deviating from a Gaussian form. (c) The kurtosis $\kappa(t)$ of $P(x, t)$ (cf. Eq. (8.30)) plotted as a function of time saturates at a value different from that of Brownian motion (line at $\kappa = 3$). The other two lines represent $\kappa(t)$ obtained from the model (Eq. (8.31)) [43].

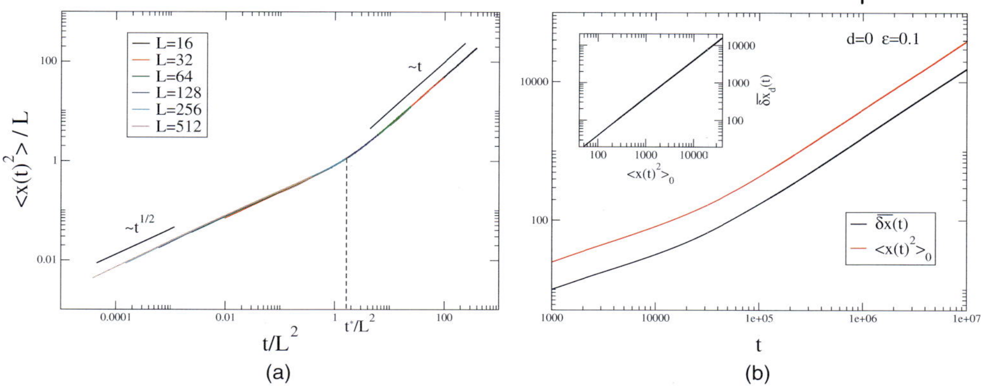

Figure 9.1 (a) $\langle x^2(t)\rangle_0/L$ versus t/L^2 plotted for several values of L in the comb model. (b) $\langle x^2(t)\rangle_0$ and the response function $\overline{\delta x}(t)$ for $L = 512$. *Inset*: The parametric plot of $\overline{\delta x}(t)$ versus $\langle x^2(t)\rangle_0$.

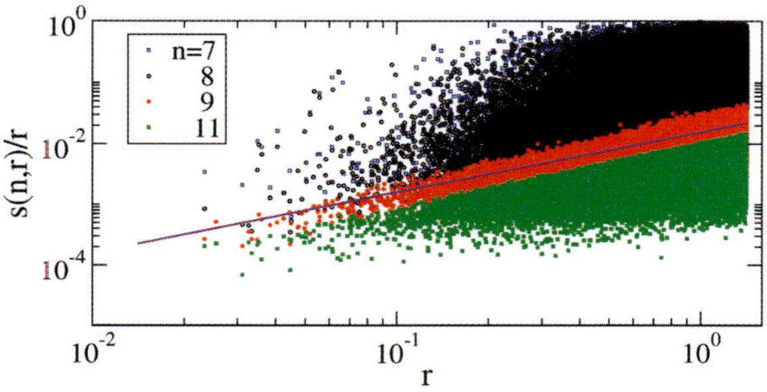

Figure 12.18 Variation of the normalized projection error $s(n, r)/r$ with the distance to the reference state r with $n = 7$, 8, 9, and 11, respectively. To improve the statistics, data from 200 reference states are presented together. A line with slope 1 is shown to guide the eyes.

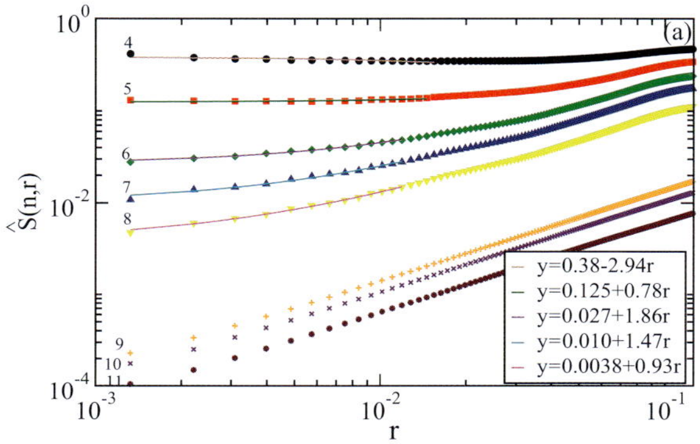

Figure 12.19 The normalized average projection error $S(n, r)/r$ versus the distance to the reference state r with n from 4 to 11. Linear fittings of data for $n < M$ confirms the saturation of $S(n, r)/r$ to a nonzero constant value.

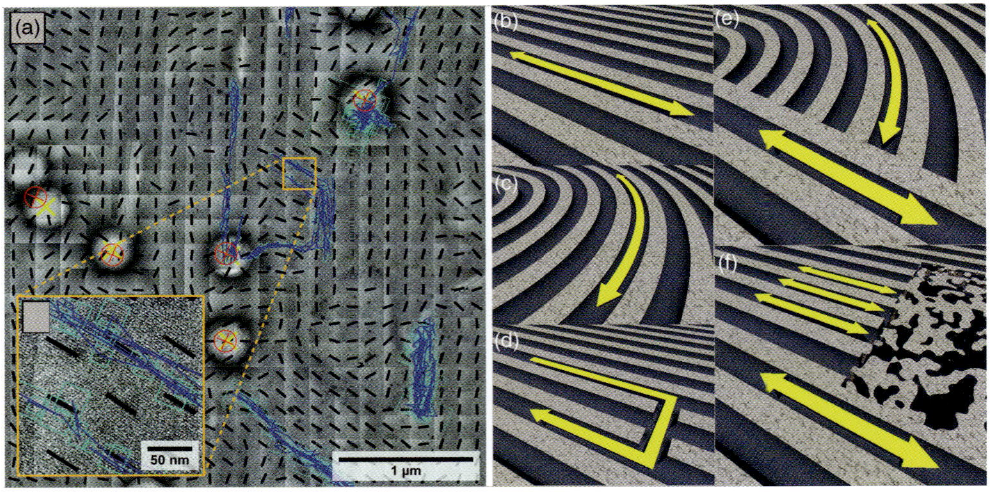

Figure 13.3 (a) Overlay of TEM image (gray) and FFT directors (black bars) with single-molecule trajectories (dark blue). The polystyrene beads, used for the overlay, are indicated by the yellow (TEM) and red (SMM) crosses. The light blue boxes show the positioning error of the SM trajectories. (b–f) Possible movement patterns of a single molecule in various structural features found in the hexagonal mesoporous films.

Part I
Fluctuation Relations

The contributions to Part I of this book are organized into three clusters of chapters, beginning with *theoretical foundations*. Spinney and Ford's opening chapter provides a pedagogical overview of fluctuation relations, summarizing key results in the field and emphasizing their connection with the second law of thermodynamics. These results are derived within the framework of continuous Markovian stochastic dynamics, in which the effects of thermal surroundings on the evolution of a system of interest are modeled by random noise. The next chapter, by Reid, Williams, Searles, Rondoni, and Evans, takes a complementary approach, using fully deterministic equations of motion to model the system's evolution. The authors show that a number of related results follow directly from the consideration of an appropriately defined *dissipation function*. These results are then illustrated using a conceptually simple and experimentally accessible system: a micron-size bead trapped with laser tweezers. Finally, Rondoni and Jepps' chapter makes a distinction between physically motivated fluctuation relations, such as those considered in the first two contributions, and the study of similar results within the framework of dynamical systems theory, where the emphasis is on generality and mathematical rigor rather than specific physical realizations. By developing a rigorous formalism for physically motivated fluctuation relations, Rondoni and Jepps explore this distinction in detail, and illuminate a number of issues such as the relationship between transient and steady-state fluctuation relations.

The second cluster of chapters within Part I considers *experimental foundations*. The contribution by Bellon, Gomez-Solano, Petrosyan, and Ciliberto reviews experiments in which the fluctuations of a system away from thermal equilibrium are measured and compared with theory. The experiments include a torsional pendulum, a polystyrene particle trapped optically in a double well, and a cantilever used for atomic force microscopy (AFM), and together they provide a set of experimental platforms for testing a variety of fluctuation relations. Next, Alemany, Ribezzi-Crivellari, and Ritort provide an introduction and up-to-date review of an

Nonequilibrium Statistical Physics of Small Systems: Fluctuation Relations and Beyond, First Edition.
Edited by Rainer Klages, Wolfram Just, and Christopher Jarzynski.

important application of fluctuation relations, namely, the recovery of equilibrium free energy differences from out-of-equilibrium single-molecule pulling experiments. Among other issues, this contribution discusses and illustrates the importance of using the appropriate microscopic definition of work.

Further developments are discussed in the third and final set of chapters of Part I. Sagawa and Ueda review the thermodynamics of feedback control, in which an external observer uses information about the fluctuations of a small system to guide its subsequent evolution, as in Maxwell's famous thought experiment. They discuss how fluctuation relations and the second law of thermodynamics itself are modified in this setting. Gaspard then gives an overview of the time-symmetry relations that are obeyed by out-of-equilibrium systems. His contribution focuses on both quantum and stochastic dynamics, and discusses applications to electron transport in mesoscopic circuits. In the closing chapter of Part I, Klages, Chechkin, and Dieterich investigate and extend fluctuation relations in the context of anomalous dynamics, as modeled by Lévy flights, long-time correlated Gaussian stochastic processes, and time-fractional kinetics. Such anomalous dynamics arise in physically relevant situations such as cell migration.

Theoretical Foundations

1
Fluctuation Relations: A Pedagogical Overview

Richard Spinney and Ian Ford

1.1
Preliminaries

Ours is a harsh and unforgiving universe, and not just in the little matters that conspire against us. Its complicated rules of evolution seem unfairly biased against those who seek to predict the future. Of course, if the rules were simple, then there might be no universe of any complexity worth considering. Perhaps richness of behavior emerges only because each component of the universe interacts with many others and in ways that are very sensitive to details: this is the harsh and unforgiving nature. In order to predict the future, we have to take into account all the connections between the components, since they might be crucial to the evolution; furthermore, we need to know everything about the present in order to predict the future: both of these requirements are in most cases impossible. Estimates and guesses are not enough: unforgiving sensitivity to the details very soon leads to loss of predictability. We see this in the workings of a weather system. The approximations that meteorological services make in order to fill gaps in understanding, or initial data, eventually make the forecasts inaccurate.

So a description of the dynamics of a complex system is likely to be incomplete and we have to accept that predictions will be uncertain. If we are careful in the modeling of the system, the uncertainty will grow only slowly. If we are sloppy in our model building or initial data collection, it will grow quickly. We may expect the predictions of any incomplete model to tend toward a state of general ignorance, whereby we cannot be sure about anything: rain, snow, heat wave, or hurricane. We must expect there to be a spread, or fluctuations, in the outcomes of such a model.

This discussion of the growth of uncertainty in predictions has a bearing on another matter: the apparent *irreversibility* of all but the most simple physical processes. This refers to our inability to drive a system exactly backward by reversing the external forces that guide its evolution. Consider the mechanical work required to compress a gas by a piston in a cylinder. We might hope to see the expended energy returned when we stop pushing and allow the gas to drive the piston all the way back to the starting point: but not all will be returned. The system

Nonequilibrium Statistical Physics of Small Systems: Fluctuation Relations and Beyond, First Edition.
Edited by Rainer Klages, Wolfram Just, and Christopher Jarzynski.

seems to mislay some energy to the benefit of the wider environment. This is the familiar process of friction. The one-way dissipation of energy during mechanical processing is an example of the famous second law of thermodynamics. But the process is actually rather mysterious: What about the underlying reversibility of Newton's equations of motion? Why is the leakage of energy one way?

We may suspect that a failure to engineer the exact reversal of a compression is simply a consequence of a lack of control over all components of the gas and its environment: the difficulty in setting things up properly for the return leg implies the virtual impossibility of retracing the behavior. So we might not expect to be able to retrace exactly. But why do we not sometimes see "antifriction?" A clue might be seen in the relative size and complexity of the system and its environment. The smaller system is likely to evolve in a more complicated fashion as a result of the coupling, while we may expect the larger environment to be much less affected. There is a disparity in the effect of the coupling on each participant, and it is believed that this is responsible for the apparent one-way nature of friction. It is possible to implement these ideas by modeling the behavior of a system using uncertain or stochastic dynamics. The probability of observing a reversal of the behavior on the return leg can be calculated explicitly and it turns out that the difference between probabilities of observing a particular compression and seeing its reverse on the return leg leads to a measure of the irreversibility of natural processes. The second law is then a rather simple consequence of the dynamics. A similar asymmetric treatment of the effect on a system of coupling to a large environment is possible using deterministic and reversible nonlinear dynamics. In both cases, Loschmidt's paradox, the apparent breakage of time reversal symmetry for thermally constrained systems, is evaded, although for different reasons.

This chapter describes the so-called *fluctuation relations*, or *theorems* [1–5], that emerge from the analysis of a physical system interacting with its environment and that provide the structure that leads to the conclusion just outlined. They can quantify unexpected outcomes in terms of the expected. They apply on microscopic as well as macroscopic scales, and indeed their consequences are most apparent when applied to small systems. They can be derived on the basis of a rather natural measure of irreversibility, just alluded to, that offers an interpretation of the second law and the associated concept of entropy production. The dynamical rules that control the universe might seem harsh and unforgiving, but they can also be charitable and from them have emerged fluctuation relations that seem to provide a better understanding of entropy, uncertainty, and the limits of predictability.

This chapter is structured as follows. In order to provide a context for the fluctuation relations suitable for newcomers to the field, we begin with a brief summary of thermodynamic irreversibility and then describe how stochastic dynamics might be modeled. We use a framework based on stochastic rather than deterministic dynamics, since developing both themes here might not provide the most succinct pedagogical introduction. Nevertheless, we refer to the deterministic framework briefly later on to emphasize its equivalence. We discuss the identification of entropy production with the degree of departure from dynamical reversibility and then take a careful look at the developments that follow, which include the various fluctuation relations, and consider how the second law might not operate as we expect. We illustrate the fluctuation relations using simple analytical models as an aid to understanding. We conclude with some final remarks, but the broader implications are to be found elsewhere in this book, for which we hope this chapter will serve as a helpful background.

1.2
Entropy and the Second Law

Ignorance and uncertainty has never been an unusual state of affairs in human perception. In mechanics, Newton's laws of motion provided tools that seemed to dispel some of the haze: here were mathematical models that enabled the future to be foretold! They inspired attempts to predict future behavior in other fields, particularly in thermodynamics, the study of systems through which matter and energy can flow. The particular focus in the early days of the field was the heat engine, a device whereby fuel and the heat it can generate can be converted into mechanical work. Its operation was discovered to produce a quantity called entropy that could characterize the efficiency with which energy in the fuel could be converted into motion. Indeed, entropy seemed to be generated whenever heat or matter flowed. The second law of thermodynamics famously states that the total entropy of the evolving universe is always increasing. But this statement still attracts discussion, more than 150 years after its introduction. We do not debate the meaning of Newton's second law anymore, so why is the second law of thermodynamics so controversial?

Well, it is hard to understand how there can be a physical quantity that never decreases. Such a statement demands the breakage of the principle of time reversal symmetry, a difficulty referred to as Loschmidt's paradox. Newton's equations of motion do not specify a preferred direction in which time evolves. Time is a coordinate in a description of the universe and it is just a convention that real-world events take place while this coordinate increases. Given that we cannot actually run time backward, we can demonstrate this symmetry in the following way. A sequence of events that take place according to time reversal symmetric equations can be inverted by instantaneously reversing all the velocities of all the participating components and then proceeding forward in time once again, suitably reversing any external protocol of driving forces, if necessary. The point is that any evolution can be imagined in reverse, according to Newton. We therefore do not expect to observe any quantity ever-increasing with time. This is the essence of Loschmidt's objection to Boltzmann's [6] mechanical interpretation of the second law.

Nobody, however, has been able to initiate a heat engine such that it sucks exhaust gases back into its furnace and combines them into fuel. The denial of such a spectacle is empirical evidence for the operation of the second law, but it is also an expression of Loschmidt's paradox. Time reversal symmetry is broken by the apparent illegality of entropy-consuming processes and that seems unacceptable. Perhaps we should not blindly accept the second law in the sense that has traditionally been ascribed to it. Or perhaps there is something deeper going on. Furthermore, a law that only specifies the sign of a rate of change sounds rather incomplete.

But what has emerged in the past two decades or so is the realization that Newton's laws of motion, when supplemented by the acceptance of uncertainty in the way systems behave, brought about by roughly specified interactions with the environment, can lead quite naturally to a quantity that grows with time, that is, uncertainty itself. It is reasonable to presume that incomplete models of the evolution of a physical system will generate additional uncertainty in the reliability of the description of the system as they are evolved. If the velocities were all instantaneously reversed, in the hope that a previous sequence of events might be reversed, uncertainty would continue to grow within such a model. We shall, of course, need to quantify this vague notion of uncertainty. Newton's laws on their own are time reversal symmetric, but intuition suggests that the injection and evolution of configurational uncertainty would break the symmetry. Entropy production might therefore be equivalent to the leakage of our confidence in the predictions of an incomplete model: an interpretation that ties in with prevalent ideas of entropy as a measure of information.

Before we proceed further, we need to remind ourselves about the phenomenology of irreversible classical thermodynamic processes [7]. A system possesses energy E and can receive additional incremental contributions in the form of heat $\mathrm{d}Q$ from a heat bath at temperature T and work $\mathrm{d}W$ from an external mechanical device that might drag, squeeze, or stretch the system. It helps perhaps to view $\mathrm{d}Q$ and $\mathrm{d}W$ roughly as increments in kinetic and in potential energy, respectively. We write the first law of thermodynamics (energy conservation) in the form $\mathrm{d}E = \mathrm{d}Q + \mathrm{d}W$. The second law is then traditionally given as Clausius' inequality:

$$\oint \frac{\mathrm{d}Q}{T} \leq 0, \tag{1.1}$$

where the integration symbol means that the system is taken around a cycle of heat and work transfers, starting and ending in thermal equilibrium with the same macroscopic system parameters, such as temperature and volume. The temperature of the heat bath might change with time, though by definition and in recognition of its presumed large size it always remains in thermal equilibrium, and the volume and shape imposed upon the system during the process might also be time dependent. We can also write the second law for an incremental thermodynamic process as

$$\mathrm{d}S_{\mathrm{tot}} = \mathrm{d}S + \mathrm{d}S_{\mathrm{med}}, \tag{1.2}$$

where each term is an incremental entropy change, the system again starting and ending in equilibrium. The change in system entropy is denoted $\mathrm{d}S$ and the change in entropy of the heat bath, or the surrounding medium, is defined as

$$\mathrm{d}S_{\mathrm{med}} = -\frac{\mathrm{d}Q}{T}, \tag{1.3}$$

such that $\mathrm{d}S_{\mathrm{tot}}$ is the total entropy change of the two combined (the "universe"). We see that Eq. (1.1) corresponds to the condition $\oint \mathrm{d}S_{\mathrm{tot}} \geq 0$, since $\oint \mathrm{d}S = 0$. A more powerful reading of the second law is that

$$\mathrm{d}S_{\mathrm{tot}} \geq 0, \tag{1.4}$$

for any incremental segment of a thermodynamic process, as long as it starts and ends in equilibrium. An equivalent expression of the law would be to combine these statements to write $dW - dE + TdS \geq 0$, from which we conclude that the *dissipative* work (sometimes called irreversible work) in an isothermal process,

$$dW_d = dW - dF, \tag{1.5}$$

is always positive, where dF is a change in Helmholtz free energy. We may also write $dS = dS_{tot} - dS_{med}$ and regard dS_{tot} as a contribution to the change in entropy of a system that is not associated with a flow of entropy from the heat bath, the dQ/T term. For a thermally isolated system, where $dQ = 0$, we have $dS = dS_{tot}$ and the second law then says that the system entropy increase is due to "internal" generation; hence, dS_{tot} is sometimes [7] denoted dS_i.

Boltzmann tried to explain what this ever-increasing quantity might represent at a microscopic level [6]. He considered a thermally isolated gas of particles interacting through pairwise collisions within a framework of classical mechanics. The quantity

$$H(t) = \int f(\mathbf{v}, t) \ln f(\mathbf{v}, t) d\mathbf{v}, \tag{1.6}$$

where $f(\mathbf{v}, t)d\mathbf{v}$ is the population of particles with a velocity in the range of $d\mathbf{v}$ about $\mathbf{v}$, can be shown to decrease with time, or remain constant if the population is in a Maxwell–Boltzmann distribution characteristic of thermal equilibrium. Boltzmann obtained this result by assuming that the collision rate between particles at velocities $\mathbf{v}_1$ and $\mathbf{v}_2$ is proportional to the product of populations at those velocities, that is, $f(\mathbf{v}_1, t)f(\mathbf{v}_2, t)$. He proposed that H was proportional to the negative of system entropy and that his so-called H-theorem provides a sound microscopic and mechanical justification for the second law. Unfortunately, this does not hold up. As Loschmidt pointed out, Newton's laws of motion cannot lead to a quantity that always decreases with time: $dH/dt \leq 0$ would be incompatible with the principle of time reversal symmetry that underlies the dynamics. The H-theorem does have a meaning, but it is statistical: the decrease in H is an expected, but not guaranteed, result. Alternatively, it is a correct result for a dynamical system that does not adhere to time reversal symmetric equations of motion. The neglect of correlation between the velocities of colliding particles, both in the past and in the future, is where the model departs from Newtonian dynamics.

The same difficulty emerges in another form when, following Gibbs, it is proposed that the entropy of a system might be viewed as a property of an ensemble of many systems, each sampled from a probability density $P(\{\mathbf{x}, \mathbf{v}\})$, where $\{\mathbf{x}, \mathbf{v}\}$ denotes the positions and velocities of all the particles in a system. Gibbs wrote

$$S_{Gibbs} = -k_B \int P(\{\mathbf{x}, \mathbf{v}\}) \ln P(\{\mathbf{x}, \mathbf{v}\}) \prod d\mathbf{x}d\mathbf{v}, \tag{1.7}$$

where k_B is Boltzmann's constant and the integration is over all phase space. The Gibbs representation of entropy is compatible with classical equilibrium thermodynamics. But the probability density P for an isolated system should evolve in time according to Liouville's theorem, in such a way that S_{Gibbs} is a *constant* of the motion. How, then, can the entropy of an isolated system, such as the universe, increase? Either equation (1.7) is valid only for equilibrium situations, something has been left out, or too much has been assumed.

The resolution of this problem is that Gibbs' expression can represent thermodynamic entropy, but only if P is not taken to provide an exact representation of the state of the universe or, if you wish, of an ensemble of universes. At the very least, practicality requires us to separate the universe into a system about which we may know and care a great deal and an environment with which the system interacts, which is much less precisely monitored. This indeed is one of the central principles of thermodynamics. We are obliged by this incompleteness to represent the probability of environmental details in a so-called coarse-grained fashion, which has the effect that the probability density appearing in Gibbs' representation of the *system* entropy evolves not according to Liouville's equations, but according to versions with additional terms that represent the effect of an uncertain environment upon an open system. This then allows S_{Gibbs} to change, the detailed nature of which will depend on exactly how the environmental forces are represented.

For an isolated system however, an increase in S_{Gibbs} will emerge only if we are obliged to coarse-grained aspects of the system itself. This line of development could be considered rather unsatisfactory, since it makes the entropy of an isolated system grain-size dependent, and alternatives may be imagined where the entropy of an isolated system is represented by something other than S_{Gibbs}. The reader is directed to the literature [8] for further consideration of this matter. However, in this chapter, we shall concern ourselves largely with entropy generation brought about by systems in contact with coarse-grained environments described using stochastic forces, and within such a framework the Gibbs' representation of system entropy will suffice.

We shall discuss a stochastic representation of the additional terms in the system's dynamical equations in the next section, but it is important to note that a deterministic description of environmental effects is also possible, and it might perhaps be thought more natural. On the other hand, the development using stochastic environmental forces is in some ways easier to present. But it should be appreciated that some of the early work on fluctuation relations was developed using deterministic so-called thermostats [1, 9], and that this theme is represented briefly in Section 1.9, and elsewhere in this book.

1.3 Stochastic Dynamics

1.3.1 Master Equations

We pursue the assertion that sense can be made of the second law, its realm of applicability and its failings, when Newton's laws are supplemented by the explicit inclusion of a developing configurational uncertainty. The deterministic rules of evolution of a system need to be replaced by rules for the evolution of the *probability* that the system should take a particular configuration. We must first discuss what we mean by probability. Traditionally, it is the limiting frequency that an event might occur among a large number of trials. But there is also a view that probability represents a distillation, in numerical form, of the best judgment or belief about the state of a system: our information [10]. It is a tool for the evaluation of *expectation* values of system properties, representing what we expect to observe based on information about a system. Fortunately, the two interpretations lead to laws for the evolution of probability that are of similar form.

So let us derive equations that describe the evolution of probability for a simple case. Consider a random walk in one dimension, where a step of variable size is taken at regular time intervals [11–13]. We write the *master equation* describing such a *stochastic process*:

$$\mathcal{P}_{n+1}(x_m) = \sum_{m'=-\infty}^{\infty} T_n(x_m - x_{m'}|x_{m'})\mathcal{P}_n(x_{m'}), \tag{1.8}$$

where $\mathcal{P}_n(x_m)$ is the probability that the walker is at position x_m at timestep n, and $T_n(\Delta x|x)$ is the transition probability for making a step of size Δx in timestep n given a starting position of x. The transition probability may be considered to represent the effect of the environment on the walker. We presume that Newtonian forces cause the move to be made, but we do not know enough about the environment to model the event any better than this. We have assumed the Markov property such that the transition probability does not depend on the previous history of the walker, but only on the position x prior to making the step. It is normalized such that

$$\sum_{m=-\infty}^{\infty} T_n(x_m - x_{m'}|x_{m'}) = 1, \tag{1.9}$$

since the total probability that *any* transition is made, starting from $x_{m'}$, is unity. The probability that the walker is at position m at time n is a sum of probabilities of all possible previous histories that lead to this situation. In the Markov case, the master equation shows that these *path* probabilities are products of transition probabilities and the probability of an initial situation, a simple viewpoint that we shall exploit later.

1.3.2
Kramers–Moyal and Fokker–Planck Equations

The Kramers–Moyal and Fokker–Planck equations describe the evolution of *probability density functions*, denoted P, which are continuous in space (KM) and additionally in time (FP). We start with the Chapman–Kolmogorov equation, an integral form of the master equation for the evolution of a probability density function that is continuous in space:

$$P(x, t+\tau) = \int T(\Delta x|x - \Delta x, t)P(x - \Delta x, t)\mathrm{d}\Delta x. \tag{1.10}$$

We have swapped the discrete time label n for a parameter t. The quantity $T(\Delta x|x, t)$ describes a jump from x through distance Δx in a period τ starting from time t. Note that T now has dimensions of inverse length (it is really a Markovian transition probability *density*), and is normalized according to $\int T(\Delta x|x, t)\mathrm{d}\Delta x = 1$.

We can turn this integral equation into a differential equation by expanding the integrand in Δx to get

$$P(x, t+\tau) = P(x, t) + \int \mathrm{d}\Delta x \sum_{n=1}^{\infty} \frac{1}{n!}(-\Delta x)^n \frac{\partial^n (T(\Delta x|x, t)P(x, t))}{\partial x^n} \tag{1.11}$$

and define the Kramers–Moyal coefficients, proportional to moments of T,

$$M_n(x, t) = \frac{1}{\tau}\int \mathrm{d}\Delta x (\Delta x)^n T(\Delta x|x, t), \tag{1.12}$$

to obtain the (discrete time) Kramers–Moyal equation:

$$\frac{1}{\tau}(P(x, t+\tau) - P(x, t)) = \sum_{n=1}^{\infty} \frac{(-1)^n}{n!} \frac{\partial^n (M_n(x, t)P(x, t))}{\partial x^n}. \tag{1.13}$$

Sometimes the Kramers–Moyal equation is defined with a time derivative of P on the left-hand side instead of a difference.

Equation (1.13) is rather intractable, due to the infinite number of higher derivatives on the right-hand side. However, we might wish to confine attention to evolution in continuous time and consider only stochastic processes that are continuous in space in this limit. This excludes processes that involve discontinuous jumps: the allowed step lengths must go to zero as the timestep goes to zero. In this limit, every KM coefficient vanishes except the first and second, consistent with the Pawula theorem. Furthermore, the difference on the left-hand side of Eq. (1.13) becomes a time derivative and we end up with the Fokker–Planck equation (FPE):

$$\frac{\partial P(x, t)}{\partial t} = -\frac{\partial (M_1(x, t)P(x, t))}{\partial x} + \frac{1}{2}\frac{\partial^2 (M_2(x, t)P(x, t))}{\partial x^2}. \tag{1.14}$$

We can define a probability current,

$$J = M_1(x, t)P(x, t) - \frac{1}{2}\frac{\partial (M_2(x, t)P(x, t))}{\partial x}, \tag{1.15}$$

and view the FPE as a continuity equation for probability density:

$$\frac{\partial P(x, t)}{\partial t} = -\frac{\partial}{\partial x}\left(M_1(x, t)P(x, t) - \frac{1}{2}\frac{\partial (M_2(x, t)P(x, t))}{\partial x}\right) = -\frac{\partial J}{\partial x}. \tag{1.16}$$

The FPE reduces to the familiar diffusion equation if we take M_1 and M_2 to be zero and $2D$, respectively. Note that it is probability that is diffusing, not a physical property like gas concentration. For example, consider the limit of the symmetric Markov random walk in one dimension as timestep and spatial step go to zero: the so-called Wiener process. The probability density $P(x, t)$ evolves according to

$$\frac{\partial P(x,t)}{\partial t} = D\frac{\partial^2 P(x,t)}{\partial x^2}, \tag{1.17}$$

with an initial condition $P(x, 0) = \delta(x)$, if the walker starts at the origin. The statistical properties of the process are represented by the probability density that satisfies this equation, that is,

$$P(x,t) = \frac{1}{(4\pi Dt)^{1/2}} \exp\left(-\frac{x^2}{4Dt}\right), \tag{1.18}$$

representing the increase in positional uncertainty of the walker as time progresses.

1.3.3 Ornstein–Uhlenbeck Process

We now consider a very important stochastic process describing the evolution of the velocity of a particle v. We shall approach this from a different viewpoint: a treatment of the dynamics where Newton's equations are supplemented by environmental forces, some of which are stochastic. It is proposed that the environment introduces a linear damping term together with random noise:

$$\dot{v} = -\gamma v + b\xi(t), \tag{1.19}$$

where γ is the friction coefficient, b is a constant, and ξ has statistical properties $\langle\xi(t)\rangle = 0$, where the angle brackets represent an expectation over the probability distribution of the noise, and $\langle\xi(t)\xi(t')\rangle = \delta(t - t')$, which states that the so-called "white" noise is sampled from a distribution with no autocorrelation in time. The singular variance of the noise might seem to present a problem, but it can be accommodated. This is the Langevin equation. We can demonstrate that it is equivalent to a description based on a Fokker–Planck equation by evaluating the KM coefficients, considering Eq. (1.12) in the form

$$M_n(v,t) = \frac{1}{\tau}\int \mathrm{d}\Delta v (\Delta v)^n T(\Delta v|v,t) = \frac{1}{\tau}\langle (v(t+\tau) - v(t))^n\rangle, \tag{1.20}$$

and in the continuum limit where $\tau \to 0$. This requires an equivalence between the average of $(\Delta v)^n$ over a transition probability density T and the average over the statistics of the noise ξ. We integrate Eq. (1.19) for small τ to get

$$v(t+\tau) - v(t) = -\gamma\int_t^{t+\tau} v\mathrm{d}t + b\int_t^{t+\tau} \xi(t')\mathrm{d}t' \approx -\gamma v(t)\tau + b\int_t^{t+\tau} \xi(t')\mathrm{d}t', \tag{1.21}$$

and according to the properties of the noise, this gives $\langle \mathrm{d}v\rangle = -\gamma v\tau$ with $\mathrm{d}v = v(t+\tau) - v(t)$, such that $M_1(v) = \langle\dot{v}\rangle = -\gamma v$. We also construct $(v(t+\tau) - v(t))^2$ and using the appropriate statistical properties and the continuum limit, we get $\left\langle (\mathrm{d}v)^2 \right\rangle = b^2\tau$ and $M_2 = b^2$. We have therefore established that the FPE equivalent to the Langevin equation (Eq. (1.19)) is

$$\frac{\partial P(v,t)}{\partial t} = \frac{\partial(\gamma v P(v,t))}{\partial v} + \frac{b^2}{2}\frac{\partial^2 P(v,t)}{\partial v^2}. \tag{1.22}$$

The stationary solution to this equation ought to be the Maxwell–Boltzmann velocity distribution $P(v) \propto \exp\left(-mv^2/2k_{\mathrm{B}}T\right)$ of a particle of mass m in thermal equilibrium with an environment at temperature T, so b must be related to T and γ in the form $b^2 = 2k_{\mathrm{B}}T\gamma/m$, where k_{B} is Boltzmann's constant. This is a connection known as a fluctuation–dissipation relation: b characterizes the fluctuations and γ the dissipation or damping in the Langevin equation. Furthermore, it may be shown that the time-dependent solution to Eq. (1.22), with initial condition $\delta(v - v_0)$ at time t_0, is

$$P_{\mathrm{OU}}^{T}[v,t|v_0,t_0] = \sqrt{\frac{m}{2\pi k_{\mathrm{B}}T(1-\mathrm{e}^{-2\gamma(t-t_0)})}}\exp\left(-\frac{m\left(v - v_0\mathrm{e}^{-\gamma(t-t_0)}\right)^2}{2k_{\mathrm{B}}T(1-\mathrm{e}^{-2\gamma(t-t_0)})}\right). \tag{1.23}$$

This is a Gaussian with time-dependent mean and variance. The notation $P_{\mathrm{OU}}^{T}[\cdots]$ characterizes this as a transition probability density for the so-called Ornstein–Uhlenbeck process starting from initial value v_0 at initial time t_0, and ending at the final value v at time t.

The same mathematics can be used to describe the motion of a particle in a harmonic potential $\phi(x) = \kappa x^2/2$, in the limit where the frictional damping coefficient γ is very large. The Langevin equations that describe the dynamics are $\dot{v} = -\gamma v - \kappa x/m + b\xi(t)$ and $\dot{x} = v$, which reduce in this so-called overdamped limit to

$$\dot{x} = -\frac{\kappa}{m\gamma}x + \frac{b}{\gamma}\xi(t), \tag{1.24}$$

which then has the same form as Eq. (1.19), but for position instead of velocity. The transition probability (1.23), recast in terms of x, can therefore be employed.

In summary, the evolution of a system interacting with a coarse-grained environment can be modeled using a stochastic treatment that includes time-dependent random external forces. However, these really represent the effect of uncertainty in the *initial* conditions for the system and its environment: indefiniteness in some of these initial environmental conditions might only have an impact upon the system at a later time. For example, the uncertainty in the velocity of a particle in a gas increases as particles that were initially far away, and that were poorly specified at the initial time, have the opportunity to move closer and interact. The evolution equations are not time reversal symmetric since the principle of causality is assumed: the probability of a system configuration depends upon events that precede it in time, and not on events in the future. The evolving probability density can capture the growth in configurational uncertainty with time. We can now explore how growth of uncertainty in system configuration might be related to entropy production and the irreversibility of macroscopic processes.

1.4 Entropy Generation and Stochastic Irreversibility

1.4.1 Reversibility of a Stochastic Trajectory

The usual statement of the second law in thermodynamics is that it is impossible to observe the reverse of an entropy-producing process. Let us immediately reject this version of the law and recognize that nothing is impossible. A ball might roll off a table and land at our feet. But there is never stillness at the microscopic level and, without breaking any law of mechanics, the molecular motion of the air, ground, and ball might conspire to reverse their macroscopic motion, bringing the ball back to rest on the table. This is not ridiculous: it is an inevitable consequence of the time reversal symmetry of Newton's laws. All we need for this event to occur is to create the right initial conditions. Of course, that is where the problem lies: it is virtually impossible to engineer such a situation, but virtually impossible is not absolutely impossible.

This of course highlights the point behind Loschmidt's paradox. If we were to time reverse the equations of motion of every atom that was involved in the motion of the ball at the end of such an event, we *would* observe the reverse behavior. Or rather more suggestively, we would observe both the forward *and* the reverse behavior with probability 1. This of course is such an overwhelmingly difficult task that one would never entertain the idea of its realization. Indeed, it is also not how one typically considers irreversibility in the real world, whether that be in the lab or through experience. What one may in principle be able to investigate is the explicit time reversal of just the motion of the particle(s) of interest to see whether the previous history can be reversed. Instead of reversing the motion of all the atoms of the ground, the air, and so on, we just attempt to roll the ball back toward the table at the same speed at which it landed at our feet. In this scenario, we would certainly not expect the reverse behavior. Now because the reverse motion is not inevitable, we have somehow, for the system we are considering, identified (or perhaps constructed) the concept of irreversibility albeit on a somewhat anthropic level: events do not easily run backward.

How have we evaded Loschmidt's paradox here? We failed to provide the initial conditions that would ensure reversibility: we left out the reversal of the motion of all other atoms. If they act upon the system differently under time reversal, then irreversibility is (virtually) inevitable. This is not so very profound, but what we have highlighted here is one of the principal paradigms of thermodynamics, the separation of the system of interest and its environment, or for our example the ball and the rest of the surroundings. Given then that we expect such irreversible behavior when we ignore the details of the environment in this way, we can ask what representation of that environment might be most suitable when establishing a measure of the irreversibility of the process? The answer to which is when the environment explicitly interacts with the system in such a way that time reversal is irrelevant. While never strictly true, this can hold as a limiting case that can be represented in a model, allowing us to determine the extent to which the reversal of just the velocities of the system components can lead to a retracing of the previous sequence of events.

Stochastic dynamics can provide an example of such a model. In the appropriate limits, we may consider the collective influence of all the atoms in the environment to act on the system in the same inherently unpredictable and dissipative way regardless of whether their coordinates are time reversed or not. In the Langevin equation, this is achieved by ignoring a quite startling number of degrees of freedom associated with the environment, idealizing their behavior as noise along with a frictional force that slows the particle regardless of which way it is traveling. If we consider now the motion of our system of interest according to this Langevin scheme, its forward and reverse motion both are no longer certain and we can attribute a probability to each path under the influence of the environmental effects. How can we measure irreversibility given these dynamics? We ask the question, what is the probability of observing some forward process compared to the probability of seeing that forward process undone? Or perhaps, to what extent has the introduction of stochastic behavior violated Loschmidt's expectation? This section is largely devoted to the formulation of such a quantity.

Intuitively, we understand that we should be comparing the probability of observing some forward and reverse behavior, but these ideas need to be made concrete. Let us proceed in a manner that allows us to make a more direct connection between irreversibility and our consideration of Loschmidt's paradox. First, let us imagine a system that evolves under some suitable stochastic dynamics. We specifically consider a realization or trajectory that runs from time $t = 0$ to $t = \tau$. Throughout this process, we imagine that any number of system parameters may be subject to change. This could be, for example under suitable Langevin dynamics, the temperature of the heat bath, or perhaps the nature of a confining potential. The changes in the parameters alter the probabilistic behavior of the system as time evolves. Following the literature, we assume that any such change in these system parameters occurs according to some protocol $\lambda(t)$ that itself is a function of time. We note that a particular realization is not guaranteed to take place, since the system is stochastic; so, consequently, we associate with it a probability of occurring that is entirely dependent on the exact trajectory taken, for example, $x(t)$, and the protocol $\lambda(t)$.

We can readily compare probabilities associated with different paths and protocols. To quantify an irreversibility in the sense of the breaking of Loschmidt's expectation however, we must consider one specific path and protocol. Recall now our definition of the paradox. In a deterministic system, a time reversal of all the variables at the end of a process of length τ leads to the observation of the reverse behavior with probability 1 over the same period τ. It is the probability of the trajectory that corresponds to this reverse behavior within a stochastic system that we must address. To do so, let us consider what we mean by time reversal. A time reversal can be thought of as the operation of the time reversal operator $\hat{T}$ on the system variables and distribution. Specifically, for position x, momentum p, and some protocol λ, we have $\hat{T}x = x$, $\hat{T}p = -p$, and $\hat{T}\lambda = \lambda$. If we were to do this after time τ for a set of Hamilton's equations of motion in which the protocol was time independent, the trajectory would be the exact time-reversed retracing of the forward trajectory. We shall call this trajectory the *reversed trajectory* and is phenomenologically the "running backward" of the forward behavior. Similarly, if we were to consider a motion in a deterministic system that was subject to some protocol (controlling perhaps some external field), we would observe the reversed trajectory only if the original protocol were performed symmetrically backward. This running of the protocol backward we shall call the *reversed protocol*.

We now are in a position to construct a measure of irreversibility in a stochastic system. We do so by comparing the probability of observing the forward trajectory under the forward protocol with the probability of observing the reversed trajectory under the reversed protocol following a time reversal at the end of the forward process. We literally attempt to undo the forward process and measure how likely that is. Since the quantities we have just defined here are crucial to this chapter, we shall make their nature absolutely clear before we proceed. To reiterate, we wish to consider the following:

- **Reversed trajectory:** Given a trajectory $X(t)$ that runs from time $t = 0$ to $t = \tau$, we define the reversed trajectory $\bar{X}(t)$ that runs forward in time explicitly such that $\bar{X}(t) = \hat{T}X(\tau - t)$. Examples are for position $\bar{x}(t) = x(\tau - t)$ and for momentum $\bar{p}(t) = -p(\tau - t)$.
- **Reversed protocol:** The protocol $\lambda(t)$ behaves in the same way as the position variable x under time reversal and so we define the reversed protocol $\bar{\lambda}(t)$ such that $\bar{\lambda}(t) = \lambda(\tau - t)$.

Given these definitions, we can construct the path probabilities we seek to compare. For notational clarity, we label path probabilities that depend upon the forward protocol $\lambda(t)$ with the superscript F to denote the forward process and probabilities that depend upon the reversed protocol $\bar{\lambda}(t)$ with the superscript R to denote the reverse process. The probability of observing a given trajectory X, $\mathcal{P}^F[X]$, has two components. First, the probability of the path given its starting point $X(0)$, which we shall write as $\mathcal{P}^F[X(\tau)|X(0)]$; second, the initial probability of being at the start of the path, which we write as $\mathcal{P}_{\text{start}}(X(0))$ since it concerns the distribution of variables at the start of the forward process. The probability of observing the forward path is then given as

$$\mathcal{P}^F[X] = \mathcal{P}_{\text{start}}(X(0))\mathcal{P}^F[X(\tau)|X(0)]. \tag{1.25}$$

It is intuitive to proceed if we imagine the path probability as being approximated by a sequence of jumps that occur at distinct times. Since continuous stochastic behavior can be readily approximated by jump processes, but not the other way round, this simultaneously allows us to generalize any statements for a wider class of Markov processes. We shall assume for brevity that the jump processes occur in discrete time. By repeated application of the Markov property for such a system, we can write

$$\begin{aligned}\mathcal{P}^F[X] = &\mathcal{P}_{\text{start}}(X_0)\mathcal{P}(X_1|X_0, \lambda(t_1)) \times \mathcal{P}(X_2|X_1, \lambda(t_2)) \times \cdots \\ &\times \mathcal{P}(X_n|X_{n-1}, \lambda(t_n)).\end{aligned} \tag{1.26}$$

Here, we consider a trajectory that is approximated by the jump sequence between $n+1$ points $X_0, X_1, \ldots, X_n$ such that there are n distinct transitions that occur at discrete times $t_1, t_2, \ldots, t_n$, and where $X_0 = X(0)$ and $X_n = X(\tau)$. $\mathcal{P}(X_i|X_{i-1}, \lambda(t_i))$ is the probability of a jump from X_{i-1} to X_i using the value of the protocol evaluated at time t_i.

Continuing with our description of irreversibility, we construct the probability of the reversed trajectory under the reversed protocol. Approximating as a sequence of jumps as before, we may write

$$\begin{aligned}\mathcal{P}^R[\bar{X}] &= \hat{T}\mathcal{P}_{\text{end}}(\bar{X}(0))\mathcal{P}^R[\bar{X}(\tau)|\bar{X}(0)] \\ &= \mathcal{P}^R_{\text{start}}(\bar{X}(0))\mathcal{P}^R[\bar{X}(\tau)|\bar{X}(0)] \\ &= \mathcal{P}^R_{\text{start}}(\bar{X}_0)\mathcal{P}(\bar{X}_1|\bar{X}_0, \bar{\lambda}(t_1)) \times \cdots \times \mathcal{P}(\bar{X}_n|\bar{X}_{n-1}, \bar{\lambda}(t_n)).\end{aligned} \tag{1.27}$$

There are two key concepts here. First, in accordance with our definition of irreversibility, we attempt to "undo" the motion from the end of the forward process and so the initial distribution is formed from the distribution to which $\mathcal{P}_{\text{start}}$ evolves under $\lambda(t)$, such that for continuous probability density distributions we have

$$P_{\text{end}}(X(\tau)) = \int \mathrm{d}X P_{\text{start}}(X(0))P^F[X(\tau)|X(0)], \tag{1.28}$$

so named because it is the probability distribution at the end of the forward process. For our discrete model, the equivalent is given by

$$\mathcal{P}_{\text{end}}(X_n) = \sum_{X_0} \cdots \sum_{X_{n-1}} \prod_{i=0}^{n-1} \mathcal{P}(X_{i+1}|X_i, \lambda(t_{i+1}))\mathcal{P}_{\text{start}}(X_0). \tag{1.29}$$

Second, to attempt to observe the reverse trajectory starting from $X(\tau)$, we must perform a time reversal of our system to take advantage of the reversibility in Hamilton's equations. However, when we time reverse the variable X, we are obliged to transform the distribution $\mathcal{P}_{\text{end}}$ as well, since the likelihood of starting the reverse trajectory with variable $\hat{T}X$ after we time reverse X is required to be the same as the likelihood of arriving at X before the time reversal. This transformed $\mathcal{P}_{\text{end}}$, $\hat{T}\mathcal{P}_{\text{end}}$, is the initial distribution for the reverse process and is thus labeled $\mathcal{P}^R_{\text{start}}$. Analogously, evolution under $\bar{\lambda}(t)$ takes the system distribution to $\mathcal{P}^R_{\text{end}}$. The forward process and its relation to the reverse process are illustrated for both coordinates x and v, which do and do not change sign following time reversal, respectively, in Figure 1.1, along with illustrations of the reversed trajectories and protocols.

Let us form our prototypical measure of the irreversibility of the path X, which for now we denote I:

$$I[X] = \ln\left[\frac{\mathcal{P}^F[X]}{\mathcal{P}^R[\bar{X}]}\right]. \tag{1.30}$$

There are some key points to note about such a quantity. First, since $\bar{X}$ and X are simply related, I is a functional of the trajectory X and accordingly will

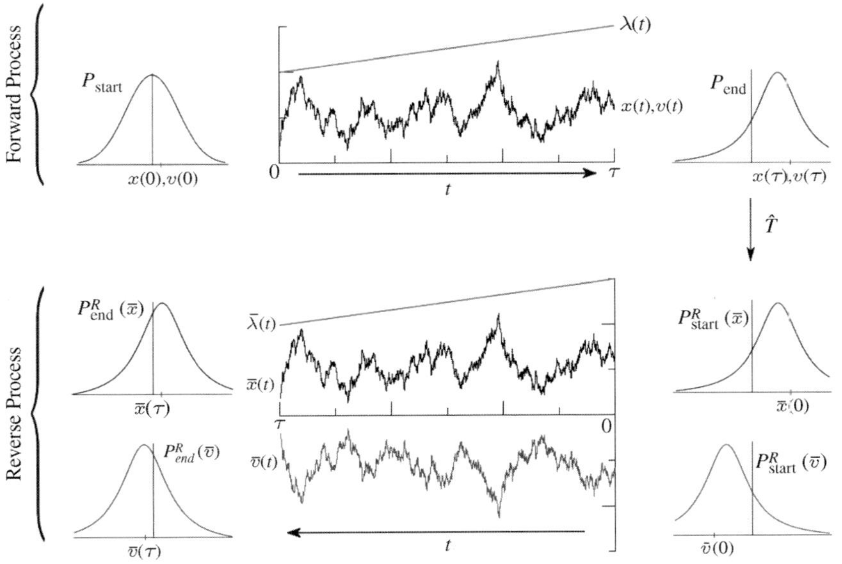

Figure 1.1 An illustration of the definition of the forward and reverse processes. The forward process consists of an initial probability *density* P_{start} that evolves forward in time under the forward protocol $\lambda(t)$ over a period τ, at the end of which the variable is distributed according to P_{end}. The reverse process consists of evolution from the distribution P^R_{start}, which is related to P_{end} by a time reversal, under the reversed protocol $\bar{\lambda}(t)$ over the same period τ, at the end of which the system will be distributed according to some final distribution P^R_{end}, which in general is not related to P_{start} and does not explicitly feature in assessment of the irreversibility of the forward process. A particular realization of the forward process is characterized by the forward trajectory $X(t)$, illustrated here as being $x(t)$ or $v(t)$. To determine the irreversibility of this realization, the reversed trajectories, $\bar{x}(t)$ or $\bar{v}(t)$, related by a time reversal, need to be considered as realizations in the reverse process.

take a range of values over all possible "realizations" of the dynamics: as such it will be characterized by a probability distribution. Furthermore, there is nothing in its form that disallows negative values. Finally, the quantity vanishes if the reversed trajectory occurs with the same probability as the forward trajectory under the relevant protocols: a process is deemed reversible if the forward process can be "undone" with equal probability. We can simplify this form since we know how the time-reversed protocols and trajectories are related. Given the step sequence laid out for the approximation to a continuous trajectory, we can transform X and t according to $\bar{X}_i = \hat{T}X_{n-i}$ and $\bar{\lambda}(t_i) = \lambda(t_{n-i+1})$ giving

$$\begin{aligned} \mathcal{P}^R[\bar{X}] &= \mathcal{P}^R_{\text{start}}(\hat{T}X(\tau))\mathcal{P}^R[\hat{T}X(0)|\hat{T}X(\tau)] \\ &= \mathcal{P}^R_{\text{start}}(\hat{T}X_n)\mathcal{P}(\hat{T}X_{n-1}|\hat{T}X_n, \lambda(t_n)) \times \cdots \times \mathcal{P}(\hat{T}X_0|\hat{T}X_1, \lambda(t_1)). \end{aligned} \tag{1.31}$$

Pointing out that $\mathcal{P}^R_{\text{start}}(\hat{T}X_n) = \hat{T}\mathcal{P}_{\text{end}}(\hat{T}X_n) = \mathcal{P}_{\text{end}}(X_n)$, we thus have

$$\begin{aligned} \ln\left[\frac{\mathcal{P}^F[X]}{\mathcal{P}^R[\bar{X}]}\right] &= \ln\left(\frac{\mathcal{P}_{\text{start}}(X(0))}{\mathcal{P}_{\text{end}}(X(\tau))}\right) + \ln\left[\frac{\mathcal{P}^F[X(\tau)|X(0)]}{\mathcal{P}^R[\hat{T}X(0)|\hat{T}X(\tau)]}\right] \\ &= \ln\left[\frac{\mathcal{P}_{\text{start}}(X_0)}{\mathcal{P}_{\text{end}}(X_n)}\prod_{i=1}^{n}\frac{\mathcal{P}(X_i|X_{i-1},\lambda(t_i))}{\mathcal{P}(\hat{T}X_{i-1}|\hat{T}X_i,\lambda(t_i))}\right], \end{aligned} \tag{1.32}$$

noting that strictly this is for a model in discrete space and time. Let us study this quantity for a specific model to understand its meaning in physical terms. Consider the continuous stochastic process described by the Langevin equation from Section 1.3.3, where $X = v$ and we have

$$\dot{v} = -\gamma v + \left(\frac{2k_B T(t)\gamma}{m}\right)^{1/2}\xi(t), \tag{1.33}$$

where $\xi(t)$ is white noise. The equivalent Fokker–Planck equation is given by

$$\frac{\partial P(v,t)}{\partial t} = \frac{\partial(\gamma v P(v,t))}{\partial v} + \frac{k_B T(t)\gamma}{m}\frac{\partial^2 P(v,t)}{\partial v^2}. \tag{1.34}$$

where P is a probability density. By inserting probability densities and associated infinitesimal volumes into Eq. (1.32) and canceling the latter, we observe that we may use probability densities to represent the quantity $I[X]$ for this continuous behavior without a loss of generality. To introduce a distinct forward and reverse process, let us allow the temperature to vary with a protocol $\lambda(t)$. We choose for simplicity a protocol that consists only of step changes such that

$$T(\lambda(t_i)) = T_j, \quad t_i \in [(j-1)\Delta t, j\Delta t], \tag{1.35}$$

where j is an integer in the range of $1 \leq j \leq N$, such that $N\Delta t = \tau$. Because the process is simply the combination of different Ornstein–Uhlenbeck processes, each of which is characterized by defined solution Eq. (1.23), we can represent the path probability in a piecewise fashion. Consolidating with our notation, the continuous Langevin behavior at some fixed temperature can be considered to be the limit, $\mathrm{d}t = (t_{i+1} - t_i) \to 0$, of the discrete jump process, so that

$$\begin{aligned} &\lim_{\mathrm{d}t\to 0}\prod_{t_i=(j-1)\Delta t}^{t_i=j\Delta t}\mathcal{P}(v_i|v_{i-1},\lambda(t_i)) = P^{T_j}_{\text{OU}}[v(j\Delta t)|v((j-1)\Delta t)]\mathrm{d}v(j\Delta t) \\ &= \left(\frac{m}{2\pi k_B T_j(1-\mathrm{e}^{-2\gamma\Delta t})}\right)^{1/2}\exp\left(-\frac{m(v(j\Delta t) - v((j-1)\Delta t)\mathrm{e}^{-\gamma\Delta t})^2}{2k_B T_j(1-\mathrm{e}^{-2\gamma\Delta t})}\right)\mathrm{d}v(j\Delta t). \end{aligned} \tag{1.36}$$

The total conditional path probability density (with units equal to the inverse dimensionality of the path) over N of these step changes in temperature is then by application of the Markov property

$$P^F[v(\tau)|v(0)] = \prod_{j=1}^{N}\left(\frac{m}{2\pi k_B T_j(1-\mathrm{e}^{-2\gamma\Delta t})}\right)^{1/2}\exp\left(-\frac{m(v(j\Delta t) - v((j-1)\Delta t)\mathrm{e}^{-\gamma\Delta t})^2}{2k_B T_j(1-\mathrm{e}^{-2\gamma\Delta t})}\right) \tag{1.37}$$

and since $\hat{T}v = -v$,

$$P^R[-v(0)| - v(\tau)] = \prod_{j=1}^{N} \left(\frac{m}{2\pi k_B T_j (1 - e^{-2\gamma\Delta t})} \right)^{1/2} \exp\left(- \frac{m(-v((j-1)\Delta t) + v(j\Delta t) e^{-\gamma\Delta t})^2}{2k_B T_j (1 - e^{-2\gamma\Delta t})} \right). \tag{1.38}$$

Taking the logarithm of their ratio explicitly and abbreviating $v(j\Delta t) = v_j$ yields

$$\ln \left[\frac{P^F[v(\tau)|v(0)]}{P^R[-v(0)| - v(\tau)]} \right] = -\frac{1}{k_B} \sum_{j=1}^{N} \frac{m}{2T_j} \left(v_j^2 - v_{j-1}^2 \right), \tag{1.39}$$

which is quite manifestly equal to the sum of negative changes of the kinetic energy of the particle scaled by k_B and the environmental temperature to which the particle is exposed. Our model consists only of the particle and the environment and so each negative kinetic energy change of the particle, $-\Delta Q$, must be associated with a positive flow of heat ΔQ_{med} into the environment such that we define $\Delta Q_{\text{med}} = -\Delta Q$. For the Langevin equation, the effect of the environment is idealized as a dissipative friction term and a fluctuating white noise characterized by a defined temperature that is entirely independent of the behavior of the particle. This is the idealization of a large equilibrium heat bath for which the exchanged heat is directly related to the entropy change of the bath through the relation $\Delta Q_{\text{med}} = T\Delta S$. It may be argued that changing between N temperatures under such an idealization is equivalent to exposing the particle to N separate equilibrium baths each experiencing an entropy change according to $\Delta Q_{\text{med},j} = T_j \Delta S_j$. We consequently assert, for this particular model at least, that

$$k_B \, ln \left[\frac{P^F[v(\tau)|v(0)]}{P^R[-v(0)| - v(\tau)]} \right] = \sum_j \frac{\Delta Q_{\text{med},j}}{T_j} = \sum_j \Delta S_j = \Delta S_{\text{med}}, \tag{1.40}$$

where the entropy production in all N baths can be denoted as a total entropy production ΔS_{med} that occurs in a generalized medium.

Let us now examine the remaining part of our quantification of irreversibility, which here is given in Eq. (1.32) by the logarithm of the ratio of $P_{\text{start}}(v(0))$ and $P_{\text{end}}(v(\tau))$. Given an arbitrary initial distribution, one can write this as the change in the logarithm of the dynamical solution to P as given by the Fokker–Planck equation (Eq. (1.34)). Consequently, we can write

$$\ln \left(\frac{P_{\text{start}}(v(0))}{P_{\text{end}}(v(\tau))} \right) = \ln \frac{P(v,0)}{P(v,\tau)} = -(\ln P(v,\tau) - \ln P(v,0)). \tag{1.41}$$

If we now characterize the *mean* entropy of our Langevin particle or "system" using a Gibbs entropy that we allow to be time dependent such that

$$\langle S_{\text{sys}} \rangle = S_{\text{Gibbs}} = -k_B \int dv \, P(v,t) \ln P(v,t), \tag{1.42}$$

one can make the conceptual leap that it is an individual value for the entropy of the system for a given v and time t that is being averaged in the above integral[1)] [14], $S_{\rm sys} = -k_{\rm B} \ln P(v, t)$. If we accept these assertions, we find that our measure of irreversibility for any one individual trajectory is formed as

$$k_{\rm B} I[X] = \Delta S_{\rm sys} + \Delta S_{\rm med}. \tag{1.43}$$

Since our model consists only of the Langevin particle (the system) and a heat bath (the medium), we therefore regard this sum as the total entropy production associated with such a trajectory and make the assertion that our measure of irreversibility is identically the increase in the total entropy of the universe, in this model at least:

$$\Delta S_{\rm tot}[X] = \Delta S_{\rm sys} + \Delta S_{\rm med} = k_{\rm B} \ln \left[\frac{P^F[X]}{P^R[\bar{X}]}\right]. \tag{1.44}$$

However, we have already stated that nothing prevents this quantity from taking negative values. If this is to be the total entropy production, how is this permitted given our knowledge of the second law of thermodynamics? In essence, describing the way in which a quantity that looks like the total entropy production can take both positive and negative values, but obeys well-defined statistical requirements such that, for example, it is compatible with the second law, is the subject matter of the so-called fluctuation theorems or fluctuation relations. These relations are disarmingly simple, but allow us to make predictions far beyond those possible in classical thermodynamics. For this class of system in fact, they are so simple that we can derive in a couple of lines a most fundamental relation and immediately reconcile the second law in terms of our irreversibility functional. Let us consider the average, with respect to all possible forward realizations, of the quantity $\exp(-\Delta S_{\rm tot}[X]/k_{\rm B})$, which we write $\langle \exp(-\Delta S_{\rm tot}[X]/k_{\rm B})\rangle$ and where the angle brackets denote a weighted path integration. Performing the average yields,

$$\begin{aligned} \langle e^{-\Delta S_{\rm tot}[X]/k_{\rm B}} \rangle &= \int dX P^F[X] e^{-\Delta S_{\rm tot}[X]/k_{\rm B}} \\ &= \int dX P^F[X] \frac{P^R[\bar{X}]}{P^F[X]} \\ &= \int d\bar{X} P^R[\bar{X}], \end{aligned} \tag{1.45}$$

where we assume the path integral measures are equivalent, $dX = d\bar{X}$, such that the Jacobian associated with the transformation between the paths is unity (this is guaranteed for any involutive transformation). Or perhaps more transparently, in the discrete approximation, multiple summations over $X_0, \ldots, X_n$ yield the same

1) Strictly, $P(v, t)$ is a probability density and so for Eq. (1.42) to be consistent with the entropy arising from the combinatorial arguments of statistical mechanics and dimensionally correct, it may be argued that we should be considering $\ln(P(v, t)dv)$. However, for relative changes, this issue is irrelevant.

result as summation over $X_n, \ldots, X_0$. The expression above now trivially integrates to unity that allows us to write the so-called [14]:

Integral Fluctuation Theorem

$$\langle e^{-\Delta S_{\text{tot}}[X]/k_B} \rangle = 1. \tag{1.46}$$

This remarkably simple relation holds for all times, protocols, and initial conditions[2)] and implies that the possibility of negative total entropy change is obligatory. Furthermore, if we make use of Jensen's inequality

$$\langle \exp(z) \rangle \geq \exp\langle z \rangle, \tag{1.47}$$

we can directly infer

$$\langle \Delta S_{\text{tot}} \rangle \geq 0. \tag{1.48}$$

Since this holds for any initial condition, we may also state that the mean total entropy monotonically increases for any process. This statement, under the stochastic dynamics we consider, is the second law. It is a replacement or reinterpretation of Eq. (1.4). The expected entropy production rate is always positive, but this is not necessarily found in detail for individual realizations. The second law, when correctly understood, is statistical in nature and we have now obtained an expression that places a fundamental bound on these statistics.

1.5 Entropy Production in the Overdamped Limit

We have formulated a quantity that we assert to be the total entropy production, though it is for a very specific system and importantly has no ability to describe the application of work. To broaden the scope of application, it is instructive to obtain a general expression like that obtained in Eq. (1.39), but for a class of stochastic behavior where we can formulate and verify the total entropy production without the need for an exact analytical result. This is straightforward for systems with detailed balance [15]; however, we can generalize further. The class of stochastic behavior we shall consider will be the simple overdamped Langevin equation that we discussed in Section 1.3.3 involving a position variable described by

$$\dot{x} = \frac{\mathcal{F}(x)}{m\gamma} + \left(\frac{2k_B T}{m\gamma}\right)^{1/2} \xi(t), \tag{1.49}$$

2) We do though assume that nowhere in the initial available configuration space we have $P_{\text{start}}(X) = 0$. This is a paraphrasing of the so-called ergodic consistency requirement found in deterministic systems [9] and insists that there must be a trajectory for every possible reversed trajectory and vice versa, so that all possible paths, $\bar{X}(t)$, are included in the integral in the final line of Eq. (1.45).

along with an equivalent Fokker–Planck equation:

$$\frac{\partial P(x,t)}{\partial t} = -\frac{1}{m\gamma}\frac{\partial(\mathcal{F}(x)P(x,t))}{\partial x} + \frac{k_B T}{m\gamma}\frac{\partial^2 P(x,t)}{\partial x^2}. \tag{1.50}$$

The description includes a force term $\mathcal{F}(x)$ that allows us to model most simple thermodynamic processes including the application of work. We describe the force as a sum of two contributions that arise respectively from a potential $\phi(x)$ and an external force $f(x)$ that is applied directly to the particle, both of which we allow to vary in time through application of a protocol such that

$$\mathcal{F}(x,\lambda_0(t),\lambda_1(t)) = -\frac{\partial\phi(x,\lambda_0(t))}{\partial x} + f(x,\lambda_1(t)). \tag{1.51}$$

The first step in characterizing the entropy produced in the medium according to this description is to identify the main thermodynamic quantities, including the heat exchanged with the bath. To do this, we paraphrase Sekimoto and Seifert [14, 16, 17] and start from basic thermodynamics and the first law:

$$\Delta E = \Delta Q + \Delta W, \tag{1.52}$$

which must hold rigorously despite the stochastic nature of our model. To proceed, let us consider the change in each of these quantities in response to evolving our system by a small time $\mathrm{d}t$ and corresponding displacement $\mathrm{d}x$. We can readily identify that the system energy for overdamped conditions is equal to the value of the conservative potential such that

$$\mathrm{d}E = \mathrm{d}Q + \mathrm{d}W = \mathrm{d}(\phi(x,\lambda_0(t))). \tag{1.53}$$

However, at this point, we reach a subtlety in the mathematics originating in the stochastic nature of x. Where normally we could describe the small change in ϕ using the usual chain rule of calculus, when ϕ is a function of the stochastic variable x, we must be more careful. The peculiarity is manifest in an ambiguity of expressing the multiplication of a continuous stochastic function by a stochastic increment. The product, which strictly should be regarded as a stochastic integral, is not uniquely defined because both function and increment cannot be assumed to behave smoothly on any timescale. The mathematical details [12] are not of our concern for this chapter and so we shall not rigorously discuss stochastic calculus or go beyond the following steps of reasoning and assumption. First, we assume that in order to work with thermodynamic quantities in the traditional sense, as in undergraduate physics, we require a small change to resemble that of normal calculus, and this requires, in all instances, multiplication to follow so-called Stratonovich rules. These rules, denoted in this chapter by the symbol $\circ$, are taken to mean evaluation of the preceding stochastic function at the midpoint of the following increment. Employing this procedure, we may write

$$\begin{aligned}\mathrm{d}E &= \mathrm{d}(\phi(x(t),\lambda_0(t))) \\ &= \frac{\partial\phi(x(t),\lambda_0(t))}{\partial\lambda_0}\frac{\mathrm{d}\lambda_0(t)}{\mathrm{d}t}\mathrm{d}t + \frac{\partial\phi(x(t),\lambda_0(t))}{\partial x}\circ\mathrm{d}x.\end{aligned} \tag{1.54}$$

Next, we can explicitly write down the work from basic mechanics as contributions from the change in potential and the operation of an external force:

$$\mathrm{d}W = \frac{\partial\phi(x(t),\lambda_0(t))}{\partial\lambda_0}\frac{\mathrm{d}\lambda_0(t)}{\mathrm{d}t}\mathrm{d}t + f(x(t),\lambda_1(t)) \circ \mathrm{d}x. \tag{1.55}$$

Accordingly, we directly have an expression for the heat transfer to the system in response to a small change $\mathrm{d}x$:

$$\begin{aligned} \mathrm{d}Q &= \frac{\partial\phi(x(t),\lambda_0(t))}{\partial x} \circ \mathrm{d}x - f(x(t),\lambda_1(t)) \circ \mathrm{d}x \\ &= -\mathcal{F}(x(t),\lambda_0(t),\lambda_1(t)) \circ \mathrm{d}x. \end{aligned} \tag{1.56}$$

We may then integrate these small increments over a trajectory of duration τ to find

$$\Delta E = \int_0^\tau \mathrm{d}E = \int_0^\tau \mathrm{d}(\phi(x(t),\lambda_0(t))) = \phi(x(\tau),\lambda_0(\tau)) - \phi(x(0),\lambda_0(0)) = \Delta\phi, \tag{1.57}$$

$$\Delta W = \int_0^\tau \mathrm{d}W = \int_0^\tau \frac{\partial\phi(x(t),\lambda_0(t))}{\partial\lambda_0}\frac{\mathrm{d}\lambda_0(t)}{\mathrm{d}t}\mathrm{d}t + \int_0^\tau f(x(t),\lambda_1(t)) \circ \mathrm{d}x, \tag{1.58}$$

and

$$\Delta Q = \int_0^\tau \mathrm{d}Q = \int_0^\tau \frac{\partial\phi(x(t),\lambda_0(t))}{\partial x} \circ \mathrm{d}x - \int_0^\tau f(x(t),\lambda_1(t)) \circ \mathrm{d}x. \tag{1.59}$$

Let us now verify what we expect; the ratio of conditional path probability densities that we use in Eq. (1.32) will be equal to the negative heat transferred to the system divided by the temperature of the environment. We no longer have a means for representing the transition probabilities in general and so we proceed using the so-called "short time propagator" [11, 12, 18], which to first order in the time between transitions, $\mathrm{d}t$, describes the probability of making a transition from x_i to x_{i+1}. We may then consider the analysis valid in the limit $\mathrm{d}t \to 0$. The short time propagator can also be thought of as a short time Green's function; it is a solution to the Fokker–Planck equation subject to a delta function initial condition, valid as the propagation time is taken to zero.

The basic form of the short time propagator is helpfully rather intuitive and most simply adopts a general Gaussian form that reflects the fluctuating component of the force about the mean due to the Gaussian white noise. Abbreviating $\mathcal{F}(x,\lambda_0(t),\lambda_1(t))$ as $\mathcal{F}(x,t)$, we may write the propagator as

$$P[x_{i+1}, t_i + \mathrm{d}t|x_i, t_i] = \sqrt{\frac{m\gamma}{4\pi k_B T\mathrm{d}t}}\exp\left[-\frac{m\gamma}{4k_B T\mathrm{d}t}\left(x_{i+1} - x_i - \frac{\mathcal{F}(x_i,t_i)}{m\gamma}\mathrm{d}t\right)^2\right]. \tag{1.60}$$

However, one must be very careful. For reasons similar to those discussed above, a propagator of this type is not uniquely defined, with a family of forms being available depending on the spatial position at which one chooses to evaluate the force $\mathcal{F}$ of which Eq. (1.60) is but one example [18]. In the same way we had to choose certain multiplication rules; it is not enough to write $\mathcal{F}(x(t),t)$

without further comment since $x(t)$ has not been fully specified. This leaves a certain mathematical freedom in how to write the propagator and we must consider which is most appropriate. Of crucial importance is that all are correct in the limit $dt \to 0$ (all lead to the correct solution of the Fokker–Planck equation), meaning our choice must rest solely on ensuring the correct representation of the entropy production. We can proceed heuristically: as we take time $dt \to 0$, we steadily approach a representation of transitions as jump processes, from which we can proceed with confidence since jump processes are the more general description of stochastic phenomena. In this limit, therefore, we are obliged to faithfully represent the ratio that appears in Eq. (1.32). In this description, the forward and reverse jump probabilities have the same functional form and to emulate this we must evaluate the short time propagators at the same position x for both the forward and reverse transitions.[3] Mathematically, the most convenient way of doing this is to evaluate all functions in the propagator midway between initial and final points. Evaluating the functions at the midpoint x' such that $2x' = x_{i+1} + x_i$ and $dx = x_{i+1} - x_i$ introduces a propagator of the form

$$P[x_{i+1}, t_i + dt | x_i, t_i] = \sqrt{\frac{m\gamma}{4\pi k_B T dt}} \exp\left[-\frac{m\gamma}{4k_B T dt}\left(dx - \frac{\mathcal{F}(x', t_i)}{m\gamma} dt\right)^2 - \frac{1}{2}\frac{\partial}{\partial x'}\left(\frac{\mathcal{F}(x', t_i)}{m\gamma}\right) dt\right] \tag{1.61}$$

and similarly

$$P[x_i, t_i + dt | x_{i+1}, t_i] = \sqrt{\frac{m\gamma}{4\pi k_B T dt}} \exp\left[-\frac{m\gamma}{4k_B T dt}\left(-dx - \frac{\mathcal{F}(x', t_i)}{m\gamma} dt\right)^2 - \frac{1}{2}\frac{\partial}{\partial x'}\left(\frac{\mathcal{F}(x', t_i)}{m\gamma}\right) dt\right]. \tag{1.62}$$

The logarithm of their ratio, in the limit $dt \to 0$, simply reduces to

$$\begin{aligned}
\lim_{dt \to 0} \ln\left[\frac{P[x_{i+1}, t_i + dt | x_i, t_i]}{P[x_i, t_i + dt | x_{i+1}, t_i]}\right] &= \ln\left(\frac{P(x_{i+1} | x_i, \lambda(t_i))}{P(x_i | x_{i+1}, \lambda(t_i))}\right) \\
&= \frac{\mathcal{F}(x', \lambda_0(t_i), \lambda_1(t_i))}{k_B T} dx \\
&= \frac{\mathcal{F}(x(t_i), \lambda_0(t_i), \lambda_1(t_i))}{k_B T} \circ dx \\
&= -\frac{dQ}{k_B T},
\end{aligned} \tag{1.63}$$

where we get to the result by recognizing that line 2 obeys our definition of Stratonovich multiplication rules since x' is the midpoint of dx and that line 3 contains the definition of an increment in the heat transfer from Eq. (1.56). We can

3) For the reader aware of the subtleties of stochastic calculus, we mention that for additive noise as considered here, this point is made largely for completeness: if one constructs the result using the relevant stochastic calculus, the ratio is independent of the choice. However, to be a well-defined quantity for cases involving multiplicative noise, this issue becomes important.

then obtain the entropy production of the entire path by constructing the integral limit of the summation over contributions for each t_i such that

$$k_B \ln \left[\frac{P^F[X(\tau)|X(0)]}{P^R[\bar{X}(\tau)|\bar{X}(0)]} \right] = -\frac{1}{T}\int_0^\tau \mathrm{d}Q = -\frac{\Delta Q}{T} = \frac{\Delta Q_{\mathrm{med}}}{T} = \Delta S_{\mathrm{med}}, \tag{1.64}$$

giving us the expected result noting that the identification of such a term from the ratio of path probabilities can also readily be achieved in full phase space [19].

1.6
Entropy, Stationarity, and Detailed Balance

Let us consider the functional for the total entropy production once more, specifically with a view to understanding when we expect an entropy change. Specifically, we aim to identify two conceptually different situations where entropy production occurs. If we consider a system evolving without external driving, it will typically, for well-defined system parameters, approach some stationary state. That stationary state is characterized by a time-independent probability density P^{st} such that

$$\frac{\partial P^{\mathrm{st}}(x,t)}{\partial t} = 0. \tag{1.65}$$

Let us write down the entropy production for such a situation. Since the system is stationary, we have $P_{\mathrm{start}} = P_{\mathrm{end}}$, but we also have a time-independent protocol, meaning we need not consider distinct forward and reverse processes such that we write path probability densities $P^R = P^F = P$. In this situation, the total entropy production for overdamped motion is given as

$$\Delta S_{\mathrm{tot}}[x] = k_B \ln \left[\frac{P^{\mathrm{st}}(x(0))P[x(\tau)|x(0)]}{P^{\mathrm{st}}(x(\tau))P[x(0)|x(\tau)]} \right]. \tag{1.66}$$

We can then ask what in general are the properties required for entropy production, or indeed no entropy production in such a situation. Clearly, there is no entropy production when the forward and reverse trajectories are equally likely and so we can write the condition for zero entropy production in the stationary state as

$$P^{\mathrm{st}}(x(0))P[x(\tau)|x(0)] = P^{\mathrm{st}}(x(\tau))P[x(0)|x(\tau)], \quad \forall\, x(0), x(\tau). \tag{1.67}$$

Written in this form, we emphasize that this is equivalent to the statement of *detailed balance*. Transitions are said to balance because the average number of all transitions to and from any given configuration $x(0)$ exactly cancel; this leads to a constant probability distribution and is the condition required for a stationary state. However, to have no entropy production in the stationary state, we require all transitions to balance in detail: we require the total number of transitions between every possible combination of two configurations $x(0)$ and $x(\tau)$ to cancel. This is also the condition required for zero probability current and for the system to be at thermal equilibrium where we understand the entropy of the universe to be maximized.

We may then quite generally place any dynamical scheme into one of two broad categories. The first is where detailed balance (Eq. (1.67)) holds and the stationary state is the thermal equilibrium.[4] Under such dynamics, systems left unperturbed will relax toward equilibrium where there is no observed preferential forward or reverse behavior, no observed thermodynamic arrow of time or irreversibility, and therefore no entropy production. Thus, all entropy production for these dynamics is the result of driving and subsequent relaxation to equilibrium or more generally a consequence of systems being out of their stationary states.

The other category therefore is where detailed balance does not hold. In these situations, we expect entropy production even in the stationary state, which by extension must have origins beyond that of driving out of and relaxation back to stationarity. So, when can we expect detailed balance to be broken? We can first identify the situations where it does hold and for overdamped motion, the requirements are well defined. To have all transitions balancing in detail is to have zero probability current, $J^{\mathrm{st}}(x,t)=0$, in the stationary state, where the current is related to the probability density according to

$$\frac{\partial P^{\mathrm{st}}(x,t)}{\partial t}=-\frac{\partial J^{\mathrm{st}}(x,t)}{\partial x}=0. \tag{1.68}$$

Utilizing the form of the Fokker–Planck equation that corresponds to the dynamics, we would thus require

$$J^{\mathrm{st}}(x,t)=\frac{1}{m\gamma}\left(-\frac{\partial\phi(x,\lambda_0(t))}{\partial x}+f(x,\lambda_1(t))\right)P^{\mathrm{st}}(x,t)-\frac{k_{\mathrm{B}}T}{m\gamma}\frac{\partial P^{\mathrm{st}}(x,t)}{\partial x}=0. \tag{1.69}$$

We can verify the consistency of such a condition by inserting the appropriate stationary distribution:

$$P^{\mathrm{st}}(x,t)\propto\exp\left[\int^{x}\mathrm{d}x'\frac{m\gamma}{k_{\mathrm{B}}T}\left(-\frac{\partial\phi(x',\lambda_0(t))}{\partial x'}+f(x',\lambda_1(t))\right)\right], \tag{1.70}$$

which is clearly of a canonical form. How can one break this condition? We would require a nonvanishing current and this can be achieved when the contents of the exponential in Eq. (1.70) are not integrable. In general, this can be achieved by using an external force that is nonconservative. However, in one dimension with natural, that is reflecting, boundary conditions, any force acts conservatively since the total distance between initial and final positions and thus work done are always path independent. To enable such a nonconservative force, one can implement periodic boundary conditions. This can be realized physically by considering motion on a ring since when a constant force acts on the particle, the work done will depend on the number of times the particle traverses the ring. If the system relaxes to its stationary state, there will be a nonzero, but constant current that

4) One can build models that have stationary states with zero entropy production where equilibrium is only local, but there is no value in distinguishing between the two situations or highlighting such cases here.

arises due to the nonconservative force driving the motion in one direction. In such a system with steady flow, it is quite easy to understand that the transitions in each direction between two configurations will not cancel and thus detailed balance is not achieved. Allowing these dynamics to relax the system to its stationary state creates a simple example of a *nonequilibrium steady state*. Generally, such states can be created by placing some constraint upon the system that stops it from reaching a thermal equilibrium. This results in a system that is perpetually attempting and failing to maximize the total entropy by equilibrating. By remaining out of equilibrium, it constantly dissipates heat to the environment and is thus associated with a constant entropy generation. As such, a system with these dynamics gives rise to irreversibility beyond that arising from driving and relaxation and possesses an underlying breakage of time reversal symmetry, leading to an associated entropy production, manifest in the lack of detailed balance.

Detailed balance may be broken in many ways and the nonequilibrium constraint that causes it may be, as we have seen, a nonconservative force or it might be an exposure to particle reservoirs with unequal chemical potentials or heat baths with unequal temperatures. The steady states of such systems in particular are of great interest in statistical physics, not only because of their qualitatively different behavior but also because they provide cases where analytical solution is feasible out of equilibrium. As we shall see later, the distribution of entropy production in these states also obeys a particular powerful symmetry requirement.

1.7
A General Fluctuation Theorem

So far, we have examined a particular functional of a path and argued from a number of perspectives that it represents the total entropy production of the universe. We have also seen that it obeys a remarkably simple and powerful relation that guarantees its positivity on average. However, we can exploit the form of the entropy production further and derive a number of fluctuation theorems that explicitly relate distributions of general entropy-like quantities. They are numerous and the differences can appear rather subtle; however, it is quite simple to derive a very general equality that we can rigorously and systematically adapt to different situations and arrive at these different relations. To do so, let us once again consider the functional that represents the total entropy production:

$$\Delta S_{\text{tot}}[X] = k_{\text{B}} \ln \left[\frac{P_{\text{start}}(X(0)) P^F[X(\tau)|X(0)]}{P^R_{\text{start}}(\bar{X}(0)) P^R[\bar{X}(\tau)|\bar{X}(0)]} \right]. \tag{1.71}$$

We are able to construct the probability distribution of this quantity for a particular process. Mathematically, the distribution of entropy production over the forward process can be written as

$$P^F(\Delta S_{\text{tot}}[X] = A) = \int \mathrm{d}X P_{\text{start}}(X(0)) P^F[X(\tau)|X(0)] \delta(A - \Delta S_{\text{tot}}[X]). \tag{1.72}$$

To proceed, we follow Harris and Schütz [20] and consider a new functional, but one that is very similar to the total entropy production. We shall generally refer to it as R and it can be written as

$$R[X] = k_B \ln \left[\frac{P^R_{\text{start}}(X(0)) P^R[X(\tau)|X(0)]}{P_{\text{start}}(\bar{X}(0)) P^F[\bar{X}(\tau)|\bar{X}(0)]} \right]. \tag{1.73}$$

Imagine that we evaluate this new quantity over the reverse trajectory, that is, we consider $R[\bar{X}]$. It will be given by

$$R[\bar{X}] = k_B \ln \left[\frac{P^R_{\text{start}}(\bar{X}(0)) P^R[\bar{X}(\tau)|\bar{X}(0)]}{P_{\text{start}}(X(0)) P^F[X(\tau)|X(0)]} \right] = -\Delta S_{\text{tot}}[X], \tag{1.74}$$

which is explicitly the negative value of the functional that represents the total entropy production in the forward process. We can similarly construct a distribution for $R[\bar{X}]$ over the reverse process. This in turn would be given as

$$P^R(R[\bar{X}] = A) = \int d\bar{X} P^R_{\text{start}}(\bar{X}(0)) P^R[\bar{X}(\tau)|\bar{X}(0)] \delta(A - R[\bar{X}]). \tag{1.75}$$

We now seek to relate this distribution to that of the total entropy production over the forward process. To do so, we consider the value the probability distribution takes for $R[\bar{X}] = -A$. By the symmetry of the delta function, we may write

$$P^R(R[\bar{X}] = -A) = \int d\bar{X} P^R_{\text{start}}(\bar{X}(0)) P^R[\bar{X}(\tau)|\bar{X}(0)] \delta(A + R[\bar{X}]). \tag{1.76}$$

We now utilize three substitutions. First, $dX = d\bar{X}$ denoting the equivalence of the path integrals owing to the Jacobian of unity. Next, we use the definition of the entropy production functional to substitute

$$P^R_{\text{start}}(\bar{X}(0)) P^R[\bar{X}(\tau)|\bar{X}(0)] = P_{\text{start}}(X(0)) P^F[X(\tau)|X(0)] e^{-\Delta S_{\text{tot}}[X]/k_B} \tag{1.77}$$

and finally the definition that $R[\bar{X}] = -\Delta S_{\text{tot}}[X]$. Performing the above substitutions, we find

$$\begin{aligned} P^R(R[\bar{X}] = -A) &= \int dX P_{\text{start}}(X(0)) P^F[X(\tau)|X(0)] e^{-\Delta S_{\text{tot}}[X]/k_B} \delta(A - \Delta S_{\text{tot}}[X]) \\ &= e^{-(A/k_B)} \int dX P_{\text{start}}(X(0)) P^F[X(\tau)|X(0)] \delta(A - \Delta S_{\text{tot}}[X]) \\ &= e^{-(A/k_B)} P^F(\Delta S_{\text{tot}}[X] = A) \end{aligned} \tag{1.78}$$

and yields the following theorem [20]:

Transient Fluctuation Theorem

$$P^R(R[\bar{X}] = -A) = e^{-(A/k_B)} P^F(\Delta S_{\text{tot}}[X] = A). \tag{1.79}$$

This is a fundamental relation and holds for all protocols and initial conditions and is of a form referred to in the literature as a finite time, transient, or detailed fluctuation theorem depending on where you look. In addition, if we integrate over all values of A on both sides, we obtain the integral fluctuation theorem:

$$1 = \langle e^{-\Delta S_{\text{tot}}/k_B} \rangle, \tag{1.80}$$

with its name now being self-explanatory. These two relations shall now form the basis of all relations we consider. However, upon returning to the transient fluctuation theorem, a valid question is what does the functional $R[\bar{X}]$ represent? In terms of traditional thermodynamic quantities, there is scant physical interpretation. It is more helpful to consider it as a related functional of the path and to understand that in general it is *not* the entropy production of the reverse path in the reverse process. It is important now to look at why. To construct the entropy production under the reverse process, we need to consider a new functional that we shall call $\Delta S_{\text{tot}}^{R}[\bar{X}]$, which is defined in exactly the same way as for the forward process. We consider an initial distribution, this time P_{start}^{R} that evolves to P_{end}^{R}, and compare the probability density for a trajectory starting from the initial distribution, this time under the reverse protocol $\bar{\lambda}(t)$, with the probability density of a trajectory starting from the time-reversed final distribution, $\hat{T}P_{\text{end}}^{R}$, so that

$$\Delta S_{\text{tot}}^{R}[\bar{X}] = k_{\text{B}} \ln \left[\frac{P_{\text{start}}^{R}(\bar{X}(0))P^{R}[\bar{X}(\tau)|\bar{X}(0)]}{\hat{T}P_{\text{end}}^{R}(X(0))P^{F}[X(\tau)|X(0)]} \right] \neq -\Delta S_{\text{tot}}[X]. \tag{1.81}$$

Crucially there is an inequality in Eq. (1.81) in general because

$$\hat{T}P_{\text{start}}(X(0))) \neq P_{\text{end}}^{R}(\bar{X}(\tau)) = \int \mathrm{d}\bar{X} P_{\text{start}}^{R}(\bar{X}(0))P^{R}[\bar{X}(\tau)|\bar{X}(0)]. \tag{1.82}$$

This is manifest in the irreversibility of the dynamics of the systems we are looking at, as is illustrated in Figure 1.1. If the dynamics were reversible, as for Hamilton's equations and Liouville's theorem, then Eq. (1.82) would hold in equality. So, examining Eqs. (1.79) and (1.81), if we wish to compare the distribution of entropy production in the reverse process with that for the forward process, we need to have $R[\bar{X}] = \Delta S_{\text{tot}}^{R}[\bar{X}]$ such that $\Delta S_{\text{tot}}[X] = -\Delta S_{\text{tot}}^{R}[\bar{X}]$. This is achieved by having $P_{\text{start}}(X(0)) = \hat{T}P_{\text{end}}^{R}(\bar{X}(\tau))$. When this condition is met, we may write

$$P^{R}(\Delta S_{\text{tot}}^{R}[\bar{X}] = -A) = \mathrm{e}^{-(A/k_{\text{B}})} P^{F}(\Delta S_{\text{tot}}[X] = A), \tag{1.83}$$

which now relates distributions of the same physical quantity, entropy change. If we assume that arguments of a probability distribution for the reverse protocol P^{R} implicitly describe the quantity over the reverse process, we may write it in its more common form:

$$P^{R}(-\Delta S_{\text{tot}}) = \mathrm{e}^{-\Delta S_{\text{tot}}/k_{\text{B}}} P^{F}(\Delta S_{\text{tot}}). \tag{1.84}$$

This will hold when the protocol and initial distributions are chosen such that evolution under the forward process followed by the reverse process together with the appropriate time reversals brings the system back into the same initial statistical distribution. This sounds somewhat challenging and indeed does not occur in any generality, but there are two particularly pertinent situations where the above does hold and has particular relevance in a discussion of thermodynamic quantities.

1.7.1 Work Relations

The first and most readily applicable example that obeys the condition $P_{\text{start}}(X(0)) = \hat{T}P^F_{\text{end}}(\bar{X}(\tau))$ involves changes between equilibrium states where one can trivially obtain the required condition by exploiting the fact that unperturbed, the dynamics will steadily bring the system into a stationary state that is invariant under time reversal. We start by defining the equilibrium distribution that represents the canonical ensemble where, as before, we consider the system energy for an overdamped system to be entirely described by the potential $\phi(x, \lambda_0(t))$ such that

$$p^{\text{eq}}(x(t), \lambda_0(t)) = \frac{1}{Z(\lambda_0(t))} \exp\left[-\frac{\phi(x(t), \lambda_0(t))}{k_B T}\right] \tag{1.85}$$

for $t = 0$ and τ, where Z is the partition function, uniquely defined by $\lambda_0(t)$, which can in general be related to the Helmholtz free energy through the relation

$$F(\lambda_0(t)) = -k_B T \ln Z(\lambda_0(t)). \tag{1.86}$$

To clarify, the corollary of these statements is to say that the directly applied force $f(x(t), \lambda_1(t))$ does not feature in the system's Hamiltonian.[5] Let us now choose the initial and final distributions to be given by the respective equilibriums defined by the protocol at the start and finish of the forward process and the same temperature:

$$\begin{aligned} P_{\text{start}}(x(0), \lambda_0(0)) &\propto \exp\left[\frac{F(\lambda_0(0)) - \phi(x(0), \lambda_0(0))}{k_B T}\right], \\ P_{\text{end}}(x(\tau), \lambda_0(\tau)) &\propto \exp\left[\frac{F(\lambda_0(\tau)) - \phi(x(\tau), \lambda_0(\tau))}{k_B T}\right]. \end{aligned} \tag{1.87}$$

We are now in a position to construct the total entropy change for a given realization of the dynamics between these two states. From the initial and final distributions, we can immediately construct the system entropy change ΔS_{sys} as

$$\begin{aligned} \Delta S_{\text{sys}} &= k_B \ln\left(\frac{P_{\text{start}}(x(0), \lambda_0(0))}{P_{\text{end}}(x(\tau), \lambda_0(\tau))}\right) = k_B \ln\left(\frac{\exp\left[(F(\lambda_0(0)) - \phi(x(0), \lambda_0(0)))/k_B T\right]}{\exp\left[(F(\lambda_0(\tau)) - \phi(x(\tau), \lambda_0(\tau)))/k_B T\right]}\right) \\ &= \frac{1}{T}(-F(\lambda_0(\tau)) + F(\lambda_0(0)) + \phi(x(\tau), \lambda_0(\tau)) - \phi(x(0), \lambda_0(0))) \\ &= \frac{\Delta\phi - \Delta F}{T}. \end{aligned} \tag{1.88}$$

The medium entropy change is as we defined previously and can be written

$$\Delta S_{\text{med}} = -\frac{\Delta Q}{T} = \frac{\Delta W - \Delta\phi}{T}, \tag{1.89}$$

5) That is not to say it may not appear in some generalized Hamiltonian. For further insight into this issue, we refer the interested reader to Refs [21, 22], noting that the approach here and elsewhere [23] best resembles the extended relation used in Ref. [21].

where ΔW is the work given earlier in Eq. (1.58), but we now emphasize that this term contains contributions due to changes in the potential and due to the external force f. We thus further define two new quantities ΔW_0 and ΔW_1 such that $\Delta W = \Delta W_0 + \Delta W_1$ with

$$\Delta W_0 = \int_0^\tau \frac{\partial \phi(x(t), \lambda_0(t))}{\partial \lambda_0} \frac{d\lambda_0(t)}{dt} dt \tag{1.90}$$

and

$$\Delta W_1 = \int_0^\tau f(x(t), \lambda_1(t)) \circ dx. \tag{1.91}$$

W_0 and W_1 are not defined in the same way with W_0 being found more often in thermodynamics and W_1 being a familiar definition from mechanics: one may therefore refer to these definitions as thermodynamic and mechanical work, respectively. The total entropy production in this case is simply given by

$$\Delta S_{\mathrm{tot}}[x] = \frac{\Delta W - \Delta F}{T}. \tag{1.92}$$

In addition, since we have established that $P^R_{\mathrm{end}}(\bar{x}(\tau)) = \hat{T} P_{\mathrm{start}}(x(0))$, we can also write

$$\Delta S^R_{\mathrm{tot}}[\bar{x}] = -\frac{\Delta W - \Delta F}{T}. \tag{1.93}$$

1.7.1.1 The Crooks Work Relation and Jarzynski Equality

The derivation of several relations follows now by imposing certain constraints on the process we consider. First let us imagine the situation where the external force $f(x, \lambda_1) = 0$ and so all work is performed conservatively through the potential such that $\Delta W = \Delta W_0$. To proceed, we should clarify the form of the protocol that would take an equilibrium system to a new equilibrium such that its reversed counterpart would return the system to the same initial distribution. This would consist of a waiting period, in principle of infinite duration, where the protocol is constant, followed by a period of driving where the protocol changes, followed by another infinitely long waiting period. Such a protocol is given and explained in Figure 1.2.

For such a process, we write the total entropy production:

$$\Delta S_{\mathrm{tot}} = \frac{\Delta W_0 - \Delta F}{T}. \tag{1.94}$$

This changes its sign for the reverse trajectory and reverse protocol and so we may construct the appropriate fluctuation relation that is now simply read off Eq. (1.79) as

$$P^F((\Delta W_0 - \Delta F)/T) = \exp\left[\frac{\Delta W_0 - \Delta F}{k_B T}\right] P^R(-(\Delta W_0 - \Delta F)/T). \tag{1.95}$$

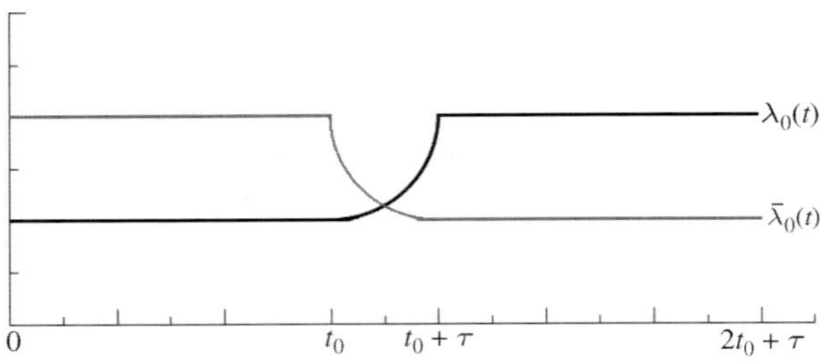

Figure 1.2 A protocol $\lambda_0(t)$, of duration $2t_0 + \tau$, which evolves a system from one equilibrium to another, defined so that the reversed protocol $\bar{\lambda}_0(t)$ returns the system to the original equilibrium. There is a period of no driving of length t_0 that corresponds to the relaxation for the reverse process, followed by a time τ of driving, followed by another relaxation period of duration t_0. As we take t_0 to infinity, we obtain a protocol that produces the condition $P^R_{\text{end}}(\bar{x}(2t_0 + \tau)) = \hat{T}P_{\text{start}}(x(0))$. We note here that $f(x(t), \lambda_1(t)) = 0$.

Since F and T are independent of the trajectory, we can simplify and find the following [5]:

The Crooks Work Relation

$$\frac{P^F(\Delta W_0)}{P^R(-\Delta W_0)} = \exp\left[\frac{\Delta W_0 - \Delta F}{k_B T}\right]. \tag{1.96}$$

Rearranging and integrating over all ΔW on both sides and taking the deterministic ΔF out of the path integral then yields an expression for the average over the forward process [4, 15, 24]:

The Jarzynski Equality

$$\langle \exp(-\Delta W_0/k_B T)\rangle = \exp(-\Delta F/k_B T). \tag{1.97}$$

The power of these statements is clarified in one very important conceptual point. In their formulation, the relations are *constructed* using the values of entropy change for a process that, after starting in equilibrium, is isolated for a long time, driven, and then left for a long time again to return to a stationary state. However, this does not mean that these quantities have to be *measured* over the whole of such a process. Why is this the case? It is because the entropy production for the whole process can be written in terms of the mechanical work and free energy change that are delivered exclusively during the driving phase when the protocol $\lambda_0(t)$ is changing. Since the work and free energy change are independent of the intervals where the protocol is constant and because we had no constraint on $\lambda_0(t)$ during the driving phase, we can therefore consider them to be valid for any protocol assuming the system is in equilibrium to start with. We can therefore state that the Crooks work relation and Jarzynski equality hold *for all times* for systems that start in equilibrium.[6] Historically, this has had one particularly important consequence: the

6) Although we have shown that this is the case for Langevin dynamics, it is important to note that these expressions can be obtained for other general descriptions of the dynamics. See Ref. [25].

results hold for driving, in principle, arbitrarily far from equilibrium. This is widely summed up as the ability to obtain equilibrium information from nonequilibrium averaging since, upon examining the form of the Jarzynski equality, we can compute the free energy difference by taking an average of the exponentiated work done in the course of some nonequilibrium process. Exploiting these facts, let us clarify what these two relations mean explicitly and what the implications are in the real world.

The Crooks Relation

Statement: For any time τ, the probability of observing trajectories that correspond to an application of ΔW_0 work, starting from an equilibrium state defined by $\lambda(0)$, under dynamics described by $\lambda(t)$ in $0 \leq t \leq \tau$, is exponentially more likely in $(\Delta W_0 - \Delta F)/k_B T$ than the probability of observing trajectories that correspond to an application of $-\Delta W_0$ work from an equilibrium state defined by $\bar{\lambda}(0)$, under dynamics described by $\bar{\lambda}(t)$.
Implication: Consider an isothermal gas in a piston in contact with a heat bath at equilibrium. Classically, we know from the second law that if we compress the gas, performing a given amount of work ΔW_0 on it, then after an equilibration period, we must expect the gas to perform an amount of work that is less than ΔW_0 when it is expanded (i.e., $-\Delta W_0$ work performed on the gas). To get the same amount of work back out, we need to perform the process quasi-statically such that it is reversible. The Crooks relation, however, tells us more. For the same example, we can state that if the dynamics of our system lead to some probability of performing ΔW_0 work, then the probability of extracting the same amount of work in the reverse process differs exponentially. Indeed, they only have the same probability when the work performed is equal to the free energy difference, often called the reversible work.

The Jarzynski Equality

Statement: For any time τ, the average value, as defined by the mean over many realizations of the dynamics, of the exponential of the negative work divided by the temperature arising from a defined change in protocol from $\lambda_0(0)$ to $\lambda_0(\tau)$ is identically equal to the exponential of the negative equilibrium free energy difference corresponding to the same change in protocol, divided by the temperature.
Implication: Consider once again the compression of a gas in a piston, but let us imagine that we wish to know the free energy change without knowledge of the equation of state. Classically, we may be able to measure the free energy change by attempting to perform the compression quasi-statically, which of course can never be fully realized. However, the Jarzynski equality states that we can determine this free energy change exactly by repeatedly compressing the gas *at any speed* and taking an average of the exponentiated work that we perform over all these fast compressions. One must, however,

exercise caution; the average taken is patently dominated by very negative values of work. These correspond to very negative excursions in entropy and are often *extremely* rare. One may find that the estimated free energy change is significantly altered following one additional realization even if hundreds or perhaps thousands have already been averaged.

These relations very concisely extend the classical definition of irreversibility in such isothermal systems. In classical thermodynamics, we may identify the difference in free energy as the maximum amount of work we may extract from the system, or rather that to achieve a given free energy change, we must perform at least as much work as that free energy change, that is,

$$\Delta W_0 \geq \Delta F, \tag{1.98}$$

with the equality holding for a quasi-static "reversible" process. But since we saw that our entropy functional could take negative values, there is nothing in the dynamics that prevents an outcome where the work is less than the free energy change. We understand now that the second law is statistical, so more generally we must have

$$\langle \Delta W_0 \rangle \geq \Delta F. \tag{1.99}$$

The Jarzynski equality tells us more than this and replaces the inequality with an equality that it is valid for nonquasistatic processes where mechanical work is performed at a finite rate such that the system is driven away from thermal equilibrium and the process is irreversible.

1.7.2
Fluctuation Relations for Mechanical Work

Let us now consider a circumstance similar, but subtly different, to that of the Jarzynski and Crooks relations. We consider a driving process that again starts in equilibrium, but this time keeps the protocol $\lambda_0(t)$ held fixed such that all work is performed by the externally applied force $f(x(t), \lambda_1(t))$, meaning $\Delta W = \Delta W_1$. Once again we seek a fluctuation relation by constructing an equilibrium-to-equilibrium process, though this time we insist that the system relaxes back to the same initial equilibrium distribution. We note that since $f(x(t), \lambda_1(t))$ may act nonconservatively, in order to allow relaxation back to equilibrium we would require that the external force be "turned off." An example set of protocols is given in Figure 1.3 for a simple external force $f(x(t), \lambda_1(t)) = \lambda_1(t)$.

For such a process, we find

$$\Delta S_{\text{tot}} = \frac{\Delta W_1}{T}, \tag{1.100}$$

since the free energy difference between the same equilibrium states vanishes. We have constructed a process such that the distribution at the end of the reverse process is (with time reversal) the same as the initial distribution of the forward

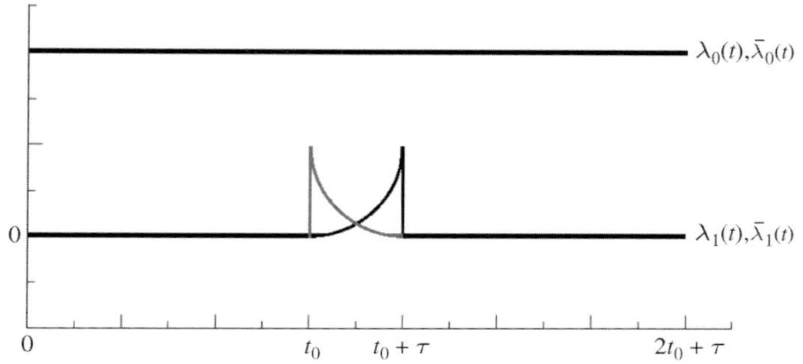

Figure 1.3 An example protocol and reversed protocol that would construct the condition $P^R_{\text{end}}(\bar{x}(2t_0 + \tau)) = \hat{T}P_{\text{start}}(x(0))$ when all work is performed through the external force $f(x(t), \lambda_1(t)) = \lambda_1(t)$ and t_0 is taken to infinity.

process and so again we are permitted to read off a set of fluctuation relations [21, 22, 26, 27] that may collectively be referred to as the following:

Fluctuation relations for mechanical work

$$\frac{P^F(\Delta W_1)}{P^R(-\Delta W_1)} = \exp\left[\frac{\Delta W_1}{k_B T}\right], \tag{1.101}$$

$$\langle \exp(-\Delta W_1/k_B T)\rangle = 1. \tag{1.102}$$

For the same reasons as in the Jarzynski and Crooks relations, they are valid for all times and thus hold as a nonequilibrium result. Taking in particular the integrated relation and comparing with the Jarzynski equality in Eq. (1.97), one may think there is an inconsistency. Both are valid for all times and arbitrary driving and concern the work done under the constraint that both start in equilibrium, yet on first inspection they seem to be saying different things. But recall our distinction between the work ΔW_0 and ΔW_1 from Eqs. (1.90) and (1.91); there are two distinct ways to describe work on such a particle. If one performs work ΔW_0, one necessarily changes the form of the system energy, whereas the application of work ΔW_1 leaves the form of the system energy unchanged. The difference is manifest in the two different integrated relations because their derivations exploit the fact that the Hamiltonian, which represents the system energy, appears in initial and final distributions. To clarify, as written the Jarzynski equality explicitly concerns driving where the application of any work also changes the Hamiltonian and thus the equilibrium state. On the other hand, the relations for W_1 concern work as the path integral of an external force such that the Hamiltonian remains unchanged for the entire process.

Of course, there is nothing in the derivation of either of these relations that precludes the possibility of both types of work to be performed at the same time and so using the same arguments, we arrive at

$$\frac{P^F(\Delta W)}{P^R(-\Delta W)} = \exp\left[(\Delta W - \Delta F)/k_B T\right] \tag{1.103}$$

and

$$\left\langle \exp\left(-\frac{\Delta W - \Delta F}{k_B T}\right)\right\rangle = 1, \tag{1.104}$$

again under the constraint that the system be initially prepared in equilibrium.

1.7.3 Fluctuation Theorems for Entropy Production

We have seen in Section 1.7.1 how relations between distributions of work can be derived from Eq. (1.84), since work can be related to the entropy production during a suitable equilibrium-to-equilibrium process. We wish now to seek situations where we can explicitly construct relations that concern the distributions of the entropy produced for forward and reverse processes that do not necessarily begin and end in equilibrium. In order to find situations where the value of the entropy production for the forward trajectory in the forward process is precisely the negative value of the entropy production for the reversed trajectory in the reverse process, we seek situations where the reverse protocol acts to return the distribution to the time-reversed initial distribution for the forward process. For the overdamped motion we have been considering, we would require

$$P_{\text{start}}(x(\tau)) = \int dx P_{\text{end}}(x(0)) P^R[x(\tau)|x(0)]. \tag{1.105}$$

This is a slightly more general situation than that previously considered of an equilibrium-to-equilibrium process, and in such cases one can expect to see a symmetry between distributions of entropy production in such forward and reverse processes along the lines of Eq. (1.84).

However, we can specify further and find an even more direct symmetry if we insist that the evolution under the forward process is indistinguishable from that under the reverse process. Mathematically, this means $P(x_{i+1}|x_i, \lambda(t_{i+1})) = P(x_{i+1}|x_i, \bar{\lambda}(t_{i+1}))$ or $P^R[x(\tau)|x(0)] = P^F[x(\tau)|x(0)]$. Given these conditions, evolution from the initial distribution will result in the final distribution and evolution under the reverse process from the final distribution will result in the initial distribution and these distributions will be the same. If we consider in more detail the requirements for such a condition, we understand there are two main ways in which this can be achieved. The first way is to require a constant protocol $\lambda(t)$. In this way, the forward process is trivially the same as the reverse process. Alternatively, we require the protocol to be time symmetric such that $\lambda(t) = \lambda(\tau - t) = \bar{\lambda}(t)$. In both situations, the forward and reverse processes are entirely indistinguishable. As such, by careful construction, we can, in these specific circumstances, relate the probability of seeing a positive entropy production to that of a negative entropy production over the *same forward process* allowing us from Eq. (1.79) to write the following [14]:

Detailed fluctuation theorem

$$P(\Delta S_{\text{tot}}) = e^{\Delta S_{\text{tot}}/k_B} P(-\Delta S_{\text{tot}}) \tag{1.106}$$

Physically, the two situations we have considered correspond to the following:

- $P_{\text{start}} = P_{\text{end}}, \quad \lambda(t) = \text{constant}$
  To satisfy such criteria, the system must be in a steady state, that is, all intrinsic system properties (probability distribution, mean system entropy, mean system energy etc.) must remain constant over the process. The simplest steady state is equilibrium that trivially has zero entropy production in detail for all trajectories. However, a *nonequilibrium* steady state can be achieved by breaking detailed balance through some constraint that prevents the equilibration, as we saw in Section 1.6. The mean entropy production rate of these states is constant, nonzero, and, as we have now shown, there is an explicit exponential symmetry in the probability of positive and negative fluctuations.
- $P_{\text{start}} = P_{\text{end}} = P, \quad \lambda(t) = \bar{\lambda}(t)$
  This condition can be achieved in a system that is being periodically driven characterized by a time symmetric $\lambda(t)$. If from some starting point we allow the system to undergo an arbitrarily large number of periods of driving, it will arrive at a so-called nonequilibrium oscillatory state such that $P(x,t) = P(x, t + t_p)$, where t_p is the period of oscillation. In this state, we can expect the above relation to hold for integer multiples of period t_p starting from a time such that $\lambda(t) = \bar{\lambda}(t)$.

1.8 Further Results

1.8.1 Asymptotic Fluctuation Theorems

In the class of system we have considered, the integral and detailed fluctuation theorems are guaranteed to hold. Indeed, it has not escaped some authors' attention that the reason they do is fully explained in the very definition of the functionals they concern [28]. There is, however, a class of fluctuation theorems that does not have this property. These are known as asymptotic fluctuation theorems. Their derivation for Langevin and then general Markovian stochastic systems is due to Kurchan [29] and Lebowitz and Spohn [30], respectively, and superficially bear strong similarities with results obtained by Gallavotti and Cohen for chaotic deterministic systems [3, 31]. They generally apply to systems that approach a steady state, and, for stochastic systems, strictly in their definition, concern a symmetry in the long-time limit of the generating function of a quantity known as an action functional or flux [30]. This quantity again relates to a trajectory that runs from x_0 through to x_n, described using jump probabilities $\sigma(x_i|x_{i-1})$ and is given as

$$\mathcal{W}(t) = \ln\left[\frac{\sigma(x_1|x_0)}{\sigma(x_0|x_1)} \cdots \frac{\sigma(x_n|x_{n-1})}{\sigma(x_{n-1}|x_n)}\right]. \tag{1.107}$$

An asymptotic fluctuation theorem, which we shall not prove and only briefly address here, states that there exists a long-time limit of the scaled cumulant generating function of $\mathcal{W}(t)$ such that

$$e(s) = \lim_{t\to\infty} -\frac{1}{t} \ln\langle\exp[-s\mathcal{W}(t)]\rangle \tag{1.108}$$

and that this quantity possesses the symmetry

$$e(s) = e(1-s). \tag{1.109}$$

From this somewhat technical definition, we can derive a fluctuation theorem closely related to those that we have already examined. The existence of such a limit implies that the distribution function $P(\mathcal{W}(t)/t)$ of the time-averaged action functional $\mathcal{W}(t)/t$ follows a large deviation behavior [32] such that in the long-time limit, we have

$$P(\mathcal{W}(t)/t) \simeq \mathrm{e}^{-t\hat{e}(\mathcal{W}/t)}, \tag{1.110}$$

where $\hat{e}$ is the Legendre transform of e defined as

$$\hat{e}(\mathcal{W}/t) = \max_s [e(s) - s(\mathcal{W}/t)], \tag{1.111}$$

maximizing over the conjugate variable s. Consequently, using the symmetry relation of Eq. (1.109), we may write

$$\begin{aligned}\hat{e}(\mathcal{W}/t) &= \max_s [e(s) - (1-s)(\mathcal{W}/t)]\\ &= \max_s [e(s) + s(\mathcal{W}/t)] - (\mathcal{W}/t)\\ &= \hat{e}(-\mathcal{W}/t) - (\mathcal{W}/t).\end{aligned} \tag{1.112}$$

Since we expect large deviation behavior described by Eq. (1.110), this implies

$$P(\mathcal{W}/t) \simeq P(-\mathcal{W}/t)\mathrm{e}^{\mathcal{W}} \tag{1.113}$$

or, equivalently,

$$P(\mathcal{W}) \simeq P(-\mathcal{W})\mathrm{e}^{\mathcal{W}}, \tag{1.114}$$

which is clearly analogous to the fluctuation theorems we have seen previously. Taking a closer look at the action functional $\mathcal{W}$, we see that it is, for the systems we have been considering, a representation of the entropy produced in the medium or a measure of the heat dissipated, up to a constant k_B. Unlike the fluctuation theorems considered earlier, this is not guaranteed for all systems. To get a basic understanding of this subtlety, we write the asymptotic fluctuation theorem for the medium entropy production for the continuous systems we have been considering in the form

$$\frac{P(\Delta S_{\mathrm{med}})}{P(-\Delta S_{\mathrm{med}})} \simeq \mathrm{e}^{\Delta S_{\mathrm{med}}/k_B}. \tag{1.115}$$

However, we know that for stationary states, the following fluctuation theorem holds for all time:

$$\frac{P(\Delta S_{\text{tot}})}{P(-\Delta S_{\text{tot}})} = \mathrm{e}^{\Delta S_{\text{tot}}/k_{\text{B}}}. \tag{1.116}$$

Since $\Delta S_{\text{tot}} = \Delta S_{\text{med}} + \Delta S_{\text{sys}}$, we may understand that the asymptotic symmetry will exist when the system entropy change is negligible compared to the medium entropy change. In nonequilibrium stationary states, we expect a continuous dissipation of heat and thus an increase of medium entropy, along with a change in system entropy that *on average* is zero. One may naively suggest that this guarantees the asymptotic symmetry since the medium entropy is unbounded and can grow indefinitely. However, if the system configuration space is unbounded, one cannot in general rule out large fluctuations to regions with *arbitrarily* low probability densities and therefore large changes in system entropy, which in principle can persist on any timescale. What is required to guarantee such a relation is the ability to neglect, in detail for all trajectories, the system entropy change compared to medium entropy change on long timescales. This can be done in general if we insist that the state space is bounded. This means that the system entropy has well-defined maximum and minimum values that can be assumed to be unimportant on long timescales and so the asymptotic symmetry necessarily follows. We note finally that systems with unbounded state space are ubiquitous and include simple harmonic oscillators [33] and so investigations of fluctuation theorems for such systems have yielded a wealth of nontrivial generalizations and extensions.

1.8.2 Generalizations and Consideration of Alternative Dynamics

What we hope the reader might appreciate following reading this chapter is the malleability of quantities that satisfy fluctuation relations. It is not difficult to identify quantities that obey relations similar to the fluctuation theorems (although it may be hard to show that they have any particular physical relevance) since the procedure simply relies on a transformation of a path integral utilizing the definition of the entropy production itself. To clarify this point, we consider some generalizations of the relations we have seen. Let us consider a new quantity that has the same form as the total entropy production:

$$G[X] = k_{\text{B}} \ln \left[\frac{P^F[X]}{P[Y]} \right]. \tag{1.117}$$

Here $P^F[X]$ is the same as before, yet we deliberately say very little about $P[Y]$ other than it is a probability density of observing some path Y related to X defined on the same space as X. Let us compute the average of the negative exponential of this quantity:

$$\begin{aligned}\langle e^{-G/k_B}\rangle &= \int dX\, P_{\text{start}}(X(0))P^F[X(\tau)|X(0)]\frac{P(Y(0))P[Y(\tau)|Y(0)]}{P_{\text{start}}(X(0))P^F[X(\tau)|X(0)]} \\ &= \int dY\, P(Y(0))P[Y(\tau)|Y(0)] \\ &= 1.\end{aligned} \tag{1.118}$$

If we have $dX = dY$ such that the Jacobian of the transformation to path Y is unity and the unspecified initial distribution $P(Y(0))$ and transition probability density $P[Y(\tau)|Y(0)]$ are normalized, then *any* such quantity $G[X]$ will obey an integral fluctuation theorem averaged over the forward process. Clearly, there are as many relations as there are ways to define $P[Y(\tau)|Y(0)]$ and $P(Y(0))$ and most will mean very little at all [28]. However, there are several such relations in the literature obtained by an appropriate choice of $P(Y(0))$ and $P[Y(\tau)|Y(0)]$ that say something meaningful, including, for example, the Seifert end point relations [34]. We will very briefly allude to just two ways that this can be achieved by first noting that one may consider an alternative dynamics known as "adjoint" dynamics, leading to conditional path probabilities written $P_{\text{ad}}[Y(\tau)|Y(0)]$, defined such that they generate the same stationary distribution as the normal dynamics, but with a probability current of the opposite sign [35]. For the overdamped motion that we have been considering, where $P^F[X] = P_{\text{start}}(x(0))$ $P^F[x(\tau)|x(0)]$, we may derive the following results:

- **Hatano–Sasa relation:** By choosing

$$P(Y(0)) = P^R_{\text{start}}(\bar{x}(0)) \tag{1.119}$$

and

$$P[Y(\tau)|Y(0)] = P^R_{\text{ad}}[\bar{x}(\tau)|\bar{x}(0)], \tag{1.120}$$

we obtain the Hatano–Sasa relation [36] or integral fluctuation theorem for the "nonadiabatic entropy production" [37–39] that concerns the so-called "excess heat" transferred to the environment [40] such that

$$\langle \exp\left[-\Delta Q_{\text{ex}}/k_B T - \Delta S_{\text{sys}}/k_B\right]\rangle = 1. \tag{1.121}$$

The exponent here is best described as the negative of the entropy production associated with movement to stationarity, which phenomenologically includes transitions between nonequilibrium stationary states for which it was first derived. This use of the P^R_{ad} adjoint dynamics is frequently described as a reversal of both the protocol and dynamics [35] to be contrasted with reversal of just the protocol for the integral fluctuation theorem.

- **Relation for the housekeeping heat:** By choosing

$$P(Y(0)) = P_{\text{start}}(x(0)) \tag{1.122}$$

and

$$P[Y(\tau)|Y(0)] = P^F_{\text{ad}}[x(\tau)|x(0)], \tag{1.123}$$

we obtain the Speck–Seifert integral relation [41] or integral fluctuation theorem for the "adiabatic entropy production" [37–39], which concerns the so-called "housekeeping heat" absorbed by the environment [40] such that

$$\langle \exp\left[-\Delta Q_{\mathrm{hk}}/k_{\mathrm{B}}T\right]\rangle = 1, \tag{1.124}$$

where $\Delta Q_{\mathrm{hk}} = \Delta Q_{\mathrm{med}} - \Delta Q_{\mathrm{ex}}$ [40] and where the negative exponent is best described as the entropy production associated with the nonequilibrium steady state. Such a consideration might be called a reversal of the dynamics, but not the protocol.

Both these relations are relevant to the study of systems where detailed balance does not hold and amount to a division in the total entropy production, or irreversibility, into the two types we considered in Section 1.6, namely, the movement toward stationarity brought about by driving and relaxation, and the breaking of time reversal symmetry that arises specifically when detailed balance is absent. Consequently, if detailed balance does hold, then the exponent in Eq. (1.124) is zero and Eq. (1.121) reduces to the integral fluctuation theorem.

1.9 Fluctuation Relations for Reversible Deterministic Systems

So far we have chosen to focus on systems that obey stochastic dynamics, whereby the interaction with the environment, and the explicit breakage of time reversal symmetry, is implemented through the presence of random forces in the equations of motion. However, there exists a framework for deriving fluctuation relations that is based on deterministic dynamical equations, whereby the environmental interaction is represented through specific nonlinear terms [9], which supplement the usual forces in Newton's equations. These have the effect of constraining some aspect of the system, such as its kinetic energy, either to a chosen constant or to a particular distribution as time progresses. Most importantly, they can be reversible, such that a trajectory driven by a specified protocol and its time-reversed counterpart driven by a time-reversed protocol are both solutions to the dynamics. In practice, these so-called thermostating terms that provide the nonlinearity are taken to act solely on the boundaries (which can be made arbitrarily remote) in order for the system to be unaffected by the precise details of the input and removal of heat. This provides a parallel framework within which the dynamics of an open system, and hence fluctuation relations, can be explored. Indeed, it was through the consideration of deterministic, reversible dynamical systems that many of the seminal insights into fluctuation relations were first obtained [1].

Given that there was a choice over the framework to employ, we opted to use stochastic dynamics to develop this pedagogical overview. This has some benefit in that the concept of entropy change can be readily attached to the idea of the growth of uncertainty in system evolution, identifying it explicitly with the intrinsic irreversibility of the stochastic dynamics. Nevertheless, it is important to review the deterministic approach as well, and explore some of the additional insight that it provides.

The main outcome of seminal and ongoing studies by Evans and coworkers [1, 2, 8] is the identification of a system property that displays a tendency to grow with time under specified non-Hamiltonian reversible dynamics. The development of the H-theorem by Boltzmann was a similar attempt to identify such a quantity. However, we shall have to confront the fact that by their very nature, deterministic equations do not generate additional uncertainty as time progresses. The configuration of a system at a time t is precisely determined given the configuration at t_0. Even if the latter were specified only through a probability distribution, all future and past configurations associated with each starting configuration are fixed, and uncertainty is therefore not changed by the passage of time. Something other than the increase in uncertainty will have to emerge in a deterministic framework if it is to represent entropy change.

Within such a framework, a system is described in terms of a probability density for its dynamical variables $x, v, \ldots$, collectively denoted Γ. An initial probability density $P(\Gamma, 0)$ evolves under the dynamics into a density $P(\Gamma, t)$. Furthermore, the starting "point" of a trajectory Γ_0 (that is, $(x(0), v(0), \ldots)$ is linked uniquely to a terminating point Γ_t (that is, $x(t), v(t), \ldots$), passing through points $\Gamma_{t'}$ in between. It may then be shown [1, 2] that

$$P(\Gamma_t, t) = P(\Gamma_0, 0) \exp\left(-\int_0^t \Lambda(\Gamma_{t'}) dt'\right), \tag{1.125}$$

where $\Lambda(\Gamma_{t'})$ is known as the phase space contraction rate associated with configuration $\Gamma_{t'}$, which may be related specifically to the terms in the equations of motion that impose the thermal constraint. For a system without constraint, and hence thermally isolated, the phase space contraction rate is therefore zero everywhere, and the resulting $P(\Gamma_t, t) = P(\Gamma_0, 0)$ is an expression of Liouville's theorem: the conservation of probability density along any trajectory followed by the system.

For typically employed thermal constraints (denoted thermostats/ergostats, depending on their nature), it may be shown that the phase space contraction rate is related to the rate of heat transfer to the system from the implied environment. For the so-called Nose–Hoover thermostat at fixed target temperature T, we are able to write $\int_0^t \Lambda(\Gamma_{t'}) dt' = \Delta Q(\Gamma_0)/k_B T$, where $\Delta Q(\Gamma_0)$ is the heat transferred to the system over the course of a trajectory of duration t starting from Γ_0.

We now consider the dissipation function $\Omega(\Gamma)$ defined through

$$\int_0^t \Omega(\Gamma_{t'}) dt' = \ln\left(\frac{P(\Gamma_0, 0)}{P(\Gamma_t^*, 0)} \exp\left(-\int_0^t \Lambda(\Gamma_{t'}) dt'\right)\right), \tag{1.126}$$

where Γ_t^* is related to Γ_t by the reversal of all velocity coordinates. Assuming that the probability density at time zero is symmetric in velocities, such that

$P(\Gamma_0, 0) = P(\Gamma_0^*, 0)$ (which ensures that the right-hand side of Eq. (1.126) vanishes when $t = 0$), we can write

$$\int_0^t \Omega(\Gamma_{t'})\mathrm{d}t' = \bar{\Omega}_t(\Gamma_0)t = \ln\left(\frac{P(\Gamma_t, t)}{P(\Gamma_t, 0)}\right), \tag{1.127}$$

defining a mean dissipation function $\bar{\Omega}_t(\Gamma_0)$ for the trajectory starting from Γ_0 and of duration t. It is a quantity that will take a variety of values for a given protocol (the specified dynamics over the period in question) depending on Γ_0, and its distribution has particular properties, just as we found for the distributions of values of functionals such as ΔS_{tot} in Eq. (1.44). For example, we have

$$\begin{aligned}\left\langle \exp\left(-\bar{\Omega}_t t\right)\right\rangle &= \int \mathrm{d}\Gamma_0 P(\Gamma_0, 0)\exp\left(-\bar{\Omega}_t(\Gamma_0)t\right)\\ &= \int \mathrm{d}\Gamma_t P(\Gamma_t, t)\exp\left(-\bar{\Omega}_t t\right)\\ &= \int \mathrm{d}\Gamma_t P(\Gamma_t, 0) = 1,\end{aligned} \tag{1.128}$$

where the averaging is over the various possibilities for Γ_0, or equivalently for Γ_t, and where we have imposed a probability conservation condition $\mathrm{d}\Gamma_0 P(\Gamma_0, 0) = \mathrm{d}\Gamma_t P(\Gamma_t, t)$, implying that $\mathrm{d}\Gamma_t$ is the region of phase space around Γ_t that contains all the end points of trajectories starting within the region $\mathrm{d}\Gamma_0$ around Γ_0. This result takes the same form as the integral fluctuation theorem obtained using stochastic dynamics, but now involves the mean dissipation function. In the deterministic dynamics literature, the result Eq. (1.128) is known as a nonequilibrium partition identity. As a consequence, we can deduce that $\langle \bar{\Omega}_t \rangle \geq 0$, again a result that resembles several already encountered.

Now let us consider a protocol that is time symmetric about its midpoint, and for simplicity, consists of time variation in the form of the system's Hamiltonian. The thermal constraint, as discussed above, is imposed through reversible non-Hamiltonian terms in the dynamics (let us say the Nose–Hoover scheme) and is explicitly time independent and therefore isothermal. For such a case, it is clear that a trajectory running from Γ_0 to Γ_t over the time period $0 \to t$ can be generated in a velocity-reversed form and in reverse sequence, by evolving for the same period forward in time under the same equations of motion, but starting from the velocity-reversed configuration at time t, that is, Γ_t^*. This evolution is precisely that which would be obtained by running a movie of the normal trajectory backward. The velocity-reversed or time-reversed counterpart to each phase space point $\Gamma_{t'}$ is visited, but in the opposite order, and the final configuration is Γ_0^*. The mean dissipation function for such a trajectory would be given by

$$\bar{\Omega}_t(\Gamma_t^*)t = \ln\left(\frac{P(\Gamma_t^*, 0)}{P(\Gamma_0, 0)}\exp\left(-\Delta Q(\Gamma_t^*)/k_B T\right)\right), \tag{1.129}$$

where $\Delta Q(\Gamma_t^*)$ is the heat transfer for this time-reversed trajectory. The symmetry of the protocol, and the symmetry of the Hamiltonian under velocity reversal, allows us to conclude that the heat transfer associated with the trajectory starting from Γ_0 is equal and opposite to that associated with starting point Γ_t^*, and hence the mean dissipation functions for the two trajectories must satisfy $\bar{\Omega}_t(\Gamma_0) = -\bar{\Omega}_t(\Gamma_t^*)$. We can then proceed to derive a specific case of the *Evans–Searles fluctuation theorem* (ESFT) associated with the distribution of values $\bar{\Omega}_t$ taken by the mean dissipation function $\bar{\Omega}_t(\Gamma_0)$:

$$\begin{aligned} P(\bar{\Omega}_t) &= \int d\Gamma_0 P(\Gamma_0, 0)\delta\left(\bar{\Omega}_t(\Gamma_0) - \bar{\Omega}_t\right) = \int d\Gamma_t P(\Gamma_t, t)\delta\left(\bar{\Omega}_t(\Gamma_0) - \bar{\Omega}_t\right) \\ &= \int d\Gamma_t P(\Gamma_t, t) \exp\left(\bar{\Omega}_t(\Gamma_0)t\right) \frac{P(\Gamma_t, 0)}{P(\Gamma_t, t)} \delta\left(\bar{\Omega}_t(\Gamma_0) - \bar{\Omega}_t\right) \\ &= \exp\left(\bar{\Omega}_t t\right) \int d\Gamma_t P(\Gamma_t, 0)\delta\left(\bar{\Omega}_t(\Gamma_0) - \bar{\Omega}_t\right) \\ &= \exp\left(\bar{\Omega}_t t\right) \int d\Gamma_t P(\Gamma_t, 0)\delta\left(-\bar{\Omega}_t(\Gamma_t^*) - \bar{\Omega}_t\right) \\ &= \exp\left(\bar{\Omega}_t t\right) \int d\Gamma_t^* P(\Gamma_t^*, 0)\delta\left(-\bar{\Omega}_t(\Gamma_t^*) - \bar{\Omega}_t\right) \\ &= \exp\left(\bar{\Omega}_t t\right) P(-\bar{\Omega}_t), \end{aligned} \tag{1.130}$$

noting that the Jacobian for the transformation of the integration variables from Γ_t to Γ_t^* is unity. Under the assumed conditions, therefore, we have obtained a relation that resembles (but historically preceded) the transient fluctuation theorem (Eq. (1.84)) or detailed fluctuation theorem (Eq. (1.106)) derived within the framework of stochastic dynamics. It only remains to make connections between the mean dissipation function and thermodynamic quantities to complete the parallel development, though it has been argued that the mean dissipation function itself is the more general measure of nonequilibrium behavior [8].

If we assume that the initial distribution is a canonical equilibrium such that $P(\Gamma_0, 0) \propto \exp\left(-H(\Gamma_0, 0)/k_B T\right)$, where $H(\Gamma_0, 0)$ is the system's Hamiltonian at $t = 0$, then we find from Eq. (1.126) that

$$\bar{\Omega}_t(\Gamma_0)t = \frac{1}{k_B T}\left(H(\Gamma_t, 0) - H(\Gamma_0, 0)\right) - \frac{1}{k_B T}\Delta Q(\Gamma_0), \tag{1.131}$$

and if the Hamiltonian at time t takes the same functional form as it does at $t = 0$, then $H(\Gamma_t, 0) = H(\Gamma_t, t)$ and we get

$$\bar{\Omega}_t(\Gamma_0)t = \frac{1}{k_B T}\left(H(\Gamma_t, t) - H(\Gamma_0, 0)\right) - \frac{1}{k_B T}\Delta Q(\Gamma_0) \tag{1.132}$$

$$\bar{\Omega}_t(\Gamma_0)t = \frac{1}{k_B T}\left(\Delta E(\Gamma_0) - \Delta Q(\Gamma_0)\right) = \frac{1}{k_B T}\Delta W(\Gamma_0), \tag{1.133}$$

where $\Delta E(\Gamma_0)$ is the change in system energy along a trajectory starting from Γ_0. Hence, the mean dissipation function is proportional to the (here solely thermodynamic) work performed on the system as it follows the trajectory starting from Γ_0. We deduce that the expectation value of this work is positive, and that the probability distribution of work for a time symmetric protocol and starting from canonical equilibrium satisfies the ESFT.

Deterministic methods may be used to derive a variety of statistical results involving the work performed on a system, including the Jarzynski equation and the Crooks relation. Nonconservative work may be included such that relations analogous to Eq. (1.101) may be obtained. However, it seems that a parallel development of the statistics of ΔS_{tot} is not possible. The fundamental problem is revealed if we try to construct the deterministic counterpart to the total entropy production defined in Eqs. (1.44) and (1.64):

$$\Delta S_{\text{tot}}^{\text{det}} = \Delta S_{\text{sys}} + \Delta S_{\text{med}} = -k_B \ln \left(\frac{P(\Gamma_t, t)}{P(\Gamma_0, 0)} \right) - \frac{\Delta Q(\Gamma_0)}{T}. \tag{1.134}$$

According to Eq. (1.125), this is identically zero. As might have been expected, uncertainty does not change under deterministic dynamics and the total entropy, in the form that we have chosen to define it, is constant. Nevertheless, the derivation of relationships involving the statistics of work performed and heat transferred, just alluded to, corresponding to similar expressions obtained using the stochastic dynamics framework, indicate that the use of deterministic reversible dynamics is an equivalent procedure for describing the behavior. Pedagogically, it is perhaps best to focus on just one approach, but a wider appreciation of the field requires an awareness of both.

1.10 Examples of the Fluctuation Relations in Action

The development of theoretical results of the kind we have seen so far is all very well, but their meaning is perhaps best appreciated by considering examples, which we do in this section. We shall consider overdamped stochastic dynamics, such that the velocities are always in an equilibrium Maxwell–Boltzmann distribution and never enter into consideration for entropy production. In the first two cases, we shall focus on the harmonic oscillator, since we understand its properties well. The only drawback of the harmonic oscillator is that it is a rather special case and some of its properties are not general [28, 42], although we deliberately avoid situations where the distributions produced are Gaussian in these examples. In the third case, we describe the simplest of nonequilibrium steady states and illustrate a detailed fluctuation theorem for the entropy, and identify its origin in the breaking of detailed balance.

1.10.1 Harmonic Oscillator Subject to a Step Change in Spring Constant

Let us consider the most simple model of the compression–expansion-type processes that are ubiquitous within thermodynamics. We start with a 1D classical harmonic oscillator subject to a Langevin heat bath. Such a system is governed by the overdamped equation of motion:

$$\dot{x} = -\frac{\kappa x}{m\gamma} + \left(\frac{2k_B T}{m\gamma}\right)^{1/2} \xi(t), \tag{1.135}$$

where κ is the spring constant. The corresponding Fokker–Planck equation is

$$\frac{\partial P(x,t)}{\partial t} = \frac{1}{m\gamma}\frac{\partial(\kappa x P(x,t))}{\partial x} + \frac{k_B T}{m\gamma}\frac{\partial^2 P(x,t)}{\partial x^2}. \tag{1.136}$$

We consider a simple work process, whereby, starting from equilibrium at temperature T, we instigate an instantaneous step change in spring constant from κ_0 to κ_1 at $t=0$ with the system in contact with the thermal bath at all times. This has the effect of compressing or expanding the distribution. We are then interested in the statistics of the entropy change associated with the process. Starting from Eqs. (1.44) and (1.64) for our definition of the entropy production, we may write

$$\Delta S_{\text{tot}} = \frac{\Delta W - \Delta\phi}{T} + k_{\text{B}} \ln\left(\frac{P_{\text{start}}(x_0)}{P_{\text{end}}(x_1)}\right), \tag{1.137}$$

utilizing notation $x_1 = x(t)$ and $x_0 = x(0)$ and $\phi(x) = \kappa x^2/2$. We also have

$$\Delta W = \frac{1}{2}(\kappa_1 - \kappa_0)x_0^2 \tag{1.138}$$

and

$$\Delta\phi = \frac{1}{2}\kappa_1 x_1^2 - \frac{1}{2}\kappa_0 x_0^2, \tag{1.139}$$

and so can write

$$\Delta S_{\text{tot}} = -\frac{\kappa_1}{2T}\left(x_1^2 - x_0^2\right) + k_{\text{B}} \ln\left(\frac{P_{\text{start}}(x_0)}{P_{\text{end}}(x_1)}\right). \tag{1.140}$$

Employing an initial canonical distribution

$$P_{\text{start}}(x_0) = \left(\frac{\kappa_0}{2\pi k_{\text{B}} T}\right)^{1/2} \exp\left(-\frac{\kappa_0 x_0^2}{2k_{\text{B}} T}\right), \tag{1.141}$$

the distribution at the end of the process will be given by

$$P_{\text{end}}(x_1) = \int_{-\infty}^{\infty} \mathrm{d}x_0\, P_{\text{OU}}^{\kappa_1}[x_1|x_0] P_{\text{start}}(x_0). \tag{1.142}$$

This is the Ornstein–Uhlenbeck process and so has transition probability density P_{OU}^{κ} given by analogy to Eq. (1.23). Hence, we may write

$$\begin{aligned} P_{\text{end}}(x_1) &= \int_{-\infty}^{\infty} \mathrm{d}x_0 \left(\frac{\kappa_1}{2\pi k_{\text{B}} T(1 - \mathrm{e}^{-(2\kappa_1 t/m\gamma)})}\right)^{1/2} \exp\left(-\frac{\kappa_1\left(x_1 - x_0 \mathrm{e}^{-(\kappa_1 t/m\gamma)}\right)^2}{2k_{\text{B}} T(1 - \mathrm{e}^{-(2\kappa_1 t/m\gamma)})}\right) \\ &\quad \times \left(\frac{\kappa_0}{2\pi k_{\text{B}} T}\right)^{1/2} \exp\left(-\frac{\kappa_0 x_0^2}{2k_{\text{B}} T}\right) \\ &= \left(\frac{\tilde{\kappa}(t)}{2\pi k_{\text{B}} T}\right)^{1/2} \exp\left(-\frac{\tilde{\kappa}(t) x_1^2}{2k_{\text{B}} T}\right), \end{aligned} \tag{1.143}$$

where

$$\tilde{\kappa}(t) = \frac{\kappa_0 \kappa_1}{\kappa_0 + \mathrm{e}^{-2\kappa_1 t/(m\gamma)}(\kappa_1 - \kappa_0)}, \tag{1.144}$$

such that P_{end} is always Gaussian. The coefficient $\tilde{\kappa}(t)$ evolves monotonically from κ_0 at $t = 0$ to κ_1 as $t \to \infty$. Substituting this into Eq. (1.140) allows us to write

$$\Delta S_{\text{tot}}(x_1, x_0, t) = -\frac{\kappa_1}{2T}\left(x_1^2 - x_0^2\right) + \frac{k_{\text{B}}}{2}\ln\left(\frac{\kappa_0}{\tilde{\kappa}(t)}\right) - \frac{\kappa_0 x_0^2}{2T} + \frac{\tilde{\kappa}(t)x_1^2}{2T} \tag{1.145}$$

for the entropy production associated with a trajectory that begins at x_0 and ends at x_1 at time t, and is not specified in between. We can average this over the probability distribution for such a trajectory to get

$$\begin{aligned}\langle \Delta S_{\text{tot}} \rangle &= \int \mathrm{d}x_0 \mathrm{d}x_1 P_{\text{OU}}^{\kappa_1}[x_1|x_0] P_{\text{start}}(x_0) \Delta S_{\text{tot}}(x_1, x_0, t) \\ &= k_{\text{B}}\left(-\frac{1}{2}\frac{\kappa_1}{\tilde{\kappa}(t)} + \frac{1}{2}\frac{\kappa_1}{\kappa_0} + \frac{1}{2}\ln\left(\frac{\kappa_0}{\tilde{\kappa}(t)}\right) - \frac{1}{2} + \frac{1}{2}\right) \\ &= \frac{k_{\text{B}}}{2}\left(\frac{\kappa_1}{\kappa_0} - \frac{\kappa_1}{\tilde{\kappa}(t)} + \ln\left(\frac{\kappa_0}{\tilde{\kappa}(t)}\right)\right),\end{aligned} \tag{1.146}$$

making full use of the separation of ΔS_{tot} into quadratic terms and the Gaussian character of the distributions. $\langle \Delta S_{\text{tot}} \rangle$ is zero at $t = 0$ and as $t \to \infty$, we find

$$\lim_{t\to\infty} \langle \Delta S_{\text{tot}} \rangle = \frac{k_{\text{B}}}{2}\left(\frac{\kappa_1}{\kappa_0} - 1 - \ln\left(\frac{\kappa_1}{\kappa_0}\right)\right), \tag{1.147}$$

which is positive since $\ln z \leq z - 1$ for all z. Furthermore,

$$\frac{\mathrm{d}\langle \Delta S_{\text{tot}} \rangle}{\mathrm{d}t} = \frac{k_{\text{B}}}{2\tilde{\kappa}^2}\frac{\mathrm{d}\tilde{\kappa}}{\mathrm{d}t}(\kappa_1 - \tilde{\kappa}), \tag{1.148}$$

and it is clear that this is positive at all times during the evolution, irrespective of the values of κ_1 and κ_0. If $\kappa_1 > \kappa_0$, then $\tilde{\kappa}$ increases with time while remaining less than κ_1, and all factors on the right-hand side of Eq. (1.148) are positive. If $\kappa_1 < \kappa_0$, then $\tilde{\kappa}$ decreases but always remains greater than κ_1 and the mean total entropy production is still positive as the relaxation process proceeds.

The work done on the system is simply the input of potential energy associated with the shift in spring constant:

$$\Delta W(x_1, x_0, t) = \frac{1}{2}(\kappa_1 - \kappa_0)x_0^2, \tag{1.149}$$

and so the mean work performed up to any time $t > 0$ is

$$\langle \Delta W \rangle = \frac{k_{\text{B}}T}{2}\left(\frac{\kappa_1}{\kappa_0} - 1\right), \tag{1.150}$$

which is greater than $\Delta F = (k_{\text{B}}T/2)\ln(\kappa_1/\kappa_0)$. The mean dissipative work is

$$\langle \Delta W_{\text{d}} \rangle = \langle \Delta W \rangle - \Delta F = \frac{k_{\text{B}}T}{2}\left(\frac{\kappa_1}{\kappa_0} - 1 - \ln\left(\frac{\kappa_1}{\kappa_0}\right)\right), \tag{1.151}$$

and this equals the mean entropy generated as $t \to \infty$ derived in Eq. (1.147), which is to be expected since the system started in equilibrium. More specifically, let us verify the Jarzynski equality:

$$\begin{aligned}\langle \exp\left(-\Delta W/k_{\mathrm{B}}T\right)\rangle &= \int \mathrm{d}x_0 P_{\mathrm{start}}(x_0) \exp\left(-(\kappa_1-\kappa_0)x_0^2/2k_{\mathrm{B}}T\right) \\ &= (\kappa_0/\kappa_1)^{1/2} = \exp\left(-\Delta F/k_{\mathrm{B}}T\right).\end{aligned} \tag{1.152}$$

Now we demonstrate that the integral fluctuation relation is satisfied. We consider

$$\begin{aligned}\langle \exp\left(-\Delta S_{\mathrm{tot}}/k_{\mathrm{B}}\right)\rangle &= \left\langle \exp\left(\frac{\kappa_1}{2k_{\mathrm{B}}T}\left(x_1^2-x_0^2\right) - \frac{1}{2}\ln\left(\frac{\kappa_0}{\tilde{\kappa}}\right) + \frac{\kappa_0 x_0^2}{2k_{\mathrm{B}}T} - \frac{\tilde{\kappa}x_1^2}{2k_{\mathrm{B}}T}\right)\right\rangle \\ &= \int \mathrm{d}x_1 \mathrm{d}x_0 P_{\mathrm{OU}}^{\kappa_1}[x_1|x_0] P_{\mathrm{start}}(x_0) \\ &\quad \times \exp\left(\frac{\kappa_1}{2k_{\mathrm{B}}T}\left(x_1^2-x_0^2\right) - \frac{1}{2}\ln\left(\frac{\kappa_0}{\tilde{\kappa}}\right) + \frac{\kappa_0 x_0^2}{2k_{\mathrm{B}}T} - \frac{\tilde{\kappa}x_1^2}{2k_{\mathrm{B}}T}\right) \\ &= \left(\frac{\tilde{\kappa}}{\kappa_0}\right)^{1/2} \int \mathrm{d}x_1 \mathrm{d}x_0 \\ &\quad \times \left(\frac{\kappa_1}{2\pi k_{\mathrm{B}}T\left(1-\mathrm{e}^{-(2\kappa_1 t/m\gamma)}\right)}\right)^{1/2} \exp\left(-\frac{\kappa_1\left(x_1 - x_0\mathrm{e}^{-(\kappa_1 t/m\gamma)}\right)^2}{2k_{\mathrm{B}}T(1-\mathrm{e}^{-(2\kappa_1 t/m\gamma)})}\right) \\ &\quad \times \left(\frac{\kappa_0}{2\pi k_{\mathrm{B}}T}\right)^{1/2} \exp\left(-\frac{\kappa_0 x_0^2}{2k_{\mathrm{B}}T}\right) \exp\left(\frac{\kappa_1}{2k_{\mathrm{B}}T}\left(x_1^2-x_0^2\right) + \frac{\kappa_0 x_0^2}{2k_{\mathrm{B}}T} - \frac{\tilde{\kappa}x_1^2}{2k_{\mathrm{B}}T}\right) = 1,\end{aligned} \tag{1.153}$$

which is a tedious integration, but the result is inevitable.

Furthermore, we can directly confirm the Crooks relation for this process. The work over the forward process is given by Eq. (1.149) and so, choosing $\kappa_1 > \kappa_0$, we can derive its distribution according to Eq. (1.154):

$$\begin{aligned}P^F(\Delta W) &= P_{\mathrm{eq}}(x_0)\frac{\mathrm{d}x_0}{\mathrm{d}\Delta W} \\ &= \left(\frac{\kappa_0}{2\pi k_{\mathrm{B}}T}\right)^{1/2} \exp\left(-\frac{\kappa_0 x_0^2}{2k_{\mathrm{B}}T}\right)\frac{1}{\kappa_1-\kappa_0}\left(\frac{2\Delta W}{\kappa_1-\kappa_0}\right)^{-1/2} H(\Delta W) \\ &= \frac{1}{\sqrt{4\pi k_{\mathrm{B}}T}}\left(\frac{\kappa_0}{\kappa_1-\kappa_0}\right)^{1/2}\Delta W^{-1/2}\exp\left(-\frac{\kappa_0}{\kappa_1-\kappa_0}\frac{\Delta W}{k_{\mathrm{B}}T}\right)H(\Delta W),\end{aligned} \tag{1.154}$$

where $H(\Delta W)$ is the Heaviside step function. Let us consider the appropriate reverse process. Starting in equilibrium defined by κ_1, where again to form the reversed protocol we must have $\kappa_1 > \kappa_0$, the work is

$$\Delta W = \frac{1}{2}(\kappa_0 - \kappa_1)x_1^2 \tag{1.155}$$

and so we can derive its distribution according to Eq. (1.156):

$$\begin{aligned}P^R(\Delta W) &= P_{\mathrm{eq}}(x_1)\frac{\mathrm{d}x_1}{\mathrm{d}\Delta W} \\ &= \left(\frac{\kappa_1}{2\pi k_{\mathrm{B}}T}\right)^{1/2} \exp\left(-\frac{\kappa_1 x_1^2}{2k_{\mathrm{B}}T}\right)\frac{1}{\kappa_0-\kappa_1}\left(\frac{2\Delta W}{\kappa_0-\kappa_1}\right)^{-1/2} H(-\Delta W) \\ &= \frac{1}{\sqrt{4\pi k_{\mathrm{B}}T}}\left(\frac{\kappa_1}{\kappa_0-\kappa_1}\right)^{1/2}\Delta W^{-1/2}\exp\left(-\frac{\kappa_1}{\kappa_0-\kappa_1}\frac{\Delta W}{k_{\mathrm{B}}T}\right)H(-\Delta W),\end{aligned} \tag{1.156}$$

so that

$$P^R(-\Delta W) = \frac{1}{\sqrt{4\pi k_B T}}\left(\frac{\kappa_1}{\kappa_0-\kappa_1}\right)^{1/2}(-\Delta W)^{-1/2}\exp\left(\frac{\kappa_1}{\kappa_0-\kappa_1}\frac{\Delta W}{k_B T}\right)H(\Delta W)$$
$$= \frac{1}{\sqrt{4\pi k_B T}}\left(\frac{\kappa_1}{\kappa_1-\kappa_0}\right)^{1/2}(\Delta W)^{-1/2}\exp\left(-\frac{\kappa_1}{\kappa_1-\kappa_0}\frac{\Delta W}{k_B T}\right)H(\Delta W). \tag{1.157}$$

Taking the ratio of these two distributions (1.154) and (1.157) gives

$$\frac{P^F(\Delta W)}{P^R(-\Delta W)} = \sqrt{\frac{\kappa_0}{\kappa_1}}\exp\left(-\frac{\kappa_0-\kappa_1}{\kappa_1-\kappa_0}\frac{\Delta W}{k_B T}\right)$$
$$= \exp\left(\frac{\Delta W-(k_B T/2)\ln(\kappa_1/\kappa_0)}{k_B T}\right) \tag{1.158}$$
$$= \exp\left((\Delta W-\Delta F)/k_B T\right),$$

which is the Crooks work relation as required.

1.10.2 Smoothly Squeezed Harmonic Oscillator

Now let us consider a process where work is performed isothermally on a particle, but this time by a continuous variation of the spring constant. We have

$$\Delta W = \int_0^\tau \frac{\partial\phi(x(t),\lambda(t))}{\partial\lambda}\frac{d\lambda}{dt}dt, \tag{1.159}$$

where $\lambda(t)=\kappa(t)$ and $\phi(x(t),\kappa(t)) = (1/2)\kappa(t)x_t^2$, where $x_t = x(t)$, such that

$$\Delta W = \int_0^\tau \frac{1}{2}\dot{\kappa}(t)x_t^2 dt. \tag{1.160}$$

Similarly, the change in system energy will be given simply by

$$\Delta\phi = \int_0^\tau \frac{d\phi(x(t),\lambda(t))}{dt}dt = \frac{1}{2}\kappa(\tau)x_\tau^2 - \frac{1}{2}\kappa_0 x_0^2. \tag{1.161}$$

Accordingly, we can once again describe the entropy production as

$$\Delta S_{\text{tot}} = \frac{1}{2T}\int_0^\tau \dot{\kappa}x_t^2 dt - \frac{1}{2T}\kappa(\tau)x_\tau^2 + \frac{1}{2T}\kappa_0 x_0^2 + k_B\ln\left(\frac{P_{\text{start}}(x_0)}{P_{\text{end}}(x_\tau)}\right). \tag{1.162}$$

For convenience, we assume the initial state to be in canonical equilibrium. The evolving distribution P satisfies the appropriate Fokker–Planck equation:

$$\frac{\partial P}{\partial t} = \frac{\kappa(t)}{m\gamma}\frac{\partial(xP)}{\partial x} + \frac{k_B T}{m\gamma}\frac{\partial^2 P}{\partial x^2}. \tag{1.163}$$

Since P is initially canonical, it retains its Gaussian form and can be written

$$P_{\text{end}}(x_\tau) = P(x_\tau,\tau) = \left(\frac{\tilde{\kappa}(\tau)}{2\pi k_B T}\right)^{1/2}\exp\left(-\frac{\tilde{\kappa}(\tau)x_\tau^2}{2k_B T}\right), \tag{1.164}$$

where $\tilde{\kappa}(t)$ evolves according to

$$\frac{\mathrm{d}\tilde{\kappa}}{\mathrm{d}t} = -\frac{2}{m\gamma}\tilde{\kappa}(\tilde{\kappa} - \kappa), \tag{1.165}$$

with initial condition $\tilde{\kappa}(0) = \kappa_0$. We can solve for $\tilde{\kappa}$: write $z = \tilde{\kappa}^{-1}$ such that

$$\frac{\mathrm{d}z}{\mathrm{d}t} = \frac{2}{m\gamma}(1 - \kappa z). \tag{1.166}$$

This has integrating factor solution:

$$z(\tau)\exp\left(\frac{2}{m\gamma}\int_0^\tau \kappa(t)\mathrm{d}t\right) = z(0) + \int_0^\tau \exp\left(\frac{2}{m\gamma}\int_0^t \kappa(t')\mathrm{d}t'\right)\frac{2}{m\gamma}\mathrm{d}t, \tag{1.167}$$

or, equivalently,

$$\frac{1}{\tilde{\kappa}(\tau)} = \frac{1}{\kappa(0)}\exp\left(-\frac{2}{m\gamma}\int_0^\tau \kappa(t)\mathrm{d}t\right) + \frac{2}{m\gamma}\int_0^\tau \exp\left(-\frac{2}{m\gamma}\int_t^\tau \kappa(t')\mathrm{d}t'\right)\mathrm{d}t. \tag{1.168}$$

Returning to the entropy production, we now write

$$\Delta S_{\mathrm{tot}} = \frac{1}{T}\int_0^\tau \frac{1}{2}\dot{\kappa}x_t^2\mathrm{d}t - \frac{1}{2T}\kappa(\tau)x_\tau^2 + \frac{1}{2T}\kappa_0 x_0^2 + \frac{k_{\mathrm{B}}}{2}\ln\left(\frac{\kappa_0}{\tilde{\kappa}(\tau)}\right) - \frac{\kappa_0 x_0^2}{2T} + \frac{\tilde{\kappa}(\tau)x_\tau^2}{2T}, \tag{1.169}$$

and we also have

$$\Delta W = \int_0^\tau \frac{1}{2}\dot{\kappa}x_t^2\mathrm{d}t. \tag{1.170}$$

We can investigate the statistics of these quantities:

$$\langle\Delta W\rangle = \int_0^\tau \frac{1}{2}\dot{\kappa}\langle x_t^2\rangle\mathrm{d}t = \int_0^\tau \frac{1}{2}\dot{\kappa}\frac{k_{\mathrm{B}}T}{\tilde{\kappa}}\mathrm{d}t, \tag{1.171}$$

and from $\Delta W_{\mathrm{d}} = \Delta W - \Delta F$, the rate of performance of dissipative work is

$$\frac{\mathrm{d}\langle\Delta W_{\mathrm{d}}\rangle}{\mathrm{d}t} = \frac{\dot{\kappa}k_{\mathrm{B}}T}{2\tilde{\kappa}} - \frac{\mathrm{d}F(\kappa(t))}{\mathrm{d}t} = \frac{\dot{\kappa}k_{\mathrm{B}}T}{2\tilde{\kappa}} - \frac{\dot{\kappa}k_{\mathrm{B}}T}{2\kappa} = \frac{k_{\mathrm{B}}T}{2}\dot{\kappa}\left(\frac{1}{\tilde{\kappa}} - \frac{1}{\kappa}\right). \tag{1.172}$$

While the positivity of $\langle\Delta W_{\mathrm{d}}\rangle$ is ensured for this process, as a consequence of the Jarzynski equation and the initial equilibrium condition, the rate of change can be both positive and negative, according to this result.

The expectation value for total entropy production in Eq. (1.169), on the other hand, is

$$\langle\Delta S_{\mathrm{tot}}\rangle = \frac{1}{T}\int_0^\tau \frac{1}{2}\dot{\kappa}\frac{k_{\mathrm{B}}T}{\tilde{\kappa}}\mathrm{d}t - \frac{1}{2T}\kappa(\tau)\frac{k_{\mathrm{B}}T}{\tilde{\kappa}} + \frac{1}{2T}k_{\mathrm{B}}T + \frac{k_{\mathrm{B}}}{2}\ln\left(\frac{\kappa_0}{\tilde{\kappa}(\tau)}\right) - \frac{k_{\mathrm{B}}T}{2T} + \frac{k_{\mathrm{B}}T}{2T} \tag{1.173}$$

and the rate of change of this quantity is

$$\begin{aligned}\frac{\mathrm{d}\langle\Delta S_{\mathrm{tot}}\rangle}{\mathrm{d}t} &= \frac{\dot{\kappa}k_{\mathrm{B}}}{2\tilde{\kappa}} - \frac{k_{\mathrm{B}}}{2}\left(\frac{\dot{\kappa}}{\tilde{\kappa}} - \frac{\kappa}{\tilde{\kappa}^2}\dot{\tilde{\kappa}}\right) - \frac{k_{\mathrm{B}}}{2}\frac{\dot{\tilde{\kappa}}}{\tilde{\kappa}} \\ &= \frac{k_{\mathrm{B}}}{2}\dot{\tilde{\kappa}}\frac{(\kappa-\tilde{\kappa})}{\tilde{\kappa}^2} \\ &= \frac{k_{\mathrm{B}}}{m\gamma}\frac{(\kappa-\tilde{\kappa})^2}{\tilde{\kappa}}.\end{aligned} \tag{1.174}$$

The monotonic increase in entropy with time is explicit. The mean dissipative work and the entropy production for a process of this kind starting in equilibrium are illustrated in Figure 1.4, where the protocol changes over a driving period followed by a subsequent period of equilibration. Note particularly that the mean entropy production never exceeds the mean dissipative work, which is delivered instantaneously, and that both take the same value as $t \rightarrow \infty$ giving insight into the operation of the Jarzynski equality, as discussed in Section 1.7.1.

It is of more interest, however, to verify that detailed fluctuation relations hold. Analytical demonstration based upon Eq. (1.169) and the probability density for a particular trajectory throughout the entire period are challenging, but a numerical

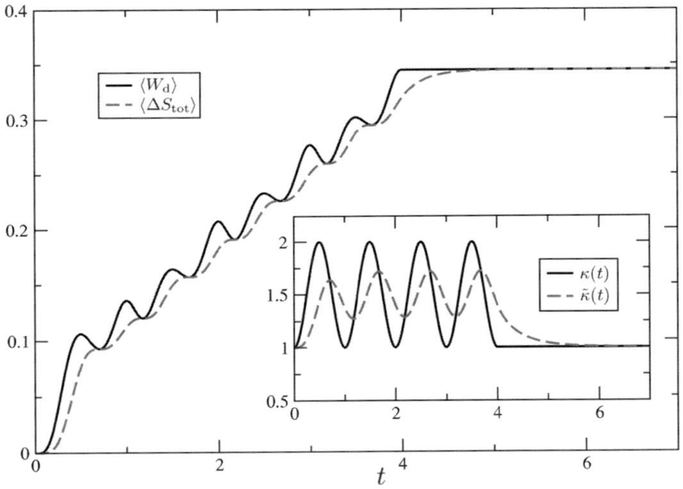

Figure 1.4 An illustration of the mean behavior of the dissipative work and entropy production for an oscillatory compression and expansion process starting in equilibrium. The mean dissipative work increases, but not monotonically, and is delivered instantly such that there is no further contribution when the protocol stops changing. The mean entropy production, however, continues to increase monotonically until it reaches the mean dissipative work after the protocol has stopped changing. The evolution of the protocol, $\kappa(t) = \sin^2(\pi t) + 1$, and the characterization of the distribution, $\tilde{\kappa}(t)$, are shown in the inset. Units are $k_{\mathrm{B}} = T = m = \gamma = 1$.

approach based upon generating sample trajectories is feasible particularly since the entire distribution can always be characterized with the known quantity $\tilde{\kappa}(t)$. As such we consider the same protocol, $\kappa(t) = \sin^2(\pi t) + 1$, wait until the system has reached a nonequilibrium oscillatory steady state, as described in Section 1.7.3, characterized here by an oscillatory $\tilde{\kappa}(t)$, as seen in Figure 1.4, and measure the entropy production over a time period across which $\kappa(t)$ is symmetric. The distribution in total entropy production over such a period and the symmetry it possesses are illustrated in Figure 1.5.

1.10.3
A Simple Nonequilibrium Steady State

Let us construct a very simple nonequilibrium steady state. We consider an overdamped Brownian motion on a ring driven in one direction by a nonconservative force. We assume a constant potential $\phi(x) = c$ such that the equation of motion is simply

$$\dot{x} = \frac{f}{m\gamma} + \left(\frac{2k_B T}{m\gamma}\right)^{1/2} \xi(t). \tag{1.175}$$

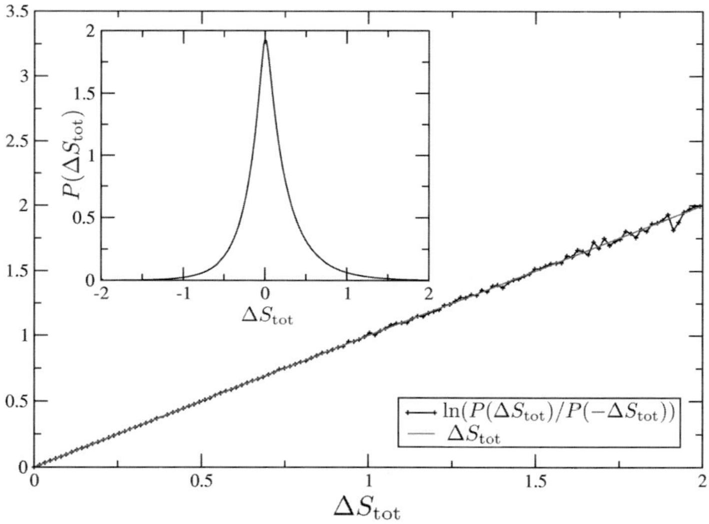

Figure 1.5 An illustration of a detailed fluctuation theorem arising for an oscillatory nonequilibrium steady state, as described in Section 1.7.3, created by compressing and expanding a particle in a harmonic potential by using protocol $\lambda(t) = \kappa(t) = \sin^2(\pi t) + 1$. The total entropy production must be measured over a time period during which the protocol is symmetric and the distribution is deemed to be oscillatory. Such a time period exists between $t = 3$ and $t = 4$, as shown in inset in Figure 1.4. Units are $k_B = T = m = \gamma = 1$.

This is just the Wiener process from Eq. (1.18) centered on a mean proportional to the external force multiplied by the time, so the probability density of a given displacement is defined by

$$P[x(\tau)|x(0)] = \sqrt{\frac{m\gamma}{4\pi k_B T\tau}} \exp\left[-\frac{m\gamma(\Delta x - (f/m\gamma)\tau)^2}{4k_B T\tau}\right], \tag{1.176}$$

where the lack of a superscript on P recognizes the constancy of the protocol, and noting that $\Delta x = x(\tau) - x(0)$ may extend an arbitrary number of times around the ring. In addition, by utilizing the symmetry of the system, we can trivially state that the stationary distribution is given by

$$P^{\mathrm{st}}(x) = L^{-1}, \tag{1.177}$$

where L is the circumference of the ring. Considering that we are in the steady state, we know that the transitions must balance in total; however, let us consider the transitions between individual configurations: comparing the probabilities of transitions, we immediately see that

$$P^{\mathrm{st}}(x(0))P[x(\tau)|x(0)] \neq P^{\mathrm{st}}(x(\tau))P[x(0)|x(\tau)]. \tag{1.178}$$

Detailed balance explicitly does not hold. For this system, not only can there be entropy production due to driving, such as is the case with expansion and compression processes, but there is also a continuous probability current in the steady state, in the direction of the force, which dissipates heat into the thermal bath. We have previously stated in Section 1.7.3 that the distribution of the entropy production in such steady states obeys a detailed fluctuation theorem for all times. We can verify that this is the case. The entropy production is rather simple and is given by

$$\begin{aligned}\Delta S_{\mathrm{tot}} &= k_B \ln \frac{P^{\mathrm{st}}(x(0))P[x(\tau)|x(0)]}{P^{\mathrm{st}}(x(\tau))P[x(0)|x(\tau)]} = k_B \ln \frac{L \exp\left[-\dfrac{m\gamma(\Delta x - (f/m\gamma)\tau)^2}{4k_B T\tau}\right]}{L \exp\left[-\dfrac{m\gamma(-\Delta x - (f/m\gamma)\tau)^2}{4k_B T\tau}\right]} \\ &= \frac{f\Delta x}{T}.\end{aligned} \tag{1.179}$$

This provides an example where the entropy production is highly intuitive. Taking $f > 0$, if the particle moves with the probability current, $\Delta x > 0$, it is carrying out the expected behavior and thus is following an entropy generating trajectory. However, if the particle moves against the current, $\Delta x < 0$, it is behaving unexpectedly and as such is performing a trajectory that destroys entropy. It follows that since an observation of the particle flowing with a current is more likely than an observation of the opposite, then on average the entropy production is positive.

Since the system is in a steady state, we expect a detailed fluctuation theorem. The transformation of the probability distribution is trivial and we have simply

$$P(\Delta S_{\text{tot}}) = \sqrt{\frac{m\gamma T}{4\pi k_B f^2 \tau}} \exp\left[-\frac{m\gamma T\left(\Delta S_{\text{tot}} - (f^2/m\gamma T)\tau\right)^2}{4k_B f^2 \tau}\right], \tag{1.180}$$

and we can verify a detailed fluctuation theorem that holds for all time. We can however probe further. While we may conceive of some fluctuations against a steady flow for a small particle, we would be quite surprised to see such deviations if we were considering a macroscopically sized object. Despite the model's limitations, let us consider an approach to macroscopic behavior while maintaining constant the ratio f/m such that the mean particle velocity is unchanged. Both the mean and variance of the distribution of entropy production increase in proportion. On the scale of the mean, the distribution of entropy change increasingly looks like a narrower and narrower Gaussian until it inevitably, for a macroscopic object, approaches a delta function where we recover the classical thermodynamic limit and are unable to perceive the fluctuations any longer.

1.11 Final Remarks

The aim of this chapter was to explore the origin, application, and limitations of fluctuation relations. We have done this within a framework of stochastic dynamics with white noise and often employing the overdamped limit in example cases where the derivations are easier: it is in the analysis of explicit examples where understanding is often to be found. Nevertheless, the results can be extended to other more complicated stochastic systems, though the details will need to be sought elsewhere. The fluctuation relations can also be derived within a framework of deterministic, reversible dynamics, which we have discussed briefly in Section 1.9. It is interesting to note that within that framework, irreversibility finds its origins in nonlinear terms that provide a contraction of phase space, in contrast to the more direct irreversibility of the equations of motion found in stochastic descriptions. Both approaches, however, are attempts to represent a dissipative environment that imposes a thermal constraint.

The fluctuation relations concern the statistics of quantities associated with thermodynamic processes, in particular the mechanical work done upon, or the heat transferred to a system in contact with a heat bath. In the thermodynamic limit, the statistics are simple: there are negligible deviations from the mean, and work and heat transfers appear to be deterministic and the second law requires entropy change to be nonnegative. But for finite size systems, there are fluctuations, and the statistics of these will satisfy one or more fluctuation relations. These can be very specific requirements, for example relating the probability of a fluctuation with positive dissipative work to the probability of a fluctuation with negative dissipative work in the reversed process. Or, the outcome can take the form of an inequality that demonstrates that the mean dissipative work over all possible realizations of the process is positive.

The core concept in the analysis, within the framework of stochastic dynamics at least, is entropy production. This no longer need to be a mysterious concept: it is a natural measure of the departure from dynamical reversibility, the loosening of the hold of Loschmidt's expectation of reversibility, when system interactions with a coarse-grained environment are taken into account. Entropy production emerges in stochastic models where there is uncertainty in initial specification. Intuitively, uncertainty in configuration in such a situation will grow with time, and the mean entropy production is this concept commodified. And it turns out that entropy production can also be related, in certain circumstances, to heat and work transfers, allowing the growth of uncertainty to be monitored in terms of thermodynamic process variables. Moreover, although it is *expected* to be positive, entropy change can be negative; and the probability of such an excursion, possibly observed by a measurement of work done or heat transferred, can be described by a fluctuation relation. In the thermodynamic limit, the entropy production appears to behave deterministically and to violate time reversal symmetry, and only then does the second law acquire its unbreakability. But for small systems interacting with a much larger environment, this status is very much diminished, and the second law is revealed to be merely a statement about what is likely to happen to the system, according to rules governing the evolution of probability that explicitly break time reversal symmetry.

References

1 Evans, D.J. Cohen, E.G.D., and Morriss, G.P. (1993) Probability of second law violations in shearing steady states. *Phys. Rev. Lett.*, **71**, 2401.

2 Evans, D.J. and Searles, D.J. (1994) Equilibrium microstates which generate second law violating steady states. *Phys. Rev. E*, **50**, 1645–1648.

3 Gallavotti, G. and Cohen, E.G.D. (1995) Dynamical ensembles in nonequilibrium statistical mechanics. *Phys. Rev. Lett.*, **74**, 2694.

4 Jarzynski, C. (1997) Nonequilibrium equality for free energy differences. *Phys. Rev. Lett.*, **78**, 2690.

5 Crooks, G.E. (1999) Entropy production fluctuation theorem and the nonequilibrium work relation for free energy differences. *Phys. Rev. E*, **60**, 2721–2726.

6 Cercignani, C. (1998) *Ludwig Boltzmann: The Man Who Trusted Atoms*, Oxford University Press.

7 Kondepudi, D. (2008) *Introduction to Modern Thermodynamics*, John Wiley & Sons, Inc., New York.

8 Evans, D.J. Williams, S.R., and Searles, D.J. (2011) A proof of Clausius' theorem for time reversible deterministic microscopic dynamics. *J. Chem. Phys.*, **134**, 204113.

9 Evans, D.J. and Searles, D.J. (2002) The fluctuation theorem. *Adv. Phys.*, **51**, 1529–1585.

10 Jaynes, E. (2003) *Probability Theory: The Logic of Science*, Cambridge University Press.

11 Risken, H. (1989) *The Fokker–Planck Equation: Methods of Solution and Applications*, 2nd edn, Springer.

12 Gardiner, C. (2009) *Stochastic Methods: A Handbook for the Natural and Social Sciences*, Springer.

13 van Kampen, N. (2007) *Stochastic Processes in Physics and Chemistry*, North-Holland.

14 Seifert, U. (2005) Entropy production along a stochastic trajectory and an

integral fluctuation theorem. *Phys. Rev. Lett.*, **95**, 040602.

15 Crooks, G.E. (1998) Nonequilibrium measurements of free energy differences for microscopically reversible Markovian systems. *J. Stat. Phys.*, **90**, 1481–1487.

16 Sekimoto, K. (1998) Langevin equation and thermodynamics. *Prog. Theor. Phys. Suppl.*, **130**, 17.

17 Sekimoto, K. (2010) *Stochastic Energetics, Lecture Notes in Physics*, vol. **799**, Springer, Berlin.

18 Wissel, C. (1979) Manifolds of equivalent path integral solutions of the Fokker–Planck equation. *Z. Phys. B*, **35**, 185–191.

19 Imparato, A. and Peliti, L. (2006) Fluctuation relations for a driven Brownian particle. *Phys. Rev. E*, **74**, 026106.

20 Harris, R.J. and Schütz, G.M. (2007) Fluctuation theorems for stochastic dynamics. *J. Stat. Mech.*, P07020.

21 Horowitz, J. and Jarzynski, C. (2007) Comparison of work fluctuation relations. *J. Stat. Mech.*, P11002.

22 Jarzynski, C. (2007) Comparison of far-from-equilibrium work relations. *C. R. Phys.*, **8** (5–6), 495–506.

23 Seifert, U. (2008) Stochastic thermodynamics: principles and perspectives. *Eur. Phys. J. B*, **64** (3–4), 423–431.

24 Jarzynski, C. (1997) Equilibrium free-energy differences from nonequilibrium measurements: a master-equation approach. *Phys. Rev. E*, **56**, 5018–5035.

25 Jarzynski, C. (2004) Nonequilibrium work theorem for a system strongly coupled to a thermal environment. *J. Stat. Mech.*, P09005.

26 Bochkov, G. and Kuzovlev, Y. (1981) Nonlinear fluctuation-dissipation relations and stochastic models in nonequilibrium thermodynamics: I. Generalized fluctuation-dissipation theorem. *Physica A*, **106**, 443–479.

27 Bochkov, G. and Kuzovlev, Y. (1981) Nonlinear fluctuation-dissipation relations and stochastic models in nonequilibrium thermodynamics: II. Kinetic potential and variational principles for nonlinear irreversible processes. *Physica A*, **106**, 480–520.

28 Shargel, B.H. (2010) The measure-theoretic identity underlying transient fluctuation theorems. *J. Phys. A*, **43**, 135002.

29 Kurchan, J. (1998) Fluctuation theorem for stochastic dynamics. *J. Phys. A*, **31**, 3719.

30 Lebowitz, J.L. and Spohn, H. (1999) A Gallavotti–Cohen type symmetry in the large deviation functional for stochastic dynamics. *J. Stat. Phys.*, **95**, 333–365.

31 Gallavotti, G. and Cohen, E. (1995) Dynamical ensembles in stationary states. *J. Stat. Phys.*, **80**, 931–970.

32 Touchette, H. (2009) The large deviation approach to statistical mechanics. *Phys. Rep.*, **478** (1–3), 1–69.

33 van Zon, R. and Cohen, E.G.D. (2003) Extension of the fluctuation theorem. *Phys. Rev. Lett.*, **91**, 110601.

34 Schmiedl, T. Speck, T., and Seifert, U. (2007) Entropy production for mechanically or chemically driven biomolecules. *J. Stat. Phys.*, **128**, 77–93.

35 Chernyak, V.Y. Chertkov, M., and Jarzynski, C. (2006) Path-integral analysis of fluctuation theorems for general Langevin processes. *J. Stat. Mech*, P08001.

36 Hatano, T. and Sasa, S. (2001) Steady-state thermodynamics of Langevin systems. *Phys. Rev. Lett.*, **86**, 3463–3466.

37 Esposito, M. and Van den Broeck, C. (2010) Three detailed fluctuation theorems. *Phys. Rev. Lett.*, **104**, 090601.

38 Esposito, M. and Van den Broeck, C. (2010) Three faces of the second law: I. Master equation formulation. *Phys. Rev. E*, **82**, 011143.

39 Van den Broeck, C. and Esposito, M. (2010) Three faces of the second law: II. Fokker–Planck formulation. *Phys. Rev. E*, **82**, 011144.

40 Oono, Y. and Paniconi, M. (1998) Steady state thermodynamics. *Prog. Theor. Phys. Suppl.*, **130**, 29–44.

41 Speck, T. and Seifert, U. (2005) Integral fluctuation theorem for the housekeeping heat. *J. Phys. A*, **38**, L581.

42 Saha, A. Lahiri, S., and Jayannavar, A.M. (2009) Entropy production theorems and some consequences. *Phys. Rev. E*, **80**, 011117.

2
Fluctuation Relations and the Foundations of Statistical Thermodynamics: A Deterministic Approach and Numerical Demonstration

James C. Reid, Stephen R. Williams, Debra J. Searles, Lamberto Rondoni, and Denis J. Evans

2.1
Introduction

Arguably, there is no more widely applicable theory than thermodynamics. However, its exact solutions are limited to equilibrium systems in the "thermodynamic limit." To treat systems outside these conditions required idealization, approximation, or some equilibrium-like approximation (linear response, local and quasi-equilibriums). In the past 20 years however, tools have been developed to treat systems of arbitrary size and also to treat systems that are not only away from equilibrium but, in fact, may also be far from equilibrium. The most well known of these tools are the fluctuation relations (FRs) [1–3].

Equilibrium statistical mechanics employs phase functions (like the Hamiltonian) and statistical weights from which equilibrium averages may be computed. The FRs use path integrals of phase functions in finite sized, nonequilibrium systems to quantify or measure their properties. For example, the Evans–Searles fluctuation theorem (ESFT) predicts the relative probability of observing opposite values of the dissipation function (Ω) averaged along a phase space trajectory of duration t. For systems that can be approximated as being in local thermodynamic equilibrium, this provides a measurement of irreversibility of the entropy production [4]. However, unlike the entropy production used in linear irreversible thermodynamics, the ESFT is exact arbitrarily far from equilibrium where entropy, or its rate of production, cannot be defined. Similarly, the Crooks relation (CR) quantifies the relative values of the work (w_t) done moving backward and forward between two equilibrium ensembles along nonequilibrium pathways [5]. The closely related Jarzynski equality (JE) refers to the ensemble average of the nonequilibrium work, $\langle e^{-\beta w_t} \rangle$, as a means of predicting equilibrium free energy differences between small equilibrium systems [6].

Since these relations were first derived at the end of the twentieth century, they have been demonstrated for a wide range of systems and in a number of experiments, including DNA stretching [7], optically trapped colloids [8], shearing

Nonequilibrium Statistical Physics of Small Systems: Fluctuation Relations and Beyond, First Edition.
Edited by Rainer Klages, Wolfram Just, and Christopher Jarzynski.

systems [9], pendulums [10], molecular motors [11], and various quantum mechanical systems [12].

Early in the twenty-first century, the Evans–Searles fluctuation relation has been used to derive a set of new interrelated theorems, each involving the dissipation function, the central argument of the ESFT. First, it was shown that for classical systems, all linear and nonlinear response theories for both driven and relaxing systems could be written in a form that involves time integrals of correlation functions involving the dissipation function: the so-called dissipation theorem (DT) [13]. Later it was proved that both thermostated and adiabatic systems, subject to a couple of simply stated conditions, relax at long times necessarily to an ergodic equilibrium state [14]. These relaxation theorems (RTs) provided a mathematical proof of Boltzmann's postulate of equal *a priori* probabilities for an isolated Newtonian system and also for the first time a proof that the Gibbs canonical distribution is the unique equilibrium distribution for systems in contact with a heat reservoir. Later a detailed mathematical proof of the Clausius inequality was given without assuming, as Clausius did in 1854, the second "law" of thermodynamics [15].

Taken together, this set of theorems constitutes a proof of the second and the zeroth "laws" of thermodynamics. These so-called "laws" of thermodynamics cease being axioms and instead become like the first "law," mathematical theorems provable from the laws of mechanics and the axiom of causality.

In this chapter, we provide a short summary of the ESFT and the CR, then outline the derivation of the relaxation theorem for systems obeying Newton's equations, and for the first time we give a mathematical proof of the relationship between the equilibrium thermodynamic entropy and thermodynamic temperature and their statistical mechanical analogs. We will also take the optical trapping system, and illustrate the application of the various theorems in a single conceptually simple system [8, 16, 17].

2.2
The Relations

The first FR to be announced was an asymptotic FR for phase space contraction in deterministic nonequilibrium steady states [9]. The first mathematical proof of an exact FR was the ESFT. The subject of the ESFT is the time integral of the dissipation function, $\Omega_t(\mathbf{\Gamma}) = \int_0^t ds\Omega(S^s\mathbf{\Gamma})$, which is defined as the probability ratio of observing in the initial distribution of states, conjugate sets of trajectories and antitrajectories:

$$\Omega_t(\mathbf{\Gamma}) = \ln\left(\frac{f(\mathbf{\Gamma}, 0)}{f(\mathbf{\Gamma}^*, 0)}\right) - \int_0^t ds\Lambda(S^s\mathbf{\Gamma}), \tag{2.1}$$

where $\mathbf{\Gamma}$ is a phase space vector consisting of all the coordinates and momenta of the N-particles comprising the system $\mathbf{\Gamma} = \{\mathbf{q}_1, \mathbf{q}_2, \ldots, \mathbf{q}_N, \mathbf{p}_1, \mathbf{p}_2, \ldots, \mathbf{p}_N\}$. $S^t\mathbf{\Gamma}$ is the phase space vector generated from the initial phase space vector $\mathbf{\Gamma}$ after a period of time t from the equations of motion $\dot{\mathbf{\Gamma}} = \mathbf{G}(\mathbf{\Gamma})$ and $\Lambda(\mathbf{\Gamma}) \equiv \partial/\partial\mathbf{\Gamma} \cdot \mathbf{G}(\mathbf{\Gamma})$.

$\boldsymbol{\Gamma}^*$ is the time reversal map (M^T) of $S^t\boldsymbol{\Gamma}$ ($\boldsymbol{\Gamma}^* = M^T(S^t\boldsymbol{\Gamma}) = \{S^t\mathbf{q}_1, S^t\mathbf{q}_2, \ldots, S^t\mathbf{q}_N, -S^t\mathbf{p}_1, -S^t\mathbf{p}_2, \ldots, -S^t\mathbf{p}_N\}$), and $f(\boldsymbol{\Gamma}, 0)$ is the phase space probability density function for the system at $\boldsymbol{\Gamma}$ and time 0. We point out here that $\Omega_t(\boldsymbol{\Gamma})$ is a function of both t and $\boldsymbol{\Gamma}$, and that it is a functional of the distribution function f. We will not normally specify this as many of the results will apply irrespective of the form of f, and in other cases the form of f is explicitly included. The number of ensemble members in any volume element is of course constant in time (conservation of probabilities in phase space), although the size and shape of the volume element will evolve. In general, the system will gain or loose heat to its surroundings. If there is positive dissipation with this loss of heat, phase space volumes contract and $\int_0^t \mathrm{d}s\Lambda(S^s\boldsymbol{\Gamma})$ is a negative number. Note that we use the notation $\bar{X}_t(\boldsymbol{\Gamma}) \equiv (1/t)X_t(\boldsymbol{\Gamma}) \equiv (1/t)\int_0^t \mathrm{d}sX(S^s\boldsymbol{\Gamma})$. Then from Eq. (2.1), it can be shown that $\Omega(\boldsymbol{\Gamma}(t)) = \mathrm{d}\Omega_t(\boldsymbol{\Gamma})/\mathrm{d}t = -\partial/\partial\boldsymbol{\Gamma} \cdot \dot{\boldsymbol{\Gamma}}(\boldsymbol{\Gamma}(t)) - \dot{\boldsymbol{\Gamma}}(\boldsymbol{\Gamma}(t)) \cdot \partial/\partial\boldsymbol{\Gamma}(\ln f(\boldsymbol{\Gamma}(t), 0))$.

Defining the dissipation function requires two conditions: ergodic consistency, that is, if $f(\boldsymbol{\Gamma},0) \neq 0$, then $f(S^t\boldsymbol{\Gamma},0) \neq 0$; and that the initial distribution is an even function of the momenta. If the system is adiabatic, time-translated infinitesimal volume elements have a unit Jacobian. Since the initial distribution is an even function of the momentum, the average momentum of the system as calculated by the observer is zero. The observer is therefore in a comoving coordinate frame.

Because the dynamics are time reversal symmetric, the denominator of the logarithm in Eq. (2.1) denotes the probability of observing at time zero the set of trajectories that are antitrajectories to those referred to in the numerator. Along the trajectories in the denominator, the value of the time-integrated dissipation will be equal in magnitude but opposite in sign to that value for those in the numerator. Therefore, we have the Evans–Searles fluctuation theorem:

$$\frac{P(\bar{\Omega}_t = A \pm \mathrm{d}A)}{P(\bar{\Omega}_t = -A \pm \mathrm{d}A)} = \exp[At \pm O(\mathrm{d}A)], \tag{2.2}$$

where $P(\Omega_t = A \pm \mathrm{d}A)$ is the probability that a system trajectory will have a value of Ω_t, which is infinitesimally close to A, and $O(\mathrm{d}A)$ is a term that goes to zero as $\mathrm{d}A$ goes to zero. To simplify the notation, in later work we will assume that $\mathrm{d}A$, and therefore this error, is infinitesimally small and can be neglected. If the dissipation is extensive, that is, if each member of the system contributes to the dissipation, then in the thermodynamic limit, it becomes infinitely more likely that the dissipation is positive rather than negative. This ultimately is a derivation of the second "law" of thermodynamics.

The dissipation function is a measure of irreversibility for sets of trajectories, making it an "entropy production-like quantity" in the sense of linear irreversible thermodynamics. In systems in the linear response regime close to equilibrium, the average dissipation is in fact equal to the average spontaneous entropy production rate of linear irreversible thermodynamics [18]. When the dissipation function is positive, the system is evolving in the way expected by the second "law." In the thermodynamic limit, the probability that the dissipation is negative goes to zero, which is exactly what the second "law" of thermodynamics states. The ESFT can thus be seen to be a generalization of the second "law" so that it applies to small systems observed for short times. In these systems, the probability that the dissipation is negative, $-A \pm dA$, is *not* actually zero, but is instead exponentially less than the probability that it is positive, $A \pm dA$. Unlike linear irreversible thermodynamics, the ESFT is valid arbitrarily far from equilibrium.

The ESFT leads to a number of corollaries such as the second law inequality (SLI):

$$\langle \bar{\Omega}_t \rangle_0 \geq 0, \quad \forall t, f(\mathbf{\Gamma}, 0). \tag{2.3}$$

Note that we have used a shortened notation when considering ensemble averages of time-dependent phase variables so that $\langle X(t) \rangle_s \equiv \int d\mathbf{\Gamma} X(S^t\mathbf{\Gamma}) f(\mathbf{\Gamma}, s)$ and $\langle \bar{X}_t \rangle_s \equiv \int d\mathbf{\Gamma} \bar{X}_t(\mathbf{\Gamma}) f(\mathbf{\Gamma}, s)$. This is also an exact result for a system of arbitrary size, arbitrarily near or far from equilibrium. On average, the ensemble of trajectories will always act as anticipated from linear irreversible thermodynamics [19].

Through the ESFT, it is seen that the dissipation function is an important quantity in nonequilibrium statistical thermodynamics. However, it is also the key property in other new relationships. The nonequilibrium partition identity (NPI) shows that an exponentially weighted average of the dissipation function is equal to unity [20]:

$$\left\langle e^{-\Omega_t} \right\rangle_0 = 1. \tag{2.4}$$

Both the ESFT and NPI can be used as diagnostic relations in simulation and experiment to check that the equations of motion used to derive the dissipation function match those of the experiment and have been used to determine unknown experimental parameters [8, 21]. The negative exponential weighting of the NPI however means that when it is applied to a system with a finite number of samples, the rarely sampled negative trajectories dominate the average, and the convergence of the relation with increasing numbers of trajectories can be poor [22]. To improve the convergence of the NPI, an alternative expression can be derived from the ESFT that removes the exponential weighting of the negative trajectories [23]:

$$\int_0^\infty \left(1 + e^{-A}\right) P(\Omega_t = A) dA = 1. \tag{2.5}$$

This is known as the partial range NPI (PRNPI).

Another innovation is to use the dissipation function to help measure or quantify the behavior of phase functions *other* than the dissipation [24]. The ESFT can be modified to consider an arbitrary phase function $D(G)$ that is odd under time reversal ($D(G) = -D(M^T G)$):

$$\frac{P(D_t = A \pm dA)}{P(D_t = -A \pm dA)} = \left\langle e^{-\Omega_t} \right\rangle^{-1}_{D_t = A \pm dA}, \tag{2.6}$$

where $\langle \cdots \rangle_{D_t = A \pm dA}$ denotes an ensemble average over all trajectories for which the time integral $D_t(\mathbf{\Gamma}) \equiv \int_0^t ds D(S^s\mathbf{\Gamma})$ takes the values $A \pm dA$. This relation (Eq. (2.6)) will be referred to as the "functional" ESFT.

Even more usefully, the phase space distribution at a time t can be related to the density at an earlier time by integrating the dissipation backward in time along a phase space trajectory:

$$f(\boldsymbol{\Gamma}, t) = \exp\left[-\int_0^{-t} ds\, \Omega(S^s\boldsymbol{\Gamma})\right] f(\boldsymbol{\Gamma}, 0). \tag{2.7}$$

This equation is true for thermostated or unthermostated systems. It is also true for both systems relaxing to equilibrium and systems driven by a dissipative external field [13].

If we compute the time-dependent average of an arbitrary phase function $B(\boldsymbol{\Gamma})$ and then differentiate in time and then reintegrate with respect to time, we obtain the following exact expression for time-dependent averages in systems of arbitrary size and arbitrarily near or far from equilibrium:

$$\langle B(t)\rangle_0 = \langle B(0)\rangle_0 + \int_0^t ds \langle \Omega(0) B(s)\rangle_0. \tag{2.8}$$

Averages of phase functions at time t are given by time integrals of the transient time correlation function between the phase function and the dissipation function. Equation (2.8) is the transient time correlation function (TTCF) form of the dissipation theorem [13, 25].

While it would appear to be easier to simply measure the value of the function at time t, $\langle B(t)\rangle$, for systems in the weak field limit, the TTCF form of the DT has better statistics from a finite number of samples. For driven systems in the weak field limit, Eq. (2.8) reduces to the well-known Green–Kubo linear response relations [26].

Equation (2.8) also allows us to introduce the definition of a *T-mixing* system. There are many subtly different forms of mixing in dynamical systems, so we introduce their different definitions here. We define a T-mixing system to be a system in which infinite time integrals of transient time correlation functions of zero mean variables $\langle A(0)\rangle_0 = 0$ converge: $\left|\int_0^\infty ds \langle A(0)B(s)\rangle_0\right| < \infty$. A special case of T-mixing is ΩT-mixing for which $\left|\int_0^\infty ds\langle \Omega(0)B(s)\rangle_0\right| < \infty$. For ΩT-mixing systems, the dissipation theorem proves that the systems *must* come to a stationary state after sufficient time has elapsed. For a proof of this statement, refer to Eq. (2.8). Because they are a special case of T-mixing, T-mixing systems are also ΩT-mixing. However, on physical grounds, it also seems highly probable that ΩT-mixing are also T-mixing systems.

In weak T-mixing systems, $\lim_{t\to\infty} \langle A(0)B(t)\rangle_0 = \langle A(0)\rangle_0 \langle B(\infty)\rangle_0$. The mixing systems met in ergodic theory have the defining property $\lim_{t\to\infty} \langle A(0)B(t)\rangle_{\text{inv}} = \langle A\rangle_{\text{inv}} \langle B\rangle_{\text{inv}}$, here the notation $\langle\cdots\rangle_{\text{inv}}$ denotes some invariant (i.e., time-independent average). In general, there is no proof of relaxation to a stationary state at long time. For that you need ΩT-mixing. *Only* in the special case of autonomous Hamiltonian systems is there a proof of relaxation using ergodic theory (in fact, to the microcanonical ensemble) (see Ref. [27]).

The next major advance in the application of dissipation to statistical mechanics is the relaxation theorem. It shows that any nonequilibrium distribution that is an even function of the momentum will eventually relax to equilibrium, where the dissipation vanishes, provided it is T-mixing and no external fields are applied to the system. This can be used to derive equilibrium distributions such as the microcanonical distribution for autonomous Hamiltonian systems [14] or the Nosé–Hoover canonical distribution [28] for Nosé–Hoover thermostated systems. Prior to the discovery of the relaxation theorems, there was no general proof of thermal relaxation to equilibrium (except for uniform ideal gases) nor was there any proof of the microscopic form of the various equilibrium distributions in the various ensembles (except for the microcanonical distribution for autonomous Hamiltonian systems).

Because the proof of the relaxation to microcanonical equilibrium using the concept of T-mixing is recent and published in a special volume [14], we give a short version of the proof here. We also give a derivation of microscopic expressions for the microcanonical entropy and temperature.

2.3 Proof of Boltzmann's Postulate of Equal *A Priori* Probabilities

It is known from ergodic theory that for an N-particle autonomous Hamiltonian system, any initial distribution will at long times relax to the equilibrium microcanonical ensemble, provided the dynamics is also mixing. For an outline of this derivation that is accessible to physicists, engineers, or chemists, see Ref. [27]. However, these ergodic theory proofs of relaxation only pertain to autonomous Hamiltonian systems. Unlike the present derivation (see below), the ergodic theory proofs seem incapable of being extended to any other form of relaxation (e.g., to nonequilibrium steady states or to canonical or isothermal isobaric equilibrium).

For autonomous Hamiltonian systems, the ergodic theory proofs are mathematically very compact, but reveal very little about the relaxation process. They only tell you that if you wait long enough, the system will relax to microcanonical equilibrium. The proof given below can also provide details about the relaxation process. Will relaxation be monotonic? Can the relaxation process be used to define nonequilibrium transport coefficients? Precisely, how do you make the connection between equilibrium thermodynamics and the microcanonical ensemble?

Consider a classical system of N-interacting particles in a periodic system with a unit cell of volume V. Initially (at $t=0$), the microstates of the system are distributed according to a normalized probability distribution function $f(\mathbf{\Gamma},0)$. As above, we assume this distribution is an even function of the momenta, $f(\mathbf{q},\mathbf{p},0)=f(\mathbf{q},-\mathbf{p},0)$. This then guarantees that the average velocity of the system is zero (i.e., we are in a comoving frame with the system).

We write the equations of motion for the N-particle system as

$$\begin{aligned} \dot{\mathbf{q}}_i &= \mathbf{p}_i/m, \\ \dot{\mathbf{p}}_i &= \mathbf{F}_i(\mathbf{q}), \end{aligned} \tag{2.9}$$

where $\mathbf{F}_i(\mathbf{q}) = -\partial\Phi(\mathbf{q})/\partial\mathbf{q}_i$ is the interatomic force on particle i, m is the particle mass, and $\Phi(\mathbf{q})$ is the interparticle potential energy. These equations of motion preserve the phase space volume: a condition known as the adiabatic incompressibility of phase space, $\Lambda = 0$ or AIΓ [26]. They also conserve energy $H_0(\mathbf{\Gamma}) \equiv \sum p_i^2/2m + \Phi(\mathbf{q}) = C$ and total momentum $\mathbf{P} = \sum_{i=1}^{N} \mathbf{p}_i = \mathbf{0}$. The equations of motion may also preserve angular momentum. However, in the interests of simplicity, that is not assumed here.

The exact equation of motion for the N-particle distribution function is the time-reversible phase continuity equation,

$$\frac{\mathrm{d}f(\mathbf{\Gamma}, t)}{\mathrm{d}t} = 0. \tag{2.10}$$

The derivation of the ESFT (2.2) considers the response of a system that is initially described by some phase space distribution. The initial distribution is not necessarily at equilibrium, however we assume $\mathbf{P} = \mathbf{0}$ because all thermodynamic quantities must be Galilean invariant. We can write any such initial distribution in the form

$$f(\mathbf{\Gamma}, 0) = \frac{\delta(H_0(\mathbf{\Gamma}) - E)\delta(\mathbf{P}) \exp[-F(\mathbf{\Gamma})]}{\int \mathrm{d}\mathbf{\Gamma}\delta(H_0(\mathbf{\Gamma}) - E)\delta(\mathbf{P}) \exp[-F(\mathbf{\Gamma})]}, \tag{2.11}$$

where $F(\mathbf{\Gamma})$ is a single-valued real function that is even in the momenta, $H_0(\mathbf{\Gamma})$ is the Hamiltonian that generates the equations of motion (2.9), and since the coordinate frame is comoving, the value of the Hamiltonian is necessarily the internal energy of the system. E is the fixed value of the internal energy of the system. For our system, (2.9), the time integral of the dissipation function $\Omega(\mathbf{\Gamma})$, (2.1), can be written as

$$\int_0^t \mathrm{d}s\, \Omega(S^s\mathbf{\Gamma}) \equiv \bar{\Omega}_t(\mathbf{\Gamma})t \equiv \Omega_t(\mathbf{\Gamma}) = \ln\left(\frac{f(\mathbf{\Gamma}, 0)}{f(S^t\mathbf{\Gamma}, 0)}\right). \tag{2.12}$$

The existence of the dissipation function only requires that the initial distribution is normalizable, even in the momenta, and that ergodic consistency holds. This means that defining the instantaneous dissipation function $\Omega(S^t\mathbf{\Gamma}) \equiv \partial\Omega_t(\mathbf{\Gamma})/\partial t$ at a phase point $S^t\mathbf{\Gamma}$ requires that $\forall \mathbf{\Gamma}, t \quad \mathrm{s.t.} \quad f(\mathbf{\Gamma}; 0) \neq 0, f(S^t\mathbf{\Gamma}; 0) \neq 0$. Since the energy is a constant of the motion, ergodic consistency requires that we only consider states on the $H_0(\mathbf{\Gamma}) = E$ hypersurface. We do not consider the shell microcanonical ensemble of states within a narrowband of energies, nor do we consider the cumulative microcanonical ensemble of all states less than a set energy. Proving ESFT (2.2) requires an additional condition: the dynamics must be time reversal symmetric, that is, any nonautonomous fields must be odd or even around $t/2$.

For the Newtonian equations of motion (2.9) that conserve energy and total momentum, consider the initial distribution:

$$f(\mathbf{\Gamma}, 0) \equiv f_{\mu C}(\mathbf{\Gamma}, 0) = \frac{\delta(H_0(\mathbf{\Gamma}) - E)\delta(\mathbf{P})}{\int \mathrm{d}\mathbf{\Gamma}\delta(H_0(\mathbf{\Gamma}) - E)\delta(\mathbf{P})}. \tag{2.13}$$

If the Hamiltonian is orientationally isotropic, then angular momentum would also be conserved. This could trivially be added into (2.13), but to simplify the notation somewhat we do not do this here. It is trivial to show that for this distribution (2.13) and the dynamics (2.9), the dissipation function $\Omega_{\mu C}(\mathbf{\Gamma})$ is identically zero:

$$\Omega_{\mu C}(\mathbf{\Gamma}) = 0, \quad \forall \mathbf{\Gamma} : H_0(\mathbf{\Gamma}) = E, \quad \mathbf{P} = \mathbf{0} \tag{2.14}$$

everywhere on the energy/zero momentum hypersurface, and from the dissipation theorem (2.7), we see that this initial distribution is preserved everywhere in the ostensible phase space:

$$f(\mathbf{\Gamma}, t) = f_{\mu C}(\mathbf{\Gamma}, 0), \quad \forall \mathbf{\Gamma}, t. \tag{2.15}$$

For ergodic systems, we call distributions that are time independent and dissipationless, *equilibrium* distributions. We shall call the distribution given in (2.13), the microcanonical distribution. We do not yet know whether the microcanonical distribution is stable with respect to small fluctuations.

Now consider an arbitrary deviation from the microcanonical distribution:

$$f(\mathbf{\Gamma}, 0) = \frac{\delta(H_0(\mathbf{\Gamma}) - E)\delta(\mathbf{P}) \exp[-g(\mathbf{\Gamma})]}{\int d\mathbf{\Gamma}\delta(H_0(\mathbf{\Gamma}) - E)\delta(\mathbf{P}) \exp[-g(\mathbf{\Gamma})]}, \tag{2.16}$$

where the deviation function $g(\mathbf{\Gamma})$ is also even in the momenta and is not a constant of the motion.

For such a system evolving under our dynamics, the time-integrated dissipation function is

$$\bar{\Omega}_t(\mathbf{\Gamma})t = g(S^t\mathbf{\Gamma}) - g(\mathbf{\Gamma}) \equiv \Delta g(\mathbf{\Gamma}, t) \tag{2.17}$$

and (2.7) becomes

$$f(\mathbf{\Gamma}, t) = \exp[-\Delta g(\mathbf{\Gamma}, -t)]f(\mathbf{\Gamma}, 0). \tag{2.18}$$

If we assume our system to be T-mixing, g is not a constant of the motion, there is dissipation, and furthermore the distribution function will not be preserved. Because the system is T-mixing, there can be no constants of the motion apart from energy and momentum (otherwise the system cannot be T-mixing!) and the only possible unchanging distribution function will be that where $g(\mathbf{\Gamma})$ is zero and in this case the distribution reduces to the equilibrium microcanonical distribution function.

If the system is not ergodic, the phase space breaks up into nonoverlapping ergodic subspaces labeled D_α and $S_\alpha(\mathbf{\Gamma}) = 1$, if $\mathbf{\Gamma} \in D_\alpha$, and $S_\alpha(\mathbf{\Gamma}) = 0$ otherwise. The set of phase functions $\{S_\alpha(\mathbf{\Gamma});\ \alpha = 1, N_\mathrm{D}\}$, where N_D is the total number of ergodic subdomains, comprise a set of constants of the motion and the system cannot be T-mixing. Within each subdomain, any deviation from local microcanonical weighting will induce dissipation and the distribution will not be preserved. The only possible unchanging distribution is microcanonical within each subdomain. The interdomain weights are, however, arbitrary. Thus, we have a new derivation of the Williams–Evans quasi-equilibrium distribution function for glassy systems [29].

The dissipation function satisfies the second law inequality (2.3):

$$\langle \Delta g(t) \rangle_0 = \int_0^\infty A(1 - e^{-A}) P(\Delta g(\mathbf{\Gamma}, t) = A) dA \geq 0, \quad \forall t, f(\mathbf{\Gamma}, 0), \tag{2.19}$$

which can only take a value of zero if $\Delta g(\mathbf{\Gamma}, t) = 0, \forall \mathbf{\Gamma}$. In (2.16), it is assumed that g is not a constant, so $\langle \Delta g(t) \rangle_0$ will only be zero when $g(\mathbf{\Gamma}) = 0, \forall \mathbf{\Gamma}$. Thus, if the system is T-mixing and the initial distribution differs from the microcanonical distribution (Eq. (2.13)), there will always be dissipation and on average this dissipation is positive. You cannot have a distribution containing trajectories of nonzero dissipation, which when averaged over all the phase space will have an average dissipation of zero: the ESFT requires that positive trajectories are exponentially more probable than negative trajectories of the same magnitude, so the average over a distribution of nonzero dissipation trajectories must be positive.

For T-mixing systems, we have derived an expression for the unique, constant energy, equilibrium state and shown that it takes on the standard form for the microcanonical distribution, modulo the facts that the momentum (possibly linear and angular) and energy are constants of the motion.

This completes our first-principles derivation of the equilibrium distribution function. We now consider the question of relaxation toward equilibrium. Again, we assume the system is T-mixing. From Eq. (2.17), $\Omega(t) = \dot{g}(t)$ and substituting this into Eq. (2.8) with $B(t) = g(t)$ gives

$$\langle g(t) \rangle_0 = \langle g(0) \rangle_0 + \int_0^t ds \langle \dot{g}(0) g(s) \rangle_0 \geq 0, \quad \forall t, \tag{2.20}$$

where the ensemble averages are with respect to the initial distribution function.

Because the system is T-mixing, at a sufficiently long times, the correlations vanish so that for $t > t_c$, where t_c is the correlation time, we can write

$$\begin{aligned} \langle g(t) \rangle_0 &= \langle g(0) \rangle_0 + \int_0^{t_c} ds \langle \dot{g}(0) g(s) \rangle_0 + \int_{t_c}^t ds \langle \dot{g}(0) \rangle_0 \langle g(s) \rangle_0 \\ &= \langle g(0) \rangle_0 + \int_0^{t_c} ds \langle \dot{g}(0) g(s) \rangle_0 \\ &= \langle g(t_c) \rangle_0, \quad \forall t > t_c. \end{aligned} \tag{2.21}$$

The final result is obtained after noting that g is an even function of the momenta, so $\langle g(t) \rangle_0$ is an even function of time. This means for $t > t_c$, the value $\langle g(t) \rangle_0$ does not change with time, that is, $\langle \dot{g}(t) \rangle_0 = 0$. Thus, at sufficiently long times, there is no dissipation. If we consider the ensemble average with respect to the distribution function at t_c, $f(\mathbf{\Gamma}, t_c)$, and its resulting dissipation integral Δg_{t_c}, we see that $\langle \Delta g_{t_c}(t) \rangle_{t_c} = 0$, and following the same arguments as above, the system can only be the dissipationless equilibrium state given by (2.13). For T-mixing systems, this dissipationless distribution is unique (i.e., such systems are ergodic). We have therefore proven the relaxation to equilibrium and Boltzmann's postulate of equal *a priori* probabilities for the equilibrium distribution of a Newtonian system.

This proof of relaxation to equilibrium is more general than Boltzmann's *H*-theorem. The *H*-theorem is valid only for uniform ideal gases. Also, Boltzmann's

H-theorem allows only a monotonic relaxation to equilibrium (which is of course valid under the circumstances treated by his theorem). Our relaxation theorems allow nonmonotonic relaxation.

It might be presumed that if the dissipation function were redefined with respect to the distribution at the time t_a, when the negative dissipation (as defined from the initial distribution) begins, then the second law inequality would be violated, $\langle \Omega_{t_a,t_b} \rangle_{t_a} < 0$. (Here $\Omega_{t_a,t_b}(\mathbf{\Gamma}) \equiv \ln \left((f(\mathbf{\Gamma}, t_a))/(f(M^T(S^{t_b}\mathbf{\Gamma}), t_a)) \right)$, so the first subscript on Ω refers to the distribution function and the second subscript refers to trajectory length). However, this is not possible. Understanding this point reveals some of the subtleties involved in the definition of the dissipation function.

To generate trajectory segments that are time-reversed conjugates of each, one can extend the time interval to both back to time zero and forward beyond the interval of interest, in a symmetric way. In the end, the dissipation satisfies the second law inequality for the extended domain: $\langle \Omega_{t_a,t_b} \rangle_{t_a} = \langle \Omega_{0,2t_a+t_b} \rangle \geq 0$ (see Ref. [30]).

The connection between equilibrium statistical mechanics and macroscopic thermodynamics is usually done extremely poorly in textbooks. We give a proof here that the microscopic expressions defined in Eqs. (2.23) and (2.24) on average are indeed equal to the thermodynamic entropy and temperature, respectively. We take as our starting point known expressions for the Galilean invariant energy and pressure to equal, on average, their thermodynamic counterparts. Energy and pressure are, as the first "law" of thermodynamics makes clear, completely mechanical in nature.

To begin with, we note that from thermodynamics we have two equations for the entropy (S), in terms of the energy (U), the volume (V), and the pressure (p):

$$\left.\frac{\partial S}{\partial U}\right|_V = \frac{1}{T}, \quad \left.\frac{\partial S}{\partial V}\right|_U = \frac{p}{T}. \tag{2.22}$$

Consider the function $\tilde{S}$ (up to an additive constant) defined as

$$\tilde{S} = -k_B \ln \int \delta(H_0(\mathbf{\Gamma}) - U)\delta(\mathbf{P})d\mathbf{\Gamma} \equiv -k_B \ln V_{\mathbf{\Gamma}}. \tag{2.23}$$

We can identify the internal energy with the value of the Hamiltonian in a comoving coordinate frame, because internal energy is the Galilean invariant mechanical energy.

By considering a vector displacement in phase space that is normal to the surface $\sum_i p_i^2/2m = K_0$, which changes the energy by an infinitesimal amount dU, it can be seen that [31]

$$\left.\frac{\partial \tilde{S}}{\partial U}\right|_V \equiv \frac{1}{\tilde{T}} = \frac{(3N-4)k_B}{2\langle K_0\rangle_{\mu C}} \equiv \frac{1}{\langle T_K(\mathbf{\Gamma})\rangle_{\mu C}}, \tag{2.24}$$

where the ensemble average is microcanonical and taken with respect to (2.13) and $T_K(\mathbf{\Gamma})$ is the instantaneous kinetic temperature.

If we now use the well-known SLLOD equations [26] (note the SLLOD equations of motion give an exact description of arbitrary homogeneous flows – see Ref. [32])

to accomplish an infinitesimal volume change at constant energy using an ergostat to fix the energy, we see that from the ergostatted equations of motion,

$$dH_0 = dU = 0 = -p\,dV - 2K_0\alpha\,dt, \tag{2.25}$$

where from the SLLOD equations, p is the microscopic expression for the pressure,

$$3pV = \sum_{i\in V} p_i^2/m - \frac{1}{2}\sum_{i\in V,\forall j} \mathbf{r}_{ij}\cdot\mathbf{F}_{ij}, \tag{2.26}$$

and α is the ergostat multiplier. As Irving and Kirkwood showed [33], this microscopic expression for p is easily identified with the microscopic mechanical force across a surface and is therefore, on average, equal to the thermodynamic pressure.

We also know from the phase continuity equation $df/dt = (3N-4)\alpha f$ that the change in phase space volume, $dV_{\boldsymbol{\Gamma}}$, caused by this constant energy volume change is

$$dV_{\boldsymbol{\Gamma}} = -(3N-4)\langle\alpha\rangle_{\mu C} V_{\boldsymbol{\Gamma}} dt. \tag{2.27}$$

From our proposed microscopic equation for the entropy, we see that

$$\left.\frac{\partial\tilde{S}}{\partial V}\right|_U = \left\langle\frac{(3N-4)k_B p}{2K_0}\right\rangle_{\mu C} = \left\langle\frac{p}{T_K}\right\rangle_{\mu C} = \frac{\langle p\rangle_{\mu C}}{\tilde{T}} + O(1/N). \tag{2.28}$$

Comparing Eqs. (2.24) and (2.28) with Eq. (2.22) and noting that the classical entropy is only defined up to an arbitrary constant, we conclude that S and $\tilde{S}$ satisfy the same differential equations with the same initial conditions:

$$\left.\frac{\partial X}{\partial V}\right|_U \Big/ \left.\frac{\partial X}{\partial U}\right|_V = p. \tag{2.29}$$

Note that $T, \tilde{T}$, which are yet unresolved, both individually cancel from the two versions of (2.29) (when $X = S, \tilde{S}$). Therefore,

$$S(V,U) = \tilde{S}(V,U) + O(1/N) + \text{constant}. \tag{2.30}$$

Substituting the thermodynamic entropy into (2.24) then shows that

$$T(V,U) = \tilde{T}(V,U) + O(1/N). \tag{2.31}$$

The $O(1/N)$ corrections disappear in the thermodynamic limit where classical thermodynamics is valid. This completes our discussion of the relaxation to equilibrium in the constant volume constant energy case.

2.4 Nonequilibrium Free Energy Relations

While the various relations based around the dissipation theorem quantify a system's behavior, the relations based upon work (or its generalizations) have generally been used for measuring free energy differences. The CR considers the work done moving between two states, here considered to be equilibrium states, along nonequilibrium paths. It gives the ratio of the probability with respect to the first equilibrium distribution that the work takes on a value $A \pm dA$ along the forward path (from state 1 to state 2) $P_1(W_F = A \pm dA)$, compared to the probability with respect to the second equilibrium distribution that it takes on a value $-A \pm dA$ along the reverse path (from state 2 to state 1) $P_2(W_R = -A \pm dA)$. The CR shows that if the equilibrium systems are canonical, this probability ratio is connected to the magnitude of the work and the free energy difference between the two systems:

$$\frac{P_1(W_F = A \pm dA)}{P_2(W_R = -A \pm dA)} = e^{\beta A - \beta \Delta F}, \tag{2.32}$$

where $\beta = 1/k_B T$, T is the temperature of the equilibrium systems, and $\Delta F \equiv F_2 - F_1$ is the difference in free energy of the two states. From Eq. (2.32), the Jarzynski equality can be obtained and this is a useful method of measuring free energy differences between systems from nonequilibrium work:

$$\left\langle e^{-\beta W} \right\rangle_1 = e^{-\beta \Delta F}. \tag{2.33}$$

In the JE, the ensemble average is evaluated in one direction only (and the work is that done starting from that initial state 1). For the JE and CR to yield correct results, one must be able to sample the antitrajectories of the most probable second "law" satisfying processes. This is practically impossible for macroscopic systems and, in fact, the JE and CR become unusable in the thermodynamic limit.

While the JE is an exact relation, it is most commonly applied to experiments/simulations where only a finite number of samples from the distribution have been taken. Similar to the NPI, the data are exponentially weighted, so that poorly sampled data from tail of the distribution where the values of the work are much smaller than the typical values of work dominate the average. There have been two main approaches to improving free energy measurements for this relation: path-based methods and bidirectional methods. Path-based methods try to choose the best ensemble to evaluate the expression through methods such as modifying the dynamics or performing umbrella sampling [34–37]. Bidirectional sampling takes advantage of the fact that the JE works in either direction between the two states. Path approaches are highly system dependent, so we will instead focus on the bidirectional methods.

If the work is measured in both the forward and reverse directions, then two JE estimates of the free energy can be produced. The most obvious method to improve sampling would be to average the two results. Generally however, the measurement in the less dissipative direction will have better statistics than either the other result on its own or the two results combined [38]. An appropriate linear sum of the two results can generate better results; however, often the best approach is to coopt an approach developed by Bennett for Monte Carlo simulations, the Bennett acceptance ratio or the maximum likelihood estimator (MLE)[39–41]. Under this approach, the work in each direction are reweighted by an estimate of the free energy, and this estimate is iteratively improved until the two sides agree:

$$\sum_{i=1}^{N_{\mathrm{F}}} \frac{1}{1+\mathrm{e}^{X(N)\left[W_{\mathrm{F},i}-\Delta F\right]}} = \sum_{i=1}^{N_{\mathrm{R}}} \frac{1}{1+\mathrm{e}^{-X(N)\left[W_{\mathrm{R},i}+\Delta F\right]}}, \tag{2.34}$$

where $X(N) = \ln\left(N_{\mathrm{F}}/N_{\mathrm{R}}\right)$, where N_{F} is the number of trajectories sampled for the forward change (from state 1 to state 2) and N_{R} is the number in the reverse change. It has been demonstrated by Shirts *et al.* [42] from the CR that this is the maximum likelihood estimator for the two distributions, that is, the free energy difference found is the one that is most likely to have generated the two observed work distributions. It has been demonstrated that for many systems this is the best free energy measurement using nonequilibrium work.

The close connection between the various FRs has enabled some approaches to cross between the two areas. It is possible to use the FT to develop an MLE-like relation for the NPI (Eq. (2.4)) [23]:

$$\sum_{\Omega_{t,i}>0} \frac{1}{1+\mathrm{e}^{\Omega_{t,i}}} \Bigg/ \sum_{\Omega_{t,j}<0} \frac{1}{1+\mathrm{e}^{\Omega_{t,j}}} = 1. \tag{2.35}$$

This method avoids the sampling problems caused by the exponential in Eq. (2.4) by weighting it by $1/(1+x)$, while still assuming that the system obeys Eq. (2.4). While this provides poorer convergence than the PRNPI (Eq. (2.5)), it exhibits much larger deviations from unity when the underlying distribution does not obey the ESFT (Eq. (2.2)). This makes it useful for checking the conformity of an experiment or simulation to these relations. Similarly, the phase function form of the CR can be extended to produce a functional form of the CR [37]:

$$\frac{P_1(D_{t,\mathrm{F}} = A \pm \mathrm{d}A)}{P_2(D_{t,\mathrm{R}} = -A \pm \mathrm{d}A)} = \mathrm{e}^{-\beta\Delta F} \left\langle \mathrm{e}^{-\beta W_{\mathrm{F}}} \right\rangle^{-1}_{1,(D_{t,\mathrm{F}}=A\pm \mathrm{d}A)}, \tag{2.36}$$

where the subscript on the ensemble average refers to the distribution function corresponding to the first state, and is over all phase points for which $D_{t,\mathrm{F}} = A \pm \mathrm{d}A$.

2.5
Simulations and Results

The FRs can be applied to a vast array of experiments and simulations using dynamics ranging in scale from quantum to molecular to coarse-grained/stochastic. A very simple system for testing the various fluctuation relations is based on an optical trapping experiment called the capture experiment. In this system, a single particle is bound to a harmonic potential created by a laser, which is equivalent to tethering the particle to a point in space with a spring. The molecules in the fluid constantly interact with the bound particle causing it to move in an apparently random fashion, while the harmonic force prevents the particle from moving too far from the trap center. These two competing forces mean that at equilibrium, the trapped particle will have a Gaussian distribution of positions and velocities, defined by the strength of the trap and the temperature of the fluid.

If the fluid is viscous with white noise, the system can be studied using either Langevin or molecular dynamics (MD) simulations, both will give the same results. However, if the solvent is viscoelastic, MD simulations are much easier [43]. The Langevin approach is difficult to apply if the system is viscoelastic (exhibits memory effects), but an even bigger problem with the Langevin or indeed any stochastic approach is that in order to write down a Langevin or stochastic equation to model dynamics, you need to assume relationships between fluctuations or noise and the friction or transport coefficients. These become complex if the noise is colored and are essentially unknown in the nonlinear response regime where the noise is field dependent and even becomes spatially anisotropic. Therefore, the MD derivation has broader applicability than the Langevin approach. The advantage of Langevin dynamics is that it is easier to perform simulations of an experimentally accessible scale, since much of the system is abstracted into the noise terms. For consistency with our theoretical approach, we will study this system using MD simulations; however, it is possible to apply the fluctuation relations to similar systems using stochastic equations of motion [17].

MD simulations integrate Newton's equations of motion to evolve a multiparticle system with time. We consider a two-dimensional system in periodic boundary conditions, comprising a fluid confined between walls. The walls are composed of two layers of thermostated particles, harmonically bound to lattice sites. In the center of the box, one of the fluid particles is bound by the harmonic trap that drives the experiment. The equations of motion for the system are

$$\begin{aligned}\dot{\mathbf{q}}_i &= \mathbf{p}_i/m,\\ \dot{\mathbf{p}}_i &= \mathbf{F}_{\mathrm{I},i} + \delta_{i,1}\mathbf{F}_{\mathrm{E},i} - \alpha S_i \mathbf{p}_i,\\ \dot{\alpha} &= \left(\frac{T_{\mathrm{K}} - T}{T}\right)\Big/ Q,\end{aligned} \tag{2.37}$$

where i is the particle index, m is the mass, $\mathbf{F}_{\mathrm{I},i}$ is the intermolecular force, and $\mathbf{F}_{\mathrm{E},i}$ is the external force (the harmonic trap), α is the thermostat multiplier, S_i is a switch controlling the thermostat that is 1 for the N_{T} wall particles and 0 for the fluid particles, T_{K} is the instantaneous kinetic temperature of the walls, and T is the thermostat temperature. The intermolecular potentials are two-body

Weeks–Chandler–Anderson (WCA) potentials, a truncation and shift of the Lennard-Jones potential to be purely repulsive [44]. For deriving the dissipation function however, the details of the potential are irrelevant. The thermostat is a Nosé–Hoover thermostat that uses integral feedback to maintain a constant temperature. This is one of the class of artificial or "fictitious" thermostats used because they are deterministic and mathematically well defined, analogous to the ergostat used in Eq. (2.25). The use of an artificial thermostat such as Nosé–Hoover means that the equations of motion for the system are no longer strictly Hamiltonian. The two important upshots of this approach are that the equilibrium distribution function for the system must be modified:

$$f(\mathbf{\Gamma}) = \frac{\exp[-[\beta H(\mathbf{\Gamma}) + D_C N_T Q\alpha^2/2]]}{Z}, \tag{2.38}$$

where Z is the partition function for the system, and that the dynamics of the system are no longer phase volume conserving in that

$$\frac{d}{dt}\ln[f(\mathbf{\Gamma},t)] = -\Lambda(\mathbf{\Gamma}) = D_c N_T \alpha, \tag{2.39}$$

where D_c is the number of Cartesian dimensions of the system.

As discussed above, the harmonic trap potential is of the form:

$$\Phi_{\text{trap}}(\mathbf{q},t) = \begin{cases} \frac{1}{2}k_i\mathbf{q}_1^2, & t \leq 0 \\ \frac{1}{2}k_f\mathbf{q}_1^2, & t > 0 \end{cases} \tag{2.40}$$

where $\mathbf{q}_1$ is the position of the trapped particle $i=1$, relative to the trap center. We can then apply Eq. (2.1) and derive both the time-integrated and instantaneous dissipation functions:

$$\begin{aligned} \Omega_t(\mathbf{\Gamma}) &= \ln\left[\frac{\exp[-\beta H_i(\mathbf{\Gamma}) - D_c N_T Q\alpha(0)^2/2]}{\exp[-\beta H_i(S^t\mathbf{\Gamma}) - D_c N_T Q\alpha(t)^2/2]}\exp\left[-\int_0^t ds\Lambda(S^s\mathbf{\Gamma})\right]\right] \\ &= \int_0^t ds[\beta\dot{H}_i(S^s\mathbf{\Gamma}) + D_c N_T Q\dot{\alpha}(s)\alpha(s) - \Lambda(S^s\mathbf{\Gamma})] \\ &= \beta\int_0^t ds[(k_i - k_f)S^s(\mathbf{q}_1\cdot\dot{\mathbf{q}}_1)] = \frac{\beta}{2}(k_i - k_f)((S^t\mathbf{q}_1)^2 - \mathbf{q}_1^2), \\ \Omega(\mathbf{\Gamma}) &= \beta(k_i - k_f)\mathbf{q}_1\cdot\dot{\mathbf{q}}_1. \end{aligned} \tag{2.41}$$

In this equation, $H_i(\mathbf{\Gamma})$ is the Hamiltonian for the system for $t < 0$. From the fluctuation theorem and the second law inequality, we know that we are likely to observe more positive dissipation results than negative results. The dissipation function has two components: the differences in the trapping constants ($k_i - k_f$) and the differences in the square of the final position of the trapped particle relative to the trap center ($\mathbf{q}_1^2(t) - \mathbf{q}_1^2(0)$). The sign of Ω_t will depend on the

Table 2.1 Sign of a particle being trapped by an optical trap when the trapping constant changes.

	$k_f > k_i$	$k_i > k_f$
$q_1^2(t) > q_1^2(0)$	$\Omega_t < 0$	$\Omega_t > 0$
$q_1^2(0) > q_1^2(t)$	$\Omega_t > 0$	$\Omega_t < 0$

The dissipation is positive when the final absolute displacement is in the same relative direction as the equipotential point of the initial absolute displacement.

signs of these two terms (see Table 2.1). We see that we have positive Ω_t for a system where the trap strength has increased and the particle finished closer to the trap center, and where the trap strength was decreased and the particle finished further from the trap center. These trajectories correspond to our physical intuition of what should happen for the system: the particle should end in a position of comparable potential energy to where it started. Furthermore, from the fluctuation theorem, we expect that the more positive the dissipation distribution, the less likely to observe antitrajectories. Looking at our first term, the magnitude of the dissipation function is controlled by the difference in trapping strengths: the greater the change in the trapping constants, the more likely it is to observe trajectories with positive dissipation. The parameters used for the simulation are contained in Table 2.2.

The system is simulated with trapping constant k_i for a sufficient time ($t > 100$) that it has equilibrated, and then periodically a snapshot is taken and used as the starting configuration for a simulation trajectory with constant k_f. Since the probability of a given system state is the same as that of the state with

Table 2.2 Parameters for molecular dynamics simulation.

Parameter	Value
N_{fluid}	50
N_{wall}	22
Δt	1×10^{-3}
m	1
ε	1
σ	1
T	1
ϱ	0.3
k_i	2
k_f	8
τ	N/A (capture) 0.1 (ramp)
Number of trajectories	100 000 (50 000 momenta reversed pairs)

opposite momenta for all the particles, a second simulation trajectory is generated by reversing the momenta of all the particles in the snapshot, minimizing the time spent running the program at equilibrium. At each time step of the simulation trajectory, the instantaneous dissipation and the time autocorrelation function of the dissipation function are calculated. The instantaneous dissipation function is used to examine the relaxation theorem and integrated to check the FT and NEPI, and the autocorrelation function is used to examine the dissipation theorem.

In addition to studying the dissipation-based fluctuation relations, we are also interested in the work-based relations. While the work can be assessed for the capture system, all the work is done at $t = 0$. To study these relations over a finite time, we modify the system so that the trapping constant changes linearly between k_i and k_f over a time τ, the (linear) ramp system. In addition, work relations are studied over forward and reverse system changes, we define the forward change as going from k_i to k_f and the reverse change as going from k_f and k_i. The ramp system is useful in that it has a known free energy, $\Delta F = \ln(k_f/k_i)$. The work function can be defined in a manner similar to the dissipation function [17]:

$$\begin{aligned} W_{\mathrm{F},\tau} &= \ln\left(\frac{f_i(\mathbf{\Gamma},0)\mathrm{d}\mathbf{\Gamma}}{f_f(\mathbf{\Gamma}^*,0)\mathrm{d}\mathbf{\Gamma}^*}\frac{Z_i}{Z_f}\right) \\ &= \ln\left(\frac{\exp\left[-(\beta H(\mathbf{\Gamma}(0),0)) - D_{\mathrm{C}}N_{\mathrm{T}}Q\alpha(0)^2/2)\right]\mathrm{d}\mathbf{\Gamma}}{\exp\left[-(\beta H(S^{\tau}\mathbf{\Gamma}^*,\tau) - D_{\mathrm{C}}N_{\mathrm{T}}Q\alpha(\tau)^2/2)\right]\mathrm{d}\mathbf{\Gamma}^*}\right) \\ &= \beta\int_0^{\tau}\mathrm{d}s\left[\dot{H}^{\mathrm{ad}}(S^s\mathbf{\Gamma},s)\right], \end{aligned} \tag{2.42}$$

where the distribution functions are equilibrium distributions associated with the systems with the initial and final trapping constants and Z_i and Z_f are the associated partition functions. $H(\mathbf{\Gamma}, t)$ is the Hamiltonian that evolves from the initial state to that of the final state over the period τ. For this system, this is the standard definition of the mechanical work modified by the thermostat. It is the total energy change minus the change due to heat transfer; hence, $\dot{H}^{\mathrm{ad}}$ in Eq. (2.42) is the work (the total change in energy = change due to heat + change due to work).

Substituting for our Hamiltonian and using the fact that $\dot{H}(\mathbf{\Gamma}, t) = \partial H(\mathbf{\Gamma}, t)/\partial t$, we derive the work function:

$$W_{\mathrm{F},\tau} = \frac{(k_f - k_i)}{2\tau}\int_0^{\tau}\mathbf{q}_1^2(s)\mathrm{d}s, \tag{2.43}$$

where τ is the ramp time. Because we change the strength of the trap at a relatively fast rate at $t = \tau$, the system is not at equilibrium. However, there is no change in the work for $t > \tau$ because $\dot{H}_0^{\mathrm{ad}}(t) = 0,\ \forall t > \tau$.

Care must be taken when applying the fluctuation relations to a finite sized data sample. To measure ensemble averages such as the NPI (2.4), we can re-express the relation in the form of a limiting sum, for example, the JE (2.33) becomes

$$\lim_{N\to\infty} \frac{1}{N}\sum_{i=1}^{N} e^{-\beta W} = e^{-\beta \Delta F}. \tag{2.44}$$

Similarly, sampling from a continuous distribution to measure the relative probabilities of points is impossible, and therefore expression in terms of probability ratios must be re-expressed in terms of finite size bins of data, for example, the ESFT (2.2) becomes

$$\lim_{\Delta A\to 0, N\to\infty} \ln\left(\frac{N_{\mathrm{Bin}}(\Omega_t = A \pm \Delta A)}{N_{\mathrm{Bin}}(\Omega_t = -A \pm \Delta A)}\right) = A, \tag{2.45}$$

where N_{Bin} is the bin population and $2\Delta A$ is the bin width. As discussed above with regard to the NPI and JE, the presence of an exponential in most of these relations tends to amplify sampling errors. Selecting an appropriate bin size for these relations is difficult: too large and the systematic error will dominate, but too small and the sampling error in the bin populations will also dominate.

In addition, measuring uncertainty in these relations is difficult. To generate uncertainties, we have repeated the simulations nine times and determined the standard error in the mean. For systems where generating multiple data sets is impossible, a block averaging method [45] or a propagating error method [46] is more appropriate. Block averaging breaks the data into subensembles containing m work values and calculates the JE estimate for each subensemble. This is done for every value of m between 1 and N, and the average free energy estimate is plotted against m. This smoothes the convergence of the work values, allowing an estimate of the limiting free energy difference and uncertainty. Regardless of the approach used, uncertainties will often be underestimated for these relations as the negative exponential means that unsampled data from the tail will have a major effect on convergence. This means that the sample distributions are highly skewed and non-Gaussian. The final approach of the sample mean to the asymptotic answer is in the vast majority of instances, one-sided.

2.6
Results Demonstrating the Fluctuation Relations

We begin by examining the ESFT (Eqs. (2.2) and (2.45)) at various times during our simulation (Figure 2.1). We see that for each of these time points, there is a linear trend close to the expected slope of 1, consistent with Eq. (2.2). We note that at extreme values of the dissipation, the number of points in each bin will drop, with exponentially fewer samples in the negative bin generally leading to a large systematic error. This problem is unavoidable as increased sampling will improve the current range of bins, but populate previously empty bins further from zero. In our

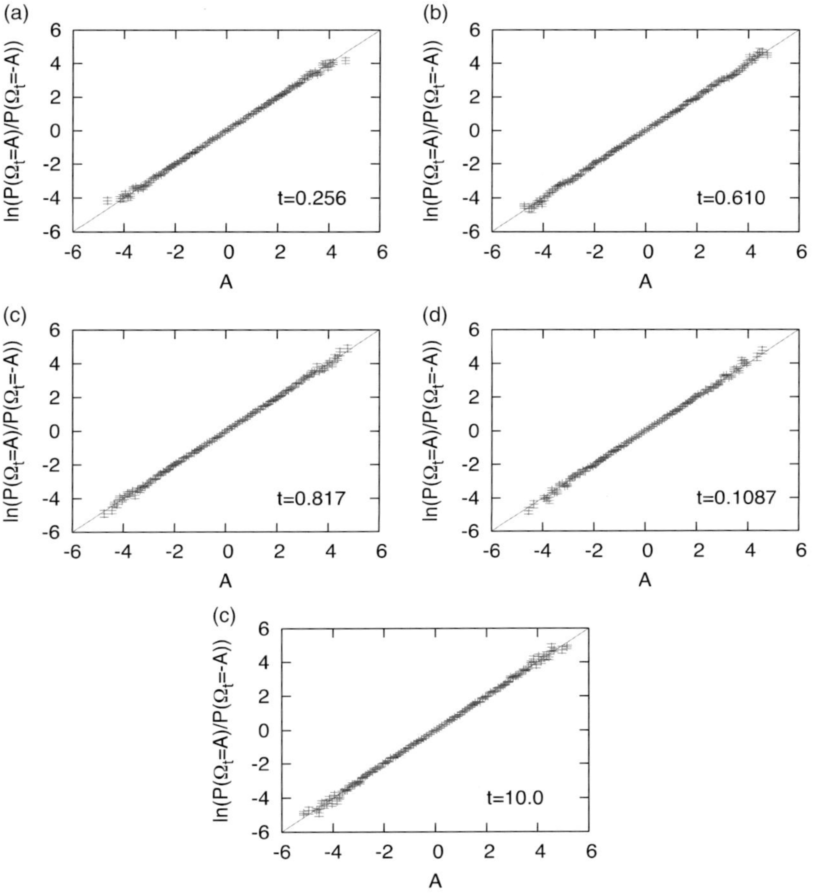

Figure 2.1 Tests of the fluctuation theorem (Eq. (2.2)). The logarithms of both sides of Eq. (2.2) are plotted for various trajectory lengths. The selected times correspond to the first maxima, zero, minima, and second zero of the instantaneous dissipation function and to the end of the simulation (see Figure 2.3). The lines represent the expected behavior [47]. (Figure 2.1e reprinted with permission from Ref. [47], Copyright 2012, American Institute of Physics.)

averaging, we discard any bin that is not populated in all nine sets of the simulations.

In Figure 2.2, we plot the ensemble average of the dissipation function $<\Omega_t>$ with time. While it fluctuates, it is always positive definite as predicted by the second law inequality (Eq. (2.3)). In Figure 2.3, we plot the ensemble average of the instantaneous dissipation function against time (left-hand side of Eq. (2.8)). In both figures, we can see that the system relaxes nonmonotonically, with zero instantaneous dissipation and a fixed total dissipation at long times. This agrees

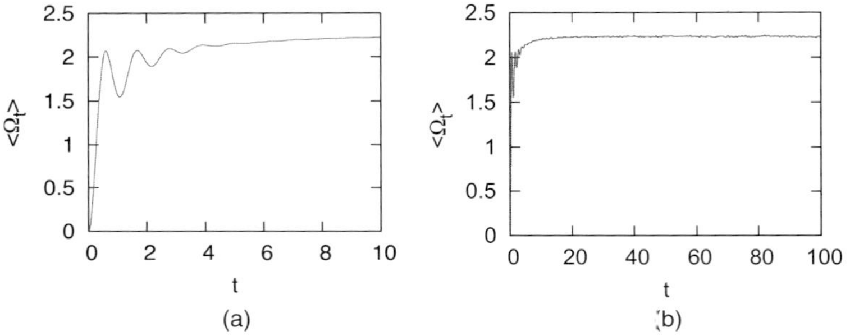

Figure 2.2 Plot of the ensemble average of dissipation versus time for two different timescales: (a) up to $t = 10$ and (b) up to $t = 100$. Note that while the value oscillates with time, it is always positive in accord with the second law inequality (Eq. (2.3)) [47]. (Figure 2.2a reprinted with permission from Ref. [47], Copyright 2012, American Institute of Physics.)

with the prediction of the relaxation theorem that as the system goes to equilibrium, the average of the dissipation function must go to zero. What is interesting about this system is that the dissipation overshoots, that is, for periods of time the average *instantaneous* behavior of the system is antidissipative (Figure 2.3). This is in contrast with the plot of the total dissipation, where the function is positive definite without being monotonic (Figure 2.2).

In Figure 2.3, we take advantage of the fact that dissipation is a phase function to plot the same function against the value derived for the instantaneous dissipation from the dissipation theorem (right-hand side of Eq. (2.8)). The results are indistinguishable on the scale of the figure. In Figure 2.3b, we plot

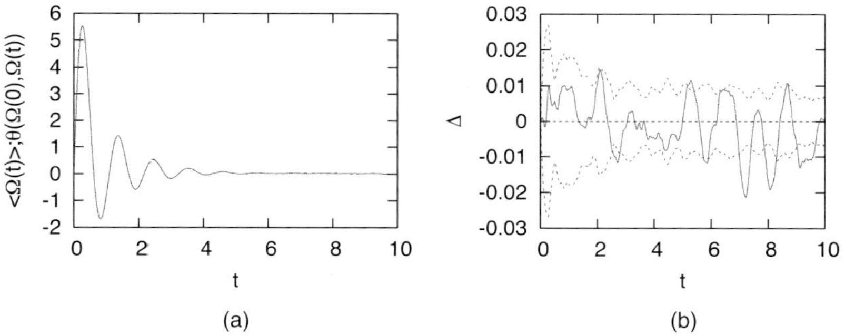

Figure 2.3 (a) Plot of the ensemble average of the instantaneous dissipation function against time (solid line) – LHS of Eq. (2.8), with a plot of the dissipation theorem versus time (dashed line, which appears to overlay the solid line on this scale): $\theta(\Omega(0),\Omega(t)) \equiv$ RHS of Eq. (2.8). (b) Plot of the difference in the average of the two functions (solid line) and the sum of the standard errors of the two functions (dashed lines) against time: $\Delta \equiv$ LHS–RHS of Eq. (2.8) [47]. (Reprinted with permission from Ref. [47], Copyright 2012, American Institute of Physics.)

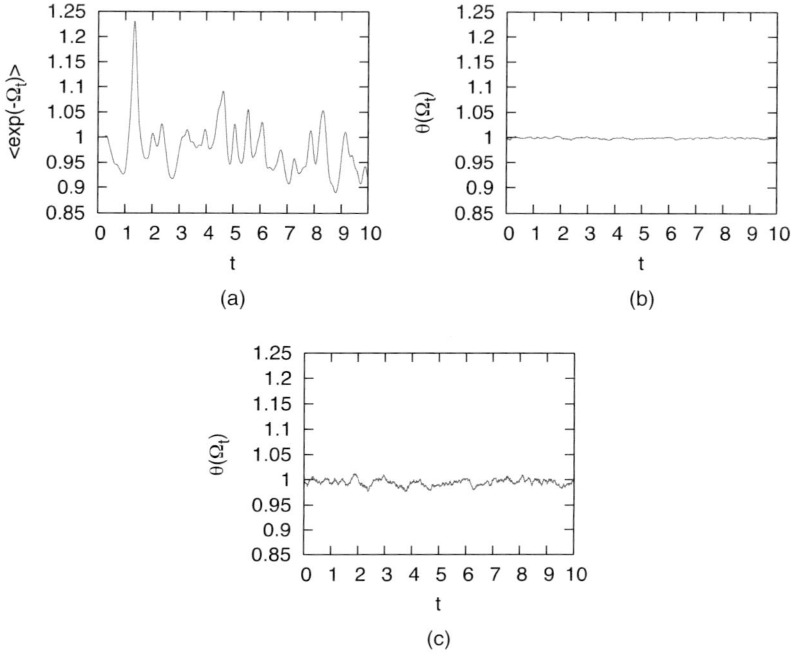

Figure 2.4 Plot of (a) NPI versus time: LHS of Eq. (2.4), (b) PRNPI versus time: $\theta(\Omega_t) \equiv$ LHS of Eq. (2.5), and (c) MLE form of the NPI versus time: $\theta(\Omega_t) \equiv$ LHS of Eq. (2.35), for a single run of the experiment. Note that the alternative methods converge significantly better than the NPI.

the difference between the two functions and the combined standard error for the two functions: even at the most extreme deviation, the two functions are in close agreement.

In Figure 2.4, we plot the NPI (Eq. (2.4)) against time, and we see that it remains close to 1 at all times, but has significant deviations. As discussed, the deviations from 1 are a common feature of NPI averages, due to the heavy weighting placed on poorly sampled antitrajectories. In Figure 2.4b and c, we plot two methods of improving NPI averages: Eqs. (2.5) and (2.35), respectively. These results are significantly better than the NPI in convergence.

In Figure 2.5, we plot the functional fluctuation theorems (Eq. (2.6)) for the change in temperature ($B_t(\mathbf{\Gamma}) = \Delta T(\mathbf{p})$) and the internal energy ($B_t(\mathbf{\Gamma}) = \Delta U(\mathbf{q})$), functions of the momenta and position, respectively. We can see good agreement with theory for both functions. Of particular interest is that the correlations of the dissipation function with temperature change, ΔT, decay with time, as the system re-equilibrates with the thermostat.

Moving to the work relations, we find that the results for the ramp system are similar to those for the capture system. In Figure 2.6, we present free energy estimates for the JE (2.33) and the MLE (2.34) as the function of the number of

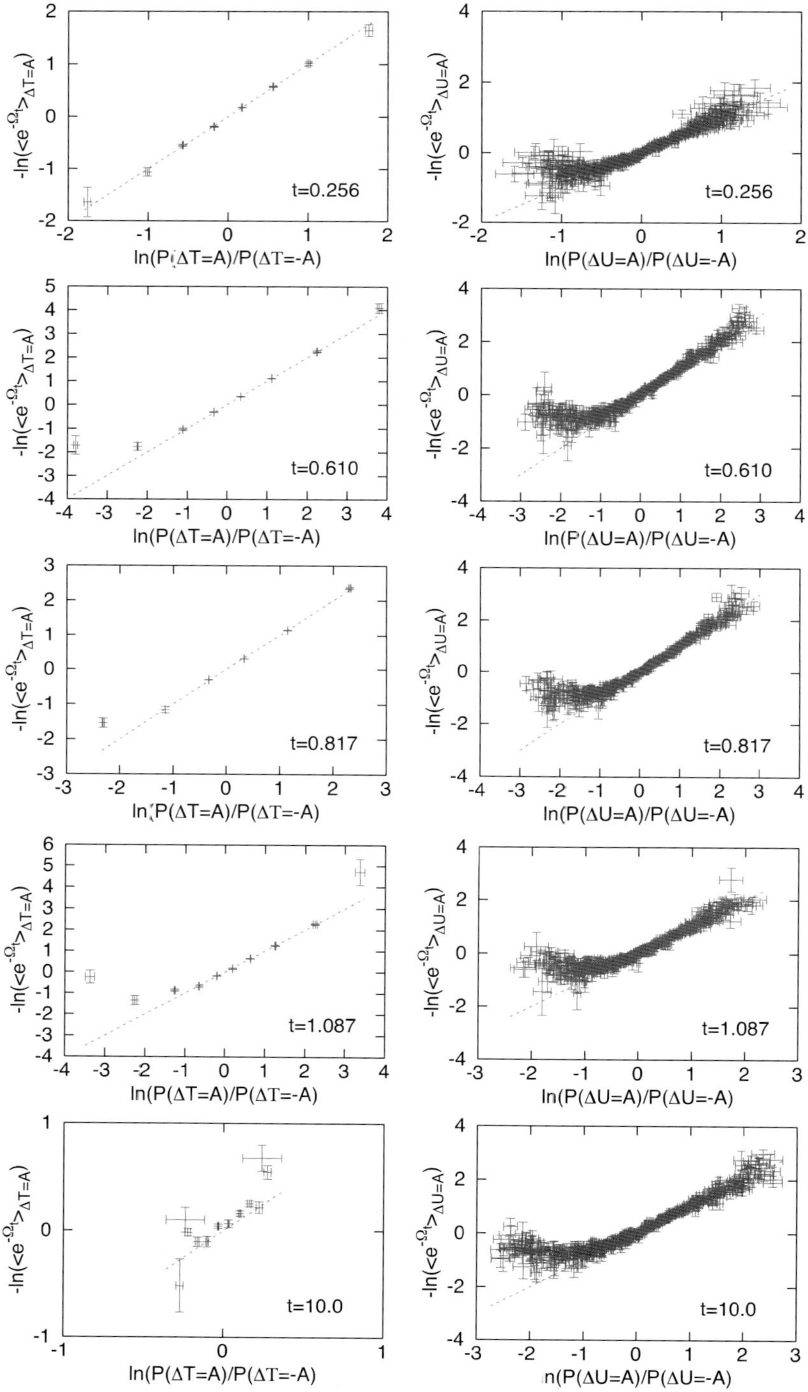
-ln(<e^{-Ω_t}>_{ΔT=A})
ln(P(ΔT=A)/P(ΔT=-A))
t=0.256
-ln(<e^{-Ω_t}>_{ΔU=A})
ln(P(ΔU=A)/P(ΔU=-A))
t=0.256
-ln(<e^{-Ω_t}>_{ΔT=A})
ln(P(ΔT=A)/P(ΔT=-A))
t=0.610
-ln(<e^{-Ω_t}>_{ΔU=A})
ln(P(ΔU=A)/P(ΔU=-A))
t=0.610
-ln(<e^{-Ω_t}>_{ΔT=A})
ln(P(ΔT=A)/P(ΔT=-A))
t=0.817
-ln(<e^{-Ω_t}>_{ΔU=A})
ln(P(ΔU=A)/P(ΔU=-A))
t=0.817
-ln(<e^{-Ω_t}>_{ΔT=A})
ln(P(ΔT=A)/P(ΔT=-A))
t=1.087
-ln(<e^{-Ω_t}>_{ΔU=A})
ln(P(ΔU=A)/P(ΔU=-A))
t=1.087
-ln(<e^{-Ω_t}>_{ΔT=A})
ln(P(ΔT=A)/P(ΔT=-A))
t=10.0
-ln(<e^{-Ω_t}>_{ΔU=A})
ln(P(ΔU=A)/P(ΔU=-A))
t=10.0

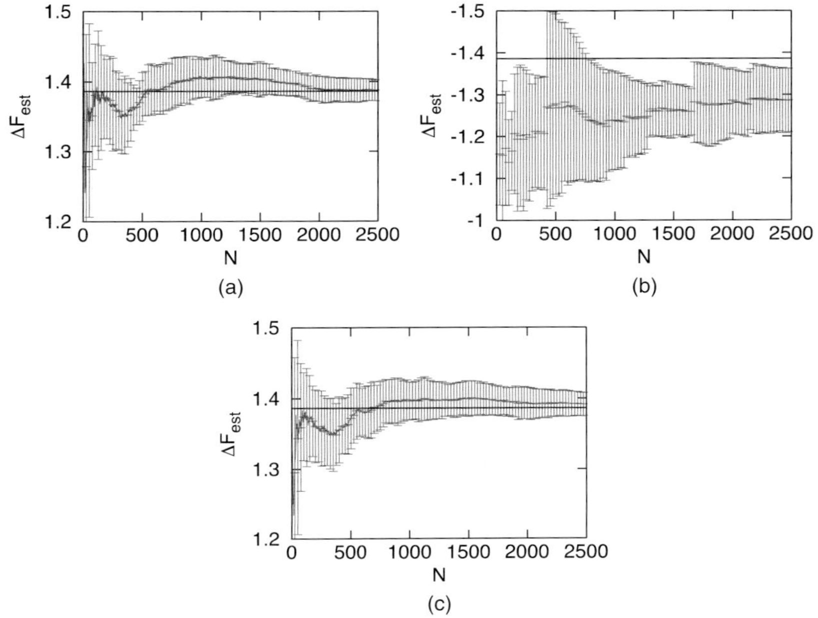

Figure 2.6 Plot of free energy estimates versus number of trajectories for (a) JE forward (Eq. (2.33)), (b) JE reverse (Eq. (2.33)), and (c) MLE (Eq. (2.34)), with the expected ΔF imposed (line). The reverse JE performs significantly worse than the other two free energy methods.

trajectories. In this system, the JE in the forward direction and the MLE converge extremely quickly toward the correct value. The JE in the reverse direction is worse as expected, due to the poor sampling of the final distribution by the initial.

In Figure 2.7, we plot the Crooks relation (2.32) and the phase function variants of Crooks (2.36) for the change in fluid temperature ($B_t(\mathbf{\Gamma}) = \Delta T(\mathbf{p})$) and internal energy ($B_t(\mathbf{\Gamma}) = \Delta U(\mathbf{q})$). The change in the kinetic temperature of the system is clustered close to zero, as expected for a thermostated system. In contrast, the change in internal energy is spread over the range. This is unsurprising as the work is effectively a time integral over the internal energy of the system.

Figure 2.5 Plots of phase function correlation functions (Eq. (2.6)) for the temperature ($B_t(\mathbf{\Gamma}) = \Delta T(\mathbf{p})$) and internal energy change ($B_t(\mathbf{\Gamma}) = \Delta U(\mathbf{q})$) of the system over various durations corresponding to the first maxima, zero, minima, and second zero of the instantaneous dissipation function and to the end of the simulation. The lines represent the expected behavior.

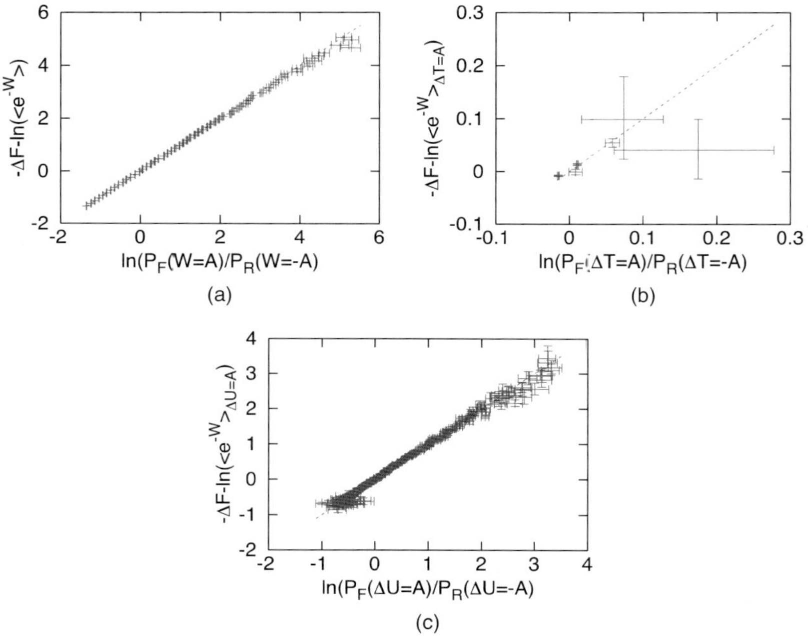

Figure 2.7 Plot of (a) Crooks relation (Eq. (2.32)), (b) Crooks for temperature of the system (Eq. (2.36)), $B_t(\mathbf{\Gamma}) = \Delta T(\mathbf{p})$), and (c) Crooks for internal energy of the system (Eq. (2.36)), $B_t(\mathbf{\Gamma}) = \Delta U(\mathbf{q})$), with expected behavior. The temperature is largely uncorrelated with the work and clusters close to zero.

2.7 Conclusion

The optical trapping system is a near-ideal system for a nontrivial test of the various FRs. The key to application of these relations however is always the derivation of the dissipation function or the work function: once derived, it simply becomes a matter of substitution into the appropriate relation. While seemingly straightforward, there is a temptation to use the mechanical work instead of the "thermodynamic" work in the work relations and also to use the instantaneous "entropy production," heat, or dissipative work instead of the dissipation function. Even if these various quantities have the same mean, the various fluctuation relations refer to *fluctuations*, and these may be very different from those of the properly defined dissipation function and work. This has caused an enormous amount of unnecessary confusion in the literature.

In practice, the biggest obstacle to applying these relations is the highly nonlinear effect of the exponential on these relations when there are a limited number of samples or the average of the dissipation or work is large. In spite of this, these relations greatly expand our understanding of nonequilibrium processes and provide useful measures of fundamental quantities such as the free energy difference between states or the nonequilibrium value of a phase function.

This work also highlights the central importance of the dissipation function to nonequilibrium statistical mechanics. It is the central function in the fluctuation theorem, the second law inequality, the dissipation theorem, and the various relaxation theorems. Although we have not discussed it here, it is also an integral part of the derivation of the Clausius inequality. Understanding each of these *nonequilibrium* theorems is essential for understanding what equilibrium is, how it is arrived at, and what mathematical forms do the various equilibrium phase space distributions take. Finally, once we have arrived at equilibrium, the dissipation is identically zero and entropy rather than dissipation assumes the central role. The material in this chapter provides a very brief synopsis of material that will appear in a forthcoming research monograph [48].

References

1 Sevick, E.M., Prabhakar, R., Williams, S.R., and Searles, D.J. (2008) *Annu. Rev. Phys. Chem.*, **59**, 603.
2 Jarzynski, C. (2008) *Eur. Phys. J. B*, **64**, 331.
3 Jarzynski, C. (2011) *Annu. Rev. Condens. Matter Phys.*, **2**, 329.
4 Evans, D.J. and Searles, D.J. (1994) *Phys. Rev. E*, **50**, 1645.
5 Crooks, G. (1998) *J. Stat. Phys.*, **90**, 14981.
6 Jarzynski, C. (1997) *Phys. Rev. Lett.*, **78**, 2690.
7 Trepagnier, E., Jarzynski, C., Ritort, F., Crooks, G., Bustamante, C., and Liphardt, J. (2004) *Proc. Natl. Acad. Sci. USA*, **101**, 15038.
8 Carberry, D.M., Reid, J.C., Wang, G.M., Sevick, E.M., Searles, D.J., and Evans, D.J. (2004) *Phys. Rev. Lett.*, **92**, 140601.
9 Evans, D.J., Cohen, E.G.D., and Morriss, G.P. (1993) *Phys. Rev. Lett.*, **71**, 2401.
10 Douarche, F., Joubaud, S., Garnier, N.B., Petrosyan, A., and Ciliberto, S. (2006) *Phys. Rev. Lett.*, **97**, 140603.
11 Seifert, U. (2005) *Eur. Phys. Lett.*, **70**, 36.
12 Monnai, T. (2005) *Phys. Rev. E*, **72**, 027102.
13 Evans, D.J., Searles, D.J., and Williams, S.R. (2008) *J. Chem. Phys.*, **128**, 014504.
14 Evans, D.J., Searles, D.J., and Williams, S.R. (2009) A simple mathematical proof of Boltzmann's equal *a priori* probability hypothesis, in *Diffusion Fundamentals III* (eds C. Chmelik, N. Kanellopoulos, J. Karger, and D. Theodorou), Leipziger Universitatsverlag, pp. 367–374.
15 Evans, D.J., Williams, S.R., and Searles, D.J. (2011) *J. Chem. Phys.*, **134**, 204113.
16 Wang, G.M., Sevick, E.M., Mittag, E., Searles, D.J., and Evans, D.J. (2002) *Phys. Rev. Lett.*, **89**, 050601.
17 Reid, J.C., Carberry, D.M., Wang, G.M., Sevick, E.M., Evans, D.J., and Searles, D.J. (2004) *Phys. Rev. E*, **70**, 016111.
18 Crooks, G. (1999) *Phys. Rev. E*, **60**, 2721.
19 Evans, D.J. and Searles, D.J. (2004) *Aust. J. Chem.*, **57**, 1119.
20 Morriss, G. and Evans, D.J. (1985) *Mol. Phys.*, **54**, 629.
21 Hayashi, K., Ueno, H., Iino, R., and Noji, H. (2010) *Phys. Rev. Lett.*, **104**, 218103.
22 Evans, D.J. and Searles, D.J. (1995) *Phys. Rev. E*, **52**, 5839.
23 Reid, J.C., Cunning, B.V., and Searles, D.J. (2010) *J. Chem. Phys.*, **133**, 154108.
24 Searles, D.J., Ayton, G., and Evans, D.J. (2000) *AIP Conf. Ser.*, **519**, 271.
25 Evans, D.J., Williams, S.R., and Searles, D.J. (2010) Thermodynamic interpretation of the dissipation function, in *Nonlinear Dynamics of Nanosystems* (eds G. Radons, B. Rumpf, and H. Schuster), Wiley-VCH Verlag GmbH, Weinheim, pp. 84–86.
26 Evans, D.J. and Morriss, G.P. (2008) *Statistical Mechanics of Nonequilibrium Liquids*, Cambridge.
27 Evans, D.J., Williams, S.R., and Rondoni, L. (2012) *J. Chem. Phys.*, submitted.
28 Evans, D.J., Searles, D.J., and Williams, S.R. (2009) *J. Stat. Mech.*, **2009**, P07029.
29 Williams, S.R. and Evans, D.J. (2007) *J. Chem. Phys.*, **127**, 184101.

30 Evans, D.J., Searles, D.J., and Williams, S.R. (2010) *J. Chem. Phys.*, **133**, 054507.
31 Butler, B.D., Ayton, G., Jepps, O.G., and Evans, D.J. (1998) *J. Chem. Phys.*, **109**, 6519.
32 Daivis, P.J. and Todd, B.D. (2006) *J. Chem. Phys.*, **124**, 194103.
33 Irving, J.H. and Kirkwook, J.G. (1950) *J. Chem. Phys.*, **18**, 817.
34 Jarzynski, C. (2002) *Phys. Rev. E*, **65**, 046122.
35 Hummer, G. (2001) *J. Chem. Phys.*, **114**, 7330.
36 Williams, S.R. and Evans, D.J. (2010) *Phys. Rev. Lett.*, **105**, 110601.
37 Williams, S.R., Evans, D.J., and Searles, D.J. (2011) *J. Stat. Phys.*, **145**, 831.
38 Jarzynski, C. (2006) *Phys. Rev. E*, **73**, 046105.
39 Bennett, C. (1976) *J. Comp. Phys.*, **22**, 245.
40 Crooks, G. (2000) *Phys. Rev. E*, **61**, 2361.
41 Shirts, M.R. and Pande, V.S. (2005) *J. Chem. Phys.*, **122**, 144107.
42 Shirts, M.R., Bair, E., Hooker, G., and Pande, V.S. (2003) *Phys. Rev. Lett.*, **91**, 140601.
43 Carberry, D.M., Baker, M.A.B., Wang, G.M., Sevick, E.M., and Evans, D.J. (2007) *J. Opt. A Pure Appl. Opt.*, **9**, S204.
44 Allen, M.P. and Tildesley, D.J. (2007) *Computer Simulation of Liquids*, Oxford.
45 Ytreberg, F.M. and Zuckerman, D.M. (2005) *J. Comput. Chem.*, **25**, 1749.
46 Dasmeh, P., Searles, D.J., Ajloo, D., Evans, D.J., and Williams, S.R. (2009) *J. Chem. Phys.*, **131**, 214503.
47 Reid, J.C., Evans, D.J. and Searles, D.J. (2012) *J. Chem. Phys.*, **136**, 021101.
48 Evans, D.J., Searles, D.J. and Williams, S. R., (2014) Dissipation and the Foundations of Classical Statistical Mechanics, Wiley-VCH, Berlin.

3
Fluctuation Relations in Small Systems: Exact Results from the Deterministic Approach

Lamberto Rondoni and O.G. Jepps

Many of the systems at the cutting edge of contemporary theoretical and/or experimental research are out of equilibrium. These can include biological processes that might be considered to take place in quasi-equilibrium, as well as transport processes that take place either in the linear (i.e., near-to-equilibrium) nonequilibrium regime or potentially far from equilibrium. As such, there is a great deal of interest in developing theories that encompass the behavior of nonequilibrium systems and processes. The fluctuation relations (FRs) comprise one of the few examples of an exact, general result that have been obtained for nonequilibrium systems to this point. The steady-state FRs describe the fluctuations of observables in nonequilibrium systems, extending Green–Kubo and Onsager relations to states that are far from equilibrium [1, 2], and providing information about fluctuations in nanoscale systems and how irreversibility can emerge from time-reversible dynamics [3, 4]. Moreover, while the classical fluctuation–dissipation relations obtain information about nonequilibrium properties of physical systems by means of equilibrium experiments, the transient fluctuation relations obtain information about equilibrium systems by means of nonequilibrium experiments, something useful when equilibrium cannot be hoped for. In a sense, this closes the circle.

One of the great challenges in recent times has been the inclusion of nanotechnological applications within the scope of these theories. The traditional path has been to extend the theoretical framework of statistical mechanics, whereby macroscopic properties are predicted from a consideration of the underlying microscopic behavior. However, the representation of macroscopic quantities within the microscopic framework usually relies (explicitly or otherwise) on a requirement that the system has a sufficiently large number of degrees of freedom (e.g., large number of constituent elements): the so-called thermodynamic limit. This condition on the system size is clearly problematic when our target systems comprise relatively few degrees of freedom, as is the case for nanotechnological applications. On the other hand, fluctuations are relevant only in (sub)systems with few degrees of freedom, unless one is prepared to observe macroscopic objects on a microscopic scale.[1)] Therefore, results such as the FRs are even more relevant to nanotechnology. The

1) Such as in the detection of gravitational waves [5].

Nonequilibrium Statistical Physics of Small Systems: Fluctuation Relations and Beyond, First Edition.
Edited by Rainer Klages, Wolfram Just, and Christopher Jarzynski.

observation of so-called anomalous transport in nanosystems (see Chapter 10) also highlights the fact that nanotechnological systems do not always behave thermodynamically, and that our theoretical development must recognize this. It is therefore of great importance to understand the consequences for theories such as the FRs away from the thermodynamic limit [6, 7].

Soon after the appearance of the FRs in the early 1990s came an explosion of interest, leading to a current, vast array of results (and their interpretation) (see Refs [8–12] and references cited therein). One of the difficulties in interpreting these results is that they tend to arise from two quite different perspectives. On the one hand is a development very much motivated by physical intuitions, using well-known properties (such as decays in correlation) of specific systems (typically systems of interacting particles) and specific observables to develop deeper theories of the fluctuations of physically interesting observables in real nonequilibrium systems. In this approach, granted the validity of the FRs for systems of physical interest (deduced empirically from numerical simulations or experimental observations), one wishes to understand the physical mechanisms underlying such results. This in turn leads toward an improved understanding of the physics of systems that obey modified FRs, or none at all. On the other hand is a much more mathematically minded development, aimed at identifying the class of dynamical systems with a (dynamical) property that enjoys fluctuations reminiscent of the FRs. This does not require the dynamical system, or the fluctuating property, to be a physically viable object. However, this approach may lead to further insight into the range of relations that could be verified for physical systems, into their generality, and into the plausibility of certain physical assumptions that would be too hard to test beyond doubt on realistic models of physical systems. For instance, by undertaking a rigorous mathematical approach, one can shed light on the role of key physical intuitions (such as decay in correlations) in the development of the FRs. Both approaches have their weaknesses and their strengths; furthermore, while at first hand they appear to lead to the same result, there are subtle yet important differences. Teasing out these differences is all the more challenging, because the methods and indeed the formalism of the two approaches are very different.

Our aim in this chapter is to consider the physically motivated FRs using a rigorous mathematical formalism. This will enable us to distinguish more clearly the various forms of the FRs (such as the transient and steady-state forms), as well as to focus in detail on the various assumptions that are required for physical approach. In doing so, we also shed some light on differences between the assumptions for the physically and mathematically motivated approaches. We will consider the relevance of the connection between the dynamics and the initial distribution, and the importance of system size for these FRs and their interpretation. It is not our intention to provide a comprehensive review of the FRs – for this, the interested reader is referred to, for example, Chapter 1 or Ref. [9].

3.1 Motivation

In the following section, we consider the merits of a formal mathematical framework from which one can explore both the physically motivated and mathematically motivated FRs. After presenting a short history of fluctuation-related theories in Section 3.1.1, we give a brief overview of the principles of nonequilibrium molecular dynamics (NEMD) simulations, the setting for the first observed FR (Section 3.1.2). An understanding of NEMD is crucial in order to understand the quantities whose fluctuations are the focus of these original FRs. In Section 3.1.3, we discuss another critical issue in the development of the FRs – the dynamical representation of dissipation. Finally, in Section 3.1.4, we highlight some of the difficulties that emerge in the development of the FRs, which will lead us to introduce a formalism in Section 3.2 that can help to clarify the conditions under which physically motivated FRs might hold.

We note that various attempts have been made to provide unifying or more general formalisms, in order to obtain further insights into both the physics and mathematics of nonequilibrium phenomena. Reference [9] sets the stage for our present perspective; Ref. [10] encompasses deterministic dissipative continuous-time dynamical systems, Langevin dynamics with nonconservative forces, and the Kraichnan model of hydrodynamic flows within the framework of diffusion stochastic processes; and Ref. [12] is mainly concerned with the minimal mathematical structure underlying the fluctuation relations, something that allows extensions to quantum statistical mechanics.

3.1.1
Why Fluctuations?

The study of fluctuations in statistical mechanics dates back to Einstein's 1905 seminal work on the Brownian motion [13], in which the first fluctuation–dissipation relation was given, and to Einstein's 1910 paper that turned Boltzmann's entropy formula into an expression for the probability of a fluctuation out of an equilibrium state [14]. Of the many authors who continued Einstein's work, we mention only a few. In 1931, Onsager obtained the transport coefficients from the observation of the thermal fluctuations [15, 16]. The fluctuation–dissipation theorem and the theory of transport coefficients received great impulse in the 1950s, thanks to the works of authors such as Callen, Welton, and R.F. Greene [17, 18], and M.S. Green and Kubo [19–22]. In 1967, Alder and Wainwright discovered long-time tails in the velocity autocorrelation functions, which implied the nonexistence of the self-diffusion coefficient, in two dimensions [23]. Anomalous divergent behavior of the transport coefficients was studied also by Kadanoff and Swift, for systems near a critical point [24]. The transient time correlation function formalism, which yields an exact relation between nonlinear steady-state response and transient fluctuations in the thermodynamic fluxes, has been developed by Visscher, Dufty, Lindenfeld, Cohen, Evans, and Morriss [25–28].

This brief and incomplete account suffices to show how the focus of studies on fluctuations of physical systems has shifted from equilibrium to nonequilibrium problems, and that it has eventually turned to anomalous phenomena, which are possible in the absence of local equilibrium. In turn, local equilibrium may be prevented by various conditions such as large gradients or large dissipative forces, low dimensionality or tight confinement, and even merely having few microscopic constituents, as in cases of bio- and nanotechnological interest. These situations often lead to anomalous phenomena (as typified by transport processes that are slower or faster than those seen in thermodynamic systems, leading to zero or infinite transport coefficients in the long-time limit), since they typically prevent a sufficiently fast decay of correlations [6, 7, 9, 11, 29].[2] Moreover, observables of small systems may have fluctuations comparable to their signal, which is one reason for which the physics of small systems and that of nonequilibrium systems may be treated within a unique setting, in which the decay of correlations, with respect to the proper probability distributions, plays a fundamental role.

3.1.2 Nonequilibrium Molecular Dynamics

One of the immediate problems encountered when developing a microscopic, dynamical theory of thermodynamic systems is that no fully satisfactory dynamics describing the evolution of the microscopic constituents of nonequilibrium systems has been developed. An analogous problem arises in equilibrium systems, where Hamiltonian dynamics forms a natural context for studying the microcanonical ensemble of equilibrium statistical mechanics, but the energy fluctuations inherent in all the other equilibrium ensembles introduce significant problems. However, the approaches toward resolving this problem in equilibrium have proven useful for nonequilibrium systems as well.

Early treatments of the equilibrium problem sometimes circumvented it by considering, for example, the canonical ensemble to comprise a distribution of microcanonical ensembles, each touring phase space on their disjoint surfaces of constant energy. Much more recently, this problem has been solved using so-called *thermostats* – artificial terms introduced into the dynamical equations that simulate an exchange of energy with an outside world (or reservoir) in accordance with the desired ensemble distribution [33–35]. While the term "thermostat" originates from the first application of simulating a canonical ensemble at fixed temperature, the method has been extended to various other ensembles including isobaric, isochoric, isoenthalpic, and constant stress models. One can derive a so-called *Gaussian* thermostat of the form

$$\dot{\mathbf{q}}_i = \mathbf{p}_i/m, \qquad \dot{\mathbf{p}}_i = \mathbf{F}_i^{\text{int}} - \alpha\mathbf{p}_i, \tag{3.1}$$

where the usual Hamiltonian equations for a system of N particles with conservative interactions $\mathbf{F}^{\text{int}}$ have been augmented to include the thermostat term $-\alpha\mathbf{p}_i$.

2) We note that anomalous transport can be studied in dynamical systems that are not fundamentally thermodynamic in nature – thermodynamic stationary states (in which local thermodynamic equilibrium holds) form a special subclass of the stationary states that dynamical systems can adopt. Our characterization of phenomena as "anomalous" here is via diverging or vanishing Green–Kubo integrals, which require a violation of local equilibrium established in thermodynamic systems (because the standard thermodynamic local balances and relations have to be violated [30–32]). For an overview of the FRs in systems with anomalous transport, see Chapter 8.

Here the thermostat multiplier α is a function of the phase variables $(\mathbf{q}, \mathbf{p})$. This style of thermostat is derived from Gauss' principle of least constraint [34, 36, 37], for which they are named. The form of α is determined by the quantity being constrained: for example, the isokinetic (IK) constraint, which fixes the kinetic energy $K = \sum_i \mathbf{p}_i^2/2m$, leads to a thermostat multiplier of the form

$$\alpha = \alpha_{IK}(\mathbf{p}, \mathbf{q}) \equiv \frac{1}{2K} \sum_{i=1}^{N} \frac{\mathbf{p}_i}{m} \cdot \mathbf{F}_i^{\text{int}}.$$

This kind of thermostat is appreciated by theorists for its simplicity and intriguing properties [34]. Furthermore, it generates a dynamics that explores the configuration space in accordance with the canonical ensemble (but not the momentum space).

A popular and (in some sense) more realistic alternative is the *Nosè–Hoover thermostat* model [38–40]. With a similar form to the Gaussian model, the Nosè–Hoover thermostat multiplier ζ is treated as an additional phase variable that can be considered to represent the external reservoir. The equations of motion are

$$\dot{\mathbf{q}}_i = \mathbf{p}_i/m, \qquad \dot{\mathbf{p}}_i = \mathbf{F}_i^{\text{int}} - \zeta \mathbf{p}_i, \qquad \zeta = \frac{1}{\tau^2}\left(\frac{K(\mathbf{p})}{K_0} - 1\right), \tag{3.2}$$

where K_0 is the target average of the kinetic energy $K(\mathbf{p})$ and τ is the relaxation time. Equations (3.2) could be considered more realistic in the sense that they generate canonical distributions for $\mathbf{q}$ *and* $\mathbf{p}$ in equilibrium, as appropriate for macroscopic isothermal systems.

At the same time, computer simulation methodologies for determining transport coefficients were being developed. As an alternative to calculating transport coefficients at equilibrium, or as the response to boundary-driven transport processes, one can consider the response to a suitably chosen fictive mechanical force $\mathbf{F}_e$, coupled to the system via phase functions $\mathbf{C}_i$ and $\mathbf{D}_i$, namely [34],

$$\begin{aligned} \dot{\mathbf{q}}_i &= \mathbf{p}_i/m + \mathbf{C}_i(\mathbf{q}, \mathbf{p}) \cdot \mathbf{F}_e, \\ \dot{\mathbf{p}}_i &= \mathbf{F}_i^{\text{int}} + \mathbf{D}_i(\mathbf{q}, \mathbf{p}) \cdot \mathbf{F}_e. \end{aligned} \tag{3.3}$$

Apart from generating the entropy necessary to drive the system out of equilibrium, the fictive forces continually add energy into the system – this energy can be removed through the same thermostat mechanism that balances energy exchange with the environment during equilibrium simulations:

$$\begin{aligned} \dot{\mathbf{q}}_i &= \mathbf{p}_i/m + \mathbf{C}_i(\mathbf{q}, \mathbf{p}) \cdot \mathbf{F}_e, \\ \dot{\mathbf{p}}_i &= \mathbf{F}_i^{\text{int}} + \mathbf{D}_i(\mathbf{q}, \mathbf{p}) \cdot \mathbf{F}_e - \alpha \mathbf{p}_i. \end{aligned} \tag{3.4}$$

The ideas underlying the introduction of these sorts of dynamics find parallels with our experience with models of equilibrium systems. Let us introduce some notation first. An observable is the time average of a phase variable $\mathcal{O} : \mathcal{M} \to \mathbb{R}$ along a trajectory $\{S^t\Gamma\}_{t\in\mathbb{R}}$ starting at the point Γ in the phase space $\mathcal{M}$. Let us denote by

$$\bar{\mathcal{O}}_{t,t+\tau}(\Gamma) := \frac{1}{\tau}\mathcal{O}_{t,t+\tau}(\Gamma) := \frac{1}{\tau}\int_t^{t+\tau} \mathcal{O}(S^s\Gamma)\mathrm{d}\,s \tag{3.5}$$

the average of $\mathcal{O}$ in the time interval $[t, t+\tau]$. The corresponding observable is defined by $\bar{\mathcal{O}} = \lim_{t\to\infty} \bar{\mathcal{O}}_{0,t}$, and represents the fact that a measurement takes a time that is virtually infinitely long, compared to the microscopic timescales. To compute such a limit is exceedingly complicated, in general, but in equilibrium the problem is commonly solved by the *ergodic hypothesis*, which states that

$$\bar{\mathcal{O}}(\Gamma) = \frac{1}{\mu(\mathcal{M})} \int_{\mathcal{M}} \mathcal{O}(\gamma) \mathrm{d}\mu(\gamma) \equiv \langle \mathcal{O} \rangle_{\mu} \tag{3.6}$$

for a suitable measure μ, and for μ almost all $\Gamma \in \mathcal{M}$.

Only a few systems of physical interest verify the mathematical statements of the ergodic theory, which requires Eq. (3.6) to hold for all integrable phase variables. Moreover, there is no hope that any many-particle system will explore its phase space as densely as suggested by Eq. (3.6), in the time that corresponds to a measurement. Nevertheless, the ergodic hypothesis is successfully applied in a very wide range of situations, because the variables of physical interest are but a few, and tend to constants in the large-N limit (cf. Chapter I of Ref. [41] and Ref. [42]). This means that the set of observables of interest is too small to probe true ergodicity, and that different, necessarily partial models of the same system may be equivalent in describing its limited set of physically interesting properties. Therefore, for an isolated system whose energy H remains within a thin shell $[E, E+\Delta E]$, it is justified to *postulate* that μ is the *microcanonical ensemble*; for a closed system in contact with a heat bath at given temperature, the *canonical ensemble* is postulated; and so on. A posteriori, one checks whether the assumption is valid or not, and finds that these classical ensembles are appropriate for the purposes of physics. In the thermodynamic limit, the different ensembles become equivalent, in the sense that the averages of local observables tend to the same values.

In modeling nonequilibrium systems by means of the thermostated equations of motion, one implicitly or explicitly assumes that such fortunate equivalences will also hold, so that certain details of the microscopic dynamics do not need to be very accurately accounted for and may be treated in the simplest effective fashion.

Indeed, for quantities not affected by how energy is removed from the system, the precise form of α is irrelevant since susceptibilities of mechanically driven processes have been shown to be similar to susceptibilities of boundary-driven processes [34, 35, 43]. Therefore, driving boundaries may be efficiently replaced by fictitious external forces and constraints *for the purpose of computing transport coefficients*, and *ad hoc* models may be devised as equivalent mechanical representations (again, in the sense that transport coefficients are equivalent) of both mechanical and thermal transport processes [7, 9, 11, 44]. While we cannot expect NEMD to reproduce the precise microscopic dynamics of a nonequilibrium system, NEMD can be rigorously shown to generate correct transport coefficients in the linear regime, and has been successfully adopted in the study of the rheology of fluids, polymers in porous media, defects in crystals, friction between surfaces, atomic clusters, biological macromolecules, and a host of other phenomena [33–35, 43].

We note that the dynamics defined in this manner are deterministic and time reversible [34].[3]

An NEMD simulation involves two choices: a choice of driving field $\mathbf{F}_e$ (motivated by the transport coefficient to be investigated) and a choice of thermostating scheme (motivated by the target ensemble). For mass diffusion, one can select $\mathbf{C}_i = 0$ and $\mathbf{D}_i = c_i E$, where E represents the field strength and c_i the response strength of particle i to the field.[4] For shear viscosity, one can select the so-called SLLOD equations of motion [34], given by

$$\dot{\mathbf{q}}_i = \mathbf{p}_i/m + \mathbf{i}\gamma y_i, \qquad \dot{\mathbf{p}}_i = \mathbf{F}_i^{\text{int}} - \mathbf{i}\gamma p_{yi} - \alpha \mathbf{p}, \tag{3.7}$$

where γ is the shear rate in the y-direction and $\mathbf{i}$ is the x-direction unit vector, and the $\alpha\mathbf{p}$ thermostat term balances the system energy in accordance with the desired constraint (constant kinetic energy, total energy, etc.).

NEMD therefore allows us to study nonequilibrium systems at the microscopic level, providing a dynamics that is consistent with the macroscopic behavior of systems away from equilibrium. The various deterministic FRs that have been derived can be understood in the context of this dynamics. In the following subsection, we will introduce some of the important functions that are the focus of the FRs, and that arise naturally from the NEMD context. We will then consider the FRs in more detail, to understand the difficulties in interpretation that lead us to a more formal presentation of the physically motivated FRs.

3.1.3 The Dissipation Function

The notion of dissipation is fundamental to the FRs. From thermodynamics, we understand equilibrium processes as those taking place without dissipation: conversely, processes in which dissipation takes place are considered to be out of equilibrium (whether relaxing to or in a nonequilibrium steady state). Yet the representation of this concept at the microscopic level is not without its difficulties.

3) This equivalence of nonequilibrium ensembles, analogous to the equivalence of equilibrium ensembles, may look puzzling from the dynamical systems point of view. Indeed, many dynamical counterexamples can be easily constructed in terms of carefully selected phase variables (see, for example, Refs [45–49]). One should then observe that equivalence of ensembles has been obtained in both equilibrium and nonequilibrium statistical mechanics only under precise conditions and for properly chosen observables of physical interest. The main conditions involve the thermodynamic limit and, in practice, the existence of local equilibrium. Equivalence is not expected in all possible dynamics, and even for thermodynamically relevant cases, it is limited to several observables under proper tuning of various parameters. This means, for instance, that the microcanonical and canonical averages of the local observables (those that depend only on the particles confined in a fixed, finite box) tend to the same values if the volume of the system grows without bounds while its density and energy density tend to a constant. Obviously, the ensembles do not agree on global quantities, such as the total energy (see, for example, Refs [11, 41, 44, 50–52]).

4) E is usually called a *color field* and c_i the *color charge*: the term “color” is used because while the response to the field is analogous to electric charges in an electric field, the particles are oblivious to one another’s charge, unlike real electric charges.

The entropy of a microcanonical ensemble, which represents an equilibrium isolated system, is given by the logarithm of its volume in phase space. As Hamiltonian dynamics preserves volumes in phase space, it is attractive to associate changes of phase space volume with changes in entropy arising from energy dissipation[5] in systems exhibiting local thermodynamic equilibrium. One can extend this idea to other ensembles, such as the canonical ensemble (which represents a closed system in contact with a heat bath), by considering disjoint sets of different energies that must also, under Hamiltonian dynamics, preserve their volume. However, as we have seen, such an equilibrium ensemble can also be modeled using the thermostated equations of motion (Eq. (3.1)), and (as we shall show below) such equations do not preserve volumes in phase space. This reflects the exchange of energy with the reservoir, which for the equilibrium system is zero *on average*, but not at every instant. It also reflects the fact that phase volumes must contract or expand in accordance with the probability distribution function (PDF) associated with that ensemble: since the dynamics must conserve probability, it cannot also preserve phase volume if the PDF is not uniform.

The changes in phase space volumes are quantified by the *phase space contraction rate*, which for a dynamical system $\dot{\Gamma} = G(\Gamma)$ is defined as

$$\Lambda := -\nabla \cdot \dot{\Gamma}(\Gamma) = -\text{div}\, G(\Gamma). \tag{3.8}$$

We may therefore extend our notion of dissipation in the following way, for the purpose of a general theory of nonequilibrium phenomena: a dynamics is called *dissipative with respect to a PDF* (of a particular equilibrium ensemble, for example) if the change in phase space volume is other than that simply predicted by the PDF. In other words, we call dissipative with respect to a PDF those dynamics for which that PDF is not invariant. Under Hamiltonian dynamics, the uniform distribution is preserved, so the Hamiltonian dynamics is not dissipative with respect to the uniform distribution. Under Gaussian IK dynamics, the configurational part of the canonical distribution is preserved: Gaussian IK dynamics is thus not dissipative with respect to the canonical distribution. Gaussian IK dynamics do not preserve the uniform distribution, however, so the Gaussian IK dynamics *is* dissipative with respect to the uniform distribution.[6]

We may then try to measure just how dissipative a dynamics is, with respect to a particular PDF. Such an approach may be formalized by introducing a *dissipation function* as the difference between the actual changes in phase space volume and the changes associated with the ensemble (effectively, the thermostat). Intuitively, if we accept the association of changes in phase volumes with changes in entropy arising from energy dissipation, and if the particular PDF corresponds to a given

5) Entropy may also change without any dissipation, as in the case of the free expansion of a gas.

6) This formal notion of dissipation does not necessarily denote physical dissipations, that is, entropy productions in the sense of irreversible thermodynamics; it is merely a useful suggestive concept, which acquires physical relevance under proper conditions. Indeed, it refers to generic dynamical systems and not to particle systems in local thermodynamic equilibrium, which is a necessary condition for entropy production to be defined.

equilibrium ensemble distribution, the dissipation function corresponds to the change in entropy relative to that required to induce the given ensemble (by the thermostat).

To make these ideas and our intuitive motivation more rigorous, consider a probability density f on the phase space $\mathcal{M}$ that is even with respect to time reversal, that is, $f(\mathcal{I}\Gamma) = f(\Gamma)$, where $\mathcal{I} : \mathcal{M} \to \mathcal{M}$ is an involution expressing the time-reversal operation on the phases $\Gamma \in \mathcal{M}$. We then introduce the following.

Definition
For a sufficiently regular probability density f, and dynamics $\dot{\Gamma} = G(\Gamma)$, the instantaneous dissipation function Ω is given by [8, 9]

$$\Omega(\Gamma) = -G(\Gamma) \cdot \frac{\mathrm{d}}{\mathrm{d}\Gamma} \ln f \bigg|_{\Gamma} + \Lambda(\Gamma) \tag{3.9}$$

provided f is ergodically consistent with the dynamics.

Ergodic consistency is the condition that the logarithmic term exists in Eq. (3.9), that is, that $f > 0$ in all regions visited by the phase space trajectories $\{S^t\Gamma\}_{t\in\mathbb{R}}$, where $S^t\Gamma$ is the point reached at time t, having started in Γ at time 0. This condition plays an important role: in physical applications f must be chosen with care, even though many phase space densities are acceptable from a mathematical point of view [9].

The time averages of Ω are then expressed by

$$\bar{\Omega}_{t,t+\tau}(\Gamma) = \frac{1}{\tau} \ln \frac{f(S^{t_0}\Gamma)}{f(\mathcal{I}S^{t+\tau}\Gamma)} + \bar{\Lambda}_{t,t+\tau}(\Gamma). \tag{3.10}$$

It has been found that Ω equals the dissipated power, divided by the kinetic temperature, in bulk thermostated systems like those of Eq. (3.4) if f is the equilibrium probability density for the given system [9, 53]. Equation (3.10) can be interpreted in two ways: either as a recipe for determining the dissipation Ω for a given background (equilibrium) distribution f, or as a recipe for determining f for a given dissipation function Ω. For a compact phase space, the uniform density $f(\Gamma) = 1/|\mathcal{M}|$ implies $\Omega = \Lambda$ – in practice, the Ω and Λ coincide only for systems whose equilibrium state is microcanonical.

We note here the condition referred to as *adiabatic incompressibility of phase space*, or AIΓ, whereby the phase space contraction rate Λ for the adiabatic equations of motion (Eq. (3.3)) (i.e., the equations of motion excluding the thermostat) is zero. This condition is met by commonly used NEMD equations (such as SLLOD or the color field) [34].

We finish by elucidating the connection between the dissipation function and the *dissipative flux* $\mathbf{J}$, which in irreversible thermodynamics represents the flow related to the entropy production. Let us assume that the initial distribution of phases f may be generated by a single field-free ($F_e = 0$) thermostated dynamics. By this we mean that not only f must be preserved by the equilibrium dynamics, but also a single field-free trajectory should explore all of its support. This is consistent with nonlinear response theory, where one is interested in the statistics of the steady-state observables, and one starts with the equilibrium ensemble dynamics [34, 54]. Consider, for instance, the Gaussian isokinetic system, where the kinetic energy is fixed at the value $K_0 = (dN - d - 1)k_B T = (dN - d - 1)/\beta$, and

$$\alpha(\Gamma) = -\frac{(\dot{H}_0(\Gamma) + \mathbf{J}(\Gamma) \cdot \mathbf{F}_e)\beta}{dN - d - 1}, \qquad \Lambda(\Gamma) = dN\alpha(\Gamma) + O_N(1), \tag{3.11}$$

where AIΓ is used to obtain the final equality and $O_N(1)$ is a correction that is order 1 in N. The equilibrium phase space distribution function takes the form $f(\Gamma) \sim e^{-\beta H_0}\delta(K(\Gamma) - K_0)$; hence, $f(\Gamma)/f(\mathcal{I}S^\tau\Gamma) = e^{\beta\int_0^\tau \dot{H}_0(\Gamma(s))ds}$ and

$$\bar{\Omega}_{0,\tau}(\Gamma) = \beta\int_0^\tau \dot{H}_0(\Gamma(s))ds + \bar{\Lambda}_{0,\tau}(\Gamma) = -\overline{(\mathbf{J} \cdot \mathbf{F}_e)}_{0,\tau}V\beta + O_N(1), \tag{3.12}$$

which shows that the dissipation function and the dissipative flux are simply related. In fact, for thermostated systems with a constant field, the (rate of) entropy production from irreversible thermodynamics $\Sigma(\Gamma) = kJ(\Gamma)$ for some constant k.

3.1.4
Fluctuation Relations: The Need for Clarification

The first FR for a nonequilibrium particle system appeared in the seminal paper by Evans, Cohen, and Morriss in 1993 [55]. It concerned fluctuations in the finite-time averages of the energy dissipation, for an isoenergetic SLLOD system simulating a shearing fluid of interacting particles. In such a system, the instantaneous energy dissipation equals the instantaneous phase space contraction rate. The form of the relation derived in Ref. [55] was

$$\frac{P_\tau(A)}{P_\tau(-A)} = e^{\tau A}, \tag{3.13}$$

where A and $-A$ are the time-averaged values of the dissipative power divided by $k_B T$, on trajectory segments of duration τ, and $P_\tau(A)$ is the steady-state probability of observing the value A, with some tolerance. In analogy with the periodic orbit expansion for dynamical systems [56–58], the relation was derived using the "Lyapunov weights" in the long-τ limit. Remarkably, Eq. (3.13) does not contain any adjustable parameter. In 1994, Evans and Searles obtained the first in a series of relations similar to Eq. (3.13), which we call transient Ω-FRs: "transient" because they describe results for ensembles of systems over finite periods of evolution, and "Ω-FRs" because they describe fluctuations in the dissipation function Ω [8, 59–65]. It was found that the transient Ω-FRs may be used to derive mathematically the steady-state Ω-FRs, by taking two large time limits (one concerning relaxation and the other concerning observation) provided a condition now known as t-mixing is verified [53, 66].[7] In 1995, Gallavotti and Cohen clarified that the argument of Ref. [55], for the steady-state fluctuations of the time-averaged phase space

7) It is important to observe that these two limits do not commute; hence, this derivation is quite a delicate one, although very informative, as will be shown below.

contraction rate Λ, required certain assumptions, such as transitivity and the Anosov property. Because most systems of practical interest are not transitive Anosov dynamical systems, they proposed the *chaotic hypothesis* – that such systems are effectively "Anosov-like" as far as their fluctuations are concerned [67–70] (see Section 3.2.5).

Of the many various developments subsequent to these original FRs, arguably one of the most interesting (especially with regard to practical application) has been the Jarzynski equality (JE), an independently obtained transient relation that connects free energy differences between two equilibrium states to nonequilibrium processes [71]. There have been various numerical and experimental tests of the FRs: the interested reader is referred to Refs [7, 9] and references cited therein.

It is important to recognize from the outset that, despite their deceptively similar forms, the transient and steady-state FRs are of totally different nature, from both theoretical and experimental points of view. Indeed, the (often) mathematically trivial transient relations, such as the transient Ω-FR, the Jarzynski, and the Crooks relations, are physically very interesting: they suggest sophisticated experiments, and shed light on the physical relevance of our current models of nonequilibrium physics. Also, they are useful in the study of nanoscale biological systems, in which no sufficiently general guiding principle has so far been firmly established. Their peculiarity is that they obtain information about equilibrium (or merely *initial* states) by means of nonequilibrium experiments, thus connecting in an intriguing fashion equilibrium states and nonequilibrium dynamics.

In the derivation of a relation like Eq. (3.13), one may follow different approaches. The one that has led to the Gallavotti–Cohen steady-state Λ-FR requires the full knowledge of the physical measure of the dynamics, which is assumed to be of the Anosov type, and hence to have a Sinai–Ruelle–Bowen (SRB) measure, μ_{SRB} [67, 68, 72]. This knowledge gives the maximum possible amount of information about the phase space distribution: once μ_{SRB} is given, all statistical properties of the dynamics can be obtained from it. However, if one is only interested in why the FR holds, less detailed knowledge of the phase space distribution is likely to suffice, and special features of the dynamics, such as the Anosov property, need not play a role. This is desirable if the FR is to apply more generally to physical systems, as observations suggest it should. Similar questions concern the relation between Khinchin's approach to the ergodic hypothesis [42] and the standard mathematical ergodic theory [11].

In the following, we will consider both transient and steady-state FRs. The transient FRs concern the time-dependent response to external drivings of ensembles of systems, or ensembles of experiments, and hold under very general conditions (time reversibility suffices for those obtained by Evans and Searles; Hamiltonian dynamics suffices for the one derived by Jarzynski). These relations hold for arbitrarily short times, while the steady-state FRs have similar applications, but are valid only asymptotically in time. The former, indeed, represent a sophisticated property of the equilibrium or initial ensemble, while the latter concern the steady state. So, while the initial ensemble can be prepared at will by the experimenter, the steady state depends on the dynamics and not on the (forgotten) initial state. This difference has physical consequences. Transient relations concern the statistical properties of the chosen initial ensembles, which may be uncovered by different nonequilibrium dynamics; steady-state relations concern the statistics of the evolution in the steady state, and therefore can be reconstructed by following the evolution of a single system: here the dynamics is crucial, while the initial ensemble is irrelevant.

Although the transient Ω-FRs were proven long ago, beginning with Refs [63, 64], their connection with steady-state relations has been clarified only recently, after it has been recognized that a particular decay of correlations must be assumed. This concept, now called *t-mixing*, deserves further consideration, also because it sheds light on the mechanisms that allow the Ω-FRs. Furthermore, in extending the ensemble derivations of the transient relations, one realizes that time reversibility and the decay of the autocorrelation of the energy dissipation imply the validity of a wide class of steady-state FRs. Some decay of correlations is always needed to reach a steady state, and to identify the statistics generated by the evolution of a single system in real space with that of an ensemble of systems in phase space. We will derive these conditions formally in the next section, taking the physically motivated transient FRs as our starting point.

3.2
Formal Development

We begin (Section 3.2.1) with a derivation of the transient FRs under the single assumption of time-reversal invariant dynamics for the system being studied. We will then consider the Jarzynski work relations in light of this formal development, in Section 3.2.2, before moving on to consider how one can extend these transient relations asymptotically in time in Section 3.2.3. Finally, we will consider the further conditions required to obtain steady-state results in Section 3.2.4. In Section 3.3, we reflect further on these conditions.

3.2.1
Transient Relations

Adopting the approach of Ref. [53], which does not require the invariant phase space probability measure to be explicitly determined, we will obtain transient fluctuation relations for Ω as well as for other functions of phase. If $\mathcal{O}$ is the observable of interest, we will speak of $\mathcal{O}$-FRs.

Recall our dynamical system of the generic form

$$\dot{\Gamma} = G(\Gamma), \qquad \Gamma \in \mathcal{M} \subset \mathbb{R}^n, \tag{3.14}$$

on the phase space $\mathcal{M}$. The vector field G is determined, for example, by particle interactions and external forces, in a system of statistical mechanical interest. We denote by $S^t\Gamma$, $t \in \mathbb{R}$, the solution of Eq. (3.14) with initial condition Γ, and adopt the finite-time-average notation of Eq. (3.5).

Consider a probability measure $\mu(\mathrm{d}\Gamma) = f(\Gamma)\mathrm{d}\Gamma$ of density f on the phase space $\mathcal{M}$. The density f need not necessarily be produced by any dynamics on $\mathcal{M}$, although it could represent, for instance, the initial equilibrium distribution of a system of interest. The probability according to μ that the phase function $\mathcal{O}$ takes values in the interval $(a, b) \subset \mathbb{R}$ is given by

$$\int_{\mathcal{O}|_{(a,b)}} \mathrm{d}\,\mu(\Gamma) = \int_{\mathcal{O}|_{(a,b)}} f(\Gamma)\mathrm{d}\Gamma, \tag{3.15}$$

where $\mathcal{O}|_{(a,b)} := \{\Gamma \in \mathcal{M} : \mathcal{O}(\Gamma) \in (a, b)\}$.

Assume that the dynamics is time-reversal invariant:

$$\mathcal{I}S^{\tau}\Gamma = S^{-\tau}\mathcal{I}\Gamma \quad \text{for all } \Gamma \in \mathcal{M} \text{ and all } \tau \in \mathbb{R}, \tag{3.16}$$

where we recall that $\mathcal{I} : \mathcal{M} \to \mathcal{M}$ is an involution representing the time inversion operator (e.g., $\mathcal{I}\Gamma \equiv \mathcal{I}(\mathbf{q}, \mathbf{p}) \equiv (\mathbf{q}, -\mathbf{p})$, for Hamiltonian dynamics), so that $\mathcal{I}^2$ is the identity transformation.

Given the odd phase variable $\mathcal{O}$ (i.e., $\mathcal{O}(\mathcal{I}\Gamma) = -\mathcal{O}(\Gamma)$), consider its averages $\bar{\mathcal{O}}_{t,t+\tau}$ defined by Eq. (3.5) with $t = 0$ and different τ. For $\delta > 0$, introduce the sets $(A)_\delta = (A - \delta, A + \delta)$, and consider the probability according to μ that $\bar{\mathcal{O}}_{0,\tau} \in (A)_\delta$ divided by the probability that $\bar{\mathcal{O}}_{0,\tau} \in (-A)_\delta$:

$$\frac{\mu(\bar{\mathcal{O}}_{0,\tau}|_{(A)_\delta})}{\mu(\bar{\mathcal{O}}_{0,\tau}|_{(-A)_\delta})} = \frac{\int_{\bar{\mathcal{O}}_{0,\tau}|(A)_\delta} f(\Gamma)\mathrm{d}\Gamma}{\int_{\bar{\mathcal{O}}_{0,\tau}|(-A)_\delta} f(\Gamma)\mathrm{d}\Gamma}. \tag{3.17}$$

To compute this quantity, observe that the points of $\mathcal{M}$ that fall in $\bar{\mathcal{O}}_{0,\tau}|_{(-A)_\delta}$ are those, and only those, obtained by doing the time inversion of the evolution, for a time τ, of the points in $\bar{\mathcal{O}}_{0,\tau}|_{(A)_\delta}$, that is,

$$\bar{\mathcal{O}}_{0,\tau}|_{(-A)_\delta} = \mathcal{I}S^{\tau}\bar{\mathcal{O}}_{0,\tau}|_{(A)_\delta}\,. \tag{3.18}$$

This allows us to compute the denominator of Eq. (3.17) in terms of $\bar{\mathcal{O}}_{0,\tau}|_{(A)_\delta}$, through the coordinate transformation[8)]

$$\Gamma = \mathcal{I}S^{\tau}X, \quad \text{whose Jacobian is } J = \left|\frac{\mathrm{d}\Gamma}{\mathrm{d}X}\right| = \mathrm{e}^{-\Lambda_{0,\tau}(X)}. \tag{3.19}$$

This leads to

$$\int_{\bar{\mathcal{O}}_{0,\tau}|_{(-A)_\delta}} f(\Gamma)\mathrm{d}\Gamma = \int_{\bar{\mathcal{O}}_{0,\tau}|_{(A)_\delta}} f(\mathcal{I}S^{\tau}X)\mathrm{e}^{-\Lambda_{0,\tau}(X)}\mathrm{d}X. \tag{3.20}$$

Then, assuming for simplicity that f is even ($f(\mathcal{I}\Gamma) = f(\Gamma)$), which is typically the case for equilibrium measures, the ratio Eq. (3.17) becomes

$$\frac{\mu(\bar{\mathcal{O}}_{0,\tau}|_{(A)_\delta})}{\mu(\bar{\mathcal{O}}_{0,\tau}|_{(-A)_\delta})} = \frac{\int_{\bar{\mathcal{O}}_{0,\tau}|_{(A)_\delta}} f(\Gamma)\mathrm{d}\Gamma}{\int_{\bar{\mathcal{O}}_{0,\tau}|_{(A)_\delta}} f(S^{\tau}X)\exp(-\Lambda_{0,\tau}(X))\mathrm{d}X}. \tag{3.21}$$

8) A similar technique is used in Ref. [73].

Using Eq. (3.10), Eq. (3.21) implies what has been referred to as the *transient $\mathcal{O}$-FR*:

$$\frac{\mu(\bar{\mathcal{O}}_{0,\tau}|_{(A)_\delta})}{\mu(\bar{\mathcal{O}}_{0,\tau}|_{(-A)_\delta})} = \langle \exp(-\Omega_{0,\tau}) \rangle^{-1}_{\bar{\mathcal{O}}_{0,\tau}\in(A)_\delta}, \tag{3.22}$$

where

$$\langle \exp[-\Omega_{0,t}] \rangle_{\bar{\mathcal{O}}_{r,s}\in(A)_\delta} := \frac{\int_{\bar{\mathcal{O}}_{r,s}|_{(A)_\delta}} \exp[-\Omega_{0,t}(\Gamma)] f(\Gamma) \mathrm{d}\Gamma}{\int_{\bar{\mathcal{O}}_{r,s}|_{(A)_\delta}} f(\Gamma) \mathrm{d}\Gamma} \tag{3.23}$$

is the ensemble average of $\exp[-\Omega_{0,t}]$, over the set of trajectories that satisfy the constraint that $\bar{\mathcal{O}}_{r,s} \in (A)_\delta$ [8, 59]. In the case that $\mathcal{O} = \Omega$, the fluctuation relation assumes the particularly elegant form, called *transient Ω-FR*:

$$\frac{\mu(\bar{\Omega}_{0,\tau}|_{(A)_\delta})}{\mu(\bar{\Omega}_{0,\tau}|_{(-A)_\delta})} = \langle \exp(-\Omega_{0,\tau}) \rangle^{-1}_{\bar{\Omega}_{0,\tau}\in(A)_\delta} = \mathrm{e}^{[A+\varepsilon(\delta,A,\tau)]\tau}, \tag{3.24}$$

where $|\varepsilon(\delta, A, \tau)| \leq \delta$. Remarkably, the time-reversal invariance is the only ingredient of this derivation.

The above relations are called "transient" because, independently of the length of τ, they express exactly a property of the initial measure μ and not of the possible steady state. The range of A's that can be observed depends on both S^τ and $\mathcal{O}$. Different choices of f are possible, which lead to different dissipation functions, and hence different transient FRs. In particular, equilibrium probability densities of many-particle systems that obey AIΓ lead to an Ω-FR that concerns the entropy production, or the energy dissipation rate Σ.

For symmetric intervals $(-\delta, \delta)$, one obtains

$$\frac{\mu(\bar{\mathcal{O}}_{0,\tau}|_{(-\delta,\delta)})}{\mu(\bar{\mathcal{O}}_{0,\tau}|_{(-\delta,\delta)})} = 1, \qquad \text{hence } \langle \mathrm{e}^{-\Omega_{0,\tau}} \rangle_{\bar{\mathcal{O}}_{0,\tau}\in(-\delta,\delta)} = 1, \tag{3.25}$$

which is a generalization of the so-called nonequilibrium partition identity [34, 64] that corresponds to the $\delta \to \infty$ limit of the second equality in Eq. (3.25). This identity can be used to test the accuracy of numerical simulations, and has even been used to calibrate experimental equipment [74–76].

In some cases, for example, IK dynamics with bounded interaction potentials, the dissipation function is bounded, $|\Omega| \leq \Omega^*$, and one obtains

$$\mathrm{e}^{-\tau\Omega^*} \leq \frac{\mu(\bar{\mathcal{O}}_{0,\tau}|_{(A)_\delta})}{\mu(\bar{\mathcal{O}}_{0,\tau}|_{(-A)_\delta})} \leq \mathrm{e}^{\tau\Omega^*}, \tag{3.26}$$

for all odd $\mathcal{O}$.

3.2.2 Work Relations: Jarzynski

Consider a finite-particle system, in equilibrium with a much larger system that constitutes a heat bath at temperature T. Assume that the overall system is described by a Hamiltonian of the form

$$\mathcal{H}(\Gamma;\lambda) = H(x;\lambda) + H_{\mathrm{E}}(y) + h_{\mathrm{i}}(x,y). \tag{3.27}$$

Here, x represents the state (positions and momenta) of the system of interest, while y represents the state of the bath. λ represents an externally controllable parameter related to the Hamiltonian of the system of interest $H(x,\lambda)$, $H_{\mathrm{E}}(y)$ represents the Hamiltonian of the bath, and $h_{\mathrm{i}}(x,y)$ represents the interaction between system and bath. Work W can be performed on the system by acting on λ according to a given protocol $\lambda(t)$, with initial and final values $\lambda(0) = A$ and $\lambda(\tau) = B$, respectively. The process can be repeated to build a PDF ϱ of the measured values of W. Unless the process is performed quasistatically, W is not the thermodynamic work done on the system, but is still a measurable quantity. Then, the Jarzynski equality predicts that [71]

$$\langle \mathrm{e}^{-\beta W} \rangle_{A\to B} = \int \mathrm{d}W \; \varrho(W)\mathrm{e}^{-\beta W} = \mathrm{e}^{-\beta[F(B)-F(A)]}, \tag{3.28}$$

where $\beta = 1/k_{\mathrm{B}}T$ and $[F(B) - F(A)]$ is the free energy difference between the initial equilibrium state, with $\lambda = A$, and the equilibrium state that could eventually be reached, if the system was allowed to relax at constant $\lambda = B$. But this final relaxation does not need to be included in the experiment.

Remark

The average over all works done in varying λ from A to B, $\langle e^{-\beta W} \rangle_{A\to B}$, is computed with respect to the initial canonical ensemble. The JE is thus a transient relation expressing a property of one such ensemble of replicas of systems, which is uncovered by performing a fixed nonequilibrium protocol.

Equation (3.28) is supposed to hold whichever protocol one follows to change λ from A to B, and hence also arbitrarily far from equilibrium (large $\mathrm{d}\lambda/\mathrm{d}t$); therefore, the presence of the equilibrium quantities $F(A)$ and $F(B)$ in Eq. (3.28) is remarkable. In particular, the externally measured work does not need to coincide with the internal work (which would not differ from experiment to experiment, if performed quasistatically). It would then seem that the JE, like the other transient relations, holds quite generally for particle systems. In reality, there are questions related to the definition of the measurable work W, and on the subdivision of the energy h_{i} between system and environment. While close to equilibrium everything is clear, because thermodynamics applies, the usefulness of the relation is hindered by the fact that all works collapse to the unique thermodynamic work.

Consequently, the most effective protocols for the applicability of the JE have been recently investigated [77] from an optimization viewpoint. This quite intriguing study concerns diffusive stochastic processes, which implicitly assume that the microscopic scales are much shorter than the shortest scales of the stochastic process at hand. The study raises further questions that are relevant on the nanoscale, to which the JE is mainly addressed: in particular, the role of the far-from-equilibrium energy dissipation must be better understood.

In any event, the JE is related to the transient Ω-FR. In the first place, the transient Ω-FR may be applied to the protocols of the Jarzynski equality [78]. Moreover, letting $A = B$, both the Ω-RF and the JE immediately yield

$$\langle e^{-\beta W} \rangle = \int P(W) e^{-\beta W} dW \int dW \, P(-W) = 1. \tag{3.29}$$

3.2.3
Asymptotic Results

One may develop further the transient $\mathcal{O}$-FRs, considering the time averages starting at any time $t > 0$, rather than at time 0, and computing

$$\frac{\mu(\bar{\mathcal{O}}_{t,t+\tau}|_{(A)_\delta})}{\mu(\bar{\mathcal{O}}_{t,t+\tau}|_{(-A)_\delta})}, \quad \text{where } \bar{\mathcal{O}}_{t,t+\tau}|_E = \{\Gamma \in \mathcal{M} : \bar{\mathcal{O}}_{t_0,t_0+\tau}(\Gamma) \in E\}. \tag{3.30}$$

The points of $\mathcal{M}$ lying in $\bar{\mathcal{O}}_{t,t+\tau}|_{(-A)_\delta}$ are given by [53]

$$\bar{\mathcal{O}}_{t,t+\tau}|_{(-A)_\delta} = \mathcal{I} S^{2t+\tau} \bar{\mathcal{O}}_{t,t+\tau}|_{(A)_\delta}, \tag{3.31}$$

from which, setting $\Gamma = \mathcal{I} S^{2t+\tau} W$, one obtains

$$\int_{\bar{\mathcal{O}}_{t,t+\tau}|_{(-A)_\delta}} f(\Gamma) d\Gamma = \int_{\bar{\mathcal{O}}_{t,t+\tau}|_{(A)_\delta}} f(S^{2t+\tau} W) \exp(-\Lambda_{0,2t+\tau}(W)) dW, \tag{3.32}$$

where we used the parity of f.

Remark
In this case, the coordinate transformation spans the whole time interval $[0, 2t + \tau]$, while $\bar{\mathcal{O}}_{t,t+\tau}$ takes only its central part $[t, t + \tau]$. In the previous cases, $t = 0$ and the two time intervals coincide. This apparently harmless fact will be crucial in the derivation of the steady-state FRs.

Using Eq. (3.10) one then obtains

$$\int_{\bar{\mathcal{O}}_{t,t+\tau}|_{(-A)_\delta}} f(\Gamma) d\Gamma = \int_{\bar{\mathcal{O}}_{t,t+\tau}|_{(A)_\delta}} f(W) e^{-\Omega_{0,2t+\tau}(W)} dW, \tag{3.33}$$

and

$$\frac{\mu(\bar{\mathcal{O}}_{t,t+\tau}|_{(A)_\delta})}{\mu(\bar{\mathcal{O}}_{t,t+\tau}|_{(-A)_\delta})} = \langle \exp(-\Omega_{0,2t+\tau}) \rangle^{-1}_{\bar{\mathcal{O}}_{t,t+\tau} \in (A)_\delta}, \tag{3.34}$$

which is the inverse of the average of $\exp(-\Omega_{0,2t+\tau})$, under the condition that $\bar{\mathcal{O}}_{t,t+\tau} \in (A)_\delta$. Taking $\mathcal{O} = \Omega$ yields

$$\begin{aligned} \frac{\mu(\bar{\Omega}_{t,t+\tau}|_{(A)_\delta})}{\mu(\bar{\Omega}_{t,t+\tau}|_{(-A)_\delta})} &= \langle \exp(-\Omega_{0,2t+\tau}) \rangle^{-1}_{\bar{\Omega}_{t,t+\tau} \in (A)_\delta} \\ &= e^{[A+\varepsilon(\delta,t,A,\tau)]\tau} \langle e^{-\Omega_{0,t} - \Omega_{t+\tau,2t+\tau}} \rangle^{-1}_{\bar{\Omega}_{t,t+\tau} \in (A)_\delta}, \end{aligned} \tag{3.35}$$

where $|\varepsilon|$ depends on δ, t_0, A, and τ but is smaller than δ.

This result is exact, and holds for all t, τ, and δ, and for any A for which both probabilities on the left-hand side do not vanish. Again, it rests only on the time reversibility of the dynamics.

If the t or $\tau \to \infty$ limits are taken, Eqs. (3.34) and (3.35) still remain transient relations because they keep referring to the initial measure μ. In particular, it is important to note that the conditional autocorrelation function

$$\begin{aligned} C(A,\delta,t,\tau) &:= \langle \mathcal{Q}_t(0)\mathcal{Q}_t(t+\tau)\rangle_{\bar{\Omega}_{t,t+\tau}\in(A)_\delta} \\ &:= \langle \mathrm{e}^{-\Omega_{0,t}-\Omega_{t+\tau,2t+\tau}}\rangle_{\bar{\Omega}_{t,t+\tau}\in(A)_\delta} \end{aligned} \tag{3.36}$$

of the observable

$$\mathcal{Q}_t(s)(\Gamma) := \exp\left[-\int_s^{s+t}\Omega(S^u\Gamma)\mathrm{d}u\right] \tag{3.37}$$

is computed with respect to the *initial* probability distribution function f, independently of how large t and τ may be. In other words C, Eqs. (3.34) and (3.35) express particular properties of the initial (possibly equilibrium) ensemble f that are accessible through the nonequilibrium dynamics. This reminds us of the Green–Kubo relations of linear response theory, which obtain a nonequilibrium property from the equilibrium distribution function and weakly nonequilibrium dynamics, because the corresponding nonequilibrium steady state is but a perturbation of the equilibrium state, negligible to first order. In our case, however, the state is not required to be close to the initial one, characterized by f: it can be arbitrarily far from that. Therefore, the quantity C should not be confused with the autocorrelation functions usually considered in the theory of dynamical systems, which are computed with respect to the invariant distribution. This distribution, indeed, is typically singular, at variance with ours, which is absolutely continuous with respect to Lebesgue. Moreover, its physical interpretation does not necessarily match with linear response theory.

However, for systems that do reach a steady state (something not required for the validity of Eqs. (3.34) and (3.35)), it may be possible to derive steady-state FRs from these transient FRs. The idea is that Eqs. (3.34) and (3.35) should concern trajectory segments of length τ distributed according to the invariant measure, or very close to that, if t, their initial instant, is sufficiently large. This may be better appreciated transferring the time evolution from the sets of phase space points to the probability distributions, using the conservation of probability that also defines the evolved probability measure μ_t, of density f_t, starting from an initial one μ_0 of density f_0:

$$\mu_t(S^tE) = \mu_0(E), \quad \text{that is,} \quad \int_{S^tE} f_t(X)\mathrm{d}X = \int_E f_0(Y)\mathrm{d}Y \tag{3.38}$$

for every (measurable) $E \subset \mathcal{M}$. Moreover, considering the coordinate transformation $Y = S^{-t}X$, one has [53]

$$\left|\frac{\partial Y}{\partial X}\right| = \mathrm{e}^{\Lambda_{-t_0,0}(X)}, \qquad f_{t_0}(X) = f_0(S^{-t_0}X)\mathrm{e}^{\Lambda_{-t_0,0}(X)} \tag{3.39}$$

or, recalling Eq. (3.10) and assuming that f_0 be even,

$$f_{t_0}(X) = f_0(X)\mathrm{e}^{-\Omega_{0,-t_0}(X)}. \tag{3.40}$$

Observe that $S^t\bar{\mathcal{O}}_{t,t+\tau}|_{(a,b)} = \bar{\mathcal{O}}_{0,\tau}|_{(a,b)}$, because a phase variable does not explicitly depend on time, and we assume autonomous dynamics. Then, one may write

$$\mu_0(\bar{\mathcal{O}}_{t,t+\tau}|_{(A)_\delta}) = \mu_t(S^t\bar{\mathcal{O}}_{t,t+\tau}|_{(A)_\delta}) = \mu_t(\bar{\mathcal{O}}_{0,\tau}|_{(A)_\delta}). \tag{3.41}$$

Taking $\mu = \mu_0$ in Eq. (3.34), this produces the *transient $\mathcal{O}$-tFR*: For every $t > 0$,

$$\frac{\mu_t(\bar{\mathcal{O}}_{0,\tau}|_{(A)_\delta})}{\mu_t(\bar{\mathcal{O}}_{0,\tau}|_{(-A)_\delta})} = \langle\exp(-\Omega_{0,2t+\tau})\rangle^{-1}_{\bar{\mathcal{O}}_{t,t+\tau}\in(A)_\delta}. \tag{3.42}$$

Relation (3.44) is exact and holds for all values of t, τ, δ, and A for which the ratio Eq. (3.34) makes sense. Also, provided the state μ_t converges to a steady-state μ_∞, the measure $\mu_t(\bar{\mathcal{O}}_{0,\tau}|_{(A)_\delta})$ approximates more and more accurately the steady-state measure of the set $\bar{\mathcal{O}}_{0,\tau}|_{(A)_\delta}$, as t grows at fixed τ.

Remark

It does make sense to speak of the evolution and possibly the limit of the real numbers expressing the time-dependent measure of the set $\mathcal{O}_{0,\tau}|_{(A)_\delta} \subset \mathcal{M}$, because this set is fixed and does not depend on t. Differently, $\mathcal{M}_{t,t+\tau}|_{(A)_\delta}$ is, in general, a very irregular object, which may wind around even more irregularly in the phase space, in particular reaching no $t \to \infty$ limit.

For $\bar{\mathcal{O}}_{t,t+\tau} = \bar{\Omega}_{t,t+\tau}$, Eq. (3.42) yields *transient Ω-tFR*: For every $t > 0$,

$$\frac{\mu_t(\bar{\Omega}_{0,\tau}|_{(A)_\delta})}{\mu_t(\bar{\Omega}_{0,\tau}|_{(-A)_\delta})} = \frac{\mathrm{e}^{[A+\varepsilon(\delta,t,A,\tau)]\tau}}{\langle\mathrm{e}^{-\Omega_{0,t}-\Omega_{t+\tau,2t+\tau}}\rangle_{\bar{\Omega}_{t,t+\tau}\in(A)_\delta}} = \frac{\mathrm{e}^{[A+\varepsilon(\delta,t,A,\tau)]\tau}}{C(A,\delta,t,\tau)}, \tag{3.43}$$

where the conditional autocorrelation function C is still computed with respect to the initial distribution f.

A direct consequence of Eq. (3.42) for $A = 0$ is that

$$\langle\exp(-\Omega_{0,2t+\tau})\rangle_{\bar{\mathcal{O}}_{t,t+\tau}\in(-\delta,\delta)} = 1. \tag{3.44}$$

In the $\delta \to \infty$ limit, this leads again to a form of nonequilibrium partition identity, for the full (unconditioned) f-ensemble average:

$$\langle\exp(-\Omega_{0,s})\rangle = 1, \quad \text{for every } s \geq 0. \tag{3.45}$$

Taking $A = 0$ in Eq. (3.43), one obtains

$$C(0,\delta,t,\tau) = \mathrm{e}^{\varepsilon(\delta,t,0,\tau)\tau}, \tag{3.46}$$

which shows that this particular conditional average is bounded in t, and tends to 1 when $\delta \to 0$, at fixed τ. Differently, the $\tau \to \infty$ limit, at fixed t and δ, may in principle vary without bounds, for unbounded Ω. The reason is that the mean of Ω may not belong to the set $(-\delta, \delta)$, especially if δ is small. In that case, $\bar{\Omega}_{t,t+\tau}|_{(-\delta,\delta)}$ may become more and more peculiar, while getting smaller and smaller, as τ grows.

Remark

If the system under investigation does not converge to a steady state with positive and negative fluctuations, no steady-state relation makes sense, but the transient FRs, being properties of the initial ensembles, hold at arbitrarily large times, and may even converge to some limit.

Next, we discuss the conditions under which such limit of the transient FRs describes not only a property of the initial ensemble, but also the statistics of trajectory segments sampled from a single steady-state trajectory.

3.2.4
Extending toward the Steady State

As discussed above, the transient FRs are exact at all times $t \geq 0$, but concern the initial and not the steady state. However, for a system that satisfies a steady-state FR such as Eq. (3.13) for the physically relevant observable Ω, this exactness implies that for large t the conditional autocorrelation function in the denominator of Eq. (3.43) deviates no more than $O(\delta)$ from unity. Let us see how this could come about for $A \neq 0$, since, in principle, this term could be of order $O(e^t)$, and t should tend to infinity.

Relation (3.13), however, is not expected to hold for all values of τ, but only for large τ. Taking the logarithm of Eq. (3.43) and dividing by τ, one obtains

$$\frac{1}{\tau}\ln\frac{\mu_t(\bar{\Omega}_{0,\tau}|_{(A)_\delta})}{\mu_t(\bar{\Omega}_{0,\tau}|_{(-A)_\delta})} = A + \varepsilon(\delta, t, A, \tau) - \frac{1}{\tau}\ln C(A, \delta, t, \tau), \tag{3.47}$$

which could turn into the steady-state FR for Ω, *provided* the conditional autocorrelation function C does not explode under the $t \to \infty$ limit, by taking large τ. There are, however, different scenarios to be considered, even under the assumption of ergodicity, without which the discussion about steady-state FRs cannot even be started. Under this assumption, the steady-state mean $\langle\Omega\rangle_\infty$ is reached by time averages starting from almost all initial conditions. Therefore, $\mu_t(\bar{\Omega}_{0,\tau}|_{(A)_\delta})$ tends to 1 if $\langle\Omega\rangle_\infty \in (A)_\delta$, while it tends to 0 if $\langle\Omega\rangle_\infty \notin (A)_\delta$.

Remark

From another perspective, this shows that the $\tau \to \infty$ limit at fixed t *usually makes no sense, as far as FRs are concerned, because in general the left-hand side of Eq. (3.47) loses its meaning in such a limit.*

Thinking of a system that does verify the steady-state FR, one is then admitting the existence of a range of values A for which the left-hand side of Eq. (3.47) preserves its meaning in the $t \to \infty$ limit, at fixed large τ. This range constitutes the domain $\mathcal{D}_{\delta,\tau}$ of the *fluctuation function* defined by

$$F_{\delta,\tau}(A) := \frac{1}{\tau}\lim_{t\to\infty}\frac{\mu_t(\bar{\Omega}_{0,\tau}|_{(A)_\delta})}{\mu_t(\bar{\Omega}_{0,\tau}|_{(-A)_\delta})}. \tag{3.48}$$

This domain is never empty, because it contains $A = 0$ at least. If a value A of the range of Ω does not belong to $\mathcal{D}_{\delta,\tau}$, it means that the function[9)]

$$M_{f_0}(A, \delta, \tau, t) := \frac{1}{\tau} \ln C(A, \delta, t, \tau) \tag{3.49}$$

has no $t \to \infty$ limit; that is, it may diverge or oscillate without ever settling on a definite value. However, if one is interested in a FR of the original form, this may be obtained in the large-τ limit if the oscillations of C are bounded with t or grow less than exponentially with τ. In that case, A may be included in the domain of the steady-state FR, and the FR is verified asymptotically in τ. If C is bounded, the error vanishes like $1/\tau$, as usual in large deviations theory. Another reason for A not to belong to the domain of the steady-state FR is that there are no fluctuations in the steady state (or possible variations on this situation). For instance, in the case that Ω tends to a nonvanishing constant, C grows exponentially with t.

If, however, the validity of the steady-state FR has been established with accuracy δ for a value A, one may ask which dynamical mechanisms allowed it. To this end, let $M_{f_0}(A, \delta, \tau) := \lim_{t\to\infty} M_{f_0}(A, \delta, \tau, t)$. We may then write

$$A - M_{f_0}(A, \delta, \tau) - \delta \le \frac{1}{\tau} \ln \frac{\mu_\infty(\bar{\Omega}_{0,\tau}|_{(A)_\delta})}{\mu_\infty(\bar{\Omega}_{0,\tau}|_{(-A)_\delta})} \le A - M_{f_0}(A, \delta, \tau) + \delta \tag{3.50}$$

and conclude that $M(A, \delta, \tau)$ vanishes as $\tau \to \infty$: A is in the domain of the steady-state Ω-FR and the standard steady-state Ω-FR (Eq. (3.13)) holds for A. This means that the condition now known as t-mixing should be verified.

Definition

A dynamical system is called t-mixing with respect to the density f with parameter $\delta > 0$ if there is $A^* > 0$, such that

$$\lim_{\tau\to\infty} \lim_{t\to\infty} M_f(A, \delta, \tau, t) = \lim_{\tau\to\infty} M_f(A, \delta, \tau) = 0, \quad \forall A \in (-A^*, A^*). \tag{3.51}$$

This is the physical mechanism underlying the validity of the steady-state Ω-FR, when it holds.[10)]

The following examples clarify this point.

Remark

If the two limits on the left-hand side of Eq. (3.51) can be exchanged, and $M_f(A, \delta, \tau, t) \to 0$ as τ grows at fixed large t, one obtains the same result, as far as the validity of the steady-state Ω-FR is concerned (cf. Section 7 of Ref. [12]). This condition is

9) The notation stresses the fact that the average is computed with the probability density f_0.

10) We note that our t-mixing condition is weaker than that adopted, for example, in Ref. [66]. There, a system is called t-mixing if $\lim_{t\to\infty} M_f(A, \delta, \tau, t)$ exists. While our weaker condition suffices for the validity of the steady-state Ω-FR, the stronger condition of Ref. [66] implies good thermodynamic behavior as well. At present, the matter of which of the two notions may be physically more relevant is still a subject of investigation; therefore, we keep the name due to Evans, Searles, and Williams for our condition (see Chapter 2).

arguably easier to obtain and, from a numerical or experimental point of view, could be hard to distinguish from the t-mixing condition. It looks interesting to investigate the relation between these two conditions.

We present three example systems here to consider the t-mixing property:

a) t-mixing is obviously verified if the Ω-autocorrelation computed with respect to the initial distribution decays instantaneously. Indeed, this implies

$$\begin{aligned} \langle e^{-\Omega_{0,t}-\Omega_{t+\tau,2t+\tau}} \rangle_{\Omega_{t,t+\tau}\in(A)_\delta} &= \langle e^{-\Omega_{0,t}-\Omega_{t+\tau,2t+\tau}} \rangle \\ &= \langle e^{-\Omega_{0,t}} \rangle \langle e^{-\Omega_{t+\tau,2t+\tau}} \rangle, \\ 1 = \langle e^{-\Omega_{0,s}-\Omega_{s,t}} \rangle &= \langle e^{-\Omega_{0,s}} \rangle \langle e^{-\Omega_{s,t}} \rangle, \qquad \langle e^{-\Omega_{s,t}} \rangle = 1, \quad \forall s,t, \end{aligned} \tag{3.52}$$

where Eq. (3.25) is used. Then, $M(A,\delta,\tau,t)$ identically vanishes for all t and τ and transient and steady-state FRs can be identified. Indeed, an instantaneous decay of correlations means an instantaneous convergence to the steady state.

b) The idealized situation (a) is not strictly verified, in general. However, tests on molecular dynamics systems [79] indicate that typically there exists a constant K (often not too far from 1), such that

$$0 < \frac{1}{K} \le \exp(M_f(A,\delta,\tau,t)\tau) \le K, \tag{3.53}$$

In the first place, M must stay limited for growing t, if A lies in the domain of the FR. Moreover, M vanishes for $\tau = 0$ and $A = 0$. Then, let us say that the Ω-autocorrelation decays with timescale t_M, the material property known as the Maxwell time in particle systems, which is of the order of magnitude of the mean free time and which is known to depend only mildly on the external field ($t_\mathrm{M}(F_e) = t_M(0) + O(F_\mathrm{e}^2) < \infty$ [4, 53]), $\ln K$ must be of order $O(t_\mathrm{M})$. For instance, take t and τ much larger than t_M and some $\hat{t} > t_M$. Then,

$$\begin{aligned} &\langle e^{-\Omega_{0,t}-\Omega_{t+\tau,2t+\tau}} \rangle_{\bar{\Omega}_{t,t+\tau}\in A_\delta^+} \approx \langle e^{-\Omega_{0,t}} \cdot e^{-\Omega_{t+\tau,2t+\tau}} \rangle_{\bar{\Omega}_{t+\hat{t},t+\tau-\hat{t}}\in A_\delta^+} \\ &\approx \langle e^{-\Omega_{0,t}} \cdot e^{-\Omega_{t+\tau,2t+\tau}} \rangle \approx \langle e^{-\Omega_{0,t}} \rangle \langle e^{-\Omega_{t+\tau,2t+\tau}} \rangle = \langle e^{-\Omega_{t+\tau,2t+\tau}} \rangle, \end{aligned} \tag{3.54}$$

with an accuracy that improves with increasing t and τ. Moreover, observe that

$$\begin{aligned} \langle e^{-\Omega_{t+\tau,2t+\tau}} \rangle &= \int e^{-\Omega_{t+\tau,2t+\tau}(\Gamma)} f_0(\Gamma) d\Gamma \\ &= \int e^{-\Omega_{0,t}(\Gamma)} f_{t+\tau}(\Gamma) d\Gamma := \langle e^{-\Omega_{0,t}} \rangle_{t+\tau} \approx \langle e^{-\Omega_{0,t}} \rangle_\infty \end{aligned} \tag{3.55}$$

because we have taken both t and τ much larger than the decorrelation time. At equilibrium, the steady-state average on the right-hand side equals 1. Then, being a material property of the system, simply related to the dissipated power, it should only have a correction of order $O(F_\mathrm{e}^2)$, when a driving external field is switched on. Indeed, close to equilibrium, the existence of the transport coefficients implies that the above must exist. Therefore, Eq. (3.53) may be assumed to hold close to equilibrium; furthermore, numerical studies show that this assumption remains valid quite far from equilibrium and for a wide range of values A [79].

c) Systems that violate the above decays of correlations may of course be conceived. Consider, for instance, a particle moving in empty space, under the action of a constant external force $\mathbf{F}_e$ and a Gaussian thermostat:

$$\dot{\mathbf{q}} = \mathbf{p}\,, \qquad \dot{\mathbf{p}} = \mathbf{F}_e - \frac{\mathbf{F}_e \cdot \mathbf{p}}{\mathbf{p} \cdot \mathbf{p}}\mathbf{p},$$

where the kinetic energy $K = \mathbf{p} \cdot \mathbf{p}/2$ is a constant of motion. For initial conditions with $\mathbf{p}$ pointing in the direction opposite to $\mathbf{F}_e$, $\mathbf{p}$ is a constant of motion; for all other initial conditions $\mathbf{p}$ tends to get parallel to $\mathbf{F}_e$, while its magnitude is constant. Therefore, there is one repeller and one attractor that are single points, and any probability density in the unit circle of velocities tends to the Dirac δ-function over the attractor. Consequently, there are no steady-state fluctuations and the domain of the steady-state Ω-FR is empty. Correspondingly, correlations never decay, and the term M_f diverges in the $t \to \infty$ limit. Nevertheless, the transient and asymptotic relations remain valid for the evolving ensembles, and express the rate at which the initial probability of observing $-A$ vanishes, compared to the initial probability of A. Similar situations, with no negative steady-state fluctuations, may be produced in more realistic models by applying extremely high fields.

Note that the steady-state Ω-FR holds if $M_f(A, \delta, \tau)$ can be made as small as one wishes, by taking sufficiently large τ and small δ. Equation (3.53) is not necessary; it suffices that M_f grows less than exponentially fast with τ. Systems of this sort, if they exist,[11] verify the steady-state FRs but may hardly obey standard thermodynamics. In all these cases, one has steady-state FRs: For any $\gamma > 0$, there is sufficiently large $\hat{\tau}$ such that

$$\begin{aligned} &\frac{1}{\tau}\left|\ln\frac{\mu_\infty(\bar{\mathcal{O}}_{0,\tau}|_{(A)_\delta})}{\mu_\infty(\bar{\mathcal{O}}_{0,\tau}|_{(-A)_\delta})} - \lim_{t\to\infty}\ln\langle \mathrm{e}^{-\Omega_{0,2t+\tau}}\rangle^{-1}_{\bar{\mathcal{O}}_{t,t+\tau}\in(A)_\delta}\right| \le \gamma, \\ &\frac{1}{\tau}\left|\ln\frac{\mu_\infty(\bar{\Omega}_{0,\tau}|_{(A)_\delta})}{\mu_\infty(\bar{\Omega}_{0,\tau}|_{(-A)_\delta})} - A\right| \le \gamma\,, \quad \text{for } \tau > \hat{\tau}. \end{aligned} \tag{3.56}$$

We conclude this subsection with some observations.

Remark

The transient FRs are very robust, since they hold very generally for time-reversal invariant dynamics. The steady-state FRs hold less generally, but still more generally than thermodynamics. Transient relations differ substantially from steady-state ones. Transient relations are exact for all observation times $\tau > 0$ (short or long ones). Steady-state relations may only be approximately valid for finite relaxation and observation times, t and τ, respectively, and hence hold only asymptotically, in the $t \to \infty$ limit followed by the $\tau \to \infty$ limit. This is because the transient relations considered here concern the initial

11) Possible candidates are objects at the nanoscale, whose correlations persist over the observation times.

distributions, while steady-state relations concern the invariant natural measure. These two kinds of relations are thus complementary.

For these reasons, asymptotic relations, which merely result from taking time to infinity in transient relations, must be distinguished from steady-state relations. However, in the case that t-mixing condition holds, one may derive steady-state relations from transient relations. This is simply a convenient way of deriving steady-state relations that could be derived without any reference to initial distributions, which are forgotten in the steady states. Indeed, the steady-state relations give no information on the initial state: if correlations decay, so that the steady state is reached, totally different initial states yield the same steady-state statistics. By contrast, changing the initial probability density leads to different results for the transient relations. For instance, in order to preserve the validity of the transient Ω-FR, one should change the Omega as well. The transient Ω-FR holds for the phase space contraction if the initial state is microcanonical: if the initial state is not microcanonical, the transient Ω-FR does not hold for the phase space contraction rate.

3.2.5
The Gallavotti–Cohen Approach

For historical and comparative purposes, it is interesting to consider the approach to steady-state FRs developed by Gallavotti and Cohen. Key to this approach is the chaotic hypothesis, the proposition that *a reversible many-particle system in a stationary state can be regarded as a transitive Anosov system for the purpose of computing its macroscopic properties.*

The idea is that dissipative, reversible, transitive Anosov maps $S^t : \mathcal{M} \to \mathcal{M}$, $t \in \mathbb{Z}$, are idealizations of nonequilibrium systems. As systems of physical interest cannot be expected to be of Anosov type, Gallavotti and Cohen assumed that their deviations from the ideal situation are not relevant in practice, that is, for their observable properties. In the Gallavotti–Cohen perspective, an important property of Anosov maps is that they admit arbitrarily fine *Markovian* partitioning of the phase space $\mathcal{M}$ into cells whose interiors are disjoint from each other, and whose boundaries are invariant sets [80]. Consequently, the interior of a cell is mapped by S into the interior of another cell. Gallavotti and Cohen further assume that the dynamics is transitive (that a typical trajectory explores $\mathcal{M}$ densely). It is this structure that guarantees that the probability (Lyapunov) weights of Eq. (1) in Ref. [55], from which the steady-state Λ-FR follows, can be assigned to the cells of a Markov partition in precise mathematical terms [67–70].

More precisely, for discrete time dynamics, let the phase space contraction rate $\Lambda(\Gamma) = -\ln J(\Gamma)$, where J is the Jacobian determinant of S. Consider the steady-state probability of the dimensionless phase space contraction rate e_τ, obtained along a trajectory segment of duration τ centered at $\Gamma \in \mathcal{M}$:

$$e_\tau(\Gamma) = \frac{1}{\tau \langle \Lambda \rangle_{\rm ss}} \sum_{k=-\tau/2}^{\tau/2-1} \Lambda(S^k \Gamma), \tag{3.57}$$

where $\langle\cdot\rangle_{ss}$ is the steady-state phase average. Let J^u be the Jacobian determinant of S restricted to the unstable manifold V^+, that is, the product of the asymptotic factors of separation of nearby points, along the directions in which distances asymptotically grow at an exponential rate. *If the system is Anosov*, the probability $\pi_\tau(B_{p,\varepsilon})$ that $e_\tau(\Gamma) \in B_{p,\varepsilon} \equiv (p-\varepsilon, p+\varepsilon)$, in the limit of fine Markov partitions and long τ, is given by the normalized sum of the weights:

$$\pi_\tau(e_\tau(\Gamma) \in B_{p,\varepsilon}) \approx \frac{1}{M} \sum_{\Gamma, e_\tau(\Gamma)\in B_{p,\varepsilon}} \left(\prod_{k=-\tau/2}^{\tau/2-1} \frac{1}{J^u(S^k\Gamma)} \right), \tag{3.58}$$

where M is a normalization constant. If the support of the physical measure is $\mathcal{M}$, which is the case if the dissipation is not exceedingly high [81], time reversibility guarantees that the support of π_τ includes an interval $[-p^*, p^*]$ for some positive p^*, and one can consider the ratio

$$\frac{\pi_\tau(B_{p,\varepsilon})}{\pi_\tau(B_{-p,\varepsilon})} \approx \frac{\sum_{\Gamma, e_\tau(\Gamma)\in B_{p,\varepsilon}} w_{\Gamma,\tau}}{\sum_{\Gamma, e_\tau(\Gamma)\in B_{-p,\varepsilon}} w_{\Gamma,\tau}}, \tag{3.59}$$

where each Γ in the numerator has a counterpart in the denominator. Denoting by $\mathcal{I}$ the involution that replaces the initial condition of one trajectory with the initial condition of the reversed trajectory, time reversibility yields

$$\Lambda(\Gamma) = -\Lambda(\mathcal{I}\Gamma), \quad w_{\mathcal{I}\Gamma,\tau} = w_{\Gamma,\tau}^{-1}, \quad \text{and} \quad \frac{w_{\Gamma,\tau}}{w_{\mathcal{I}\Gamma,\tau}} = \mathrm{e}^{-\tau\langle\Lambda\rangle_{ss}p} \tag{3.60}$$

if $e_\tau(\Gamma) = p$. Taking small ε, the division of each term in the numerator of (3.59) by its counterpart in the denominator approximately equals $\mathrm{e}^{-\tau\langle\Lambda\rangle p}$, which then equals the ratio in (3.59). In the limit of small ε, infinitely fine Markov partition and large τ, the authors of Ref. [68] obtained the following theorem.

Gallavotti–Cohen Theorem (1995)

Let $(\mathcal{M}, S)$ be dissipative (i.e., $\langle\Lambda\rangle_{ss} > 0$) and reversible and assume that the chaotic hypothesis holds. Then,

$$\frac{\pi_\tau(B_{p,\varepsilon})}{\pi_\tau(B_{-p,\varepsilon})} = \mathrm{e}^{\tau\langle\Lambda\rangle_{ss}p}, \tag{3.61}$$

with an error in the argument of the exponential that can be estimated to be p- and τ-independent.

If the Λ-FR (hence the chaotic hypothesis on which it is based) holds, the function $C(p;\tau,\varepsilon) = (1/\tau\langle\Lambda\rangle_{ss})\ln[\pi_\tau(B_{p,\varepsilon})/\pi_\tau(B_{-p,\varepsilon})]$ tends to a straight line of slope 1 for growing τ, apart from small errors.

Under the assumption that Λ coincides with the entropy production rate, the Λ-FR can be used to obtain the Green–Kubo relations and the Onsager reciprocal relations, in the limit of small external drivings [1], and the Λ-FR appears to be an extension of such relations to nonequilibrium systems. To achieve this, Gallavotti assumes that the continuous-time IK system is driven by ℓ fields F_i, and that

$\Lambda(\Gamma) = \sum_{i=1}^{\ell} F_i J_i^0(\Gamma) + O(F^2)$ (thus vanishing for zero field). In this way, the linear "currents" J_i^0, which are proportional to the forces F_i, are defined. The validity of the Λ-FR (Eq. (3.61)) implies $\langle\Lambda\rangle_{ss} = C_2/2 + O(F^3)$, for constant C_2. Now, let the full (nonlinear) currents be defined by $J_i(\Gamma) = -\partial_{F_i}\Lambda(\Gamma)$, and the transport coefficients be $L_{ij} = \partial_{F_j}\langle J_i\rangle_{ss}|_{F=0}$. Assuming differentiability with respect to the F_i [82], the validity of the Λ-FR, and time reversibility, one can write

$$\langle\Lambda\rangle_{ss} = \frac{1}{2}\sum_{i,j=1}^{\ell}(\partial_{F_j}\langle J_i\rangle_{ss} + \partial_{F_i}\langle J_j\rangle_{ss})|_{F=0}F_iF_j = \frac{1}{2}\sum_{i,j=1}^{\ell}(L_{ij} + L_{ji})F_iF_j \tag{3.62}$$

to second order in the forces. Equating this with $C_2/2$ and considering $(L_{ij} + L_{ji})/2$ with $i = j$, one recovers the Green–Kubo relations. To obtain the symmetry $L_{ij} = L_{ji}$, Gallavotti extends the Λ-FR to consider the joint distribution of Λ and its derivatives. He introduces the dimensionless current q, averaged over a long time τ, through the relation

$$\frac{1}{\tau}\int_{-\tau/2}^{\tau/2} F_j\partial_{F_j}\Lambda(S^t\Gamma)\mathrm{d}t = F_j\langle\partial_{F_j}\Lambda\rangle_{ss}q(\Gamma) \tag{3.63}$$

and considers the joint distribution $\pi_\tau(p, q)$, with corresponding large deviation functional $\zeta(p, q) = -\lim_{\tau\to\infty}(1/\tau)\ln\pi_\tau(p, q)$. The result is a relation similar to the Λ-FR:

$$\lim_{\tau\to\infty}\frac{1}{\tau\langle\Lambda\rangle_{ss}p}\ln\frac{\pi_\tau(p, q)}{\pi_\tau(-p, -q)} = 1\,. \tag{3.64}$$

This difference $\zeta(p, q) - \zeta(-p, -q)$ is thus independent of q, implying $L_{ij} = L_{ji}$ in the limit of small F. This work was refined in Ref. [3] and inspired further investigations, such as Refs [2, 44].

This approach is not without difficulties, particularly regarding physical interpretation. Λ, which is directly related to the thermostating term of nonequilibrium molecular dynamics, is only proportional to Σ/kT in very special cases [4, 9]. Moreover, it was found that the steady-state Λ-FR is hard, if not impossible, to verify in non-isoenergetic systems with singular Λ, close to equilibrium [60, 83, 84].

In Gaussian isokinetic systems, Λ is the sum of a dissipative term Ω and a conservative singular term. Ω obeys the FR, while the conservative term does not: but its averages over long time intervals are expected to become negligible with respect to the averages of Ω as the length of the interval grows [4, 84]. Thus, in the long-time limit, the Λ-FR should hold as a consequence of the validity of the Ω-FR, while the convergence times of the Λ-FR would diverge when equilibrium is approached, because Ω vanishes as the *square* of the driving forces. These observations eventually led to the conclusion that, in some cases, Λ describes heat fluxes, not entropy productions [4, 78, 85], and hence that in those cases the Λ-FR has to be modified, to mimic the heat FR of van Zon and Cohen for stochastic systems [86]. This amounts also to saying that the white noise present in the systems studied by van Zon and Cohen is reproduced to some extent by the deterministic chaos of the uniformly hyperbolic dynamical systems of Ref. [85], and that the same may be expected in sufficiently chaotic particle systems (like typical NEMD models). This provides us with one example of equivalence between stochastic and deterministic evolutions, as far as the FRs are concerned.

We note that the mathematical description of Gallavotti and Cohen and the physical approach of Evans and Searles contribute differently to our understanding of nonequilibrium physics. The mathematical approach is concerned with the identification of dynamical systems that allow a rigorous derivation of some kind of FR, for one phase function. This may appear to be of marginal physical relevance, and one could argue that the stringent Anosov conditions may conceal the true reasons for a real object to obey one FR (in the way that Khinchin argues for a weaker condition than ergodicity). Nonetheless, intriguing physical questions have been raised by the mathematics, like the (still open) question of which observables and which systems of physical interest verify the modified FR of Refs [87, 88] (see, for example, Refs [51, 62]). On the other hand, the physical approach is concerned with understanding the mechanisms by which a particular observable, of one physical system, obeys a given FR. Derivations of the FRs such as those of Refs [8, 53] may look mathematically uninteresting, because they rely on *physical* assumptions that seem too hard to prove, such as the *t*-mixing condition. Such conditions have a natural physical origin, but need to be more intensively mathematically investigated.

We believe that combining the mathematical with the physical approach is bound to lead to important advances in the field.

3.3 Discussion

Our analysis shows that the steady-state Ω-FR and its consequences can be obtained from only time reversibility, the existence of a unique steady state, and Ω-autocorrelation decay with respect to the initial probability density: the *t*-mixing property. These are the physical mechanisms underlying the validity of the steady-state Ω-FR and, indeed, they correctly identify the relevant timescales. From a purely mathematical point of view, the decay of the Ω-autocorrelation could be relaxed, but is always present if thermodynamics applies. This explains why steady-state Ω-FR is so generally verified, even though no special requirement can be expected to be satisfied by the corresponding microscopic dynamics (and despite the very restrictive conditions required by the mathematically motivated derivations).

A number of predictions follow from the analysis leading to the concept of *t*-mixing [53], most of which have still to be considered in experimental tests. Much remains to be known about this property. Could certain dynamics enjoy a decay of correlations with respect to μ_∞, but not with respect to μ_0? In our example (c), M_{f_0} diverges as t grows, so the steady-state Ω-FR does not hold, in agreement with the fact that the steady state of that system has no fluctuations: more general systems, such as those verifying axiom-C, warrant investigation, as does the limit behavior of M_f more generally.

The transient FRs connect equilibrium and nonequilibrium properties of physical systems in a striking fashion, in that they consider at once the statistical properties of equilibrium states and nonequilibrium dynamics. Their predictions describe the statistics of ensembles of experiments, are valid under extremely wide conditions, and can be verified by a large variety of physical systems. Despite their apparent mathematical simplicity, transient relations (such as the Jarzynski equality) can prove exceedingly useful, providing a route to determining properties that may otherwise be very difficult to extract in practice. Moreover, as noted earlier, transient relations close very well the circle of fluctuations: the fluctuation–dissipation relations obtain nonequilibrium properties from equilibrium experiments; the transient relations obtain equilibrium properties from nonequilibrium experiments [7, 11].

The steady-state relations, on the other hand, concern the asymptotic statistics generated by a single system evolution, if a steady state is reached. If a single steady state is attained, its statistics may be recovered from an individual trajectory, and this may have no knowledge of the initial ensemble f. If a steady state is not attained, or if the steady state is sufficiently far from equilibrium that negative values of the dissipation function cannot be observed, the steady-state FRs cannot apply; the transient FRs (including their asymptotic forms) remain valid as properties of the evolving ensembles, because they require only time reversibility, and refer to the initial state (equilibrium). The steady-state Ω-FR also cannot hold if there are distinct basins of attraction in $\mathcal{M}$: again, the transient and asymptotic relations remain valid, as they refer to the initial ensemble.

In the linear regime, absence of decay of correlations of Ω with respect to the equilibrium density f by which it is defined corresponds to the nonexistence of the transport coefficient associated with Ω. Further away from equilibrium, nonlinear response theory applies: the time correlation function is still determined with respect to f, but the dynamics used to compute the nonlinear transport coefficients is the *nonequilibrium* one [34]. As τ constitutes the time separation of the quantities that should decorrelate, the growth of τ contributes to this decorrelation. This virtually guarantees the validity of the steady-state Ω-FR in most cases concerning macroscopic physics. In mesoscopic systems, this is no longer the case, and here other predictions like those related to the $\mathcal{O}$-FRs become of interest.

The Ω-FRs refer directly to the dissipative fluxes (or related quantities) rather than to the phase space contraction rate. As discussed in Ref. [4], the PDFs of Λ and Ω may be quite different even if their steady-state averages coincide, and the convergence times of the Λ-FR typically diverge as the dissipative fields F decrease. This is due to the fact that Λ contains "undesirable" terms that average to zero: as $F \to 0$, the dissipation decreases as $o(F^2)$ while the "undesirable" terms remain $o(1)$. Longer and longer averaging times are thus required before the Λ-FR converges. The finite-time relations illustrated in this paper still hold at all times.

The importance of N is usually made fairly clearly in equilibrium statistical mechanics: in the development of FRs, it has an often subtle, but ultimately similar origin. Partly we rely on large N to justify the association of particular phase functions with thermodynamic properties (most notably the entropy production). Also, systems with large numbers of particles exhibit decays in correlation that are extremely rapid for observables of thermodynamic significance. For systems with moderate or small N, commonly correlations no longer decay on a shorter timescale than the period of observation. The analysis based on the decay of the Ω-autocorrelation attributes a lesser importance to strong chaos than to correlation decays for a small set of observables that are favored by large N [89, 90].

We conclude with a consequence of the (transient or steady-state) $\mathcal{O}$-FRs, yet to be investigated. These FRs can reveal information about experimentally inaccessible quantities, from the statistics of accessible ones. Consider Eq. (3.22), rewritten as

$$\Omega_{0,\tau}^{(\mathcal{O})}(A,\delta) = \ln\langle\exp(-\Omega_{0,\tau})\rangle_{\bar{\mathcal{O}}_{0,\tau}\in(A)_\delta} = \ln\frac{\mu(\bar{\mathcal{O}}_{0,\tau}|_{(-A)_\delta})}{\mu(\bar{\mathcal{O}}_{0,\tau}|_{(A)_\delta})}. \tag{3.65}$$

In the small-δ limit, $\Omega_{0,\tau}^{(\mathcal{O})}(A,\delta)$ approaches an average of $\Omega_{0,\tau}$ over all trajectories for which $\bar{\mathcal{O}}_{0,\tau}$ equals A. As a consequence, knowledge of the observable $\mathcal{O}$ may be used to reconstruct Ω and its statistics. Such an approach is analogous to the fundamental approach of developing physically relevant ensembles in equilibrium statistical mechanics.

The reconstruction of Ω may then be made more accurate, complementing Eq. (3.65) with similar relations concerning other observables, so as to obtain a function of a certain number of variables $\Omega_{0,\tau}^{(\mathcal{O}_1,\mathcal{O}_2,\ldots)}(A,B,\ldots)$ that gets closer and closer to the phase variable Ω, that is, to the full ensemble f as the number of observables increases. The vast separation of scales that leads to standard thermodynamic behavior would serve to prevent this detailed reconstruction of the phase space, as usual in statistical mechanics: too many observables would have to be considered. Indeed, the attraction in this approach is that it invites the possibility of shifting our focus from details of the mathematical definition of Ω toward the physical quantity (e.g., Σ) that it should represent.

3.4 Conclusions

The fluctuation relations provide one of the few exact sets of results available to us to describe nonequilibrium systems. In analogy with the study of equilibrium systems via statistical mechanics, the theoretical development of the FRs has emerged following different perspectives on the problem at hand. The mathematical approach focuses on the properties that a dynamical system should have in order to exhibit fluctuations in the nature of the computationally observed FRs. It is then asserted that real dynamical systems that exhibit FRs of this form must be like these idealized dynamical systems, in some sense. The physical approach considers typical properties of typical real dynamical systems that are known to obey the FRs, and asks what conditions are required of those properties (and those dynamical systems) for the FRs to hold. In this chapter, we have presented a mathematically rigorous framework in which these questions and their answers can be considered.

This framework is in part based upon the formal theory originally developed to support the use of NEMD in calculating transport coefficients in the linear response regime. This theory allows us to develop a connection between dynamical and thermodynamical properties of the nonequilibrium system, in particular those properties related to entropy production. More recent developments of this theory have led to a means of distinguishing fluctuations that could be attributed to non-dissipative effects (e.g., in relation to the background ensemble constraints) from those that could be attributed to dissipative effects (e.g., the nonequilibrium field).

Armed with this theory, we can formally develop the transient Ω-FRs toward a steady-state form, and consider the conditions under which this extension is possible as well as the order of error terms. The main advantage of this approach is that we obtain a formal expression for this condition, which has become known as t-mixing and which is related to the decay in correlations of the dissipation function (or relevant function in the generalized case). Current work is now focused on a detailed exploration of this condition in various systems that are known either to obey the FRs or to demonstrate anomalous FRs, as in the case of Ref. [91], or none at all.

For equilibrium systems the importance of system size is often explicit, or is implicit in the association of dynamical properties with their thermodynamic counterparts. For the FRs in nonequilibrium systems, the influence of system size is also present in relation to the decay in correlations (which, indeed, is the influence that interested Khinchin in his alternative approach to the ergodic question). Anomalous FRs have been associated with systems of few degrees of freedom, where correlations decay slowly. Again, the connection between the number of degrees of freedom, the slow decay of correlations, and the condition of t-mixing is a research topic of current investigation.

Acknowledgments

Thanks are in order to Rainer Klages for many useful observations. L.R. gratefully acknowledges financial support from the European Research Council under the European Community's Seventh Framework Programme (FP7/2007–2013)/ERC Grant Agreement No. 202680. The EC is not liable for any use that can be made of the information contained herein.

References

1 Gallavotti, G. (1996) Extension of Onsager's reciprocity to large fields and the chaotic hypothesis. *Phys. Rev. Lett.*, **77**, 4334.

2 Rondoni, L. and Cohen, E.G.D. (1998) Orbital measures in non-equilibrium statistical mechanics: the Onsager relations. *Nonlinearity*, **11**, 1395.

3 Gallavotti, G. and Ruelle, D. (1997) SRB states and nonequilibrium statistical mechanics close to equilibrium. *Commun. Math. Phys.*, **190**, 279.

4 Evans, D.J., Searles, D.J., and Rondoni, L. (2005) On the application of the Gallavotti–Cohen fluctuation relation to thermostatted steady states near equilibrium. *Phys. Rev. E*, **71**, 056120.

5 Bonaldi, M., Conti, L., Gregorio, P.D., Rondoni, L., Vedovato, G., Vinante, A., Bignotto, M., Cerdonio, M., Falferi, P., Liguori, N., Longo, S., Mezzena, R., Ortolan, A., Prodi, G.A., Salemi, F., Taffarello, L., Vitale, S., and Zendri, J.P. (2009) Nonequilibrium steady-state

fluctuations in actively cooled resonators. *Phys. Rev. Lett.*, **103**, 010601.

6 Bustamante, C., Liphardt, J., and Ritort, F. (2005) The nonequilibrium thermodynamics of small systems. *Phys. Today*, **58**, 43.

7 Marconi, U.M.B., Puglisi, A., Rondoni, L., and Vulpiani, A. (2008) Fluctuation–dissipation: response theory in statistical physics. *Phys. Rep.*, **461**, 111.

8 Evans, D.J. and Searles, D.J. (2002) The fluctuation theorem. *Adv. Phys.*, **52**, 1529.

9 Rondoni, L. and Mejia-Monasterio, C. (2007) Fluctuations in nonequilibrium statistical mechanics: models, mathematical theory, physical mechanisms. *Nonlinearity*, **20**, R1.

10 Chetrite, R. and Gawedzky, K. (2008) Fluctuation relations for diffusion processes. *Commun. Math. Phys.*, **282**, 469.

11 Jepps, O.G. and Rondoni, L. (2010) Deterministic thermostats, theories of nonequilibrium systems and parallels with the ergodic condition. *J. Phys. A*, **43**, 133001.

12 Jaksic, V., Pillet, C.A., and Rey-Bellet, L. (2011) Entropic fluctuations in statistical mechanics. I. Classical dynamical systems. *Nonlinearity*, **24**, 699.

13 Einstein, A. (1905) The motion of elements suspended in static liquids as claimed in the molecular kinetic theory of heat. *Ann. Phys.*, **17**, 549.

14 Einstein, A. (1910) Theory of opalescence of homogeneous liquids and mixtures of liquids in the vicinity of the critical state. *Ann. Phys.*, **33**, 1275.

15 Onsager, L. (1931) Reciprocal relations in irreversible processes. I. *Phys. Rev.*, **37**, 405.

16 Onsager, L. (1931) Reciprocal relations in irreversible processes. II. *Phys. Rev.*, **38**, 2265.

17 Callen, H.B. and Welton, T.A. (1951) Irreversibility and generalized noise. *Phys. Rev.*, **83**, 34.

18 Callen, H.B. and Greene, R.F. (1952) On a theorem of irreversible thermodynamics. *Phys. Rev.*, **86**, 702.

19 Green, M.S. (1951) Brownian motion in a gas of noninteracting molecules. *J. Chem. Phys.*, **19**, 1036.

20 Green, M.S. (1952) Markoff random processes and the statistical mechanics of time-dependent phenomena. *J. Chem. Phys.*, **20**, 1281.

21 Green, M.S. (1954) Markoff random processes and the statistical mechanics of time-dependent phenomena. II. Irreversible processes in fluids. *J. Chem. Phys.*, **22**, 398.

22 Kubo, R. (1957) Statistical-mechanical theory of irreversible processes. I. General theory and simple applications to magnetic and conduction problems. *J. Phys. Soc. Jpn.*, **12**, 570.

23 Alder, B.J. and Wainwright, T.E. (1967) Velocity autocorrelations for hard spheres. *Phys. Rev. Lett.*, **18**, 988.

24 Kadanoff, L.P. and Swift, J. (1968) Transport coefficients near the liquid–gas critical point. *Phys. Rev.*, **166**, 89.

25 Visscher, W.M. (1974) Transport processes in solids and linear-response theory. *Phys. Rev. A*, **10**, 2461.

26 Dufty, J.W. and Lindenfeld, M.J. (1979) Non-linear transport in the Boltzmann limit. *J. Stat. Phys.*, **20**, 259.

27 Cohen, E.G.D. (1983) Kinetic theory of non-equilibrium fluids. *Physica A*, **118**, 17.

28 Morriss, G.P. and Evans, D.J. (1987) Application of transient correlation functions to shear flow far from equilibrium. *Phys. Rev. A*, **35**, 792.

29 Jepps, O.G. and Rondoni, L. (2006) Thermodynamics and complexity of simple transport phenomena. *J. Phys. A*, **39**, 1311.

30 Jou, D., Casas-Vásquez, J., and Lebon, G. (2008) *Extended Irreversible Thermodynamics*, Springer, Berlin.

31 Kreuzer, H.J. (1981) *Nonequilibrium Thermodynamics and Its Statistical Foundations*, Clarendon Press, Oxford.

32 Chapman, S. and Cowling, T.G. (1990) *The Mathematical Theory of Non-Uniform Gases*, Cambridge University Press, Cambridge, UK.

33 Allen, M.P. and Tildesley, D.J. (1987) *Computer Simulation of Liquids*, Clarendon Press, Oxford.

34 Evans, D.J. and Morriss, G.P. (2008) *Statistical Mechanics of Nonequilibrium Liquids*, Cambridge University Press, Cambridge, UK.

35 Hoover, W.G. (1991) *Computational Statistical Mechanics*, Elsevier.

36 Gauss, K.F. (1829) Nouveau principe de mécanique. *J. Reine Angew. Math.*, **4**, 232.

37 Lanczos, C. (1979) *The Variational Principles of Mechanics*, Dover, New York.

38 Nosé, S. (1984) A unified formulation of the constant temperature molecular dynamics methods. *J. Chem. Phys.*, **81**, 511.

39 Nosé, S. (1984) A molecular dynamics method for the simulations in the canonical ensemble. *Mol. Phys.*, **52**, 255.

40 Hoover, W.G. (1985) Canonical dynamics: equilibrium phase-space distributions. *Phys. Rev. A*, **31**, 1695.

41 Gallavotti, G. (2000) *Statistical Mechanics: A Short Treatise*, Springer, Berlin.

42 Khinchin, A.I. (1949) *Mathematical Foundations of Statistical Mechanics*, Dover, New York.

43 Sarman, S., Evans, D.J., and Cummings, P.T. (1998) Recent developments in non-Newtonian molecular dynamics. *Phys. Rep.*, **305**, 1.

44 Rondoni, L. (2002) Deterministic thermostats and fluctuation relations, in *Dynamics of Dissipation*, Lecture Notes in Physics, vol. **597** (ed. R.P. Garbaczewski), Springer, Berlin.

45 Bonetto, F., Chernov, N.I., and Lebowitz, J.L. (1998) (Global and local) fluctuations of phase space contraction in deterministic stationary nonequilibrium. *Chaos*, **8**, 823.

46 Wagner, C., Klages, R., and Nicolis, G. (1999) Thermostating by deterministic scattering: heat and shear flow. *Phys. Rev. E*, **60**, 1401.

47 Wagner, C. (2000) Lyapunov instability for a hard-disk fluid in equilibrium and nonequilibrium thermostated by deterministic scattering. *J. Stat. Phys.*, **98**, 723.

48 Bonetto, F. and Lebowitz, J.L. (2001) Thermodynamic entropy production fluctuation in a two-dimensional shear flow model. *Phys. Rev. E*, **64**, 056129.

49 Bonetto, F., Daems, D., Lebowitz, J.L., and Ricci, V. (2002) Properties of stationary nonequilibrium states in the thermostatted periodic Lorentz gas: the multiparticle system. *Phys. Rev. E*, **65**, 051204.

50 Gallavotti, G. (1997) Dynamical ensemble equivalence in fluid mechanics. *Physica D*, **105**, 163.

51 Gallavotti, G., Rondoni, L., and Segre, E. (2004) Lyapunov spectra and nonequilibrium ensembles equivalence in 2D fluid mechanics. *Physica D*, **187**, 338.

52 Gallavotti, G. and Presutti, E. (2010) Nonequilibrium, thermostats, and thermodynamic limit. *J. Math. Phys.*, **51**, 015202.

53 Evans, D.J., Searles, D.J., and Rondoni, L. (2007) The steady state fluctuation relation for the dissipation function. *J. Stat. Phys.*, **128**, 1337.

54 Evans, D.J., Searles, D.J., and Williams, S.R. (2008) On the fluctuation theorem for the dissipation function and its connection with response theory. *J. Chem. Phys.*, **128**, 014504.

55 Evans, D.J., Cohen, E.G.D., and Morriss, G.P. (1993) Probability of second law violations in shearing steady flows. *Phys. Rev. Lett.*, **71**, 2401.

56 Parry, W. (1986) Synchronisation of canonical measures for hyperbolic attractors. *Commun. Math. Phys.*, **106**, 267.

57 Vance, W.N. (1992) Unstable periodic orbits and transport properties of nonequilibrium steady states. *Phys. Rev. Lett.*, **69**, 1356.

58 Cvitanović, P., Artuso, R., Mainieri, R., Tanner, G., and Vattay, G. (2005) *Chaos: Classical and Quantum*, ChaosBook.org, Niels Bohr Institute, Copenhagen, Denmark.

59 Searles, D.J., Ayton, G., and Evans, D.J. (2000) Generalised fluctuation formula. *AIP Conf. Ser.*, **519**, 271.

60 Searles, D.J. and Evans, D.J. (2000) Ensemble dependence of the transient fluctuation theorem. *J. Chem. Phys.*, **113**, 3503.

61 Williams, S.R., Searles, D.J., and Evans, D. J. (2004) Thermostat invariance of the transient fluctuation theorem. *Phys. Rev. E*, **70**, 066113.

62 Williams, S.R., Searles, D.J., and Evans, D. J. (2006) Numerical study of the steady state fluctuation relations far from equilibrium. *J. Chem. Phys.*, **124**, 194102.

63 Evans, D.J. and Searles, D.J. (1994) Equilibrium microstates which generate second law violating steady states. *Phys. Rev. E*, **50**, 1645–1648.

64 Evans, D.J. and Searles, D.J. (1995) Steady states, invariant measures, and response theory. *Phys. Rev. E*, **52**, 5839–5848.

65 Searles, D.J. and Evans, D.J. (2001) A fluctuation theorem for heat flow. *Int. J. Thermophys.*, **22**, 123.

66 Evans, D.J., Searles, D.J., and Williams, S.R. (2010) On the probability of violations of Fourier's law for heat flow in small systems observed for short times. *J. Chem. Phys.*, **132**, 024501.

67 Gallavotti, G. and Cohen, E.G.D. (1995) Dynamical ensembles in nonequilibrium statistical mechanics. *Phys. Rev. Lett.*, **94**, 2694.

68 Gallavotti, G. and Cohen, E.G.D. (1995) Dynamical ensembles in stationary states. *J. Stat. Phys.*, **80**, 931.

69 Gallavotti, G. (1995) Reversible Anosov diffeomorphisms and large deviations. *Math. Phys. Electron. J.*, **1**, 1.

70 Gallavotti, G. (2004) Comments on "On the application of the Gallavotti–Cohen fluctuation relation to thermostatted steady states near equilibrium". cond-mat/0402676.

71 Jarzynski, C. (1997) Nonequilibrium equality for free energy differences. *Phys. Rev. Lett.*, **78**, 2690.

72 Ruelle, D. (1999) Smooth dynamics and new theoretical ideas in nonequilibrium statistical mechanics. *J. Stat. Phys.*, **95**, 393.

73 Balkovsky, E., Falkovich, G., and Fouxon, A. (2001) Intermittent distribution of inertial particles in turbulent flows. *Phys. Rev. Lett.*, **86**, 2790.

74 Wang, G.M., Sevick, E.M., Mittag, E., Searles, D.J., and Evans, D.J. (2002) Experimental demonstration of violations of the second law of thermodynamics for small systems and short time scales. *Phys. Rev. Lett.*, **89**, 050601.

75 Garnier, N. and Ciliberto, S. (2005) Nonequilibrium fluctuations in a resistor. *Phys. Rev. E*, **71**, 060101.

76 Wang, G.M., Reid, J.C., Carberry, D.M., Williams, D.R.M., Sevick, E.M., and Evans, D.J. (2005) Experimental study of the fluctuation theorem in a nonequilibrium steady state. *Phys. Rev. E*, **71**, 046142.

77 Aurell, E., Mejía-Monasterio, C., and Muratore-Ginanneschi, P. (2011) Optimal protocols and optimal transport in stochastic thermodynamics. *Phys. Rev. Lett.*, **106**, 250601.

78 Evans, D.J. (2003) A non-equilibrium free energy theorem for deterministic systems. *Mol. Phys.*, **101**, 1551.

79 Searles, D.J., Johnston, B.M., Evans, D.J., and Rondoni, L. (2011) manuscript in preparation.

80 Sinai, Y.G. (1977) *Lectures in Ergodic Theory*, Lecture Notes in Mathematics, Princeton University Press.

81 Evans, D.J., Cohen, E.G.D., Searles, D.J., and Bonetto, F. (2000) Note on the Kaplan–Yorke dimension and linear transport coefficients. *J. Stat. Phys.*, **101**, 17.

82 Ruelle, D. (1997) Differentiation of SRB states. *Commun. Math. Phys.*, **187**, 227.

83 Dolowschiak, M. and Kovacs, Z. (2005) Fluctuation formula in the Nosé–Hoover thermostated Lorentz gas. *Phys. Rev. E*, **71**, 025202.

84 Zamponi, F., Ruocco, G., and Angelani, L. (2004) Fluctuations of entropy production in the isokinetic ensemble. *J. Stat. Phys.*, **115**, 1655.

85 Bonetto, F., Gallavotti, G., Giuliani, A., and Zamponi, F. (2006) Chaotic hypothesis, fluctuation theorem and singularities. *J. Stat. Phys.*, **123**, 39.

86 van Zon, R. and Cohen, E.G.D. (2003) Stationary and transient work-fluctuation theorems for a dragged Brownian particle. *Phys. Rev. E*, **67**, 046102.

87 Bonetto, F., Gallavotti, G., and Garrido, P. L. (1997) Chaotic principle: an experimental test. *Physica D*, **105**, 226.

88 Bonetto, F. and Gallavotti, G. (1997) Reversibility, coarse graining and the chaoticity principle. *Commun. Math. Phys.*, **189**, 263.

89 Giberti, C., Rondoni, L., and Vernia, C. (2007) Temporal asymmetry of fluctuations in the nonequilibrium FPU model. *Physica D*, **228**, 64.

90 Altmann, E.G. and Kantz, H. (2007) Hypothesis of strong chaos and anomalous diffusion in coupled symplectic maps. *Europhys. Lett.*, **78**, 10008.

91 Chechkin, A.V. and Klages, R. (2009) Fluctuation relations for anomalous dynamics. *J. Stat. Mech.*, L03002.

Experimental Foundations

4
Measuring Out-of-Equilibrium Fluctuations

L. Bellon, J. R. Gomez-Solano, A. Petrosyan, and Sergio Ciliberto

4.1
Introduction

In this chapter, we analyze the problem of the measure of the fluctuations of the injected and dissipated power in systems driven out of equilibrium by external forces. We study out-of-equilibrium processes in which the injected and dissipated energies are of the order of the thermal one. Therefore, thermal fluctuations cannot be neglected and play an important role in the dynamics. Our purpose is to describe several experiments performed to understand the practical applications and the limit of the theoretical approaches. We will focus on two major theoretical aspects: the fluctuation theorem (FT) and the fluctuation–dissipation theorem (FDT) in an out-of-equilibrium system. We follow an experimental approach and the connection with the theory will be done starting from the experimental measurements.

We start by an analysis of the experimental results on the energy fluctuations measured in a harmonic oscillator driven out of equilibrium by an external force. In Section 4.3, the experimental results on the harmonic oscillator are used to introduce the property of fluctuation theorems. In Section 4.4, the nonlinear case of a Brownian particle confined in a time-dependent double-well potential is presented. This is a very important case because many measurements in the literature are done for linear potentials and this is one of the few examples where FT is applied to a very nonlinear potential. In Section 4.5, we analyze the case of the application of the FT when the driving force is also random. We show that it is extremely important to compare the variance of the fluctuations imposed by the random driving with respect to those produced by the thermal forcing. The main result is that FT does not apply when the variance of the random forcing is larger than that of the thermal fluctuation. In Section 4.6, we will review the possible applications of FT and the experiments where these have been used. In Section 4.7, we will discuss another interesting aspect, which is the measure of the linear response in an out-of-equilibrium system. We will show that the new formulations of the FDT somehow related to FT are quite useful for this purpose because they

Nonequilibrium Statistical Physics of Small Systems: Fluctuation Relations and Beyond, First Edition.
Edited by Rainer Klages, Wolfram Just, and Christopher Jarzynski.

allow the estimation of the response starting from the measurement of fluctuations of different quantities in the steady state. Finally, we conclude in Section 4.7.

4.2
Work and Heat Fluctuations in the Harmonic Oscillator

The choice of discussing the dynamics of the harmonic oscillator is dictated by the fact that it is relevant for many practical applications such as the measure of the elasticity of nanotubes [1], the dynamics of the tip of an AFM [2], the MEMS, and the thermal rheometer that we developed several years ago to study the rheology of complex fluids [3, 4].

4.2.1
The Experimental Setup

This device is a very sensitive torsion pendulum as sketched in Figure 4.1a. It is composed of a brass wire (length 10 mm, width 0.5 mm, and thickness 50 μm) and a glass mirror with a golden surface (Figure 4.1c). The mirror (length 2 mm, width 8 mm, and thickness 1 mm) is glued in the middle of the brass wire. The elastic torsional stiffness of the wire is $C = 4.65 \times 10^{-4}$ N m rad^{-1}. It is enclosed in a cell (Figure 4.1d), which is filled with a fluid. We used either air or a water–glycerol mixture at 60% concentration. The system is a harmonic oscillator with resonant frequency $f_0 = \sqrt{C/I_{\text{eff}}}/2\pi = \omega_0/2\pi$ and a relaxation time $\tau_\alpha = 2I_{\text{eff}}/\nu = 1/\alpha$. I_{eff} is the total moment of inertia of the displaced masses (i.e., the mirror and the mass of displaced fluid) [5]. The damping has two contributions: the viscous damping ν of the surrounding fluid and the viscoelasticity of the brass wire.

The angular displacement of the pendulum θ is measured by a differential interferometer [6–9] that uses the two laser beams reflected by the mirror (Figure 4.1a). The measurement noise is two orders of magnitude smaller than thermal fluctuations of the pendulum. $\theta(t)$ is acquired with a resolution of 24 bits at a sampling rate of 8192 Hz, which is about 40 times f_0. We drive the system out of equilibrium by forcing it with an external torque M by means of a small electric current

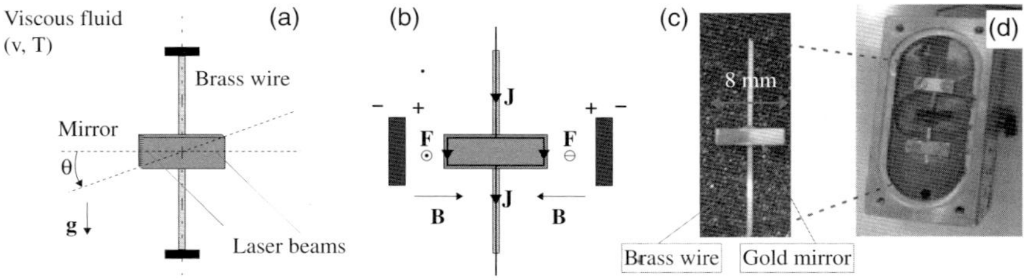

Figure 4.1 (a) The torsion pendulum. (b) The magnetostatic forcing. (c) Picture of the pendulum. (d) Cell where the pendulum is installed. (For a color version of this figure, please see the color plate at the end of this book.)

J flowing in a coil glued behind the mirror (Figure 4.1b). The coil is inside a static magnetic field. The displacements of the coil and therefore the angular displacements of the mirror are much smaller than the spatial scale of inhomogeneity of the magnetic field. So the torque is proportional to the injected current: $M = AJ$; the slope A depends on the geometry of the system. The practical realization of the montage is shown in Figure 4.1c and d. In equilibrium, the variance $\delta\theta^2$ of the thermal fluctuations of θ can be obtained from equipartition, that is, $\delta\theta = \sqrt{k_B T/C} \simeq 2$ nrad for our pendulum, where T is the temperature of the surrounding fluid.

4.2.2
The Equation of Motion

The dynamics of the torsion pendulum can be assimilated to that of a harmonic oscillator damped by the viscoelasticity of the torsion wire and the viscosity of the surrounding fluid, whose equation of motion reads in the temporal domain

$$I_{\text{eff}}\,\ddot{\theta} + \int_{-\infty}^{t} G(t-t')\dot{\theta}(t')\mathrm{d}t' + C\theta = M + \eta, \tag{4.1}$$

where G is the memory kernel and η is the thermal noise. In Fourier space (in the frequency range of our interest), this equation takes the simple form

$$[-I_{\text{eff}}\omega^2 + \hat{C}]\hat{\theta} = \hat{M} + \eta, \tag{4.2}$$

where ^ denotes the Fourier transform and $\hat{C} = C + i[C_1'' + \omega C_2'']$ is the complex frequency-dependent elastic stiffness of the system. C_1'' and C_2'' are the viscoelastic and viscous components of the damping term, respectively.

4.2.2.1 **Equilibrium**

At equilibrium, that is, $M = 0$, the fluctuation–dissipation theorem gives a relation between the amplitude of the thermal angular fluctuations of the oscillator and its response function. The response function of the system $\hat{\chi} = \hat{\theta}/\hat{M} = \hat{\theta}/A\hat{J}$ can be measured by applying a torque with a white spectrum. When $M = 0$, the amplitude of the thermal vibrations of the oscillator is related to its response function via the fluctuation–dissipation theorem. Therefore, the thermal fluctuation power spectral density (psd) of the torsion pendulum reads for positive frequencies

$$\langle|\hat{\theta}|^2\rangle = \frac{4k_{\text{B}}T}{\omega}\,\text{Im}\,\hat{\chi} = \frac{4k_{\text{B}}T}{\omega}\,\frac{C_1'' + \omega C_2''}{[-I_{\text{eff}}\omega^2 + C^2] + [C_1'' + \omega C_2'']^2}. \tag{4.3}$$

The angle brackets are ensemble averages. As an example, the spectrum of θ measured in the glycerol–water mixture is shown in Figure 4.2a. In this case, the resonance frequency is $f_0 = \sqrt{C/I_{\text{eff}}}/2\pi = \omega_0/2\pi = 217$ Hz and the relaxation time $\tau_\alpha = 2I_{\text{eff}}/\nu = 1/\alpha = 9.5$ ms. The measured spectrum is compared with that obtained from Eq. (4.3) using the measured χ. The viscoelastic component at low

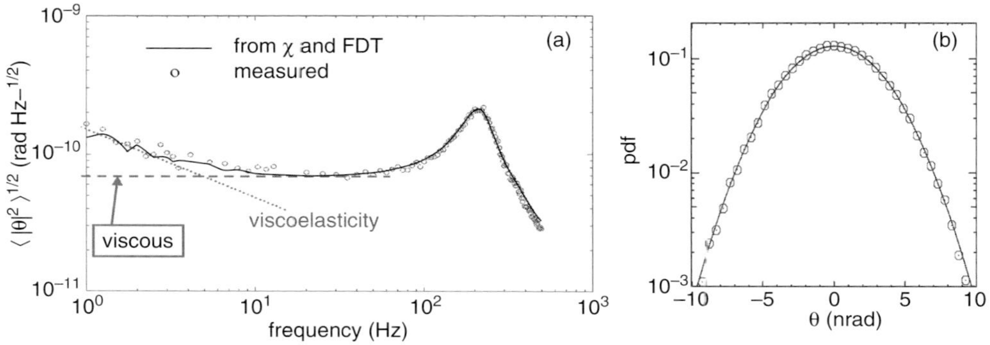

Figure 4.2 Equ librium: the pendulum inside a glycerol–water mixture with $M = 0$. (a) Square root of the power spectral density of θ. Open circles denote directly measured spectrum and the solid line is the spectrum estimated from the measure of χ and using Eq. (4.3). The dashed and dotted lines show the viscous and viscoelastic components of the damping, respectively. (b) Probability density function of θ. The continuous line is a Gaussian fit.

frequencies corresponds to a constant $C_1'' \neq 0$. Indeed, if $\omega \to 0$, then from Eq. (4.3)$\langle|\hat{\theta}|^2\rangle \propto 1/\omega$ as seen in Figure 4.2a. Instead, if $C_1'' = 0$, then for $\omega \to 0$ from Eq. (4.3) the spectrum is constant as a function of ω. It is important to stress that in the viscoelastic case the noise η is correlated and the process is not Markovian, whereas in the viscous case the process is Markovian. Thus, by changing the quality of the fluid surrounding the pendulum, one can tune the Markovian nature of the process. In the following, we will consider only the experiment in the glycerol–water mixture, where the viscoelastic contribution is visible only at very low frequencies and is therefore negligible. This allows a more precise comparison with theoretical predictions often obtained for Markovian processes. The probability density function (pdf) of θ, plotted in Figure 4.2b, is a Gaussian.

4.2.3
Nonequilibrium Steady State: Sinusoidal Forcing

We now consider a periodic forcing of amplitude M_0 and frequency ω_d, that is, $M(t) = M_0 \sin(\omega_d t)$ [8–11]. This is a very common kind of forcing that has been already studied in the case of the first-order Langevin equation [12] and the two-level system [13] and in a different context for the second-order Langevin equation [14]. Furthermore, this is a very general case because using Fourier transform, any periodical forcing can be decomposed into a sum of sinusoidal forcing. We explain here the behavior of a single mode. Experiments have been performed at various M_0 and ω_d. We present here the results for a particular amplitude and frequency: $M_0 = 0.78$ pN m and $\omega_d/2\pi = 64$ Hz. This torque is plotted in Figure 4.3a. The mean of the response to this torque is sinusoidal, with the same frequency, as can be seen in Figure 4.3b. The system is clearly in a nonequilibrium steady state (NESS).

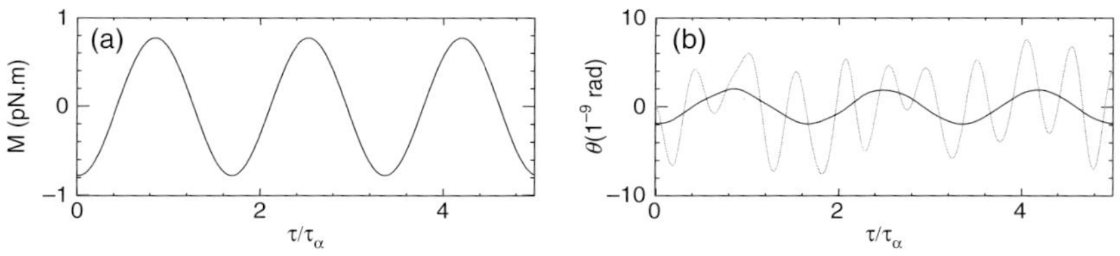

Figure 4.3 (a) Sinusoidal driving torque applied to the oscillator. (b) Response of the oscillator to this periodic forcing (gray line); the dark line represents the mean response $\langle\theta(t)\rangle$.

The work done by the torque $M(t)$ during a time $\tau_n = 2\pi n/\omega_d$ is

$$W_n = W_{\tau=\tau_n} = \int_{t_i}^{t_i+\tau_n} M(t)\frac{d\theta}{dt}dt. \tag{4.4}$$

As θ fluctuates, also W_n is a fluctuating quantity whose probability density function is plotted in Figure 4.4a for various n. This plot has interesting features. Specifically, work fluctuations are Gaussian for all values of n and W_τ takes negative values as long as τ_n is not too large. The probability of having negative values of W_τ decreases when τ_n is increased. There is a finite probability of having negative values of the work; in other words, the system may have an instantaneous negative entropy production rate although the average of the work $\langle W_n\rangle$ is of course positive ($\langle\cdot\rangle$ stands for ensemble average). In this specific example, $\langle W_n\rangle = 0.04n(k_BT)$. We now consider the energy balance for the system.

4.2.4
Energy Balance

As the fluid is rather viscous, we will take into account only the standard viscosity that is $C_1'' = 0$ and $C_2'' = \nu$. In such a case, Eq. (4.1) simplifies to

$$I_{\text{eff}}\frac{d^2\theta}{dt^2} + \nu\frac{d\theta}{dt} + C\theta = M + \eta, \tag{4.5}$$

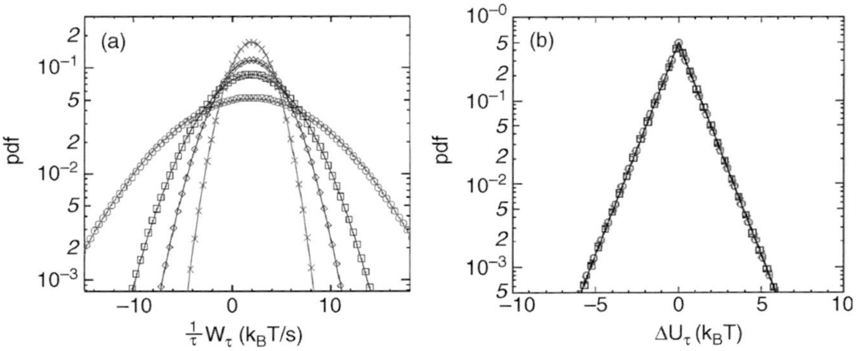

Figure 4.4 Sinusoidal forcing. (a) Pdf of W_τ: $n = 7$ (○), $n = 15$ (□), $n = 25$ (◇), and $n = 50$ (×). (b) Pdf of ΔU_τ.

where η is the thermal noise, which in this case is delta-correlated in time: $\langle \eta(t)\eta(t')\rangle = 2k_B T\nu\delta(t-t')$.

When the system is driven out of equilibrium using a deterministic torque, it receives some work and a fraction of this energy is dissipated into the heat bath. Multiplying Eq. (4.5) by $\dot{\theta}$ and integrating between t_i and $t_i+\tau$, one obtains a formulation of the first law of thermodynamics between the two states at times t_i and $t_i+\tau$ (Eq. (4.6)). This formulation has been first proposed in Ref. [15] and used in other theoretical and experimental works [12, 16]. The change in internal energy ΔU_τ of the oscillator over a time τ, starting at a time t_i, is written as

$$\Delta U_\tau = U(t_i+\tau) - U(t_i) = W_\tau - Q_\tau, \tag{4.6}$$

where W_τ is the work done on the system over a time τ:

$$W_\tau = \int_{t_i}^{t_i+\tau} M(t')\frac{d\theta}{dt}(t')dt' \tag{4.7}$$

and Q_τ is the heat dissipated by the system. The internal energy is the sum of the potential energy and the kinetic energy:

$$U(t) = \left\{\frac{1}{2}I_{\text{eff}}\left[\frac{d\theta}{dt}(t)\right]^2 + \frac{1}{2}C\theta(t)^2\right\}. \tag{4.8}$$

The heat transfer Q_τ is deduced from Eq. (4.6); it has two contributions:

$$\begin{aligned} Q_\tau &= W_\tau - \Delta U_\tau \\ &= \int_{t_i}^{t_i+\tau} \nu\left[\frac{d\theta}{dt}(t')\right]^2 dt' - \int_{t_i}^{t_i+\tau} \eta(t')\frac{d\theta}{dt}(t')dt'. \end{aligned} \tag{4.9}$$

The first term corresponds to the viscous dissipation and is always positive, whereas the second term can be interpreted as the work of the thermal noise that has a fluctuating sign. The second law of thermodynamics imposes $\langle Q_\tau\rangle$ to be positive.

4.2.5
Heat Fluctuations

The dissipated heat Q_τ cannot be directly measured because we have seen that Eq. (4.9) contains the work of the noise (the heat bath) that experimentally is impossible to measure, because η is unknown. However, Q_τ can be obtained indirectly from the measure of W_τ and ΔU_τ, whose pdfs measured during the periodic forcing are exponential for any τ, as shown in Figure 4.4b. We first give some comments on the average values. The average of ΔU_τ is obviously vanishing because the time τ is a multiple of the period of the forcing. Therefore, $\langle W_n\rangle$ and $\langle Q_n\rangle$ are equal.

We rescale the work W_τ (the heat Q_τ) by the average work $\langle W_\tau\rangle$ (the average heat $\langle Q_\tau\rangle$) and define $w_\tau = W_\tau/\langle W_\tau\rangle (q_\tau = Q_\tau/\langle Q_\tau\rangle)$. In this chapter, x_τ (respectively X_τ) stands for either w_τ or q_τ (respectively, W_τ or Q_τ).

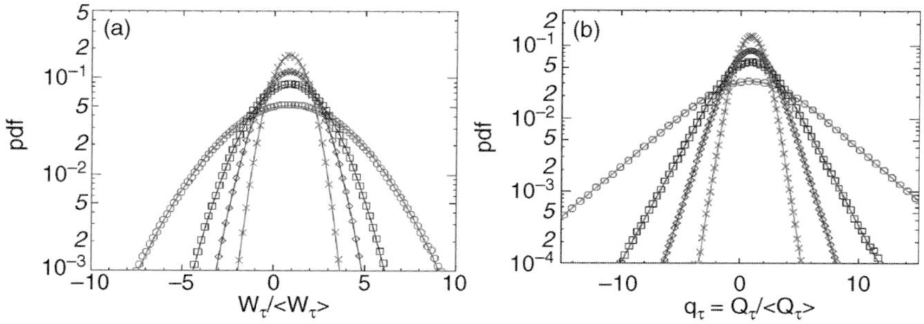

Figure 4.5 Sinusoidal forcing. (a) Pdf of W_τ and (b) pdf of Q_τ for various n: $n = 7$ (○), $n = 15$ (□), $n = 25$ (◇), and $n = 50$ (×). The continuous lines in the figure are not fits but are analytical predictions obtained from the Langevin dynamics as discussed in Section 4.3.3.

We compare now the pdfs of w_τ and q_τ in Figure 4.5. The pdfs of heat fluctuations q_n have exponential tails (Figure 4.5b). It is interesting to stress that although the two variables W_τ and Q_τ have the same mean values they have very different pdfs. The pdfs of w_τ are Gaussian, whereas those of q_τ are exponential.[1] On a first approximation, the pdfs of q_τ are the convolution of a Gaussian (the pdf of W_τ) and exponential (the pdf of ΔU_τ). In Figure 4.5, the continuous lines are analytical predictions obtained from the Langevin dynamics with no adjustable parameter (see Section 4.3.3).

4.3 Fluctuation Theorem

In the previous section, we have seen that both W_τ and Q_τ present negative values; that is, the second law is verified only on average but the entropy production can have instantaneously negative values. The probabilities of getting positive and negative entropy production are quantitatively related in nonequilibrium systems by the Fluctuation Theorems (FTs). [70–78].

There are two classes of FTs. The *stationary state fluctuation theorem* (SSFT) [18, 72–78] considers a nonequilibrium steady state. The *transient fluctuation theorem* (TFT) [72] describes transient nonequilibrium states, where τ measures the time since the system left the equilibrium state. A fluctuation relation (FR) examines the symmetry around 0 of the probability density function $p(x_\tau)$ of a quantity x_τ, as defined in the previous section. It compares the probability to have a positive event ($x_\tau = +x$) versus the probability to have a

1) This can be understood by considering that W is linear in θ and the potential U is quadratic. Thus, as θ is normally distributed, one expects for the distribution of W a Gaussian and for U and Q an exponential distribution (see Ref. [11]).

negative event ($x_\tau = -x$). We quantify the FT using a function S (symmetry function):

$$S(x_\tau) = \frac{k_B T}{\langle X_\tau \rangle} \ln \left(\frac{p(x_\tau = +x)}{p(x_\tau = -x)} \right). \tag{4.10}$$

The *transient fluctuation theorem* states that the symmetry function is linear with x_τ for any value of the time integration τ and the proportionality coefficient is equal to 1 for any value of τ:

$$S(x_\tau) = x_\tau, \quad \forall x_\tau, \quad \forall \tau. \tag{4.11}$$

Contrary to TFT, the *stationary state fluctuation theorem* holds only in the limit of infinite time (τ):

$$\lim_{\tau \to \infty} S(x_\tau) = x_\tau. \tag{4.12}$$

In the following, we will assume linearity at finite time τ [17, 18] and use the following general expression:

$$S(x_\tau) = \Sigma_x(\tau) x_\tau, \tag{4.13}$$

where for SSFT $\Sigma_x(\tau)$ takes into account the finite-time corrections and $\lim_{\tau \to \infty} \Sigma_x(\tau) = 1$, whereas $\Sigma_x(\tau) = 1$, $\forall \tau$ for TFT.

However, these claims are not universal because they depend on the kind of x_τ that is used. Specifically, we will see in the next sections that the results are not exactly the same if X_τ is replaced by any one of W_τ, Q_τ, and $T\Delta s_{\text{tot},\tau}$, defined in Section 4.3.2. Furthermore, the definitions given in this section are appropriate for stochastic systems and the differences between stochastic and chaotic systems will not be developed in this chapter. A discussion on this point can be found in Ref. [19].

4.3.1
FTs for Gaussian Variables

Let us suppose that the variable X_τ has a Gaussian distribution of mean $\langle X_\tau \rangle$ and variance $\sigma^2_{X_\tau}$. It is easy to show that in order to satisfy FTs, the variable X_τ must have the following statistical property:

$$\sigma^2_{X_\tau} = 2k_B T \langle X_\tau \rangle. \tag{4.14}$$

This is an interesting relation because it imposes that the relative fluctuations of X_τ are

$$\frac{\sigma_{X_\tau}}{X_\tau} = \sqrt{\frac{2k_B T}{\langle X_\tau \rangle}}. \tag{4.15}$$

This means that the probability of having negative events reduces by increasing X_τ; specifically, from Eq. (4.15) it follows that $P(X_\tau < 0) = \text{erfc}(\sqrt{\langle X_\tau \rangle / 2k_B T})$, where erfc is the complementary error function. It is now possible to estimate the length of the time interval $t_{\rm obs}$ needed to observe at least one negative event, which is

$$t_{obs} = \frac{\tau}{\mathrm{erfc}(\sqrt{\langle X_\tau \rangle / 2k_{\rm B}T})}, \tag{4.16}$$

where we used the fact that all the values W_τ computed on different intervals of length τ are independent, which is certainly true if τ is larger than the correlation time.

Let us consider the specific example of Section 4.2.3, that is, $\langle W_\tau \rangle = 0.04n(k_{\rm B}T)$ at $\omega_{\rm d}/2\pi = 67$ Hz, $M_0 = 0.78$ pN m, and $\tau = 2\pi n/\omega_{\rm d}$. The pdfs of W_τ are Gaussian in this case (Figure 4.4) and, as we will see in the next section, they satisfy SSFT for large τ. Therefore, Eq. (4.15) holds for $X_\tau = W_\tau$ and we may estimate t_{obs} in the asymptotic limit $\tau_\alpha \ll \tau$. For example, at $n = 200$, one obtains from the above-mentioned experimental values $\tau \simeq 3$ s $\gg \tau_\alpha$ and $\langle W_\tau \rangle = 8k_{\rm B}T$. Inserting these experimental values in Eq. (4.15), one gets roughly a negative event over an observational time $t_{\rm obs} \simeq 641$ s, which is already a rather long time for the distance between two events. For larger n and larger M_0, this time becomes exponentially large. This justifies the fact that millions of data are necessary in order to have a reliable measure of SSFT.

4.3.2 FTs for W_τ and Q_τ Measured in the Harmonic Oscillator

The questions we ask are whether for finite time FTs are satisfied for either $x_\tau = w_\tau$ or $x_\tau = q_\tau$ and what are the finite-time corrections. First, we test the correction to the proportionality between the symmetry function $S(x_\tau)$ and x_τ. In the region where the symmetry function is linear with x_τ, we define the slope $\Sigma_x(\tau)$, that is, $S(x_\tau) = \Sigma_x(\tau)x_\tau$. Second, we measure finite-time corrections to the value $\Sigma_x(\tau) = 1$, which is the asymptotic value expected from FTs.

In this chapter, we will focus on the SSFT applied to the experimental results of Section 4.2.3 and to other examples. The TFT will be not discussed here and the interested readers may look at Ref. [11].

From the pdfs of w_τ and q_τ plotted in Figure 4.5, we compute the symmetry functions defined in Eq. (4.10). The symmetry functions $S(w_n)$ are plotted in Figure 4.6a as a function of w_n. They are linear in w_n. The slope $\Sigma_w(n)$ is not equal to 1 for all n but there is a correction at finite time (Figure 4.7a). Nevertheless, $\Sigma_w(n)$ tends to 1 for large n. Thus, SSFT is satisfied for W_τ and for a sinusoidal forcing. The convergence is very slow and we have to wait a large number of periods of forcing for the slope to be 1 (after 30 periods, the slope is still 0.9). This behavior is independent of the amplitude of the forcing M_0 and consequently of the mean value of the work $\langle W_n \rangle$, which, as explained in Section 4.3.1, changes only the time needed to observe a negative event. The system satisfies the SSFT for all forcing frequencies $\omega_{\rm d}$ but finite-time corrections depend on $\omega_{\rm d}$, as can be seen in Figure 4.7a.

We now analyze the pdfs of q_τ (Figure 4.5b) and compute the symmetry functions $S(q_n)$ of q_n plotted in Figure 4.6b for different values of n. They are clearly

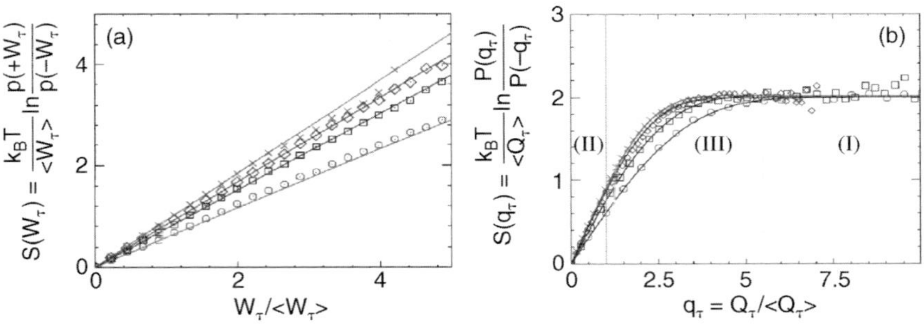

Figure 4.6 Sinusoidal forcing. Symmetry functions for SSFT. (a) Symmetry functions $S(w_\tau)$ plotted as a function of w_τ for various n: $n = 7$ (○), $n = 15$ (□), $n = 25$ (◇), and $n = 50$ (×). For all n, the dependence of $S(w_\tau)$ on w_τ is linear, with slope $\Sigma_w(\tau)$. (b) Symmetry functions $S(q_\tau)$ plotted as a function of q_τ for various n. The dependence of $S(q_\tau)$ on q_τ is linear only for $q_\tau < 1$. Continuous lines are theoretical predictions.

very different from those of w_n plotted in Figure 4.6a. For $S(q_n)$, three different regions appear:

I) For large fluctuations q_n, $S(q_n)$ equals 2. When τ tends to infinity, this region spans from $q_n = 3$ to infinity.

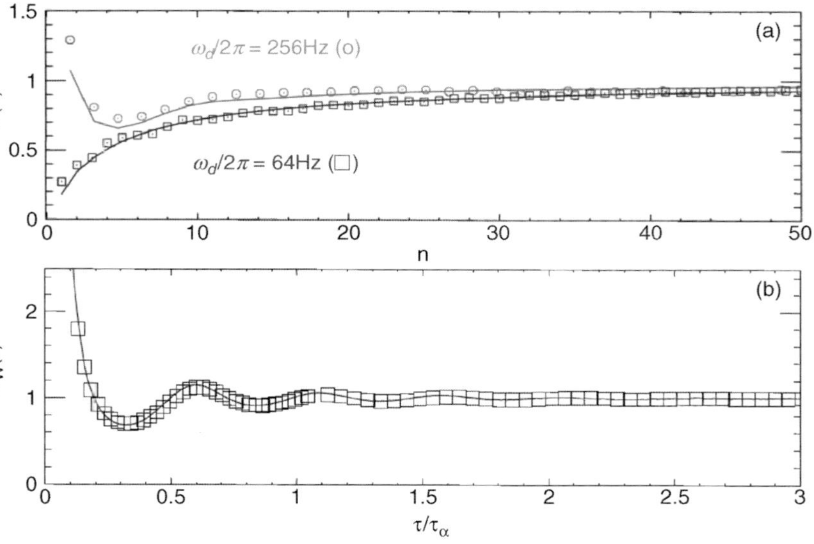

Figure 4.7 Finite-time corrections for SSFT. (a) Sinusoidal forcing. $\Sigma_w(\tau)$ as a function of n obtained from the slopes of the straight lines of Figure 4.6a (□). The circles correspond to another measurement performed at a different frequency. The finite-time corrections depend on the driving frequency. The slopes $\Sigma_q(\tau)$ measured for $q_\tau < 1$ (Figure 4.6b) have exactly the same values of $S_w(\tau)$ as a function of n. (b) Linear forcing. $\Sigma_w(\tau)$ measured as a function of τ with the driving torque M has a linear dependence on time. The finite-time corrections depend on the form of driving.

II) For small fluctuations q_n, $S(q_n)$ is a linear function of q_n. We then define $\Sigma_q(n)$ as the slope of the function $S(q_n)$, that is, $S(q_n) = \Sigma_q(n)q_n$. We have measured [11] that $\Sigma_q(n) = \Sigma_w(n)$ for all the values of n; that is, finite-time corrections are the same for heat and work. Thus, $\Sigma_q(n)$ tends to 1 when τ is increased and SSFT holds in this region that spans from $q_n = 0$ up to $q_n = 1$ for large τ. This effect has been discussed for the first time in Refs [17, 18].

III) A smooth connection between the two behaviors.

These regions define the fluctuation relation from the heat dissipated by the oscillator. The limit for large τ of the symmetry function $S(q_\tau)$ is rather delicate and has been discussed in Ref. [11].

The conclusions of this experimental analysis are that SSFT holds for work for any value of w_τ whereas for heat it holds only for $q_\tau < 1$. The finite-time corrections to FTs, described by $1 - \Sigma$, are not universal. They are the same for both w_τ and q_τ but they depend on the driving frequency as shown in Figure 4.7a. Furthermore, they depend on the kind of driving force. In Figure 4.7b, we plot $\Sigma_w(\tau)$ measured when the harmonic oscillator is driven out of equilibrium by a linear ramp.[2)] The difference with respect Figure 4.7a is quite evident.

4.3.3 Comparison with Theory

This experimental analysis allows a very precise comparison with theoretical predictions using the Langevin equation (Eq. (4.5)) and using two experimental observations: (a) the properties of heat bath are not modified by the driving and (b) the fluctuations of the W_τ are Gaussian (see also Ref. [20], where it is shown that in Langevin dynamics W_τ has a Gaussian distribution for any kind of deterministic driving force if the properties of the bath are not modified by the driving and the potential is harmonic). The observation in point (a) is extremely important because it is always assumed to be true in all the theoretical analysis. In Ref. [11], this point has been precisely checked. Using these experimental observations, one can compute the pdf of q_τ and the finite-time corrections $\Sigma(\tau)$ to SSFT (see Ref. [11]). The continuous lines in Figures 4.4,4.6, and 4.7 are not fit but analytical predictions, with no adjustable parameters, derived from the Langevin dynamics of Eq. (4.5) (see Ref. [11] for more details).

4.3.4 Trajectory-Dependent Entropy

In previous sections, we have studied the energy W_τ injected into the system during the time τ and the energy dissipated toward the heat bath Q_τ. These two quantities and the internal energy are related by the first law of thermodynamics (Eq. (4.9)).

2) The stationarity in the case of a ramp is discussed in Refs [11, 18].

Following notations of Ref. [21], we define the entropy variation in the system during a time τ as

$$\Delta s_{\mathrm{m},\tau} = \frac{1}{T} Q_\tau. \tag{4.17}$$

For thermostated systems, entropy change in medium behaves like the dissipated heat. The nonequilibrium Gibbs entropy is

$$\langle s(t)\rangle = -k_\mathrm{B} \int \mathrm{d}\vec{x}\ p(\vec{x}\,(t), t, \lambda_t) \ln p(\vec{x}\,(t), t, \lambda_t), \tag{4.18}$$

where λ_t denotes the set of control parameters at time t and $p(\vec{x}\,(t), t, \lambda_t)$ is the probability density function to find the particle at a position $\vec{x}\,(t)$ at time t, for the state corresponding to λ_t. This expression allows the definition of a "trajectory-dependent" entropy:

$$s(t) \equiv -k_\mathrm{B} \ln p(\vec{x}\,(t), t, \lambda_t). \tag{4.19}$$

The variation $\Delta s_{\mathrm{tot},\tau}$ of the total entropy s_tot during a time τ is the sum of the entropy change in the system during τ and the variation of the "trajectory-dependent" entropy during a time τ, $\Delta s_\tau \equiv s(t+\tau) - s(t)$:

$$\Delta s_{\mathrm{tot},\tau} \equiv s_\mathrm{tot}(t+\tau) - s_\mathrm{tot}(t) = \Delta s_{\mathrm{m},\tau} + \Delta s_\tau. \tag{4.20}$$

In this section, we study fluctuations of $\Delta s_{\mathrm{tot},\tau}$ computed using (4.17) and (4.19). We will show that $\Delta s_{\mathrm{tot},\tau}$ satisfies a SSFT for all τ. In Ref. [22], the relevance of boundary terms such as Δs_τ has been pointed out for Markovian processes.

We investigate the data of the harmonic oscillator described in Section 4.2.3. The probability to compute is the joint probability $p(\theta(t_\mathrm{i}+\tau_n), \dot{\theta}(t_\mathrm{i}+\tau_n), \varphi)$, where φ is the starting phase $\varphi = t_\mathrm{i}\omega_\mathrm{d}$. As the system is described by a second-order linear differential equation, $\theta(t_\mathrm{i}+\tau_n)$ and $\dot{\theta}(t_\mathrm{i}+\tau_n)$ are two independent degrees of freedom, and thus the joint probability can be factorized into a product. The expression of the trajectory-dependent entropy is

$$\Delta s_{\tau_n} = -k_\mathrm{B} \ln\left(\frac{p(\theta(t_\mathrm{i}+\tau_n), \varphi)\ p(\dot{\theta}(t_\mathrm{i}+\tau_n, \varphi))}{p(\theta(t_\mathrm{i}), \varphi)\ p(\dot{\theta}(t_\mathrm{i}, \varphi))}\right). \tag{4.21}$$

For computing correctly the trajectory-dependent entropy, we have to calculate the $p(\theta(t_\mathrm{i}), \varphi)$ and $p(\dot{\theta}(t_\mathrm{i}), \varphi)$ for each initial phase φ (see Figure 4.8a). These distributions turn out to be independent of φ and they correspond to the equilibrium fluctuations of θ and $\dot{\theta}$ around the mean trajectory defined by $\langle\theta(t)\rangle$ and $\langle\dot{\theta}(t)\rangle$. The distribution of $\theta(t_\mathrm{i})$ is plotted in Figure 4.8b, where the continuous line corresponds to the equilibrium distribution. Once the $p(\theta(t_\mathrm{i}, \varphi)$ and $p(\dot{\theta}(t_\mathrm{i}), \varphi)$ are determined, we compute the "trajectory-dependent" entropy. As fluctuations of θ and $\dot{\theta}$ are independent of φ, we can average Δs_{τ_n} over φ, which improves a lot the statistical accuracy. We stress that it is not equivalent to calculate first the pdfs over all values of φ – which would correspond here to the convolution of the pdf of the fluctuations with the pdf of a periodic signal – and then compute the trajectory-dependent entropy. The results are shown in Figure 4.9.

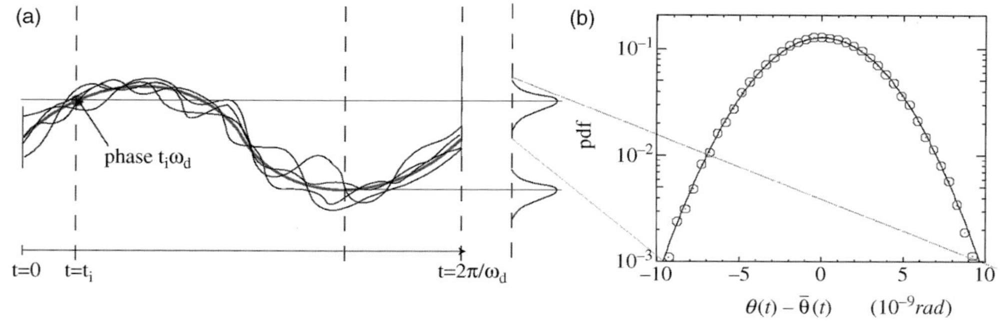

Figure 4.8 (a) Schematic diagram illustrating the method to compute the trajectory-dependent entropy. (b) Pdf of $\theta(t)$ around the mean trajectory $\langle\theta(t)\rangle$. The continuous line is the equilibrium distribution.

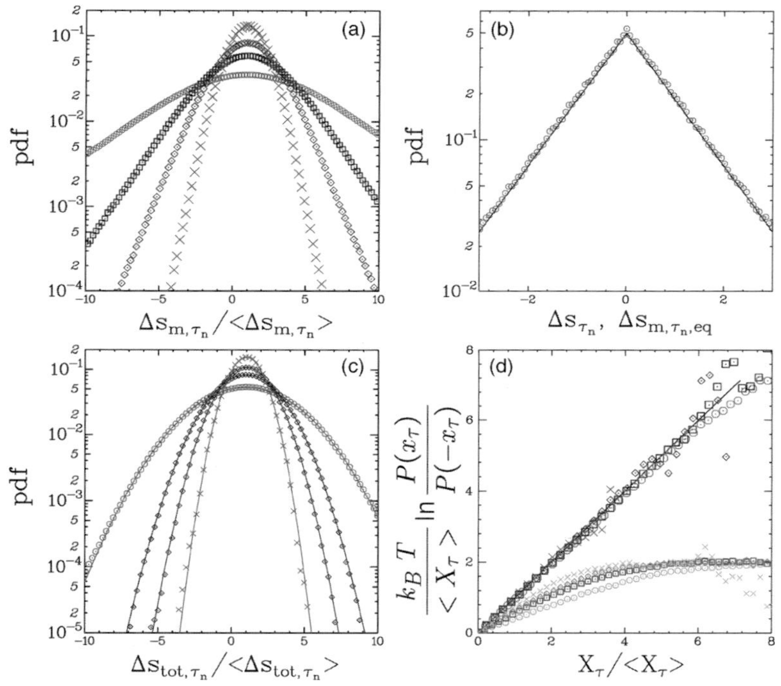

Figure 4.9 Torsion pendulum. (a) Pdfs of the normalized entropy variation $\Delta s_{m,\tau_n}/\langle\Delta s_{m,\tau_n}\rangle$ integrated over n periods of forcing, with $n = 7$ (○), $n = 15$ (□), $n = 25$ (◇), and $n = 50$ (×). (b) Pdfs of Δs_{τ_n}; the distribution is independent of n and here $n = 7$. Continuous line is the theoretical prediction for equilibrium entropy exchanged with thermal bath $\Delta s_{m,\tau_n,eq}$. (c) Pdfs of the normalized total entropy $\Delta s_{tot,\tau_n}/\langle\Delta s_{tot,\tau_n}\rangle$, with $n = 7$ (○), $n = 15$ (□), $n = 25$ (◇), and $n = 50$ (×). (d) Symmetry functions for the normalized entropy variation in the system (small symbols in light colors and X_τ stands for $T\Delta s_{m,\tau_n} = Q_\tau$) and for the normalized total entropy (large symbols in dark colors and X_τ stands for $T\Delta s_{tot,\tau_n}$) for the same values of n.

In Figure 4.9a, we recall the main results for the dissipated heat $Q_\tau = T\Delta s_{\mathrm{m},\tau_n}$. Its average value $\langle T\Delta s_{\mathrm{m},\tau_n}\rangle$ is linear in τ_n and equal to the injected work. The pdfs of $T\Delta s_{\mathrm{m},\tau_n}$ are not Gaussian and extreme events have an exponential distribution. The pdf of the "trajectory-dependent" entropy is plotted in Figure 4.9b; it is exponential and independent of n. We superpose to it the pdf of the variation of internal energy divided by T at equilibrium: the two curves match perfectly within experimental errors, so the "trajectory-dependent" entropy can be considered as the entropy that would be exchanged with the thermostat if the system were at equilibrium. The average value of Δs_{τ_n} is zero, so the average value of the total entropy is equal to the average of injected power divided by T. In Figure 4.9c, we plot the pdfs of the normalized total entropy for four typical values of integration time. We find that the pdfs are Gaussian for any time.

The symmetry functions (Eq. (4.10)) of the dissipated heat $S(T\Delta s_{\mathrm{m},\tau_n} = Q_\tau)$ and the total entropy $S(T\Delta s_{\mathrm{tot},\tau_n})$ are plotted in Figure 4.9d. As we have already seen in Figure 4.6, $S(Q_\tau)$ is a nonlinear function of $Q_\tau = T\,\Delta s_{\mathrm{m},\tau}$. The linear behavior, with a slope that tends to 1 for large time, is observed only for $\Delta s_{\mathrm{m},\tau_n} < \langle\Delta s_{\mathrm{m},\tau_n}\rangle < 1$. For the normalized total entropy, the symmetry functions are linear with $\Delta s_{\mathrm{tot},\tau_n}$ for all values of $\Delta s_{\mathrm{tot},\tau_n}$ and the slope is equal to 1 for all values of τ_n. Note that it is not exactly the case for the first values of τ_n because these are the times over which the statistical errors are the largest and the error in the slope is large.

For the harmonic oscillator, we have obtained that the "trajectory-dependent" entropy can be considered as the entropy variation in the system during a time τ that one would have if the system were at equilibrium. Therefore, the total entropy is the additional entropy due to the presence of the external forcing: *this is the part of entropy that is created by the nonequilibrium stationary process.* The total entropy (or excess entropy) satisfies the fluctuation theorem for all times and for all kinds of stationary external torque [21, 22]. More details on this problem can be found in Ref. [23].

4.4 The Nonlinear Case: Stochastic Resonance

The harmonic oscillator cannot be driven to a nonlinear regime without forcing it to such a high level where thermal fluctuations become negligible. Thus, in order to study the nonlinear effects, we change experiment and measure the fluctuations of a Brownian particle trapped in a nonlinear potential produced by two laser beams, as shown in Figure 4.10. It is very well known that a particle of small radius $R \simeq 2\,\mu\mathrm{m}$ is trapped in the focus of a strongly focused laser beam, which produces a harmonic potential for the particle, whose Brownian motion is confined inside this potential well. When two laser beams are focused at a distance $D \simeq R$, as shown in Figure 4.10a, the particle has two equilibrium positions, that is, the foci of the two beams. Thermal fluctuations may force the particle to move from one to the other. The particle feels an equilibrium potential $U_0(x) = ax^4 - bx^2 - dx$, shown in Figure 4.10b, where a, b, and d are determined by the laser intensity and

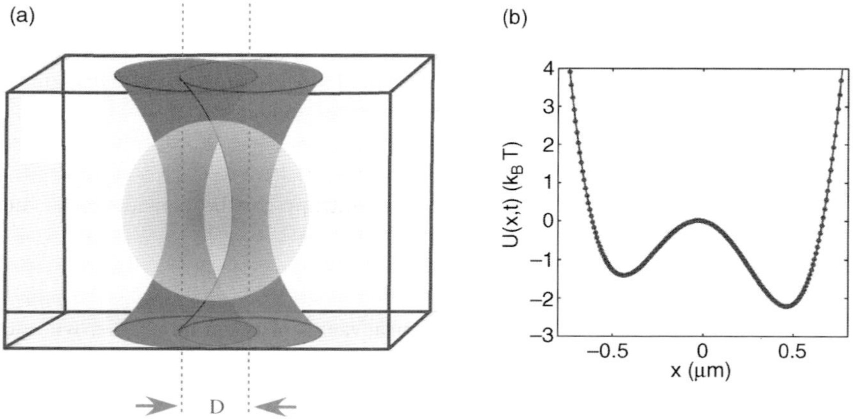

Figure 4.10 (a) Drawing of the polystyrene particle trapped by two laser beams whose axis distance is about the radius of the bead. (b) Potential felt by the bead trapped by the two laser beams. The barrier height between the two wells is about $2k_BT$.

by the distance of the two focal points. This potential has been computed from the measured equilibrium distribution of the particle $P(x) \propto \exp(U_0(x))$. The right–left asymmetry of the potential (Figure 4.10b) is induced by small unavoidable asymmetries, induced by the optics focusing the two laser beams. In our experiment, the distance between the two spots is 1.45 μm, which produces a trap whose minima are at $x_{\min} = \pm 0.45$ μm. The total intensity of the laser is 29 mW on the focal plane, which corresponds to an interwell barrier energy $\delta U_0 = 1.8\,k_BT$, $ax^4_{\min} = 1.8_BT$, $bx^2_{\min} = 3.6k_BT$, and $d|x_{\min}| = 0.44k_BT$ (see Ref. [24] for more experimental details). The rate at which the particle jumps from one potential well to the other is determined by the Kramers' rate $r_K = (1/\tau_0)\exp(-\delta U_0/k_BT)$, where τ_0 is a characteristic time. In our experiment, $r_K \simeq 0.3$ Hz at 300 K.

To drive the system out of equilibrium, we periodically modulate the intensity of the two beams at low frequency. Thus, the potential felt by the bead is the following profile:

$$U(x,t) = U_0(x) + U_p(x,t) = U_0 + cx\sin(2\pi ft), \tag{4.22}$$

with $c|x_{\min}| = 0.81k_BT$. The amplitude of the time-dependent perturbation is synchronously acquired with the bead trajectory.[3)]

An example of the measured potential for $t = 1/4f$ and $3/4f$ is shown in Figure 4.11a. This figure is obtained by measuring the probability distribution function $P(x,t)$ of x for fixed values of $c\sin(2\pi ft)$; it follows that $U(x,t) = -\ln(P(x,t))$.

3) The parameters given here are average parameters since the coefficients a, b, and c, obtained from fitted steady distributions at given phases, vary with the phase ($\delta a/a \approx 10\%$, $\delta b/b \approx \delta c/c \approx 5\%$).

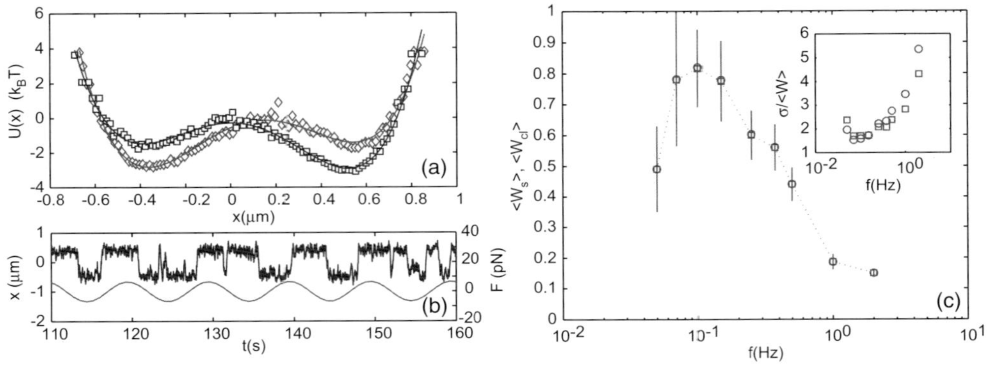

Figure 4.11 (a) The perturbed potential at $t = 1/4f$ and $t = 3/4f$. (b) Example of trajectory of the glass bead and the corresponding perturbation at $f = 0.1$ Hz. (c) Mean injected energy in the system over a single period as a function of the driving frequency. $\langle W_s \rangle$ (□) and $\langle W_{cl} \rangle$ (○) coincide as their mean values are equal within experimental errors. The error bars are computed from the standard deviation of the mean over different runs. *Inset*: Standard deviations of work distributions over a single period normalized by the average work as a function of the frequency (same symbols).

The x position of the particle can be described by a Langevin equation

$$\gamma \dot{x} = -\frac{\partial U(x,t)}{\partial x} + \eta, \tag{4.23}$$

with $\gamma = 1.61 \times 10^{-8}$ N s m^{-1} the friction coefficient and η the thermal noise delta correlated in time. When $c \neq 0$, the particle can experience a stochastic resonance [25], when the forcing frequency is close to the Kramers' rate. An example of the sinusoidal force with the corresponding position is shown in Figure 4.11b. Since the synchronization is not perfect, sometimes the particle receives energy from the perturbation, and sometimes the bead moves against the perturbation leading to a negative work on the system. Two kinds of work can be defined in these experiments [24]:

$$W_{s,n}(t) = \int_t^{t+t_f} dt \frac{\partial U(x,t)}{\partial t}, \tag{4.24}$$

$$W_{cl,n}(t) = -\int_t^{t+t_f} dt\, \dot{x} \frac{\partial U_p(x,t)}{\partial x}, \tag{4.25}$$

where in this case $t_f = n/f$ is a multiple of the forcing period. The work $W_{s,n}$ is the stochastic work (used in Jarzynski and Crooks relations [7, 26, 27]) and $W_{cl,n}$ is the classical work that will be discussed in this chapter. The results on $W_{s,n}$ are quite similar but there are subtle differences, which are discussed in Ref. [24].

We first measure the average work received over one period for different frequencies ($t_f = 1/f$ in Eq. (4.25)). Each trajectory is here recorded during 3200 s in different consecutive runs, which corresponds to 150 up to 6400 forcing periods, for the range of frequencies explored. In order to increase the statistics, we consider 10^5 different t_0. Figure 4.11c shows the evolution of the mean work per period for both definitions of the work. First, the input average work decreases to zero when the frequency tends to zero. Indeed, the bead hops randomly several times between the two wells during the period. Second, in the limit of high frequencies, the particle does not have the time to jump on the other side of the trap but rather stays in the same well during the period; thus, the input energy again decreases with increasing frequency. In the intermediate regime, the particle can almost synchronize with the periodical force and follows the evolution of the potential. The maximum of injected work is found around the frequency $f \approx 0.1$ Hz, which is comparable with half of the Kramers' rate of the fixed potential $r_K = 0.3$ Hz. This maximum of transferred energy shows that the stochastic resonance for a Brownian particle is a bona fide resonance, as it was previously shown in experiments using resident time distributions [28, 29] or directly in simulations [30, 31]. In the inset of Figure 4.11, we plot the normalized standard deviation of work distributions ($\sigma/\langle W\rangle$) as a function of the forcing frequency. The curves present a minimum at the same frequency of 0.1 Hz, in agreement again with the resonance phenomena.

In order to study FT for stochastic resonance, we choose for the external driving a frequency $f = 0.25$ Hz, which ensures a good statistic, by allowing the observation of the system over a sufficient number of periods. We compute the works and the dissipation using 1.5×10^6 different t on a time series that spans about 7500 periods of the driving.

We consider the pdf $P(W_{\text{cl}})$ that is plotted in Figure 4.12a. Notice that for small n the distributions are double peaked and very complex. They tend to a Gaussian for large n (inset of Figure 4.12a). In Figure 4.12b, we plot the normalized symmetry function of $W_{\text{cl},n}$. We can see that the curves are close to the line of slope 1. For high values of work, the dispersion of the data increases due to the lack of events. The slope tends toward 1 as expected by the SSFT. It is remarkable that straight lines are obtained even for n close to 1, where the distribution presents a very complex and unusual shape (Figure 4.12a). We do not discuss here the case of $W_{s,n}$ as

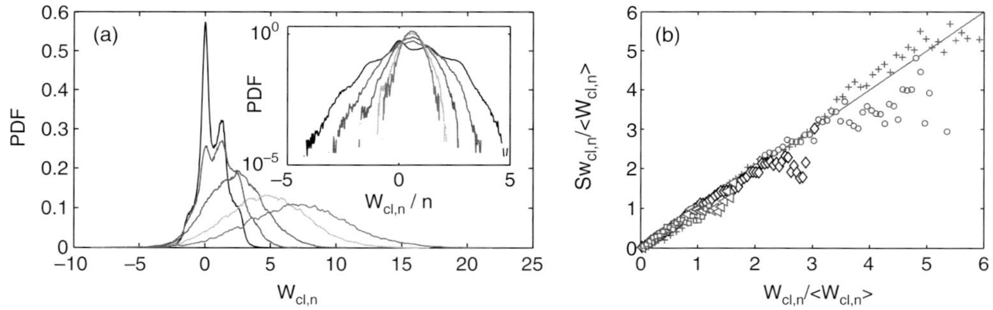

Figure 4.12 (a) Distribution of classical work W_{cl} for different numbers of period $n = 1, 2, 4, 8$, and 12 ($f = 0.25$ Hz). *Inset*: Same data in lin-log. (b) Normalized symmetry function as a function of the normalized work for $n = 1$ (+), 2 (○), 4 (◇), 8 (Δ), and 12 (□). (For a color version of this figure, please see the color plate at the end of this book.)

the behavior is quite similar to that of $W_{\mathrm{cl},n}$ [24]. The very fast convergence to the asymptotic value of the SSFT is quite striking in this example. The measurements are in full agreement with a realistic model based on the Fokker–Planck equations where the measured values of $U(x,t)$ have been inserted [32]. This example shows the application of FT in a nonlinear case where the distributions are strongly non-Gaussian.

4.5
Random Driving

In the two experiments described in the previous sections, the force that drives the system out of equilibrium is inherently deterministic. However, it has been recently argued that the nature (deterministic or stochastic) of the forcing can play an important role in the distribution of the injected work leading to possible deviations from the relation (4.10) for large fluctuations ($W_\tau/\langle W_\tau \rangle > 1$). Indeed, it has been found in experiments and simulations such as a Brownian particle in a Gaussian white [33] and colored [34] noise bath, turbulent thermal convection [35, 79], wave turbulence [36], a vibrating metallic plate [37], an RC electronic circuit [38], and a gravitational wave detector [39] that the probability density functions of the work done by a stochastic force are not Gaussian but asymmetric with two exponential tails leading to violations of the FT in the form of effective temperatures or nonlinear relations between the left- and the right-hand side of Eq. (4.12). It is important to remark that in the systems previously cited the steady-state FT is violated because in such a case the external random force acts itself as a kind of thermal bath. One question that naturally arises is what the work fluctuation relations will become when in addition to the external random forcing a true thermalization process is allowed. In this situation, there are two sources of work fluctuations: the external force and the thermal bath. As pointed out in Refs [37, 40], one is interested in the distribution of the work fluctuations done by the external random force in the presence of a thermostat and the conditions under which the FT could be valid.

In this section, we address these questions in two experimental systems: a Brownian particle in an optical trap and a microcantilever used for atomic force microscopy (AFM). Both are in contact with a thermal bath and driven by an external random force whose amplitude is tuned from a small fraction to several times the amplitude of the intrinsic thermal fluctuations exerted by the thermostat.

4.5.1
Colloidal Particle in an Optical Trap

The first system we study consists of a spherical silica bead of radius $r = 1\,\mu\mathrm{m}$ immersed in ultrapure water that acts as a thermal bath. The experiment is performed at a room temperature of $27 \pm 0.5\,^\circ\mathrm{C}$ at which the dynamic viscosity of water is $\eta = (8.52 \pm 0.10) \times 10^{-4}$ Pa s. The motion of the particle is confined by an

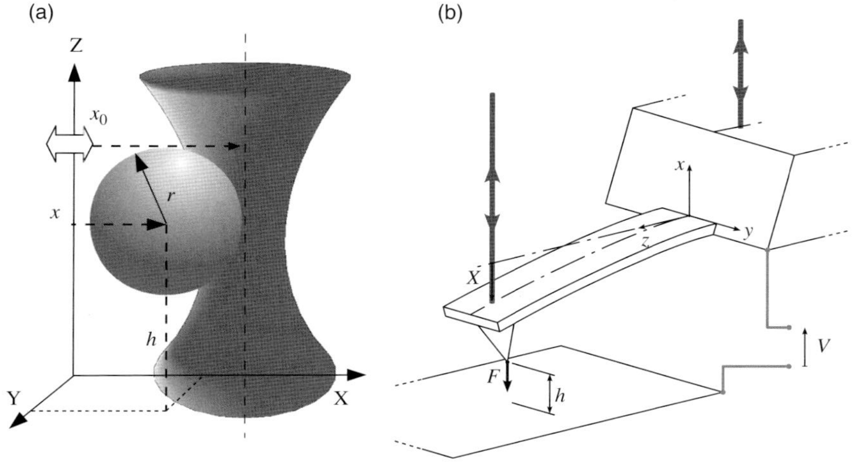

Figure 4.13 (a) Colloidal particle in the optical trap with modulated position. (b) AFM cantilever close to a metallic surface. See the text for explanation.

optical trap that is created by tightly focusing a Nd:YAG laser beam ($\lambda = 1064$ nm) by means of a high numerical aperture objective ($63\times$, NA = 1.4). The trap stiffness is fixed at a constant value of $k = 5.4$ pN μm^{-1}. The particle is kept at $h =$ 10 μm above the lower cell surface to avoid hydrodynamic interactions with the boundary. Figure 4.13a sketches the configuration of the bead in the optical trap. An external random force is applied to the particle by modulating the position of the trap $x_0(t)$ using an acousto-optic deflector, along a given direction on the plane perpendicular to the beam propagation. The modulation corresponds to a Gaussian Ornstein–Uhlenbeck noise of mean $\langle x_0(t)\rangle = 0$ and covariance $\langle x_0(s)x_0(t)\rangle = A\exp(-|t-s|/\tau_0)$. The correlation time of the modulation is set to $\tau_0 = 25$ ms whereas the value of its amplitude A is tuned to control the driving intensity. We determine the particle barycenter (x, y) by image analysis using a high-speed camera at a sampling rate of 1 kHz with an accuracy better than 10 nm. See Ref. [41] for more details about the experimental apparatus. The attractive force exerted by the optical trap on the bead at time t along the direction of the beam modulation is given by $-k(x(t) - x_0(t))$. Hence, for the experimentally accessible timescales, the dynamics of the coordinate x is described by the overdamped Langevin equation

$$\gamma\dot{x} = -kx + \zeta_T + f_0. \tag{4.26}$$

In Eq. (4.26), $\gamma = 6\pi r\eta$ is the viscous drag coefficient, ζ_T is as usual a Gaussian white noise ($\langle\zeta_T\rangle = 0$, $\langle\zeta_T(s)\zeta_T(t)\rangle = 2k_BT\gamma\delta(t-s)$) that mimics the collisions of the thermal bath particles with the colloidal bead, and $f_0(t) = kx_0(t)$ plays the role of the external stochastic force. The standard deviation of f_0 (δf_0) is chosen as the main control parameter of the system. Besides the correlation time τ_0 of f_0, there is a second characteristic timescale in the dynamics of Eq. (4.26): the viscous relaxation time in the optical trap $\tau_\gamma = \gamma/k = 3$ ms $< \tau_0$. In order to quantify the relative strength of the external force with respect to the thermal fluctuations, we introduce a dimensionless parameter that measures the distance from equilibrium:

$$\alpha = \frac{\langle x^2 \rangle}{\langle x^2 \rangle_{\text{eq}}} - 1, \tag{4.27}$$

where $\langle x^2 \rangle$ is the variance of x in the presence of $f_0 > 0$, whereas $\langle x^2 \rangle_{\text{eq}}$ is the corresponding variance at equilibrium ($f_0 = 0$). The dependence of α on δf_0 is quadratic, as shown in Figure 4.14a.

The work done by the external random force on the colloidal particle (in $k_B T$ units) is

$$w_\tau = \frac{1}{k_B T} \int_t^{t+\tau} \dot{x}(t') f_0(t') \mathrm{d}t'. \tag{4.28}$$

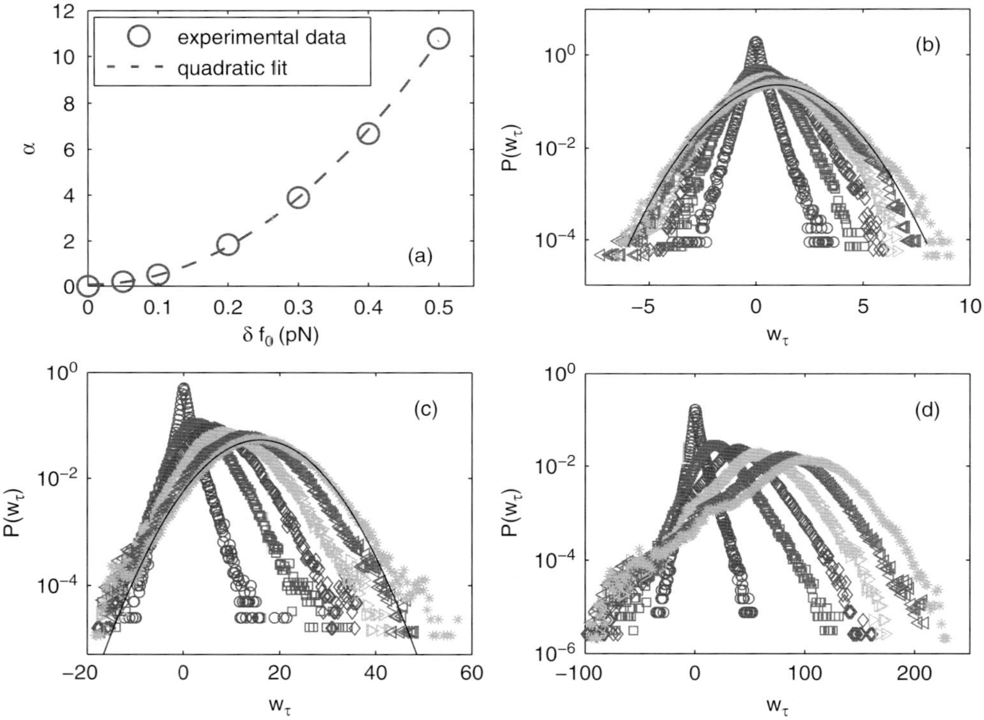

Figure 4.14 (a) Dependence of the parameter α on the standard deviation of the Gaussian exponentially correlated external force f_0 acting on the colloidal particle. Probability density functions of the work w_τ for (b) $\alpha = 0.20$, (c) $\alpha = 3.89$, and (d) $\alpha = 10.77$. The symbols correspond to integration times $\tau = 5$ ms (○), 55 ms (□), 105 ms (◇), 155 ms (◁), 205 ms (▷), and 255 ms (∗). The solid black lines in (b) and (c) are Gaussian fits. (For a color version of this figure, please see the color plate at the end of this book.)

Thus, by measuring simultaneously the time evolution of the barycenter position of the particle and the driving force, we are able to compute directly the work injected on the system by the driving. In Figure 4.14b–d, we show the probability density functions of w_τ for different values of τ and α. We observe that for a fixed value of α, the pdfs have asymmetric exponential tails at short integration times and they become smoother as the value of τ increases. For $\alpha = 0.20$ they approach an approximately Gaussian profile (Figure 4.14b), whereas asymmetric non-Gaussian tails remain for increasing values of α. As shown in Figure 4.14c and d, the asymmetry of these tails becomes very pronounced for large $\alpha > 1$ even for integration times as long as $\tau = 250\,\text{ms} = 10\tau_0$, where we have taken τ_0 because it is the largest correlation time of the dynamics. As pointed out in Ref. [38], the deviations of the linear relation of Eq. (4.12) (with respect to w_τ) can occur for extreme values of the work fluctuations located on these tails.

We define the asymmetry function of the pdf P as

$$\varrho(w) = \lim_{\tau/\tau_c \to \infty} \ln \frac{P(w_\tau = w)}{P(w_\tau = -w)}, \tag{4.29}$$

so that SSFT of Eq. (4.12) reads

$$\varrho(w) = w. \tag{4.30}$$

From the experimental pdfs of w_τ, we compute $\varrho(w)$ as the logarithm in Eq. (4.29) for integration times $\tau = 10\tau_c$. We checked that for this integration time the limit of Eq. (4.29) has been attained. Figure 4.15 shows the profile of the asymmetry functions for different values of α. We notice that for sufficiently small

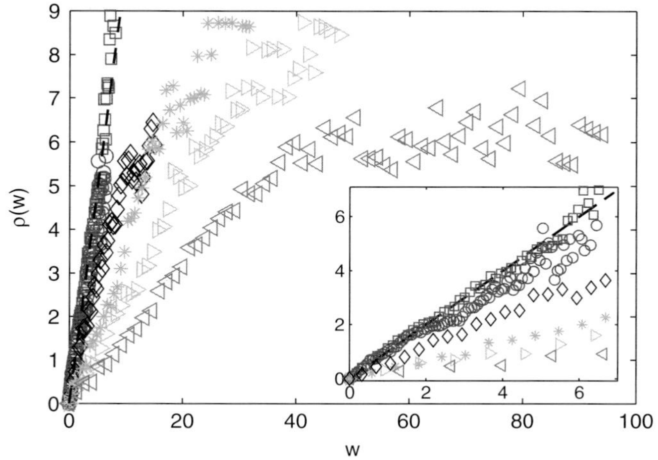

Figure 4.15 Asymmetry function of the pdf of the work done by the external force on the colloidal bead computed at $\tau = 0.25$ s for different values of the parameter α: 0.20 ($\circ$), 0.51 ($\square$), 1.84 ($\diamond$), 3.89 ($*$), 6.69 ($\triangleright$), and 10.77 ($\triangleleft$). The dashed line represents the prediction of the fluctuation theorem $\rho(w) = w$. *Inset*: Expanded view around $\rho(w) = w$.

values ($\alpha = 0.20, 0.51 < 1$), the FT given by Eq. (4.30) is verified by the experimental data (see the inset of Figure 4.15). To our knowledge, this is the first time that the FT has been verified for a random force without introducing any prefactor in the linear relation of Eq. (4.30).

It is important to point out that any deviation from the linear relation of Eq. (4.30) for extreme fluctuations is unlikely since we probed values as large as $w_\tau/\langle w_\tau \rangle \sim 5$. Indeed, it is argued [33, 34, 38, 40] that, for strongly dissipative systems driven by a random force, the deviations from FT may occur around $w_\tau/\langle w_\tau \rangle \sim 1$. Furthermore, in the present case the validity of the FT for weak driving amplitudes $\alpha < 1$ is consistent with the fact that for integration times $\tau > 25$ ms, the ratio $\varrho(w)/w$ has converged to its asymptotic value 1 for all measurable w. Note that this convergence to FT prediction is quite similar to that measured in system driven by deterministic forces [10, 11, 24, 42]. For example, in the case of a harmonic oscillator driven out of equilibrium by a sinusoidal external force, the asymptotic value of $\varrho(w)/w$ is reached for integration times larger than the forcing period [10, 11]. Sect. 4.3.2.

In contrast for $1 \lesssim \alpha$, deviations from Eq. (4.30) are expected to occur because the fluctuations of injected energy produced by the external random force become larger than those injected by the thermal bath. Indeed, Figure 4.15 shows that for values above $\alpha = 1.84$, Eq. (4.30) is not verified anymore but ϱ becomes a nonlinear function of w_τ. For small values of w_τ, it is linear with a slope that decreases as the driving amplitude increases, whereas there is a crossover to a slowest dependence around $w_\tau/\langle w_\tau \rangle \sim 1$, a qualitatively similar behavior to those reported in Refs [33, 36–39]. We finish this section by emphasizing that we have clearly found that for an experimental system whose dynamics correspond to a first-order Langevin equation subjected to both thermal and external noises, the FT can be satisfied or not depending on the relative strength of the external driving. The details about how these deviations arise and the convergence to generic work fluctuation relations will be given further. We first analyze the experiment on the AFM.

4.5.2
AFM Cantilever

A second example of a system for which thermal fluctuations are non-negligible in the energy injection process at equilibrium is the dynamics of the free end of a rectangular microcantilever used in AFM measurements. The cantilever is a mechanical clamped-free beam, which can be bended by an external force F and is thermalized with the surrounding air. The experiment is sketched in Figure 4.13b.

We use conductive cantilevers from Nanoworld (PPP-CONTPt). They present a nominal rectangular geometry: 450 μm long, 50 μm wide, and 2 μm thick, with a 25 nm $PtIr_5$ conductive layer on both sides. The deflection is measured with a homemade interferometric deflection sensor [43], inspired by the original design of Schonenberger [44] with a quadrature phase detection technique [6]: the interference between the reference laser beam reflecting on the chip of the cantilever and the sensing beam on the free end of the cantilever gives a direct measurement of the deflection X. Our detection system has a very low intrinsic noise, as low as 4 pm rms in the 100 kHz bandwidth we are probing [43, 45].

From the power spectrum of the deflection fluctuations of the free end at equilibrium ($F = 0$), we verify that the cantilever dynamics can be reasonably modeled as a stochastic harmonic oscillator with viscous dissipation [45, 46]. Hence, in the presence of the external force, the dynamics of the vertical coordinate X of the free end is described by the second-order Langevin equation

$$m\ddot{X} + \gamma\dot{X} = -kX + \zeta_T + F, \tag{4.31}$$

where m is the effective mass, γ is the viscous drag coefficient, k is the stiffness associated with the elastic force on the cantilever, and ζ_T models the thermal fluctuations. m, γ, and k can be calibrated at zero forcing using fluctuation–dissipation theorem, relating the observed power spectrum of X to the harmonic oscillator model: in our experiment we measure $m = 2.75 \times 10^{-11}$ kg, $\gamma = 4.35 \times 10^{-8}\ \mathrm{kg\,s^{-1}}$, and $k = 8.05 \times 10^{-2}\ \mathrm{N\,m^{-1}}$. The amplitude of the equilibrium thermal fluctuations of the tip position (i.e., $\sqrt{\langle x^2 \rangle} = \sqrt{k_B T/k} \simeq 2\,10^{-10}$ m) is two orders of magnitude larger than the detection noise (i.e., 4 pm rms). The signal to noise ratio is even better when the system is driven by an external force F. The characteristic timescales of the deflection dynamics are the resonance period of the harmonic oscillator $\tau_k = 2\pi\sqrt{m/k} = 116\ \mu$s and the viscous relaxation time $\tau_\gamma = m/\gamma = 632\ \mu$s, which is the longest correlation time.

When a voltage V is applied between the conductive cantilever and a metallic surface brought close to the tip ($h \sim 10\ \mu$m apart), an electrostatic interaction is created. The system behaves like a capacitor with stored energy $E_c = (1/2)C(X)V^2$, with C the capacitance of the cantilever tip/surface system. Hence, the interaction between the cantilever and the opposite charged surface gives rise to an attractive external force $F = -\partial_X E_c = -aV^2$ on the free end, with $a = \partial_X C/2$. If we apply a static voltage $\bar{V}$, the force F can be deduced from the stationary solution of Eq. (4.31): $k\bar{X} = -a\bar{V}^2$, where $\bar{X}$ is the mean measured deflection.[4] k being already calibrated, we validate this quadratic dependence of forcing in V and measure $a = 1.49 \times 10^{-11}\ \mathrm{N\,V^{-2}}$.

As the electrostatic force F is only attractive, its mean value cannot be chosen to be 0. We thus generated a driving voltage V designed to create a Gaussian white noise forcing f_0 around an offset $\bar{F}$: $F = \bar{F} + f_0$. The variance δf_0 of f_0 is the main control parameter of the system. In the absence of fluctuations ζ_T and f_0, Eq. (4.32) has the stationary solution $\bar{X} = \bar{F}/k$. This solution corresponds to the mean position attained by the free end in the presence of the zero mean fluctuating forces. Hence, we focus on the dynamics of the fluctuations $x = X - \bar{X}$ around $\bar{X}$, which are described by the equation

$$m\ddot{x} + \gamma\dot{x} = -kx + \zeta_T + f_0. \tag{4.32}$$

4) The quadratic dependence in V is valid only after taking care to compensate for the contact potential between the tip and the sample, which gives a small correction of the order of a few tens of mV.

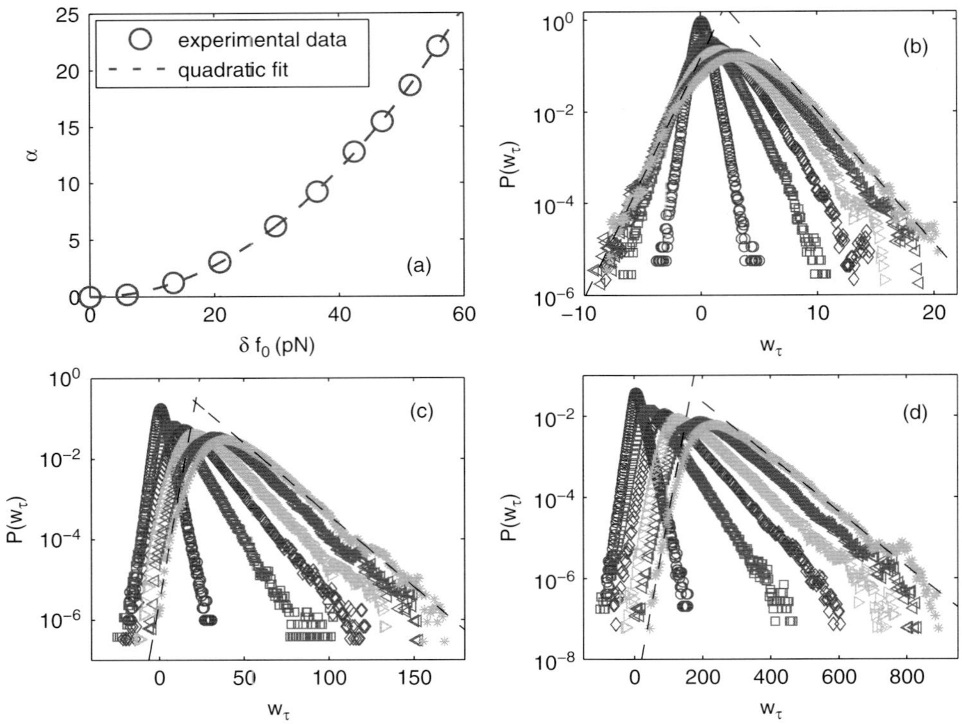

Figure 4.16 (a) Dependence of the parameter α on the standard deviation of the Gaussian white external force f_0 acting on the cantilever. Probability density functions of the work w_τ for (b) $\alpha = 0.19$, (c) $\alpha = 3.03$, and (d) $\alpha = 18.66$. The symbols correspond to integration times $\tau = 97\ \mu s$ (○), $1.074\ \mu s$ (□), 2.051 ms (◇), 3.027 ms (◁), 4.004 ms (▷), and 4.981 ms (∗). The black dashed lines in (b)–(d) represent the exponential fits of the corresponding tails. (For a color version of this figure, please see the color plate at the end of this book.)

Figure 4.16a shows the dependence between the parameter α (defined by Eq. (4.27)) for the stochastic variable x and the standard deviation of the forcing δf_0. We find that similarly to the position fluctuations of the colloidal particle, this dependence is quadratic for the cantilever. On the other hand, the work done by the external random force during an integration time τ is computed from Eq. (4.28). The corresponding pdfs are shown in Figure 4.16b–d. Unlike the case of the colloidal particle, in this case the pdfs do not converge to a Gaussian distribution but to a profile with asymmetric exponential tails even for the weakest applied driving amplitude ($\alpha = 0.19$) and for integration times as long as $\tau = 8\tau_\gamma$, as shown in Figure 4.16b–d. Surprisingly, when computing the asymmetry function for the two smallest values of α (0.19 and 1.21) for $\tau = 4\tau_\gamma$, the steady-state FT of Eq. (4.30) is perfectly verified, as shown in the inset of Figure 4.17. Once again, work fluctuations as large as five times their mean value located on the exponential tails are

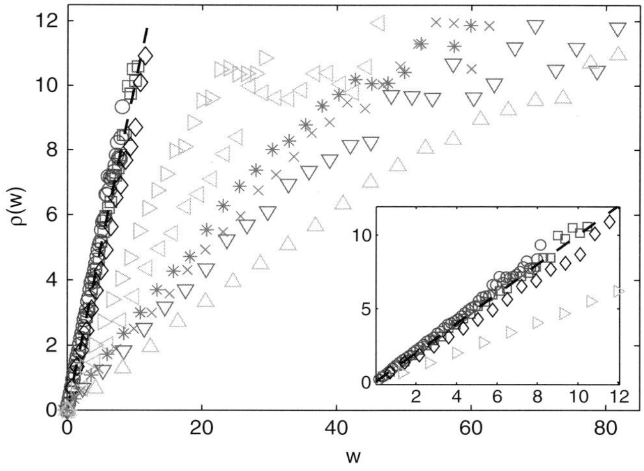

Figure 4.17 Asymmetry function of the probability density function of the work done by the external force on the AFM cantilever computed at $\tau = 4\tau_\gamma$ for different values of the parameter α: 0.19 ($\circ$), 1.21 ($\square$), 3.03 ($\diamond$), 6.18 ($\triangleright$), 9.22 ($\triangleleft$), 12.77 ($*$), 15.46 ($\times$), 18.66 (∇), and 22.10 (Δ). The dashed line corresponds to the prediction of the fluctuation theorem $\rho(w) = w$. *Inset*: Expanded view around $\rho(w) = w$.

probed and hence deviations from FT are unlikely for the same reasons discussed for the case of the Brownian particle.

In Figure 4.17, we see that for larger values of the driving amplitude, the deviations from Eq. (4.30) appear as a nonlinear relation with a linear part for small fluctuations whose slope decreases as α increases and a crossover for larger fluctuations, qualitatively similar to the behavior observed for the colloidal particle, as shown for $3.03 < \alpha$. In the following, we discuss the properties of these deviations as the energy injection process becomes dominated by the external force.

4.5.3
Fluctuation Relations Far from Equilibrium

We address now the question of how the deviations from Eq. (4.30) arise as the external stochastic force drives the system far from equilibrium. As shown previously, for $1.84 \leq \alpha$ for the colloidal particle and $3.03 \leq \alpha$ for the cantilever, the forcing amplitude is strong enough to destroy the conditions for the validity of the FT. Hence, we note that there are two well-defined limit regimes depending on the driving amplitude: one occurring at small values of α for which the steady-state FT is valid, and the limit $\alpha \gg 1$ for which the role of the thermal bath is negligible in the energy injection process, which is completely dominated by the external stochastic force. In order to investigate whether the transition between these two regimes is abrupt or not, we proceed by noting that for the latter the stochastic force term ζ_T in Eqs. (4.26) and (4.32) will be negligible compared to f_0. This implies that the resulting statistical time-integrated properties of the corresponding non-equilibrium steady state will be invariant under a normalization of the timescales and the temperature of the system. In particular, the resulting work fluctuation relations for w_τ must lead to a master curve for the asymmetry function in the far-from-equilibrium limit $\alpha \gg 1$. The information about the transition of the fluctuation relations to this regime is given by the convergence to the master curve.

We introduce the normalized work w_τ^* in the following way:

$$w_\tau^* = \frac{\tau_c}{\tau} \frac{w_\tau}{1+\alpha}. \tag{4.33}$$

The physical idea behind this normalization is that for $\alpha \gg 1$, the thermal bath alone works as a heat reservoir for viscous dissipation whereas its coupling with the external forcing plays the role of a nonequilibrium thermal bath at an effective temperature: $\kappa\langle x^2\rangle/k_B = (1+\alpha)T \approx \alpha T$. The prefactor τ_c/τ is introduced in such a way that w_τ^* represents the average normalized work done during the largest correlation time of the system. Accordingly, the asymmetry function must be redefined as

$$\varrho^*(w^*) = \lim_{\tau/\tau_c \to \infty} \frac{\tau_c}{\tau} \ln \frac{P(w_\tau^* = w^*)}{P(w_\tau^* = -w^*)}. \tag{4.34}$$

Figure 4.18a shows the asymmetry function ϱ^* for the normalized work w_τ^* on the colloidal particle at large values of α for which Eq. (4.30) is violated. The timescale τ_c in the computation of (4.33) and (4.34) is taken as the correlation time ($\tau_0 = 25$ ms) of the exponentially correlated stochastic forcing of Eq. (4.26). For comparison, we also show the corresponding curves at $\alpha = 0.20$ and 0.51 as circles and squares, respectively, for which Eq. (4.30) holds. The convergence to a master curve is verified, which means that for a sufficiently strong forcing the thermal bath acts only as a passive reservoir for the energy dissipation without providing any important contribution to the energy injection process in the system. Evidently, the normalized asymmetry function for the values α that verify the FT lies far from the master curve. We point out that the transition to the limit $\alpha \gg 1$ is rather continuous since intermediate regimes occur, as observed for $\alpha = 1.84$. In this case, neither the FT is satisfied as shown previously in Figure 4.15 nor the master curve is attained, since in this case the strength of thermal noise is still comparable to that of the external noise.

The results for the normalized asymmetry function of the work done on the cantilever by the external force are shown in Figure 4.18b. The curves corresponding to the verification of the FT for $\alpha = 0.19$ and 1.21 are also plotted for comparison. The convergence to a master curve is also checked as the value of α increases. Indeed, when comparing our normalized experimental curves with the analytical expression carried out by Ref. [33] for the asymmetry function of the work distribution on a Brownian particle driven entirely by a Gaussian white noise

$$\varrho^*(w^*) = \begin{cases} 4w^*, & w^* < 1/3, \\ \dfrac{7}{4}w^* + \dfrac{3}{2} - \dfrac{1}{4w^*}, & w^* \geq 1/3, \end{cases} \tag{4.35}$$

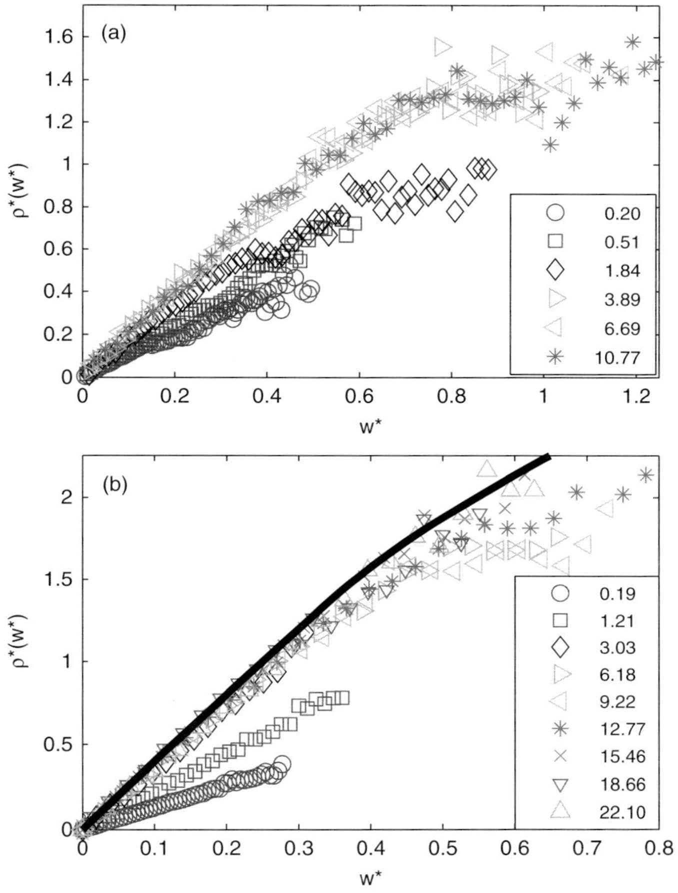

Figure 4.18 (a) Asymmetry function of the pdf of the normalized work done by the Gaussian Ornstein–Uhlenbeck force on the colloidal particle for different values of the parameter α. (b) Asymmetry function of the pdf of the normalized work done by the Gaussian white force on the cantilever for different values of the parameter α. The thick solid line represents the analytical expression given by Eq. (4.35).

we check that the assumption of the convergence of the energy injection process in the cantilever to that of a Langevin dynamics for a harmonic oscillator entirely dominated by the external noise is valid. Finite α corrections can be detected for large values of the normalized work indicating that the thermal bath still influences the energy injection into the cantilever. These corrections seem to vanish as the system is driven farther from equilibrium, as observed in Figure 4.18 for $\alpha = 22.10$.

Finally, we point out that the profile of the master curve strongly depends on the kind of stochastic force: a Gaussian Ornstein–Uhlenbeck process in the first example and a Gaussian white noise in the second one. Non-Gaussian extensions of the external random force are expected to lead to striking modification of the fluctuation relations in the limit $\alpha \gg 1$, as recently investigated for an asymmetric Poissonian shot noise [40].

4.5.4
Conclusions on Randomly Driven Systems

We have studied the FT in two experimental systems in contact with a thermal bath and driven out of equilibrium by a stochastic random force. The main results of our study are that the validity of FT is controlled by the parameter α. For small α, we have shown that the validity of the steady-state FT is a very robust result regardless of the details of the intrinsic dynamics of the system (first- and second-order Langevin dynamics) and the statistical properties of the forcing (white and colored random Gaussian noise). Indeed, these specific features vanish when the integration of w_τ is performed for τ much larger than the largest correlation time of the system.

In contrast for very large $\alpha \geq 1$, when the randomness of the system is dominated by the random forcing, we have shown that the symmetry relation given by the FT is violated. We remark that this happens because in such a case the injected work is not a good measure of the total entropy production of the system. Then the hypotheses for the validity of the FT are not satisfied. For $\alpha \gg 1$, the results at different driving amplitudes can be set on a master curve by defining a suitable effective temperature that is a function of α. We have shown that this master curve is system dependent. Therefore, some care must be taken when trying to apply directly the FT to systems completely driven by random or chaotic forcing, such as in turbulent fluids and granular matter. For small systems, this is not an issue as the stochasticity is in general dominated by the thermal fluctuations of the bath.

4.6
Applications of Fluctuation Theorems

The fluctuation theorems have several important consequences such as the Jarzynski and Crooks equalities [26, 27, 47], which are useful to compute the free energy difference between two equilibrium states using any kind of transformation [7, 9, 48, 49]. The Hatano–Sasa [16] relations and the recently derived fluctuation–dissipation theorems [50] are related to FTs and are useful to compute the response of a NESS using the steady-state fluctuations of the NESS. As we have seen, the FT allows the calculation of tiny amount of heat, which can be useful in many applications in aging systems [41, 51, 52] and biological systems.

The FTs for Langevin systems can be used to measure an unknown average power. This idea has been discussed first in the context of electrical circuits [42] and in Ref. [53] it has been applied for the first time to the measure of the torque of a molecular motor.

In the next section, we will discuss the problem related to the applications of FDT in a nonequilibrium steady state.

4.6.1 Fluctuation–Dissipation Relations for NESS

As we have seen in the previous section, current theoretical developments in nonequilibrium statistical mechanics have led to significant progress in the study of systems around states far from thermal equilibrium. Systems in nonequilibrium steady states are the simplest examples because the dynamics of their degrees of freedom x under fixed control parameters λ can be statistically described by time-independent probability densities $\varrho_0(x,\lambda)$. NESS naturally occur in mesoscopic systems such as colloidal particles dragged by optical tweezers, Brownian ratchets, and molecular motors because of the presence of nonconservative or time-dependent forces [54]. At these length scales fluctuations are important, so it is essential to establish a quantitative link between the statistical properties of the NESS fluctuations and the response of the system to external perturbations. Around thermal equilibrium, this link is provided by the fluctuation–dissipation theorem [55].

The generalization of the fluctuation–dissipation theorem around NESS for systems with Markovian dynamics has been achieved in recent years from different theoretical approaches [56–64]. The different generalized formulations of the fluctuation–dissipation theorem link correlation functions of the fluctuations of the observable of interest $O(x)$ in the unperturbed NESS with the linear response function of $O(x)$ due to a small external time-dependent perturbation around the NESS. The observables involved in such relations are not unique but they are equivalent in the sense that they lead to the same values of the linear response function. These theoretical relations may be useful in experiments and simulations to know the linear response of the system around NESS. Indeed, the response can be obtained from measurements entirely done at the unperturbed NESS of the system of interest without any need to perform the actual perturbation. Nevertheless, the theoretical equivalence of the different observables involved in those relations does not translate into equivalent experimental accessibility: for example, strongly fluctuating observables such as instantaneous velocities may lead to large statistical errors in the measurements [65]. Besides, NESS quantities themselves such as local mean velocities, joint stationary densities, and the stochastic entropy are not in general as easily measurable as dynamical observables directly related to the degrees of freedom [66]. Hence, before implementing the different fluctuation–response formulas in real situations, it is important to test its experimental validity under very well-controlled conditions and to assess the influence of finite data analysis. The experimental test of some fluctuation–dissipation relations has been recently done in Refs [65–68] for colloidal particles in toroidal optical traps.

In this chapter, we discuss the effects of the finite number of experimental data on the determination of the linear response function around a NESS for a micron-sized system with Markovian dynamics: a Brownian particle in a toroidal optical trap. For this purpose, we perform the respective data analysis on the measurements reported in Refs [66, 68]. In Section 4.6.1.1, we briefly describe a generalized fluctuation–dissipation relation that has been derived for Markovian dynamics around a NESS exploiting the properties of the stationary density $\varrho_0(x, \lambda)$. In Section 4.6.1.2, we recall the main features of a previous experiment that we use in this chapter for the data analysis. In Section 4.6.2, we discuss the two different kinds of finite sampling effects that can appear in the computation of the different terms involved in the fluctuation–dissipation relation. We show that the generalized fluctuation–dissipation relation is verified experimentally when performing a careful data analysis, which takes into account these effects. Finally, we present the conclusion.

4.6.1.1 Hatano–Sasa Relation and Fluctuation–Dissipation Around NESS

The Hatano–Sasa relation provides a general identity for the transitions between either equilibrium or nonequilibrium steady states of Markovian systems [16]. In the following, we will focus on a Langevin system with Markovian dynamics described by a steady-state probability density $\varrho_0(x, \lambda)$. When the system is subjected to a time-dependent variation of the control parameter $\lambda(t)$ between an initial time t_i and a final time t_f, the Hatano–Sasa identity reads

$$\left\langle \exp\left(-\int_{t_i}^{t_f} \mathrm{d}t\, \dot{\lambda}_\alpha(t) \frac{\partial \phi(x_t, \lambda(t))}{\partial \lambda_\alpha} \right) \right\rangle = 1, \tag{4.36}$$

where $\phi(x, \lambda) = -\ln \varrho_0(x, \lambda)$ and the average $\langle \cdot \rangle$ is performed over an infinite number of realizations of a prescribed time-dependent protocol $\lambda(t)$. From Eq. (4.36), Joanny and coworkers have directly derived a generalized fluctuation–dissipation relation that holds in the linear response regime around a NESS [50]:

$$R_{\alpha\gamma}(t-s) = \frac{\mathrm{d}}{\mathrm{d}t} \left\langle \frac{\partial \phi(x_t, \lambda_{\mathrm{SS}})}{\partial \lambda_\alpha} \frac{\partial \phi(x_s, \lambda_{\mathrm{SS}})}{\partial \lambda_\gamma} \right\rangle_0 . \tag{4.37}$$

In Eq. (4.37), $R_{\alpha\gamma}(t-s) = \delta \langle O(x_t) \rangle_h / \delta h_s |_{h=0}$ is the linear response function of the observable $O(x) = \partial \phi(x, \lambda_{\mathrm{SS}}) / \partial \lambda_\alpha$ due to a small external time-dependent perturbation $h_s = \lambda(s) - \lambda_{\mathrm{SS}}$ around $\lambda_{\mathrm{SS}} = \lambda(t_i)$ fixing an initial NESS at time t_i. The averages $\langle \cdot \rangle_h$ and $\langle \cdot \rangle_0$ are performed over the perturbed and unperturbed processes, respectively.

In experiments, the formal average involved in Eq. (4.36) is not perfectly computed because of the finite number of independent realizations of $\lambda(t)$. Hence, Eq. (4.36) allows one to estimate the experimental precision of (4.37) computed from a given number of experimental data provided that one can measure the observable $\partial \phi(x, \lambda_{\mathrm{SS}}) / \partial \lambda_\alpha$. In the next section, we tackle this problem for the experimental trajectories of a colloidal particle in a toroidal optical trap.

4.6.1.2 Brownian Particle in a Toroidal Optical Trap

The Brownian motion of a colloidal particle in a toroidal optical trap has become an experimental model to study the generalization of the fluctuation–dissipation theorem around a NESS [65–68]. This is because it is a system with a single translational degree of freedom where one can easily tune its relevant control parameters.

Our experiment has been already described in detail in Refs [66, 68], so here we only explain it briefly. The Brownian motion of a spherical silica particle (radius $r = 1\,\mu\text{m}$) immersed in water is confined on a thin torus of major radius $a = 4.12\,\mu\text{m}$ by a tightly focused laser beam rotating at 200 Hz. The water reservoir acts as a thermal bath at fixed temperature ($T = 20 \pm 0.5\,^\circ\text{C}$) providing thermal fluctuations to the particle. The viscous drag coefficient at this temperature is $\gamma = 1.89 \times 10^{-8}\,\text{kg s}^{-1}$. The rotation frequency of the laser is so high that it is not able to trap continuously the particle in the focus because the viscous drag force of the surrounding water quickly exceeds the optical trapping force. Consequently, at each rotation the beam only kicks the particle a small distance along the circle of radius a exerting a nonconservative force $f = 66\,\text{fN}$ on it in the direction of the rotation. Thus, the particle motion is effectively confined in a circle: the angular position θ of its barycenter is the only relevant degree of freedom. In addition, a static light intensity profile (amplitude about 5% of the total laser intensity [66, 68]) is created along the circle acting as a periodic potential $U(\theta) = U(\theta + 2\pi)$ of amplitude $68.8k_BT$. Figure 4.19a depicts this experimental configuration. We track the 2D particle position by video microscopy in order to measure the time evolution of θ. Thus, for the experimentally accessible length and timescales, the dynamics of θ_t is modeled by the first-order Langevin equation [65–69]

$$\dot{\theta} = -\partial_\theta H(\theta) + F + \xi, \tag{4.38}$$

where $H(\theta) = U(\theta)/(\gamma a)$ with amplitude $A = \max\{H(\theta)\} = 68.8k_BT/(\gamma a^2)$, $F = f/(\gamma a)$, and ξ is a white noise process of zero mean and covariance $\langle \xi_t \xi_s \rangle = 2[k_BT/(\gamma a^2)]\delta(t - s)$. Under these fixed conditions, the dynamics of θ_t settles into a NESS whose probability density function $\varrho_0(\theta, A)$ is plotted in Figure 4.19b (solid black line).

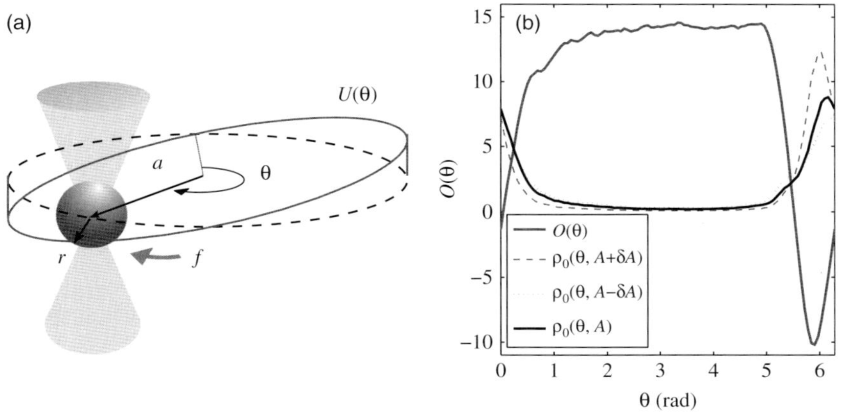

Figure 4.19 (a) Sketch of a Brownian particle in a toroidal optical trap subjected to a nonconservative force f and undergoing a periodic potential $U(\theta)$. (b) Experimental profile of the observable $O(\theta)$ defined in Eq. (4.39) computed using the NESS densities $\rho_0(\theta, A + \delta A)$ and $\rho_0(\theta, A - \delta A)$ around $\rho_0(\theta, A)$ at fixed F.

4.6.2 Generalized Fluctuation–Dissipation Relation

In the following analysis, we take $x = \theta$ as the single degree of freedom and $\lambda = A$ as the main control parameter of the system. In this case, the observable of interest involved in the fluctuation–dissipation relation (4.37) is

$$O(\theta) = -\frac{\partial \ln \varrho_0(\theta, A)}{\partial A}. \tag{4.39}$$

The experimental profile of $O(\theta)$, computed as a discrete three-point derivative of $-\ln \varrho_0(\theta, A)$ at two different NESS around A, is shown in Figure 4.19b.

We focus on the response of the system after applying a small Heaviside perturbation δA to A at time t_i: $A \rightarrow A + \delta A$. In the experiment, this dynamical procedure is done by suddenly switching the laser power modulation as explained in Refs [66, 68]. This procedure yields the integrated response function, defined as

$$\chi(t - t_\mathrm{i}) = \int_{t_\mathrm{i}}^{t} R(t - s)\mathrm{d}s = \frac{\langle O(\theta_t)\rangle_{\delta A}}{\delta A}, \tag{4.40}$$

where the average $\langle\cdot\rangle_{\delta A}$ must be performed over the perturbed process at time t. Then the integrated version of the generalized fluctuation–dissipation relation (4.37) is in this case

$$\chi(t - t_\mathrm{i}) = \langle O(\theta_t)O(\theta_{t_\mathrm{i}})\rangle_0 - \langle O(\theta_t)O(\theta_t)\rangle_0. \tag{4.41}$$

We now study the effects of a finite number of realizations of the perturbation δA and the finite number of trajectories used to compute the averages $\langle\cdot\rangle_{\delta A}$ and $\langle\cdot\rangle_0$ in Eqs. (4.40) and (4.41), respectively.

4.6.2.1 Statistical Error

As discussed in Section 4.6.1.1, the Hatano–Sasa relation (4.36) can be used to estimate the error of the experimental computation of Eq. (4.41) when performing $N < \infty$ independent realizations of δA around the NESS. In the case of the dynamical process defined by the Heaviside perturbation $A \rightarrow A + \delta A$ at time t_i, as done in the experiment, Eq. (4.36) reads

$$\langle \exp[-\delta A O(\theta_{t_\mathrm{i}})]\rangle = 1. \tag{4.42}$$

Equation (4.42) depends only on the initial values $O(\theta_{t_\mathrm{i}})$ when the system is still in NESS. Therefore, for a finite number of trajectories N, we introduce an estimator of the error of Eq. (4.42):

$$\Delta(N) = \left| \frac{1}{N}\sum_{j=1}^{N} \exp\left[-\delta A O_j(\theta_{t_\mathrm{i}})\right] - 1 \right|, \tag{4.43}$$

where $O_j(\theta_{t_\mathrm{i}})$ is the jth sampled NESS initial condition. Figure 4.20a shows the behavior of the error $\Delta(N)$ computed for $\delta A = 0.05A$ (the value realized in the dynamical experiment) using N experimental values of $O(\theta)$ drawn from the NESS

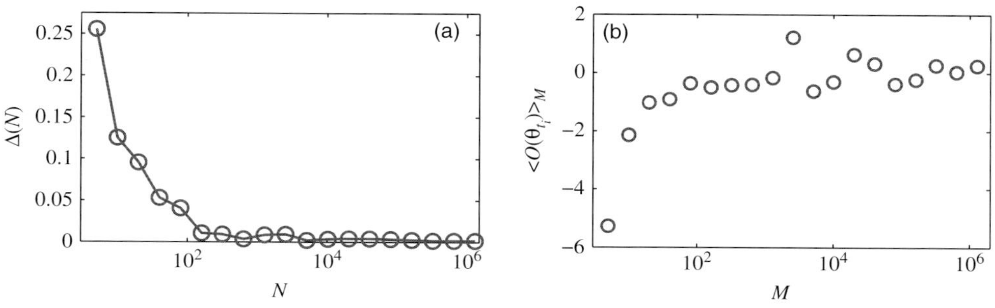

Figure 4.20 (a) Estimate of the error of the Hatano–Sasa relation (4.42) for a finite number N of realizations of δA. (b) Estimate of the error of $\langle O(\theta_{t_i})\rangle_0$ using Eq. (4.45) for M NESS data.

distribution. For small $N \lesssim 100$, the error is non-negligible, $\Delta(N) \geq 4\%$, and this must be taken into account in the final accuracy of the generalized fluctuation–dissipation relation when comparing the left-hand side with the right-hand side of Eq. (4.41) using the experimental data. Then as N increases $\Delta(N)$ quickly converges to 0: for $N \geq 500$ the precision of Eq. (4.42) found in the experiment is better than 1%.

4.6.2.2 **Effect of the Initial Sampled Condition**

According to the Heaviside procedure for δA, in Eqs. (4.40) and (4.41) the initial condition $\theta(t_i)$ for the perturbed process is sampled from the NESS density $\varrho_0(\theta, A)$. Then the integrated linear response function must formally satisfy the initial condition

$$\chi(0) = \frac{\langle O(\theta_{t_i})\rangle_0}{\delta A} = 0, \tag{4.44}$$

where the last equality is due to the normalization of $\varrho_0(\theta_{t_i}, A)$. We are interested in the effect of a finite number $M < \infty$ of initial values $\theta(t_i)$ drawn from the initial NESS on the estimate of $\chi(t - t_i)$. It should be noted that in practice a small M may significantly affect the computation of $\langle O(\theta_{t_i})\rangle_0$ in Eq. (4.44) because most of the positive values of $O(\theta)$ lie in the region where $\varrho_0(\theta, A)$ is rarely sampled, as shown in Figure 4.19b. In Figure 4.20b, we plot some values of the finite average

$$\langle O(\theta_{t_i})\rangle_M = \frac{1}{M}\sum_{j=1}^{M} O_j(\theta_{t_i}), \tag{4.45}$$

where $O_j(\theta_{t_i})$ is the jth initial condition at NESS. As expected, for small M, $\langle O(\theta_{t_i})\rangle_M < 0$ due to the fact that one samples mostly the negative values around the maximum of $\varrho_0(\theta, A)$. The convergence to the theoretical value $\langle O(\theta_{t_i})\rangle_0 = 0$ is very slow: as M increases, $\langle O(\theta_{t_i})\rangle_M$ becomes very sensitive to M and large positive values of $\langle O(\theta_{t_i})\rangle_M$ can be obtained. The general trend is around $\langle O(\theta_{t_i})\rangle_0 = 0$, though. Then even for large M one must be careful with the computation of the integrated response function since a large initial error of $\chi(0)$ due to the use of the average $\langle\cdot\rangle_M$ may significantly propagate as t increases.

In order to avoid the problem of the sensitivity to the initial condition, instead of using directly the average $\langle\cdot\rangle_M$ in Eq. (4.40), one can define an estimator $\chi_M(t - t_{t_i})$ satisfying the initial condition $\chi_M(0) = 0$ as required ideally by Eq. (4.44). In this way, the propagation of the initial error given by $\langle O(\theta_{t_i})\rangle_M$ is suppressed at the beginning. An intuitive way to define χ_M can be outlined from the usual protocol to compute the integrated response function

$$\chi(t - t_i) = \frac{\langle O^{\delta A}(\theta_t)\rangle_{\delta A} - \langle O(\theta_{(t-t_i+t^*)})\rangle_0}{\delta A}, \tag{4.46}$$

where the time t^* is chosen such that $O(\theta_{t^*}) = O^{\delta A}(\theta_{t_i})$ and $O^{\delta A}(\theta_t)$ denotes the observable measured during the perturbed process. Notice that Eq. (4.46) is justified by the fact that in the case of an infinite number of samples $\langle O(\theta_t)\rangle_0 = 0$, $\forall t$, because of the time translational invariance of the NESS. In contrast, when M is finite, it is useful to take into account, in Eq. (4.46), that $\langle\cdot\rangle_{\delta A}$ and $\langle\cdot\rangle_0$ are performed independently first on the perturbed trajectory $O^{\delta A}(\theta_t)$ and then on the unperturbed ones $O(\theta_{(t-t_i+t^*)})$, specifically

$$\chi_M(t - t_i) = \frac{1}{\delta A}\left[\frac{1}{M}\sum_{j=1}^{M} O_j^{\delta A}(\theta_t) - \frac{1}{L}\sum_{k=1}^{L} O_k(\theta_{(t-t_i+t^*)})\right], \tag{4.47}$$

where L is the number of unperturbed trajectories such that $O_k(\theta_{t^*}) = O_j^{\delta A}(\theta_{t_i})$. Therefore, Eq. (4.47) can be rewritten as

$$\chi_M(t - t_i) = \frac{1}{\delta A}\frac{1}{M}\sum_{j=1}^{M}\left\{\frac{1}{L}\sum_{k=1}^{L}\delta O_{jk}(\theta_t)\right\}, \tag{4.48}$$

where $\delta O_{jk}(\theta_t) \equiv O_j^{\delta A}(\theta_t) - O_k(\theta_{(t-t_i+t^*)})$ is the instantaneous difference between a perturbed trajectory $O_j^{\delta A}(\theta_t)$ and an unperturbed one $O_k(\theta_t)$. An example of this procedure is depicted in Figure 4.21a, where t_i has been set equal to zero. We see that for a given perturbed trajectory $O_j^{\delta A}(\theta_t)$ (thick dashed black line), obtained after δA has been applied, one should look for an unperturbed NESS trajectory $O_k(\theta(t))$ such that $O_k(\theta_{t^*}) = O_j^{\delta A}(\theta_{t_i})$ like the four unperturbed trajectories shown by the solid lines.

In this way, $\delta O_{jk}(\theta_{t_i}) = 0$ by construction and the estimator defined by Eq. (4.48) satisfies the condition $\chi_M(0) = 0$. For $M, L \to \infty$, $\chi_M(t - t_i)$ converges to χ defined by Eq. (4.40), because $\langle\langle O(\theta_t)\rangle_L\rangle_M \to 0$. In Figure 4.21b, we show $\chi_M(t)$, with t_i redefined as $t_i = 0$, computed using Eq. (4.48) for different values of M and for fixed $L = 1$ (solid lines) and $L = 200$ (dashed line). As M increases for $L = 1$ the curves converge to a single profile that must correspond to that of $\chi(t)$ ideally given by Eq. (4.40). The additional conditional average done for $L = 200$ smooths out the slightly fluctuating profile for $M = 500$ (thick solid black line) resulting in the thick dashed solid line. For comparison we also show in the inset of Figure 4.21b the raw estimate of χ obtained using the average of Eq. (4.45) for the same $M = 500$ perturbed trajectories without correcting the effect of the initial sampling. In this case,

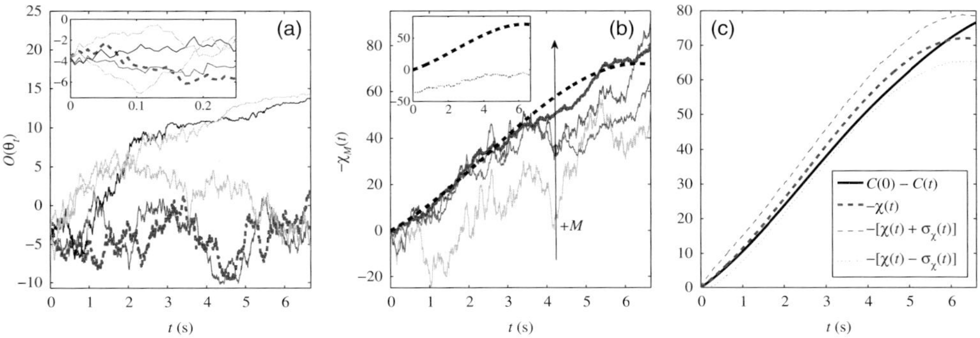

Figure 4.21 (a) Examples of perturbed (thick dashed black line) and unperturbed (solid lines) trajectories used to estimate $\chi(t)$ using Eq. (4.48). *Inset*: Expanded view at short time. (b) Estimate of $-\chi(t)$ for $M =$ 50, 100, 250, 500 and $L = 1$ (solid lines) and $M = 500$ and $L = 200$ (dashed line). *Inset*: Comparison of $\chi_M(t)$ for $M = 500$ and $L = 200$ with the poor estimate done by the uncorrected average of Eq. (4.45) for $M = 500$. (c) Comparison between the experimental $C(0) - C(t)$ and the best estimate of $-[\chi(t) \pm \sigma_\chi(t)]$.

the propagation of the initial large error of $\langle O(\theta_0)\rangle_0$ gives rise to a very poor estimate of the integrated response function for $t > 0$.

4.6.2.3 **Experimental Test**

Finally, we proceed to test the theoretical fluctuation–dissipation relation (4.41) for the experimental unperturbed NESS trajectories of the Brownian particle and those perturbed around NESS. For this purpose, we compare the best estimate $\chi_M(t)$ of $\chi(t)$ done for $M = 500$ and $L = 200$ with the right-hand side of Eq. (4.41). The involved correlation function $C(t) = \langle O(\theta_t)O(\theta_0)\rangle_0$ on the right-hand side is computed using unperturbed NESS trajectories. In Figure 4.21c, we compare $\chi_M(t)$ with $C(t) - C(0)$. Besides, one can estimate the statistical error of the experimental $\chi(t)$ at each $t \geq 0$ by computing the standard deviation of $\langle O^{\delta A}(\theta_t) - O(\theta_t)\rangle_M$ over the L possible choices of the unperturbed $O_k(\theta_t)$. The standard deviation $\pm\sigma_\chi(t)$ obtained in this way is also shown in Figure 4.21c showing that after following the careful procedure to estimate χ the relation $\chi(t) = C(t) - C(0)$ is verified by the finite experimental data. Note that without the finite correction $\langle\langle O(\theta_t)\rangle\rangle_M$ of the initial condition in Eq. (4.48), one would largely underestimate the direct measurement of the integrated response function leading to an apparent violation of Eq. (4.41). The results of this chapter are consistent with those of Refs [66, 68], where two fluctuation–dissipation formulas equivalent to Eq. (4.41) but involving different observables from the one studied here are checked experimentally.

4.6.3 **Discussion on FDT**

We have studied the influence of finite sampling in the computation of the linear response function of a Brownian particle in a toroidal optical trap around a NESS. We have shown that there are two different effects that may lead to a very poor estimate of the experimental linear response function when the data analysis is not performed carefully. This is an important point that must be assessed in general when applying in experiments and numerical simulations the different generalized fluctuation–dissipation formulas recently derived for NESS.

4.7 Summary and Concluding Remarks

In this chapter, we have reviewed several experimental results on systems coupled with a heat bath and driven out of equilibrium by external forces either deterministic or random. We have seen that in both cases we observe that the external forces may produce a negative work because of fluctuations. The probability of these negative events has been analyzed in the framework of fluctuation theorem.

We have mainly discussed the stochastic systems described by Langevin equations, with both harmonic and anharmonic potentials. We have seen the injected and dissipated powers exhibit present different behaviors. FTs are valid for any value of W_τ whereas can be applied only for $Q_\tau < \langle Q_\tau \rangle$ in the case of the heat. We have also seen that the finite-time corrections to SSFT depend on the driving and on the properties of the system. We have introduced the total entropy takes into account both contributions, that due to the external force (the entropy of the medium) and the equilibrium-like one (stochastic entropy). For the total entropy, FTs are valid for all the times. We discussed the applications of FTs to extract important physical properties of a stochastic system. Thus, one may conclude that for Markovian systems driven by a deterministic force, FT do not present any major problem and can be safely applied. The case of random driving is more complex and we have seen that several problems may arise when the variance of the driving becomes larger than the fluctuations induced by the thermal bath.

We have finally discussed the measure of the response in a system driven in a NESS. Among the various formulations tested experimentally, we have chosen the one derived in Ref. [50], which seems to us the easiest to apply in experiments. We pointed out several statistical problems that may arise on the direct application of the proposed relation and we have shown how to overcome them. The real problem of this formulation is that the variables in some cases do not have a direct physical interpretation as they come from a nonlinear transformation. Finally, we pointed out that there is the problem of the application of FDT in NESS when the dynamics is non-Markovian. Several theoretical approaches, that we have not discussed in this chapter, have been used to model this problem in experiments on NESS in complex fluids with non-newtonian behavior (see for example Refs. [80–82]).

Acknowledgments

Several results reviewed in this chapter have been obtained in collaboration with N. Garnier, S. Joubaud, and P. Jop. We acknowledge useful discussions and collaboration during the past 10 years with D. Andrieux, E. Cohen, R. Chetrite, G. Gallavotti, P. Gaspard, K. Gawedzky, and A. Imparato.

References

1 Kis Andras, A. and Zettl, A. (2008) *Philos. Trans. R. Soc. A*, **366**, 1591.

2 Gomez-Solano, J.R., Bellon, L., Petrosyan, A., and Ciliberto, S. (2010) *Europhys. Lett.*, **89**, 60003.

3 Bellon, L., Buisson, L., Ciliberto, S., and Vittoz, F. (2002) *Rev. Sci. Instrum.*, **73** (9), 3286.

4 Bellon, L. and Ciliberto, S. (2002) *Physica D*, **168**, 325.

5 Lamb, H. (1945) *Hydrodynamics*, 6th edn, Dover, republication of the 1932 edition, Cambridge University Press.

6 Bellon, L., Ciliberto, S., Boubaker, H., and Guyon, L. (2002) *Opt. Commun.*, **207**, 49–56.

7 Douarche, F., Ciliberto, S., and Petrosyan, A. (2005) *J. Stat. Mech.*, 09011.

8 Douarche, F., Joubaud, S., Garnier, N., Petrosyan, A., and Ciliberto, S. (2006) *Phys. Rev. Lett.*, **97**, 140603.

9 Douarche, F., Buisson, L., Ciliberto, S., and Petrosyan, A. (2004) *Rev. Sci. Instrum.*, **75** (12), 5084.

10 Joubaud, S., Garnier, N.B., Douarche, F., Petrosyan, A., Ciliberto, S., and Physique, C.R. (2007) *Work, Dissipation, and Fluctuations in Nonequilibrium Physics* (eds B. Derrida, P. Gaspard, and C. Van den Broeck), Elsevier, cond-mat/0703695.

11 Joubaud, S., Garnier, N.B., and Ciliberto, S. (2007) *J. Stat. Mech.*, P09018.

12 Blickle, V., Speck, T., Helden, L., Seifert, U., and Bechinger, C. (2006) *Phys. Rev. Lett.*, **96**, 070603.

13 Schuler, S., Speck, T., Tietz, C., Wrachtrup, J., and Seifert, U. (2005) *Phys. Rev. Lett.*, **94**, 180602.

14 Zamponi, F., Bonetto, F., Cugliandolo, L., and Kurchan, J. (2005) *J. Stat. Mech.*, P09013.

15 Sekimoto, K. (1998) *Prog. Theor. Phys.*, (Suppl. 130), 17.

16 Hatano, T. and Sasa, S. (2001) *Phys. Rev. Lett.*, **86**, 3463.

17 van Zon, R., Ciliberto, S., and Cohen, E.G.D. (2004) *Phys. Rev. Lett.*, **92** (13), 130601.

18 (a) van Zon, R. and Cohen, E.G.D. (2003) *Phys. Rev. Lett.*, **91** (11), 110601. (b) van Zon, R. and Cohen, E.G.D. (2003) *Phys. Rev. E*, **67**, 046102. (c) van Zon, R. and Cohen, E.G.D. (2004) *Phys. Rev. E*, **69**, 056121.

19 Ciliberto, S., Joubaud, S., and Petrosyan, A. (2010) *J. Stat. Mech.*, P12003.

20 Speck, T. and Seifert, U. (2005) *Eur. Phys. J. B*, **43**, 521.

21 Seifert, U. (2005) *Phys. Rev. Lett.*, **95**, 040602.

22 Puglisi, A., Rondoni, L., and Vulpiani, A. (2006) *J. Stat. Mech.*, P08010.

23 Joubaud, S., Garnier, N., and Ciliberto, S. (2008) *Eur. Phys. Lett.*, **82**, 30007.

24 Jop, P., Ciliberto, S., and Petrosyan, A. (2008) *Eur. Phys. Lett.*, **81**, **5**, 50005.

25 Benzi, R., Parisi, G., Sutera, A., and Vulpiani, A. (1983) *SIAM J. Appl. Math.*, **43**, 565.

26 Jarzynski, C. (1997) *Phys. Rev. Lett.*, **78** (14), 2690.

27 Crooks, G.E. (1999) *Phys. Rev. E*, **60** (3), 2721.

28 Gammaitoni, L., Marchesoni, F., and Santucci, S. (1995) *Phys. Rev. Lett.*, **74** (7), 1052–1055.

29 Schmitt, C., Dybiec, B., Hänggi, P., and Bechinger, C. (2006) *Europhys. Lett.*, **74** (6), 937.

30 Iwai, T. (2001) *Physica A*, **300**, 350–358.

31 Dan, D. and Jayannavar, A.M. (2005) *Physica A*, **345**, 404–410.

32 Imparato, A., Jop, P., Petrosyan, A., and Ciliberto, S. (2008) *J. Stat. Mech.*, P10017.

33 Farago, J. (2002) *J. Stat. Phys.*, **107**, 781.

34 Farago, J. (2004) *Physica A*, **331**, 69.

35 Shang, X.-D., Tong, P., and Xia, K.-Q. (2005) *Phys. Rev. E*, **72**, 015301(R).

36 Falcon, E., Aumaître, S., Falcón, C., Laroche, C., and Fauve, S. (2008) *Phys. Rev. Lett.*, **100**, 064503.

37 Cadot, O., Boudaoud, A., and Touz, C. (2008) *Eur. Phys. J. B*, **66**, 399.

38 Falcón, C. and Falcon, E. (2009) *Phys. Rev. E*, **79**, 041110.

39 Bonaldi, M. *et al.* (2009) *Phys. Rev. Lett.*, **103**, 010601.

40 Baule, A. and Cohen, E. (2009) *Phys. Rev. E*, **80**, 011110.

41 Jop, P., Gomez Solano, R., Petrosyan, A., and Ciliberto, S. (2009) *J. Stat. Mech.*, P04012.

42 Garnier, N. and Ciliberto, S. (2005) *Phys. Rev. E*, **71**, 060101(R).

43 Paolino, P. and Bellon, L. (2009) *NANOTECHNOLOGY*, **20** (40), 405705, DOI: 10.1088/0957-4484/20/40/405705, Published: OCT 7 2009.

44 Schonenberger, C. and Alvarado., S.F. (1989) *Rev. Sci. Instrum.*, **60**, 3131.

45 Paolino, P. and Bellon, L. (2009) *Nanotechnology*, **20**, 405705.

46 Bellon, L. (2008) *J. Appl. Phys.*, **104**, 104906.

47 Jarzynski, C. (2000) *J. Stat. Phys.*, **98** (1/2), 77.

48 Ritort, F. (2003) *Sémin. Poincaré*, **2**, 63.

49 Liphardt, J., Dumont, S., Smith, S.B., Ticono, I., Jr., and Bustamante, C. (2002) *Science*, **296**, 1832.

50 Prost, J., Joanny, J.F., and Parrondo, J.M.R. (2009) *Phys. Rev. Lett.*, **103**, 090601.

51 Crisanti, A. and Ritort, F. (2004) *Europhys. Lett.*, **66**, 253.

52 Gomez-Solano, J.R., Petrosyan, A., and Ciliberto, S. (2011) *Phys. Rev. Lett.*, **106**, 200602.

53 Hayashi, K., Ueno, H., Iino, R., and Noji, H. (2010) *Phys. Rev. Lett.*, **104**, 218103.

54 Reimann, P. (2002) *Phys. Rep.*, **57**, 361.

55 Marini Bettolo Marconi, U. *et al.* (2008) *Phys. Rep.*, **461**, 111.

56 Harada, T. and Sasa, SI. (2005) *Phys. Rev. Lett.*, **95**, 130602.

57 Lippiello, E., Corberi, F., and Zannetti, M. (2005) *Phys. Rev. E*, **71**, 036104.

58 Speck, T. and Seifert, U. (2006) *Europhys. Lett.*, **74**, 391.

59 Chetrite, R., Falkovich, G., and Gawedzki, K. (2008) *J. Stat. Mech.*, P08005.

60 Chetrite, R. and Gawedzki, G. (2009) *J. Stat. Phys.*, **137**, 890.

61 Baiesi, M., Maes, C., and Wynants, B. (2009) *Phys. Rev. Lett.*, **103**, 010602.

62 Baiesi, M., Maes, C., and Wynants, B. (2009) *J. Stat. Phys.*, **137**, 1094.

63 Baiesi, M., Boksenbojm, E., Maes, C., and Wynants, B. (2010) *J. Stat. Phys.*, **139**, 492.

64 Seifert, U. and Speck, T. (2010) *Europhys. Lett.*, **89**, 10007.

65 Mehl, J., Blickle, V., Seifert, U., and Bechinger, C. (2010) *Phys. Rev. E*, **82**, 032401.

66 Gomez-Solano, J.R., Petrosyan, A., Ciliberto, S., and Maes, C. (2011) *J. Stat. Mech.*, P01008.

67 Blickle, V. *et al.* (2007) *Phys. Rev. Lett.*, **98**, 210601.

68 Gomez-Solano, J.R., Chetrite, R., Gawedzki, G., Petrosyan, A., and Ciliberto, S. (2009) *Phys. Rev. Lett.*, **103**, 040601.

69 Blickle, V., Speck, T., Seifert, U., and Bechinger, C. (2007) *Phys. Rev. E*, **75**, 060101(R).

70 Evans, D.J., Cohen, E.G.D., and Morriss, G.P. (1993) *Phys. Rev. Lett.*, **71**, 2401.

71 (a) Evans, D.J., and Searles, D.J., (1994) *Phys. Rev. E*, **50**, 1645. (b) Evans, D.J., and Searles, D.J., (2002) *Adv. Phys.*, **51** (7), 1529. (c) Evans, D.J., Searles, D.J., and Rondoni, L. (2005) *Phys. Rev. E*, **71** (5), 056120.

72 (a) Gallavotti, G., and Cohen, E.G.D., (1995) *Phys. Rev. Lett.*, **74**, 2694. (b) Gallavotti, G., and Cohen, E.G.D. (1995) *J. Stat. Phys.*, **80** (5–6), 931.

73 Lebowitz, J.L. and Spohn, H. (1999) *J. Stat. Phys.*, **95**, 333.

74 Kurchan, J. (1998) *J. Phys. A: Math. Gen.*, **31**, 3719.

75 Searles, D.J., Rondoni, L., and Evans, D.J. (2007) *J. Stat. Phys.*, **128**, 1337.

76 Rondoni, L. and Meőja-Monasterio, C. (2007) *Nonlinearity*, **20**, R1.

77 Zamponi, F. (2007) *J. Stat. Mech.*, P02008.

78 Gallavotti, G., Bonetto, F., and Gentile, G. (2004) *Aspects of the Ergodic, Qualitative and Statistical Theory of Motion*, Springer, Berlin.

79 Ciliberto, S. and Laroche, C. (1998) *J. Phys. IV (France)*, **8**, 215.

80 Ohkuma, T. and Ohta, T. (2007) *J. Stat. Mech.*, P10010.

81 Toyabe, S. and Sano, M. (2008) PRE 77, 041403.

82 arXiv:1203.3571.

5
Recent Progress in Fluctuation Theorems and Free Energy Recovery

Anna Alemany, Marco Ribezzi-Crivellari, and Felix Ritort

5.1
Introduction

Free energy (FE) is a key quantity in thermodynamics that combines the effects of two opposite trends in nature: the minimization of energy led by deterministic or ordering forces and the maximization of entropy induced by the randomness or noise of the environment. Free energy minimization determines the spontaneous evolution and final equilibrium state attained by thermal systems. The knowledge of free energy differences between different states of a system provides a deep insight into its response to external perturbations. This applies to systems ranging from single molecules to the entire universe.

One classical approach to free energy measurements is to quasi-statically (i.e., infinitely slow) change a control parameter and measure the net amount of energy exchanged between the system and the environment. The ultimate goal of these measurements is the determination of the free energy branches, which describe the dependence of the free energy of the system on the control parameter. At macroscopic scales, the experimental outcomes of thermodynamic manipulations, be it quasi-static or not, do not significantly change over different repetition of the same protocol (meaning same system, source, initial state, and process). Yet, this is true only for large enough systems. The situation is very different at microscopic scales, where outcomes from repetitions of an identical experimental protocol can be different. The origins of these fluctuations are thermal random forces caused by the environment that deliver uncontrolled amounts of energy to the system as the transformation is carried out.

Recent theoretical developments in nonequilibrium physics have shown how, using these fluctuations, it is possible to recover free energy differences and free energy landscapes from irreversible work measurements in thermodynamic transformations of small systems. These developments go under the name of fluctuation theorems and are reviewed in detail in other chapters of this book. From an experimental viewpoint, these theoretical results provide a powerful method to obtain thermodynamic quantities from experiments carried out in nonequilibrium conditions. In this chapter, we will succinctly review some of the main

Nonequilibrium Statistical Physics of Small Systems: Fluctuation Relations and Beyond, First Edition.
Edited by Rainer Klages, Wolfram Just, and Christopher Jarzynski.

developments in this field in the context of single-molecule experiments and explain how fluctuation theorems can be applied to the measurement of free energies of folding in molecular structures. A few topics reviewed in this chapter have been already discussed in a previous paper by us [1].

The plan of this chapter is as follows: Section 5.2 contains a brief description of equilibrium-based bulk and computational methods to recover free energy differences. Section 5.3 describes the main experimental techniques used to manipulate small systems with special emphasis on DNA stretching experiments with optical tweezers (OTs). Section 5.4 overviews fluctuation relations (FRs) and their usefulness to recover free energy differences from irreversible work measurements. Section 5.5 discusses issues related to the correct definition of control parameters, configurational variables, and mechanical work in small systems. The right definition of mechanical work has generated some debate and confusion during the past years and we found this volume a good opportunity to clarify this point. In Section 5.6 we analyze an important extension of the Crooks fluctuation relation to the case where states are in partial (rather than global) equilibrium. Such extension allows a full recovery of free energy branches and partially equilibrated states. Finally, we discuss the main limitations of free energy recovery when dissipation is too large and free energy estimates obtained from bidirectional and unidirectional methods too biased. The chapter ends with some conclusions.

5.2 Free Energy Measurement Prior to Fluctuation Theorems

The understanding of many biophysical processes does often require the knowledge of the underlying free energy differences. For instance, this is the case in the study of the folding behavior of proteins and nucleic acids or ligand binding by proteins. Given the interest of these phenomena, several methods were developed to reliably measure FE changes in experiments or to estimate them through the combination of statistical mechanics and computational techniques. Before discussing the recently introduced nonequilibrium methods based on fluctuation relations, we review the classical experimental and computational approaches to FE recovery.

5.2.1 Experimental Methods for FE Measurements

Equilibrium thermodynamics offers several equalities between FE differences and directly measurable quantities: The most famous of these equalities links the FE difference between two equilibrium states (A and B) to the reversible work $W_{\mathrm{rev}}^{\mathrm{A}\to\mathrm{B}}$ needed to bring the system from state A to state B through a series of equilibrium states:

$$\Delta G_{\mathrm{AB}} = W_{\mathrm{rev}}^{\mathrm{A}\to\mathrm{B}}, \tag{5.1}$$

where G is the Gibbs free energy, as we have in mind a situation in which temperature T and pressure P are controlled. A second important relation, suitable for the study of FE changes in chemical reactions, concerns the FE difference between reagents (R) and products (P) to the equilibrium constant K of a reaction [2]:

$$\Delta G_{\mathrm{RP}} = -N_{\mathrm{A}} k_{\mathrm{B}} T \log K = -N_{\mathrm{A}} k_{\mathrm{B}} T \log \frac{[P]}{[R]}, \tag{5.2}$$

where $[\cdot]$ denote the molar concentrations, N_{A} is the Avogadro number ($6.02 \times 10^{23}\ \mathrm{mol}^{-1}$), and k_{B} is the Boltzmann constant ($1.38 \times 10^{-23}\ \mathrm{J\,K}^{-1}$). It is important to stress that these equations pertain to the domain of equilibrium thermodynamics and the validity of any result obtained through them is to be questioned whenever the assumption of thermodynamic equilibrium is violated.

The denaturation transition in proteins and nucleic acids (DNA and RNA) can be used to illustrate some of the equilibrium FE measurement methods. Denaturation, that is, the loss of the native folded configuration, in proteins and nucleic acids can be induced by different means. Prominent examples are thermal and chemical denaturation. In thermal denaturation experiments, unfolding is induced raising the temperature of the solution in which the molecules are stored. In chemical denaturation experiments, unfolding is instead obtained raising the concentration of a denaturant (e.g., urea or guanidine), which interferes with the noncovalent interactions (hydrogen bonds and van der Waals forces) that stabilize the native configuration. The FE difference between the native and the unfolded states can be estimated by measuring the fraction of unfolded molecules as a function of the temperature or denaturant concentration and using Eq. (5.2). The following are some of the most common techniques available to measure these populations:

- **UV absorbance spectroscopy:** This uses the sample's absorbance at 287 nm as a probe of the relative population of the folded and unfolded states. This is possible because some amino acid residues change their absorbance according to the surrounding environment. The optical density of trypsin, for example, is decreased when it passes from an hydrophobic environment (e.g., inside a folded protein) to an hydrophilic environment (e.g., exposed to a polar solvent).
- **Circular dichroism (CD) spectroscopy:** This compares the differences in absorption of left-handed versus right-handed circularly polarized light. The double helix of nucleic acids and the different types of secondary structure in proteins (α-helix, β-sheet, and random coil) have specific CD signatures in the far UV region. From the CD spectrum of a sample, it is possible to estimate how the fraction of each of these structures changes at different denaturation conditions [3].
- **Fluorescence spectroscopy:** This uses the natural fluorescence of tryptophan, an amino acid residue. Tryptophan is excited by a monochromatic light source with $\simeq 290$ nm wavelength and the resulting fluorescence spectrum is analyzed through a monochromator (e.g., diffraction grating). Both the yield of tryptophan fluorescence and the peak of the spectrum are sensitive to the environment. A tryptophan buried in the hydrophobic core of a folded protein has a high yield

and gives intense fluorescence, while in an hydrophilic environment, it gives lower intensity. The fraction of folded proteins contained in a sample can be estimated measuring its fluorescence intensity.

Another common method for FE measurements is calorimetry. Most calorimetric methods rest on the thermodynamic equality relating FE to the specific heat at constant pressure c_P:

$$\left(\frac{\partial^2 G}{\partial T^2}\right)_P = -\frac{c_P}{T}, \tag{5.3}$$

which allows FE recovery via the Gibbs–Helmholtz equation.

- **Differential scanning calorimetry (DSC):** This is one of the most usual calorimetric techniques. DSC determines the constant pressure heat capacity (c_P) of a protein or nucleic acid in solution as function of the temperature [4]. The heat capacity is obtained as the ratio between the heat flux toward the sample and the temperature increase in the sample. Unfolding of the solute (protein or nucleic acids) induces a peak in the heat capacity, which is due to the extra heat needed to disrupt the noncovalent interactions that stabilize the native structure. The enthalpy or FE changes across the transition are measured by integration of the c_P.

FE recovery from both thermal and chemical denaturation experiments is based on a series of assumptions. These include matters such as the order of the denaturation transition, the reversibility of the denaturation process, and the number of states in the transition region [5]. Moreover, the interpretation of denaturation experiments requires the formulation of a model for the effect of the denaturant on the FE, an issue that is often dealt with by a linear extrapolation. Nevertheless, it is the need for reversibility that puts fundamental limits on the applicability of these classical techniques. For example, in the last years, many kinetically stable proteins, that is, proteins whose native is not thermodynamically but just kinetically stable, have been discovered [4]. The stability of these proteins is difficult to be investigated by the aforementioned techniques.

5.2.2
Computational FE Estimates

The combination of statistical mechanics and the continuously growing computational power have paved the way for numerical techniques to estimate FE changes. We will review only two classical approaches of free energy calculations that are a limit case of the more general techniques based on fluctuation relations: thermodynamic integration (TI) [6] and the free energy perturbation method (FEP) [7].

Let us consider a system of N particles contained in a three-dimensional box at temperature T. The Hamiltonian of this system will be denoted by H_λ, where λ is a control parameter that distinguishes different states of the system (e.g., charge or dipole moment of the particles). The Helmholtz FE, ($F(N, V, T)$), which is the natural thermodynamic potential when temperature (T) and volume (V) are controlled, is linked to the canonical partition function $Z_\lambda(N, V, T)$ by

$$F_\lambda(N, V, T) = -k_B T \ln Z_\lambda(N, V, T), \tag{5.4}$$

$$Z_\lambda(N, V, T) = \int \exp\left(-\frac{H_\lambda}{k_B T}\right) d^{3N}x d^{3N}p. \tag{5.5}$$

The FE difference ΔF between two states of a system characterized by two different values of the control parameter $\lambda(\lambda_A, \lambda_B)$ is according to Eq. (5.4):

$$\Delta F = -k_B T \ln \frac{Z_{\lambda_B}}{Z_{\lambda_A}}. \tag{5.6}$$

Using the definition of partition function introduced in Eqs. (5.5) and (5.6), it can be transformed as follows:

$$\begin{aligned}\Delta F &= -k_B T \log \frac{\int \exp(-(H_{\lambda_B}/k_B T)) d^{3N}x d^{3N}p}{\int \exp(-(H_{\lambda_A}/k_B T)) d^{3N}x d^{3N}p} \\ &= -k_B T \log \frac{\int \exp(-((H_{\lambda_B} - H_{\lambda_A})/k_B T)) \exp(-(H_{\lambda_A}/k_B T)) d^{3N}x d^{3N}p}{\int \exp(-(H_{\lambda_A}/k_B T)) d^{3N}x d^{3N}p} \\ &= -k_B T \log \left\langle \exp\left(-\frac{H_{\lambda_B} - H_{\lambda_A}}{k_B T}\right)\right\rangle_A\end{aligned} \tag{5.7}$$

The FE difference between two states can thus be obtained by averaging the exponential of the Hamiltonian difference, with respect to the Boltzmann distribution corresponding to one state. This could be done with Monte Carlo simulations. The use of Eq. (5.7) is limited by a practical problem: If the Hamiltonians defining the two states are very different, then the "typical" configurations for the two states will also be very different. As a consequence, the convergence of a numerical calculation can be impractically slow. One possible solution is to use the TI method, which uses a slightly modified version of Eq. (5.7). Taking the derivative of F_λ with respect to λ, we get

$$\frac{dF}{d\lambda} = \frac{\int \partial_\lambda H_\lambda \exp(-(H_\lambda/k_B T)) d^{3N}x d^{3N}p}{\int \exp(-(H_\lambda/k_B T)) d^{3N}x d^{3N}p} = \langle \partial_\lambda H_\lambda \rangle_\lambda, \tag{5.8}$$

so the calculation of the FE difference between two states characterized by very different values of λ can be reduced to a series of smaller steps. The interested reader will find rich information on free energy calculations, from the initial stages to the recent developments in Ref. [8].

5.3 Single-Molecule Experiments

The boost experienced by nanotechnologies from 1990 to the present day have made it possible to manipulate a single molecule at a time by exerting forces in the range of piconewtons and measure molecular distances with nanometric resolution.

In traditional bulk experiments, like calorimetry or UV absorbance, samples under study contain a large number N of molecules. Experimental measurements are the result of an average over the behavior of all the molecules. Fluctuations are on the order of $1/\sqrt{N}$ and become negligible. In contrast, single-molecule systems provide a powerful example of the so-called *small systems*, where fluctuations play an important role as they are on the same order of magnitude as the energy exchanged by the system with the environment [9, 10].

Small systems are not restricted to single molecules. On the one hand, fluctuations in the measurement of energy exchanges with the surroundings of any macroscopic system operating at short enough timescales could play an important role. On the other hand, if measurements in a molecular system are carried out over long times (compared to the heat diffusion time of the system), fluctuations become small and irrelevant. In this chapter, we will focus our attention on single-molecule experiments, a canonical example of small system manipulation.

5.3.1
Experimental Techniques

There are several techniques that allow the manipulation of a single molecule at a time. Mostly used are the atomic force microscopy (AFM), magnetic tweezers (MTs), and laser optical tweezers. Differences between the three techniques appear when considering the operating range in force and distance and the choice of the control parameter to carry out experiments (see Section 5.5).

- The AFM is based on the principle that a cantilever with a tip can sense the roughness of the surface and deflect by an amount that is proportional to the proximity of the tip to the surface [11, 12]. The typical experimental setup in force spectroscopy experiments is as follows: A surface is coated with the desired molecules and the AFM tip is coated with molecules that can bind (either specifically or nonspecifically) to the molecules on the substrate. By moving the tip to the substrate, a contact between the tip and one of the molecules adsorbed on the substrate is made. Retraction of the tip at constant speed allows to measure the deflection of the cantilever in real time, providing (when its stiffness in known) the force acting on the molecule as a function of its end-to-end distance. The main drawback in AFM experiments is the presence of undesired interactions or the interaction of many molecules between the tip and the substrate. Different strategies, like the design of polyproteins, have been specifically developed to overcome these limitations.

  The AFM covers forces between 20–1000 pN depending on the stiffness of the cantilever. Typical values of the stiffness are in the range of 10–1000 pN nm^{-1}. Resolution in the AFM is limited by thermal fluctuations. When the cantilever stage is held at a constant position, the force acting on the tip and the extension between tip and substrate fluctuate. The respective r.m.s.d. are given by

$\sqrt{\langle \delta x^2 \rangle} = \sqrt{k_B T/k} \sim 2$ Å and $\sqrt{\langle \delta f^2 \rangle} = \sqrt{k_B T k} \sim 10$ pN, where k_B is the Boltzmann constant, T is the absolute temperature of the environment, and k is the stiffness of the cantilever, taken equal to 100 pN nm^{-1}. AFMs are ideal to investigate strong intermolecular and intramolecular interactions, for example, pulling experiments in biopolymers such as polysaccharides, proteins, and nucleic acids [13].

- MTs are based on the principle that a magnetized bead with dipole moment μ experiences a force f when immersed in a magnetic field gradient B equal to $f = \mu \nabla B$. The typical experimental setup is as follows: A bead is trapped in the magnetic field gradient generated by two strong magnets and a molecule is attached between the surface of the magnetic bead and a glass surface. Molecules are pulled or twisted by moving the translation stage that supports the magnets. A microscope objective with a CCD camera is used to determine the position of the bead, equal to the extension of the molecule x. Force is measured by using the expression $f = k_B T x / \langle \delta x^2 \rangle$ [14, 15].

  Typical values of the magnetic trap stiffness are 10^{-4} pN nm^{-1}, which induces large fluctuations in the extension of the molecule, on the order of 20 nm. The typical range of operating forces is 10^{-2}–10 pN, where the maximum value of the force depends on the size of the magnetic bead. When the magnet stage is kept fixed, the force acting on the bead can be kept constant because the spatial region occupied by the bead is small enough for the magnetic field gradient to be considered uniform. Therefore, although the bead position fluctuates, the force is always constant. MTs have been extensively used to investigate elastic and torsional properties of DNA molecules and processes involving molecular motors [16].

- The principle of OT is based on the optical gradient force generated by a focused beam of light acting on an object with an index of refraction higher than that of the surrounding medium [17, 18]. The typical experimental setup is as follows: A micrometer-sized polystyrene or silica bead is captured in the optical trap, while another bead is immobilized by air suction in the tip of a micropipette. Molecular biology tools are employed to insert a molecule between molecular handles, and the whole molecular complex is inserted between the two beads (Figure 5.1b). In a more recent setup, dual traps are also used.

  The operating force range in OT is (0.5–100) pN, depending on the size of the bead (on the order of 1–3 μm) and the power of the laser (few hundred milliwatts). To a very good approximation, the trapping potential is harmonic, $f = -kx$, where k is the stiffness constant of the trap and x is the distance of the trapped bead to the center of the optical trap. Typical values of the stiffness of the optical trap are 10^{-2}–10^{-4} times smaller than that for AFM tips; therefore, force resolution is at least 10 times better, on the order of 0.1 pN. Depending on the experimental setup, the spatial resolution can reach the nanometer level only in carefully isolated environments (absence of air currents, mechanical and acoustic vibrations, and temperature oscillations). OTs have been widely used to investigate nucleic acids and molecular motors.

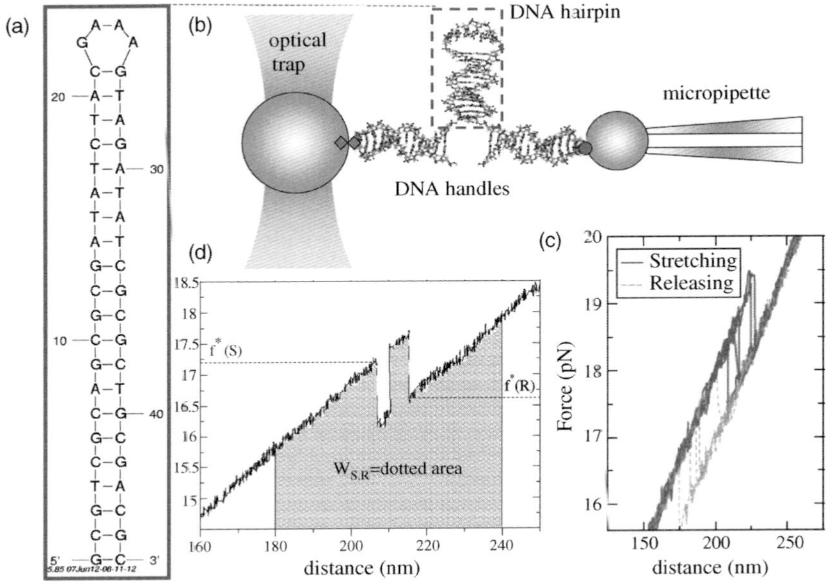

Figure 5.1 Mechanical folding/unfolding of DNA hairpins. (a) The sequence of a DNA hairpin with a 21 bp stem ending in a tetraloop. (b) Experimental setup (figure not to scale). A molecular construct made of the hairpin shown in (a) flanked by two dsDNA handles (29 bp each) is tethered between two micrometer-sized beads. In the experiments, the trap is moved relative to the pipette at speeds ranging from 10 to 1000 nm s^{-1}. (c) Different force cycles recorded at 300 nm s^{-1}. Dark-colored curves indicate the stretching parts of the cycle, whereas light-colored curves indicate the releasing parts of the cycle. Note that the forces of unfolding and refolding are random due to the stochastic nature of the thermally activated unfolding/folding process. The marked hysteresis is a signature of an irreversible process. (d) Measurement of work for a single trajectory. It is given by the area below the force–distance curve integrated between two trap positions. Trap distances are relative. Note that there might be more than one unfolding or refolding events along each trajectory. $f^*(S)(f^*(R))$ defines the first rupture force in the unfolding (refolding) process. (Figure taken from Ref. [19].)

5.3.2
Pulling DNA Hairpins with Optical Tweezers

Nucleic acids stretched under force are excellent model systems to explore the non-equilibrium physics of small systems. DNA hairpins are one of the most simple and versatile molecular systems one can think about. They consist of a stable DNA double helix, called stem, ended by a loop (Figure 5.1a). When force is applied to both ends of the stem by pulling them apart, the base pairs establishing the stem become progressively disrupted until the DNA hairpin unfolds completely, around 15 pN (depending on the salt concentration, the temperature, and the pulling speed).

The force range accessible with an optical tweezers instrument makes it an appropriate tool to study the behavior of DNA and RNA hairpins under mechanical force. In a pulling experiment with optical tweezers, the two ends of a DNA hairpin are linked to double-stranded DNA handles [20, 21]. One handle is attached to a polystyrene bead that is immobilized by air suction in the tip of a micropipette, whereas the other handle is attached to another polystyrene bead that is captured in the optical trap (Figure 5.1b). Handles are used to avoid spurious interactions between the beads and the molecule under study. The relative distance λ between the center of the trap and the tip of the micropipette is the control parameter in these experiments. At the beginning of an unzipping experiment, both λ and the force applied to the system are low and the hairpin is in its folded conformation. Then, λ is increased at a constant pulling speed until a maximum force is achieved, typically around 30 pN, where the hairpin is unfolded. In a rezipping experiment, we apply the time-reversed protocol for $\lambda(t)$ and the DNA hairpin recovers its native conformation. When unzipping short hairpins, a jump in force is generally observed in the force–distance curve (FDC), characteristic of two-states systems: either the hairpin is folded or it is unfolded (Figure 5.1d). In contrast, large hairpins tend to display different levels of rupture force corresponding to different intermediates. The unraveling of rupture process is thermally activated, and the force at which the hairpin unfolds or refolds changes in each cycle, as shown in Figure 5.1c. Depending on the pulling speed, FDCs reveal different degree of hysteresis.

5.4 Fluctuation Relations

Fluctuation relations establish fundamental equalities between the work applied in nonequilibrium systems and FE differences. As such they overcome some of the limitations of classical FE measurement techniques, introduced in Section 5.2.

To better understand FR, consider a generic system in thermal equilibrium that is transiently driven out of equilibrium during the time interval $[0,t_B]$ by varying λ according to a protocol $\lambda(t)$ (hereafter referred to as the forward protocol) from an initial value $\lambda(0) = \lambda_A$ to a final value $\lambda(t_B) = \lambda_B$. The work W measured along the trajectory significantly changes in each realization because it is not possible to control the configurations explored by the system: work fluctuations are on the same order of magnitude as the energy exchanges with the environment. According to the second law of thermodynamics, the average of the work over multiple independent realizations of the experiment is larger than or equal (if the perturbation is slow enough) to the FE difference between the initial and the final states, $\langle W \rangle = W_{\text{rev}} + \langle W \rangle_{\text{dis}} \geq \Delta G$. Thus, the faster the perturbation, the larger the dissipated work $\langle W \rangle_{\text{dis}}$. Let us also consider the reversed process where the system starts in equilibrium at λ_B and λ is varied according to the time reversal protocol, $\lambda(t_B - t)$, until reaching the final value λ_A (Figure 5.2).

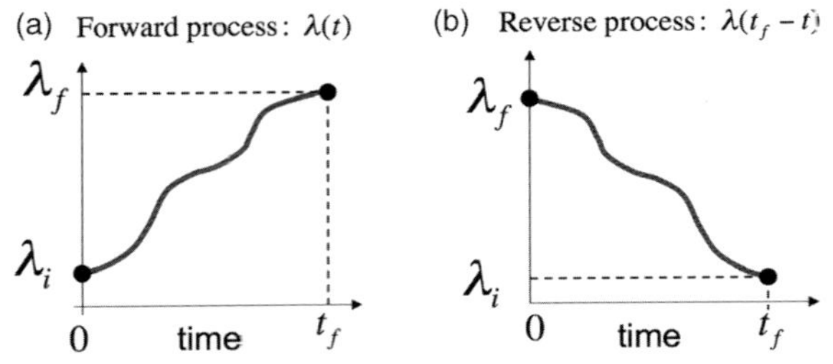

Figure 5.2 Forward and reverse paths. (a) An arbitrary forward protocol. The system starts in equilibrium at λ_A and is transiently driven out of equilibrium until λ_B. At λ_B, the system may or may not be in equilibrium. (b) The reverse protocol of (a). The system starts in equilibrium at λ_B and is transiently driven out of equilibrium until λ_A. At λ_A, the system may or may not be in equilibrium.

An important FR is the Crooks fluctuation relation (CFR) that reads as [22, 23]

$$\frac{P_F(W)}{P_R(-W)} = \exp\left(\frac{W - \Delta G}{k_B T}\right), \tag{5.9}$$

where $P_F(W)$ is the probability density function of the work performed along the forward protocol, $P_R(-W)$ is the probability density function of the work performed along the reversed protocol, and $\Delta G = G(\lambda_B) - G(\lambda_A)$ is equal to the Gibbs free energy difference between the equilibrium states at λ_B and λ_A. In single-molecule experiments, pressure and temperature are usually controlled, so the Gibbs free energy is the relevant quantity. An important consequence of Eq. (5.9) is that the FE difference corresponds to the work equally probable in both the forward and the reversed protocol, which is always the same no matter how far from equilibrium the system is brought under the perturbative protocol. An equally important result is the prediction of the existence of trajectories where the dissipated work is negative, meaning that heat from the environment is absorbed and converted into useful work. It must be stressed that there is no violation of the second law because thermodynamics does not focus on a single trajectory but on the average over an infinite number of them.

Particular corollaries of the CFR can be obtained by integration of Eq. (5.9). For instance, if we multiply and divide the LHS of Eq. (5.9) by a generic function $\phi(W)$ and integrate over W, we get

$$e^{-(\Delta G/k_B T)} = \frac{\langle e^{-(W/k_B T)}\phi(W)\rangle_F}{\langle \phi(-W)\rangle_R}, \tag{5.10}$$

where $\langle \cdots \rangle_{F(R)}$ denote the average over forward(reversed) trajectories. When $\phi(W) = 1$, we get the well-known Jarzynski equality, $e^{-\Delta G/k_B T} = \langle e^{-W/k_B T}\rangle$ [24], that has been used for FE recovery [25, 26]. However, this expression is strongly biased for a finite number of measurements [27, 28]. More information is given in Section 5.7. Bidirectional methods that combine information from the forward and reverse processes and use the CFR have proven to be more predictive [29–31].

5.4.1 Experimental Validation of the Crooks Equality

The CFR was experimentally tested in 2005 in RNA pulling experiments with laser tweezers [32] showing this to be a reliable and useful methodology to extract FE differences between states that could not be measured with bulk methods. The forward(reversed) process is identified with the unzipping(rezipping) of the hairpin that starts in $\lambda_A(\lambda_B)$ where the hairpin is folded(unfolded). The work is evaluated as the area below the experimental FDC (Figure 5.1d):

$$W = \int_{\lambda_A}^{\lambda_B} f \, d\lambda. \tag{5.11}$$

The definition of work is not straightforward. For more information, see Section 5.5. In Figure 5.3a, we show the experimental $P_F(W)$ and $P_R(-W)$ obtained in pulling experiments at three different pulling speeds for the hairpin sketched in Figure 5.1a (results obtained in the Small Biosystems Lab in Barcelona using dual beam miniaturized optical tweezers) [19, 33, 34]. Note that the crossing point of the histograms, equal to the FE difference between the final and the initial states, does not depend on the pulling speed, whereas hysteresis effects (and therefore dissipation) increase.

A first validation of Eq. (5.9) is shown in Figure 5.3b, where the logarithm of the ratio between the probabilities $P_F(W)$ and $P_R(-W)$ is represented versus $W - \Delta G$. Note that data collapse in a single straight line with slope almost equal to 1.

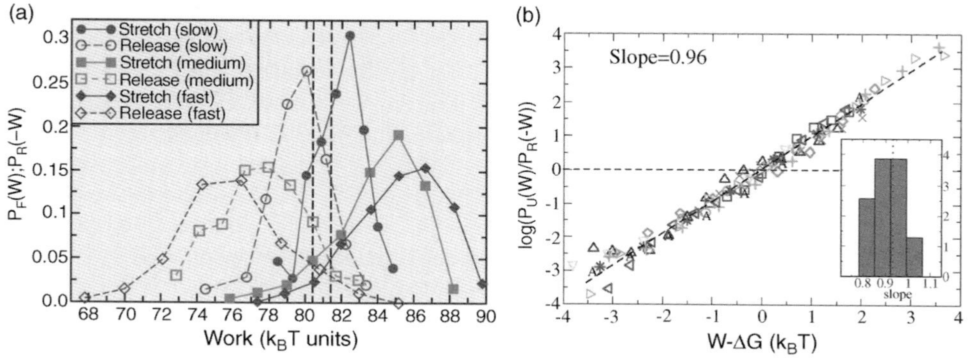

Figure 5.3 The Crooks fluctuation relation. (a) Work distributions for the hairpin shown in Figure 5.1 measured at three different pulling speeds: 50 nm s^{-1} (diamonds), 100 nm s^{-1} (squares), and 300 nm s^{-1} (circles). Unfolding or forward (continuous lines) and refolding or reverse work distributions (dashed lines) cross each other at a value of 81.0 ± 0.2 k_BT independent of the pulling speed. (b) Experimental test of the CFR for 10 different molecules pulled at different speeds. The log of the ratio between the unfolding and refolding work distributions is equal to $(W - \Delta G)$ in k_BT units. The inset shows the distribution of slopes for the different molecules that are clustered around an average value of 0.96. (Figure taken from Ref. [19].) (For a color version of this figure, please see the color plate at the end of this book.)

Therefore, an estimation of the FE of the system is measured by identifying the value of work at which $\log[P_F(W)/P_R(-W)] = 0$.

However, a proper estimation of the FE using Eq. (5.9) is limited by the number of unzipping and rezipping cycles. If we do not have enough data, the tail of the work distributions is not well captured and a bias is introduced in the FE estimation. Several procedures have been proposed in literature to beat this inconvenience (more information is given in Section 5.7). Here, we present the Bennett acceptance ratio method [29] as a first approach to estimate FE differences. This method consists of using the following function $\phi(W)$ in Eq. (5.10):

$$\phi(W) = \frac{1}{1 + (n_F/n_R)\exp\left((W - \Delta G)/k_B T\right)}, \tag{5.12}$$

where n_F and n_R are the number of measured forward and reversed trajectories. This function is known to minimize the statistical variance of the estimation of the FE difference ΔG. Therefore, we estimate the FE difference as the solution to the transcendental equation:

$$\begin{aligned} \frac{u}{k_B T} &= z(u) \\ &= -\log\left\langle \frac{e^{-(W/k_B T)}}{1 + (n_F/n_R)e^{(W-u)/k_B T}} \right\rangle_F + \log\left\langle \frac{1}{1 + (n_F/n_R)e^{-((W+u)/k_B T)}} \right\rangle_R . \end{aligned} \tag{5.13}$$

The FE estimate is the FE difference of the total system (hairpin, handles, and bead) between the initial and the final states. In order to obtain the FE of formation of the hairpin, we need to subtract the contributions coming from the work performed by the trap to stretch the handles, pull on the bead, and stretch the elastic ssDNA released when the hairpin unfolds [19]. Once we do that, we obtain that the FE of the hairpin sketched in Figure 5.1 and whose experimental data are shown in Figure 5.3 is $61.0 \pm 0.5\ k_B T$, which is in fair agreement with the predicted FE obtained from calorimetric melting experiments, 60.53 $k_B T$ [35, 36].

5.5
Control Parameters, Configurational Variables, and the Definition of Work

In small systems, it is crucial to make a distinction between controlled parameters and noncontrolled or fluctuating variables. Controlled parameters are macroscopic variables that are imposed on the system by the external sources (e.g., the thermal environment) and do not fluctuate with time. In contrast, noncontrolled variables are microscopic quantities describing the internal configuration of the system and do fluctuate in time because they are subject to Brownian forces. Let us consider a typical single-molecule experiment where a protein is pulled by an AFM. In this case, the control parameter is given by the position of the cantilever that determines the degree of stretching and the average tension applied to the ends of the protein. Also, the temperature and the pressure inside the fluidic chamber are controlled parameters. However, the height of the tip with respect to the substrate or the force acting on the protein is fluctuating variable that describes the molecular extension of the protein tethered between tip and substrate. Also, the position of each of the residues along the polypeptide chain is fluctuating variable. Both molecular extension or force and the residues' positions define different types of configurational variables. However, only the former is subject to experimental measurement and therefore we will restrict our discussion throughout this chapter to such kind of experimentally accessible configurational variables. Figure 5.4 illustrates other examples of control parameters and configurational variables. In what follows, we will denote by λ the set of controlled (i.e., nonfluctuating) parameters and by x the set of configurational (i.e., fluctuating) variables. The definition of what are controlled parameters or configurational variables is broad. For example, a force can be a configurational variable and a molecular extension can be a controlled parameter, or vice versa, depending on the experimental setup (Figure 5.4c).

The energy of a system acted by external sources can be generally described by a Hamiltonian or energy function $U(x, \lambda)$. The net variation of U is given by the conservation law:

$$dU = \left(\frac{\partial U}{\partial x}\right) dx + \left(\frac{\partial U}{\partial \lambda}\right) d\lambda = đQ + đW, \tag{5.14}$$

where $đQ$ and $đW$ stand respectively for the infinitesimal heat and work transferred to the system. The previous mathematical relation has a simple physical interpretation. Heat accounts for the energy transferred to the system when the configurational variables change at fixed value of the control parameter. Work is

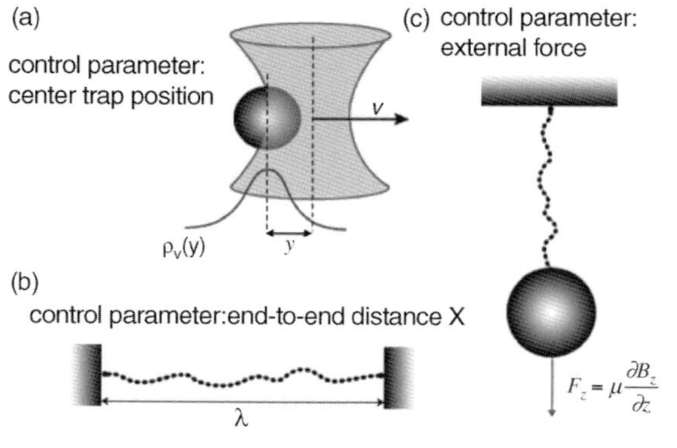

Figure 5.4 Control parameter and configurational variables. Different experimental setups corresponding to different types of control parameters (denoted as λ) or configurational variables (denoted as x). (a) A micrometer-sized bead dragged through water. λ could be the center of the trap measured in the lab (i.e., fixed to the water) frame, whereas x is the displacement of the bead, indicated as y, with respect to the center of the trap. (b) A polymer tethered between two surfaces. λ is the distance between the surfaces and x is the force acting on the polymer. (c) A polymer stretched with magnetic tweezers. λ is the force acting on the magnetic bead and x is the molecular extension of the tether.

the energy delivered to the system by the external sources upon changing the control parameter for a given configuration. The total work performed by the sources on the system when the control parameter is varied from λ_A to λ_B is given by

$$W = \int_{\lambda_A}^{\lambda_B} đW = \int_{\lambda_A}^{\lambda_B} \left(\frac{\partial U}{\partial \lambda}\right) d\lambda = \int_{\lambda_A}^{\lambda_B} F(x,\lambda) d\lambda, \tag{5.15}$$

where $F(x, \lambda)$ is a generalized force defined as

$$F(x,\lambda) = \left(\frac{\partial U}{\partial \lambda}\right). \tag{5.16}$$

It is important to stress that the generalized force is not necessarily equal to the mechanical force acting on the system. In other words, $F(x, \lambda)$ is a configurational dependent variable conjugated to the control parameter λ and has dimensions of [energy]/[λ], which are not necessarily Newtons. In the example shown in Figure 5.4c, the control parameter is the magnetic force $\lambda \equiv f$ and the configurational variable x is the molecular extension of the polymer. The total Hamiltonian of the system is then given by $U(x,f) = U_0(x) - fx$, where $U_0(x)$ is the energy of the system at $\lambda = f = 0$. In other words, the external force f shifts all energy levels (defined by x) of the original system by the amount $-fx$. The generalized force is then given by $F(x,f) = -x$ (i.e., it has the dimensions of a length) and $đW = -xdf$. The fact that $đW$ is equal to $-xdf$ and not equal to fdx has generated some controversy over the past years [37]. Below we show how this distinction is already important for the simplest case of a bead in the optical trap. In Section 5.5.1, we also show how the physically sound definition of mechanical work is amenable to experimental test.

5.5.1
About the Right Definition of Work: Accumulated versus Transferred Work

In a pulling experiment with OT (Section 5.3.2), there are two possible representations of the pulling curves (Figure 5.5b). In one representation, the force is plotted versus the relative trap–pipette distance (λ), the so-called force–distance curve (hereafter referred to as FDC). In the other representation, the force is plotted versus the relative molecular extension (x), the so-called force–extension curve (hereafter referred to as FEC). In the optical tweezers setup, $\lambda = x + y$, where y is the distance between the bead and the center of the trap. The measured force is given by $F = ky$, where k is the stiffness of the trap. The areas below the FDC and the FEC define two possible work quantities, $W = \int_{\lambda_A}^{\lambda_B} F\, d\lambda$ and $W' = \int_{x_A}^{x_B} F dx$. If the bead and the molecular system are in mechanical equilibrium, then the average forces acting on the bead and the molecule are identical. From the relation $d\lambda = dx + dy$, we get

$$W = W' + W_b = W' + \frac{F_B^2 - F_A^2}{2k}, \tag{5.17}$$

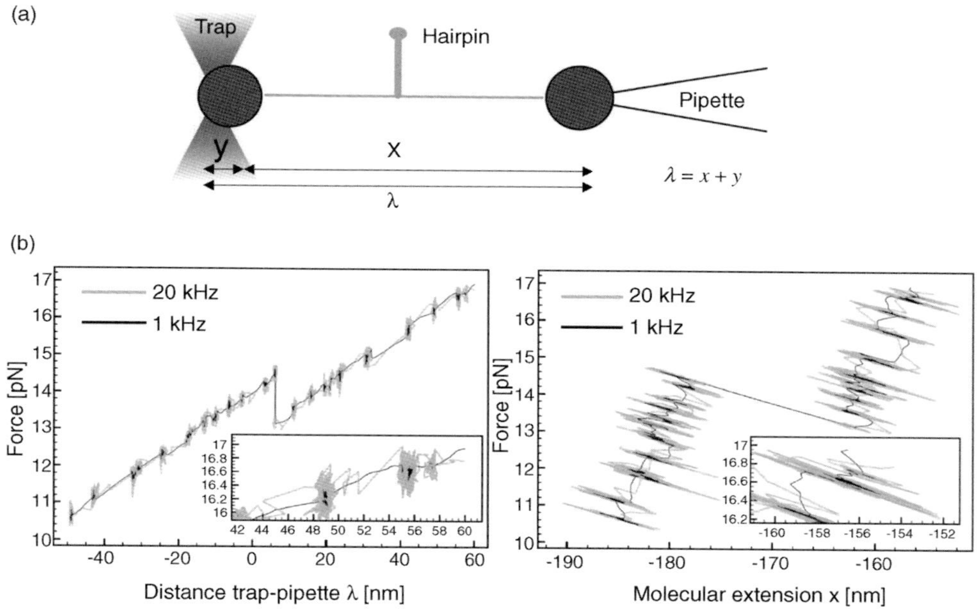

Figure 5.5 FDC versus FEC. (a) Experimental setup and different variables. (b) The FDC and FEC are defined as the curves obtained by plotting the force versus the trap position or the molecular extension, respectively. Although force fluctuations in both types of curves show a dependence on the bandwidth of the measurement (black, 1 kHz; gray, 20 kHz) only in the FEC, the measurement of the work is very sensitive to such fluctuations. (Figure taken from Ref. [38].)

where F_A, F_B are the initial and final forces along a given trajectory. W is often called the total accumulated work and contains the work exerted to displace the bead in the trap, W_b, and the work transferred to the molecular system, W' (therefore, receiving the name of transferred work) [38, 39]. The term W_b appearing in Eq. (5.17) implies that W, W' cannot simultaneously satisfy the CFR. What is the right definition of the mechanical work? In other words, which work definition satisfies the CFR? The answer to our question is straightforward if we correctly identify which are the control parameters and which are the configurational variables. In the lab frame defined by the pipette (or by the fluidic chamber to which the pipette is glued), the control parameter λ is given by the relative trap–pipette distance, whereas the molecular extension x stands for the configurational variable. Note that due to the noninvariance property of the CFR under Galilean transformations, y cannot be used as configurational variable because it is defined with respect to the comoving frame defined by the trap. The total energy of the molecular system is then given by $U(x,\lambda) = U_m(x) + (k/2)(\lambda - x)^2$, where $U_m(x)$ is the energy of the molecular system. From Eq. (5.16) and using $\lambda = x + y$, we get $F = ky = k(\lambda - x)$. From Eq. (5.15), we then conclude that the mechanical work that satisfies the CFR is the accumulated work W rather than the transferred work W'. We remark a few relevant facts:

1) **The transferred work W' does not satisfy the CFR and depends on the bandwidth of the measurement:** The FDC and FEC are sensitive to the bandwidth or data acquisition rate of the measurement (Figure 5.5b). Although W is insensitive to the bandwidth, W' is not (Figure 5.6a). This difference is very important because it implies that the bandwidth dependence implicit in the boundary term in Eq. (5.17) (the power spectrum of the force depends on the bandwidth if this is much smaller than the corner frequency of the bead) is fully contained in W'. Operationally it is much easier to use W rather than W'. As shown in Figure 5.6b, W satisfies the CFR, whereas W' does not. The logarithm of the ratio $\log(P_F(W')/P_R(-W'))$ plotted versus W'/k_BT is strongly

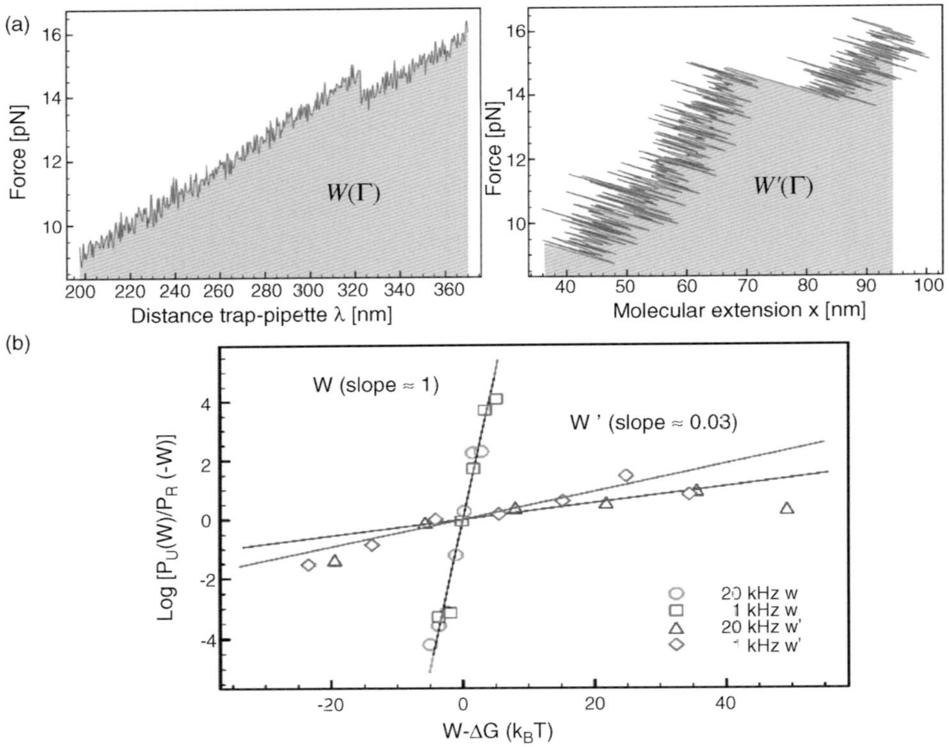

Figure 5.6 Accumulated (W) versus transferred (W') work. (a) The two work quantities for a given experimental trajectory. Note that the effect of bandwidth-dependent force fluctuations is much larger for W' compared to W, showing the importance of the boundary term Eq. (5.17). (b) Experimental test of the CFR. When using W, the CFR is satisfied at all bandwidths. However, when we use W', the CFR is strongly violated and depends on the measurement bandwidth. (Figure taken from Ref. [38].)

bandwidth dependent and exhibits a slope 30 times smaller than 1 (i.e., the slope expected for W from the CFR) [38].

2) **How big is the error made in recovering free energy differences by using W' rather than W?** Despite that W and W' differ only by a boundary term (cf. Eq. (5.17)), one can show that, for the case of the mechanical folding/unfolding of the hairpin, the error in recovering FE differences using the Jarzynski equality can be as large as 100% [38]. The error or discrepancy increases with the bandwidth. Interestingly enough, for small enough bandwidths (but always larger than the coexistence kinetic rates between the folded and unfolded states, otherwise the folding/unfolding transitions are smeared out), fluctuations in the boundary term in Eq. (5.17) are negligible and both W and W' are equally good. This explains why previous experimental tests of the CFR that used W' instead of W produced satisfactory results [32].
3) **Nonequivalence between moving the trap and moving the pipette or chamber:** The noninvariance of the CFR under Galilean transformations suggests that moving the optical trap inside the fluidic chamber should not be necessarily equivalent to moving the pipette glued to the fluidic chamber. We have to distinguish two cases, depending on whether the fluid inside the chamber is dragged (*stick* conditions) or not (*slip* conditions) by the moving chamber. The two scenarios are physically different because in the former case the bead in the trap is subject to an additional Stokes force due to the motion of the fluid. If the fluid is not dragged by the moving chamber (*slip* conditions), then y is the right configurational variable. In this case, $U(y, \lambda) = U_{\mathrm{m}}(\lambda - y) + (k/2)(y)^2$ and the generalized force is equal to $F = U'_{\mathrm{m}}(\lambda - y)$. Note that this F is not equal to the instantaneous force measured by the optical trap, but to the instantaneous force acting on the molecule. Even in case of mechanical equilibrium, the difference between the two instantaneous forces, $U'_{\mathrm{m}}(\lambda - y)$ and ky, produces a net nonnegligible difference term. If the fluid does move with the chamber (*stick* conditions), then y is again the right configurational variable and we recover the main results of this section. Interestingly, all experiments done until now that use motorized stages to move chambers operate in *stick* conditions, so we do not expect experimental discrepancies for the definition of the work.
4) **Other cases where the work definition matters:** As we showed in Figure 5.6, both the CFR and the right definition of work W are amenable to experimental test. Another interesting example where the boundary term is relevant is when the force f (rather than the trap position) is controlled. As we saw in Section 5.5, the work in that case is given by $W_{X_0} = -\int_{f_{\mathrm{A}}}^{f_{\mathrm{B}}} X \mathrm{d}f$, where $X = y + x + X_0$ is the absolute trap–pipette distance. Because X_0 stands for an arbitrary origin, the work W_{X_0} is also a quantity that depends on X_0. This may seem unphysical, but it is not [37]. The CFR is invariant with respect to the value of X_0 as it can be easily checked by writing, $W_{X_0} = W_{X_0=0} - X_0(f_{\mathrm{B}} - f_{\mathrm{A}})$, and using Eq. (5.9) gives $\Delta G_{X_0} = \Delta G_{X_0=0} - fX_0$. If the force f is controlled, then the other work-related quantities such as $W' = \int_{x_{\mathrm{A}}}^{x_{\mathrm{B}}} f \mathrm{d}x$ or $W'' = \int_{X_{\mathrm{A}}}^{X_{\mathrm{B}}} f \mathrm{d}X$ differ from W by finite boundary terms. Again, these terms make the CFR not to be satisfied for W' and W''. These predictions are amenable to experimental test in magnetic

tweezers (where the force is naturally controlled) or in optical tweezers operating in the force clamp mode with infinite bandwidth [40], or even in a force feedback mode with finite bandwidth.

5.6 Extended Fluctuation Relations

The growing theoretical understanding of fluctuation theorems has led to the development of different tools, especially suited to measure FE differences under special requirements. For example, the CFR can be generalized to cases where the system is initially in partial, rather than global, equilibrium in both the forward and the reverse protocols [41]. Suppose we take a system at fixed control parameter λ in thermal equilibrium with a bath at temperature T. The probability distribution over configurational variables x is Gibbsian over the whole phase space S, meaning $P_\lambda^{\mathrm{eq}}(x) = \exp(-E_\lambda(x)/k_\mathrm{B}T)/Z_\lambda$ where Z_λ is the partition function, $Z_\lambda = \sum_{x\in S}\exp(-E_\lambda(x)/k_\mathrm{B}T)$, where $E_\lambda(x)$ is the energy function of the system for a given pair λ, x. We refer to this condition as global thermodynamic equilibrium. However, we might consider a case where the initial state is Gibbsian, but restricted over a subset of configurations $S' \subseteq S$. We refer to this case as partial thermodynamic equilibrium. Partially equilibrated states satisfy $P_{\lambda,S'}^{\mathrm{eq}}(x) = P_\lambda^{\mathrm{eq}}(x)\chi_{S'}(x)Z_\lambda/Z_{\lambda,S'}$, where $\chi_{S'}$ is the characteristic function defined over the subset $S' \subseteq S$ ($\chi_{S'} = 1$ if $x \in S'$ and zero otherwise) and $Z_{\lambda,S'}$ is the partition function restricted to the subset S', that is, $Z_{\lambda,S'} = \sum_{x\in S'}\exp(-E_\lambda(x)/k_\mathrm{B}T)$. The partial FE is then given by $G_{\lambda,S'} = -k_\mathrm{B}T\log Z_{\lambda,S'}$. Let us suppose the following scenario: along the forward process, the system is initially in partial equilibrium in S_A at λ_A and along the reverse process, the system is initially in partial equilibrium in S_B at λ_B. The extended CFR reads

$$\frac{p_\mathrm{F}^{S_\mathrm{A}\to S_\mathrm{B}}}{p_\mathrm{R}^{S_\mathrm{A}\leftarrow S_\mathrm{B}}}\frac{P^{S_\mathrm{A}\to S_\mathrm{B}}(W)}{P^{S_\mathrm{A}\leftarrow S_\mathrm{B}}(-W)} = \exp\left[\beta\left(W - \Delta G_{S_\mathrm{A},\lambda_\mathrm{A}}^{S_\mathrm{B},\lambda_\mathrm{B}}\right)\right], \tag{5.18}$$

where the direction of the arrow distinguishes forward from reverse, $p_\mathrm{F}^{S_\mathrm{A}\to S_\mathrm{B}}$ ($p_\mathrm{R}^{S_\mathrm{A}\leftarrow S_\mathrm{B}}$) stands for the probability to be in S_B (S_A) at the end of the forward (reverse) process, and $\Delta G_{S_\mathrm{A},\lambda_\mathrm{A}}^{S_\mathrm{B},\lambda_\mathrm{B}} = G_{S_\mathrm{B}}(\lambda_\mathrm{B}) - G_{S_\mathrm{A}}(\lambda_\mathrm{A})$ is the FE difference between the partially equilibrated states S_A and S_B. Partially equilibrated states appear in many cases, from protein and peptide–nucleic acid binding to intermediate and misfolded molecular states.

The usefulness of the extended CFR relies on our possibility to experimentally distinguish the different kinetic states along any trajectory. For example, a molecule pulled by stretching forces can be in partial equilibrium when it stays either in the folded or in the unfolded state until it transits to the other state. If S_A stands for the folded state and S_B stands for the unfolded state, the extended CFR makes possible to extract the free energies $G_{S'}(\lambda)$ of the folded and unfolded states

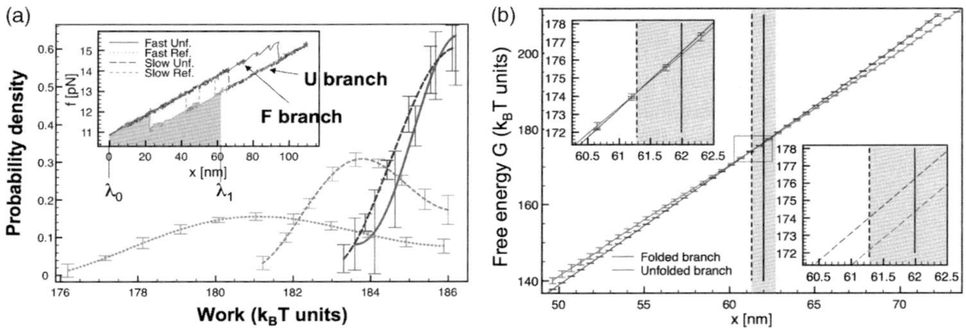

Figure 5.7 The extended Crooks fluctuation relation. (a) Constrained work distributions measured in a 20 bp hairpin at two different pulling speeds: 300 nm s^{-1} (dark-straight line unfolding; light-straight lines refolding) and 40 nm s^{-1} (dark-dashed lines unfolding; light-dashed refolding). (inset in part (a)). The forward trajectories we consider are those where the hairpin starts partially equilibrated in the folded (F) state at λ_A and ends in the unfolded (U) state (partially equilibrated or not) at λ_B. Note that due to the correction term $p_F^{F\to U}/p_R^{F\leftarrow U}$ appearing in Eq. (5.18), the restricted unfolding and refolding work distributions should not cross each other at an energy value that is independent of the pulling speed. (b) Reconstruction of the folded (dark color) and unfolded (light) free energy branches by applying the extended CFR (Eq. (5.18)) as shown in (a) and by varying the parameter $x \equiv \lambda_B$. The two branches cross each other around $x_c \simeq 62$ nm corresponding to the coexistence transition. For $x < x_c$ ($x > x_c$), the F (U) state is the minimum free energy state. The upper left inset shows an enlarged view of the crossing region. The lower left region shows the importance of the correction term $p_F^{F\to U}/p_R^{F\leftarrow U}$ appearing in Eq. (5.18). If that term was not included in Eq. (5.18), the coexistence transition disappears. (Figure taken from Ref. [41].)

$S' = S_A, S_B$ along the λ-axis, that is, the folded and unfolded branches. Figure 5.7 shows an experimental verification of this result for a two-state molecule.

It is worth mentioning that Eq. (5.18) leads to an extended Jarzynski equality for kinetic states:

$$\left\langle \exp\left(-\frac{W - \Delta G_{S_A,\lambda_A}^{S_B,\lambda_B}}{k_B T}\right)\right\rangle = \frac{p_R^{S_A\leftarrow S_B}}{p_F^{S_A\to S_B}}. \tag{5.19}$$

This expression is formally equivalent to a recently proposed generalization of the Jarzynski equality to systems with a feedback control [42, 43]:

$$\left\langle \exp\left(-\frac{W - \Delta G_{S_A,\lambda_A}^{S_B,\lambda_B}}{k_B T}\right)\right\rangle = \gamma, \tag{5.20}$$

where γ quantifies how efficiently the obtained information is used for the feedback control. Note that both equations use more information about the response of the system during the perturbative protocol than just the total work. In Eq. (5.19), the trajectories are classified according to the initial and final states to measure free energy differences, an operation that can be carried out *a posteriori*. In contrast, in Eq. (5.20), the response of the system influences the protocol itself in a feedback loop. The formal similarity of these two equations leaves the open question of whether the operation of classifying trajectories can be considered as a feedback.

5.6.1
Experimental Measurement of the Potential of Mean Force

Another interesting physical quantity is the so-called *potential of mean force*, which is the free energy as a function of a reaction coordinate. This specific case was considered by Hummer and Szabo [25], with regard to unidirectional measurements, using the molecular extension as reaction coordinate in pulling experiments:

$$G(x) = -k_B T \log \left\langle \delta(x - x_t) \exp\left(-\frac{W}{k_B T}\right)\right\rangle. \tag{5.21}$$

Here, x_t is the molecular extension at the end of the pulling process and $G(x)$ is the total free energy at fixed molecular extension. The free energy surface (FES) of the molecule can then be obtained by subtracting the contributions due to beads and handles from $G(x)$. The theory was later extended by Minh and Adib [30] to take into account bidirectional measurements. These methods have been applied to study the FES of protein domains [44] and of DNA hairpins [45].

The recovery of FESs or free energy branches (Figure 5.7b) can be seen as two different applications of extended fluctuation relations. In the first case (Eq. (5.18)), the method provides the FE of a system partially equilibrated at a given molecular extension, while in the second case (Eq. (5.21)), the system is partially equilibrated in a given state. Indeed, Eq. (5.18) can be cast in a form very similar to Eq. (5.21). Imagine a molecule whose initial state is definite ($S_A = A$), but which can be in different states at the end of the forward protocol ($S_B \in \{B_1, B_2, \ldots\}$). In this case, according to the notation of the previous section, we have

$$p_F^{S_A \to S_B} = \langle \delta(S_B = B) \rangle, \tag{5.22}$$

$$p_R^{S_A \leftarrow S_B} = 1. \tag{5.23}$$

Now, we introduce Eqs. (5.22) and (5.23) in Eq. (5.18) and after rearranging the different terms and integrating both sides over work values, we get

$$\left\langle \exp\left(-\frac{W}{k_B T}\right)\right\rangle^{A \to B} \langle \delta(S_B = B) \rangle = \exp\left(-\frac{\Delta G_{A,\lambda_A}^{B,\lambda_B}}{k_B T}\right). \tag{5.24}$$

To cast this last equation in the form of Eq. (5.21), we need to transform the restricted average over trajectories going from A to B to an average over all trajectories. This can be done using the following identity:

$$\langle \mathcal{O} \rangle^{A \to B} = \frac{\langle \mathcal{O}\delta(S_B = B) \rangle}{\langle \delta(S_B = B) \rangle}, \tag{5.25}$$

where $\langle \cdots \rangle$ is the unrestricted average over all trajectories and $\mathcal{O}$ represents any observable of the system. Finally,

$$\Delta G_{A,\lambda_A}^{B,\lambda_B} = -k_B T \log \left\langle \exp\left(-\frac{W}{k_B T}\right) \delta(S_B = B) \right\rangle. \quad (5.26)$$

If now the final condition is a given molecular extension, then the method by Hummer and Szabo (Eq. (5.21)) is recovered. Otherwise, if the final condition is a given state (native, intermediate, and misfolded), we shall fall back to the case discussed in the previous section.

5.7 Free Energy Recovery from Unidirectional Work Measurements

Free energy recovery from irreversible pulling experiments might be done by applying the Jarzynski equality or the Crooks fluctuation relation to unidirectional or bidirectional work measurements, respectively. By combining information from the forward and reverse processes, the bidirectional methods provide less biased free energy estimates than the unidirectional methods. However, in many situations, dissipation is high and the bias obtained by applying any of the two methods is so large that a theory for the bias is needed to reduce the uncertainty. Highly irreversible conditions are found in mechanically unfolding molecular objects that exhibit low thermodynamic stability but high kinetic stability. In this case, a large hysteresis is observed between the forward and reverse force–distance curves in pulling cycles. Consequently, the forward and reverse work distributions tend to appear separated by a large gap of empty events along the work axis and the error in predicting free energy differences appears too large using any standard statistical methods. Examples of highly dissipative systems are found in many proteins that are stabilized in their native state by intramolecular contacts that often display a large kinetic stability. Other cases include RNA or DNA molecules stabilized by ligands or cases in which the experimental technique makes it difficult to see the reverse process, for example, stretching polyproteins with AFM [44]. In general, any molecular complex exhibiting free energy landscapes with a high-energy activation state along the reaction pathway (e.g., during unfolding) will tend to exhibit large differences between work values measured along the forward and reverse protocols. In these conditions, it might be useful to improve free energy predictions from unidirectional work measurements using theory-based methods that predict how the free energy estimator depends on the number of experiments (N). Despite the large amount of work along this direction, only a few analytical results are available. Recently it was shown how it is possible to apply disordered systems theory based on the random energy model to predict the dependence of the bias of the Jarzynski estimator on the number of work measurements [46]. By assuming a generalized compressed exponential work distribution with unbounded lower tail, it has been possible to extract formulas for the N-dependence of the Jarzynski free energy estimator that match reasonably well the experimental results obtained in

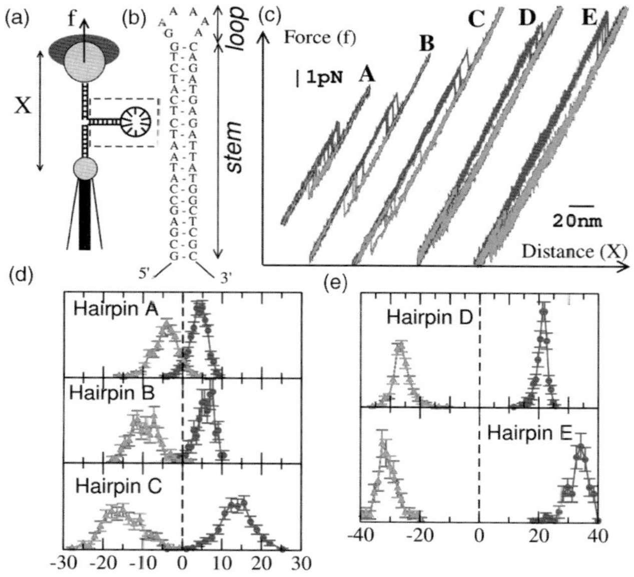

Figure 5.8 Work distributions in DNA hairpins of varying loop size. (a) Experimental setup. The DNA hairpin (squared region) is linked to the two microspheres by short (29 bp) dsDNA handles [20]. (b) Hairpin sequences. The stem is fixed and the size of the loop (sequence 5′-GAAA . . . -3′) takes the values of 4, 6, 12, 16, and 20 (hairpins A, B, C, D, and E, respectively). (c) FDC for the different DNA hairpin sequences. For each sequence, different unfolding (dark curves) and folding (light curves) cycles are shown. As the loop size increases, FDC tends to exhibit larger hysteresis, indicating a more irreversible unfolding/folding process. (d and e) Work distributions measured for all sequences (dark unfolding; light folding). In the horizontal axis, the dissipated work, $W - \Delta G$, is shown in $k_B T$ units with ΔG estimated from combining unidirectional and bidirectional estimates [46]. For hairpins A, B, and C, a crossing at $W - \Delta G = 0$ is apparent; whereas for hairpins D and E, it must be guessed. In the latter strongly dissipative regime, it is very difficult to extract the value of ΔG from unidirectional work measurements, unless the mathematical form of the work tails spanning the whole range of unfolding and folding work values is known. (Panels (a)–(c) are taken from Ref. [46].)

DNA hairpins with varying loop sizes (Figure 5.8). The main drawback of this methodology is the assumption of a specific form for the unbounded tail of the work distribution. Beyond the measurement of a few hundreds or thousands of work values, accurate information about the mathematical form of the unbounded tail of the work distribution is generally missing. Indeed, exact analytical results obtained in two-level systems [47] and molecular dynamics simulations of mechanical unfolding of peptides [48] show that the unbounded left tail of work distributions are rather described by functions more complex than just compressed exponentials. These distributions have tails interpolating between Gaussian fluctuations around the most probable work and an exponential decay further out in the tail. A thorough analysis of the Jarzynski estimator in strongly dissipative systems should make assumptions about the shape of work tails, possibly obtained from exactly solvable models. Otherwise, the error committed in estimating free energy differences from unidirectional work measurements can be too large.

5.8 Conclusions

The use of fluctuation relations for free energy recovery in single-molecule experiments is now a well-established methodology. Future perspectives in this field aim to extend the applicability of this technique from the now classical example of two-state DNA hairpins to more complex situations. For instance, the technique described in Section 5.6 can be applied to study the thermodynamics of intermediates and misfolded states. Such kinetic states, which are hardly observed in classical experiments (Section 5.2), are crucial, for example, to understand the folding behavior of proteins. Moreover, the same method could be applied to study the energetics of binding of proteins or drugs to nucleic acids.

The applicability of fluctuation relations is limited in systems with large dissipation, where free energy estimations can be strongly biased. This is the case, for instance, of many proteins under pulling experiments [49]. A boost in the investigation of protein stability by single-molecule manipulation and nonequilibrium experiments could be provided by the improvement and application of the technique discussed in Section 5.7.

Finally, the recent results in thermodynamics of feedback protocols [42, 43, 50] have not found their application to single-molecule experiments yet. For instance, they could be useful in designing new experimental protocols to reduce the bias and improve free energy estimates.

Overall, the application of fluctuation theorems to free energy recovery is a rapidly evolving field, from both the theoretical and experimental perspectives. Many exciting developments are to be expected.

References

1 Alemany, A., Ribezzi, M., and Ritort, F. (2011) Recent progress in fluctuation theorems and free energy recovery. *AIP Proceedings conference*, vol. 1332, 96–110.

2 Atkins, P. and de Paula, J. (2009) *Elements of Physical Chemistry*, Oxford University Press.

3 Greenfield, N.J. (2006) Using circular dichroism spectra to estimate protein secondary structure. *Nat. Protoc.*, **1** (6), 2876.

4 Sánchez-Ruiz, J.M. (2011) Probing free-energy surfaces with differential scanning calorimetry. *Annu. Rev. Phys. Chem.*, **62** (1), 231–255.

5 Y., Min and Bolen, D.W. (1995.) How valid are denaturant-induced unfolding free energy measurements? Level of conformance to common assumptions over an extended range of ribonuclease A stability. *Biochemistry*, **34** (11), 3771–3781.

6 Kirkwood, J.G. (1935) Statistical mechanics of fluid mixtures. *J. Chem. Phys.*, **3**, 300.

7 Zwanzig, R.W. (1954) High-temperature equation of state by a perturbation

method: I. Nonpolar gases. *J. Chem. Phys.*, **22**, 1420.
8 Chipot, C. and Pohorille, A. (2007) *Free Energy Calculations: Theory and Applications in Chemistry and Biology*, vol. **86**, Springer.
9 Bustamante, C., Liphard, J., and Ritort, F. (2005) The nonequilibrium thermodynamics of small systems. *Phys. Today*, **58**, 43–38.
10 Ritort, F. (2008) Nonequilibrium fluctuations in small systems: from physics to biology. *Adv. Chem. Phys.*, **137**, 31–123.
11 Hugel, T. and Seitz, M. (2001) The study of molecular interactions by AFM force spectroscopy. *Macromol. Rapid Commun.*, **22** (13), 989–1016.
12 Zlatanova, J., Lindscay, S.M., and Leuba, S.H. (2008) Single molecule force spectroscopy in biology using the atomic force microscope. *Prog. Biophys. Mol. Biol.*, **74**, 37–61.
13 Galera-Prat, A., Gómez-Sicilia, A., Oberhauser, A.F., Cieplak, M., and Carrión-Vázquez, M. (2010) Understanding biology by stretching proteins: recent progress. *Curr. Opin. Struct. Biol.*, **20** (1), 63–69.
14 Gossel, C. and Croquette, V. (2002) Magnetic tweezers: micromanipulation and force measurement at the molecular level. *Biophys. J.*, **82** (6), 3314–3329.
15 Zlatanova, J. and Leuba, S.H. (2003) Magnetic tweezers: a sensitive tool to study DNA and chromatin at the single-molecule level. *Biochem. Cell Biol.*, **81** (3), 151–159.
16 Mosconi, F., Allemand, J.F., Bensimon, D., and Croquette, V. (2009) Measurement of the torque on a single stretched and twisted DNA using magnetic tweezers. *Phys. Rev. Lett.*, **201**, 078301.
17 Moffitt, J.R., Chemla, Y.R., Smith, S.B., and Bustamante, C. (2008) Recent advances in optical tweezers. *Annu. Rev. Biochem.*, **77**, 205–228.
18 Ritort, F. (2006) Single-molecule experiments in biological physics: methods and applications. *J. Phys. Condens. Mater.*, **18**, R531.
19 Mossa, A., Manosas, M., Forns, N., Huguet, J.M., and Ritort, F. (2009) Dynamic force spectroscopy of DNA hairpins: I. force kinetics and free energy landscapes. *J. Stat. Mech.*, P02060.
20 Forns, N., de Lorenzo, S., Manosas, S., Hayashi, K., Huguet, J.M., and Ritort, F. (2011) Single-molecule experiments using molecular constructs with short handles. *Biophys. J.*, **100**, 1765–1774.
21 Woodside, M.T., Behnke-Parks, W.M., Larizadeh, K., Travers, K., Herschlag, D., and Block, S.M. (2006) Nanomechanical measurements of the sequence-dependent folding landscapes of single nucleic acid hairpins. *Proc. Natl. Acad. Sci. USA*, **103** (16), 6190–6195.
22 Crooks, G.E. (1999) Entropy production fluctuation theorem and the nonequilibrium work relation for free energy differences. *Phys. Rev. E*, **60**, 2721.
23 Crooks, G.E. (2000) Path-ensemble averages in systems driven far from equilibrium. *Phys. Rev. E*, **61**, 2361–2366.
24 Jarzynski, C. (1997) Nonequilibrium equality for free energy differences. *Phys. Rev. Lett.*, **78** (14), 2690.
25 Hummer, G. and Szabo, A. (2001) Free energy reconstruction from nonequilibrium single-molecule pulling experiments. *Proc. Natl. Acad. Sci. USA*, **98** (7), 3658.
26 Liphardt, J., Dumont, S., Smith, S.B., Tinoco, I., and Bustamante, C. (2002) Equilibrium information from nonequilibrium measurements in an experimental test of Jarzynski's equality. *Science*, **296** (5574), 1832.
27 Ritort, F. Bustamante, C., and Tinoco, I. (2002) A two-state kinetic model for the unfolding of single molecules by mechanical force. *Proc. Natl. Acad. Sci. USA*, **99** (21), 13544.
28 Zuckerman, D.M. and Woolf, T.B. (2002) Overcoming finite-sampling errors in fast-switching free-energy estimates: extrapolative analysis of a molecular system. *Chem. Phys. Lett.*, **351** (5–6), 445–453.
29 Bennett, C.H. (1976) Efficient estimation of free energy differences from Monte Carlo data. *J. Comput. Phys.*, **22** (2), 245–268.
30 Minh, D.D.L. and Adib, A.B. (2008) Optimized free energies from

bidirectional single-molecule force spectroscopy. *Phys. Rev. Lett.*, **100** (18), 180602.

31 Shirts, M.R., Bair, E., Hooker, G., and Pande, V.S. (2003) Equilibrium free energies from nonequilibrium measurements using maximum-likelihood methods. *Phys. Rev. Lett.*, **91** (14), 140601.

32 Collin, D., Ritort, F., Jarzynski, C., Smith, S.B., Tinoco, I., and Bustamante, C. (2005) Verification of the Crooks fluctuation theorem and recovery of RNA folding free energies. *Nature*, **437** (7056), 231–234.

33 Huguet, J.M., Bizarro, C.V., Forns, N., Smith, S.B., Bustamante, C., and Ritort, F. (2010) Single-molecule derivation of salt dependent base-pair free energies in DNA. *Proc. Natl. Acad. Sci. USA*, **107**, 15431–15436.

34 Manosas, M., Mossa, A., Forns, N., Huguet, J.M., and Ritort, F. (2009) Dynamic force spectroscopy of DNA hairpins: II. Irreversibility and dissipation. *J. Stat. Mech.*, P02061.

35 Santalucia, J., Jr. (1998) A unified view of polymer, dumbbell, and oligonucleotide DNA nearest-neighbor thermodynamics. *Proc. Natl. Acad. Sci. USA*, **95**, 1460–1465.

36 Zuker, M. (2003) Mfold web server for nucleic acid folding and hybridization prediction. *Nucleic Acids Res.*, **31**, 3406–3415.

37 Peliti, L. (2008) On the work: Hamiltonian connection in manipulated systems. *J. Stat. Mech.*, P05002.

38 Mossa, A., de Lorenzo, S., Huguet, J.M., and Ritort, F. (2009) Measurement of work in single molecule experiments. *J. Chem. Phys.*, **130**, 234116.

39 Schurr, J.M. and Fujimoto, S.B. (2003) Equalities for the nonequilibrium work transferred from an external potential to a molecular system: analysis of single-molecule extension experiments. *J. Phys. Chem. B*, **107**, 14007.

40 Greenleaf, J.W., Woodside, M.T., Abbondanzieri, E.A., and Block, S.M. (2005) Passive all-optical force clamp for high-resolution laser trapping. *Phys. Rev. Lett.*, **95**, 218102.

41 Junier, I., Mossa, A., Manosas, M., and Ritort, F. (2009) Recovery of free energy branches in single molecule experiments. *Phys. Rev. Lett.*, **102**, 070602.

42 Sagawa, T. and Ueda, M. (2010) Generalized Jarzynski equality under nonequilibrium feedback control. *Phys. Rev. Lett.*, **204**, 090602.

43 Toyabe, S., Sagawa, T., Ueda, M., Muneyuki, E., and Sano, M. (2010) Experimental demonstration of information-to-energy conversion and validation of the generalized Jarzynski equality. *Nat. Phys.*, **6**, 988–992.

44 Harris, N.C., Song, Y., and Kiang, C.H. (2007) Experimental free energy surface reconstruction from single-molecule force spectroscopy using Jarzynski's equality. *Phys. Rev. Lett.*, **99** (6), 68101.

45 Grupta, A.N., Vincent, A., Neupane, K., Yu, H., Wang, F., and Woodside, M.T. (2011) Experimental validation of free-energy-landscape reconstruction from non-equilibrium single-molecule force spectroscopy measurements. *Nat. Phys.*, **7**, 631–634.

46 Palassini, M. and Ritort, F. (2011) Improving free-energy estimates from unidirectional work measurements: theory and experiment. *Phys. Rev. Lett.*, **107**, 060601.

47 Ritort, F. (2004) Work and heat fluctuations in two-state systems. *J. Stat. Mech.*, P10016.

48 Procacci, P. and Marsili, S. (2010) Energy dissipation asymmetry in the non equilibrium folding/unfolding of the single molecule alanine decapeptide. *Chem. Phys.*, **375**, 8–15.

49 Shank, E.A., Cecconi, C., Dill, J.W., Marqusee, S., and Bustamante, C. (2010) The folding cooperative of a protein is controlled by its chain topology. *Nature*, **465**, 634.

50 Horowitz, J.M. and Vaikuntanathan, S. (2010) Nonequilibrium detailed fluctuation theorem for repeated discrete feedback. *Phys. Rev. E*, **82**, 061120.

Further Developments

6
Information Thermodynamics: Maxwell's Demon in Nonequilibrium Dynamics

Takahiro Sagawa and Masahito Ueda

6.1
Introduction

The profound interrelationship between information and thermodynamics was first brought to light by Maxwell [1] in his *gedankenexperiment* about a hypothetical being of intelligence, which was later christened by William Thomson as Maxwell's demon. Since then, numerous discussions have been spurred as to whether and how Maxwell's demon is compatible with the second law of thermodynamics [2–4].

To understand the roles of Maxwell's demon, let us consider a situation in which a gas is confined in a box surrounded by adiabatic walls. A barrier is inserted at the center of the box to divide it into two. The temperatures of the gases in the two boxes are assumed to be initially the same. We assume that the demon is present near the barrier and can close or open a small hole in the barrier. If a faster-than-average (slower-than-average) molecule comes from the right (left) box, the demon opens the hole. Otherwise, the demon keeps it closed. By doing so over and over again, the temperature of the left gas becomes higher than that of the right one, in apparent contradiction to the second law of thermodynamics. This example illustrates the two essential roles of the demon:

- The demon observes individual molecules and obtains the information about their velocities.
- The demon opens or closes the hole based on each measurement outcome, which is the feedback control.

In general, feedback control implies that a control protocol depends on the measurement outcome, or equivalently, we control a system based on the obtained information [5, 6]. The crucial point here is that the measurements are performed at the level of thermal fluctuations (i.e., the demon can distinguish the velocities of individual molecules). Therefore, Maxwell's demon can be characterized as a feedback controller that utilizes information about a thermodynamic system at the level of thermal fluctuations (Figure 6.1).

In the nineteenth century, it was impossible to observe and control individual atoms and molecules and, therefore, it was not necessary to take into account the

Nonequilibrium Statistical Physics of Small Systems: Fluctuation Relations and Beyond, First Edition.
Edited by Rainer Klages, Wolfram Just, and Christopher Jarzynski.

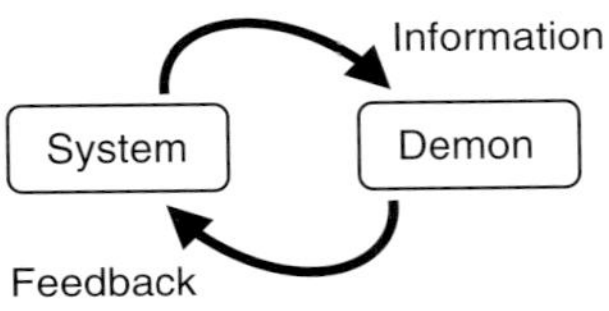

Figure 6.1 Maxwell's demon as a feedback controller. The demon performs feedback control based on the information obtained from measurement at the level of thermal fluctuations.

effect of feedback control on the formulation of thermodynamics. However, due to the recent advances in manipulating microscopic systems, the effect of feedback control on thermodynamic systems has become relevant to real experiments. We can simulate the role of Maxwell's demon in real experiments and can reduce the entropy of small thermodynamic systems.

In this chapter, we review a general theory of thermodynamics that involves measurements and feedback control [7–46]. We generalize the second law of thermodynamics by including information contents concerning the thermodynamics of feedback control. By "demon," we mean a type of devices that perform feedback control at the level of thermal fluctuations.

This chapter is organized as follows. In Section 6.2, we discuss the Szilard engine which is a prototypical model of Maxwell's demon, and examine the consistency between the demon and the second law. In Section 6.3, we review information contents that are used in the following sections. In Section 6.4, we discuss a generalized second law of thermodynamics with feedback control, which is the main part of this chapter. In Section 6.5, we generalize nonequilibrium equalities such as the fluctuation theorem and the Jarzynski equality to the case with feedback control. In Section 6.6, we discuss the energy cost (work) that is needed for measurement and information erasure. Section 6.7 concludes this chapter.

6.2
Szilard Engine

In 1929, Szilard proposed a simple model of Maxwell's demon that illustrates the quantitative relationship between information and thermodynamics [47]. In this section, we briefly review the model, which is called the Szilard engine, and discuss its physical implications.

The Szilard engine consists of a single-particle gas that is in contact with a single heat bath at temperature T. By a measurement, we obtain 1 bit of information about the position of the particle and use that information to extract work from the engine via feedback control. While the engine eventually returns to the initial equilibrium, the total amount of the extracted work is positive. The details of the control protocol are as follows (Figure 6.2).

Step 1: Initial state. We prepare a single-particle gas in a box of volume V_0, which is at thermal equilibrium with temperature T.

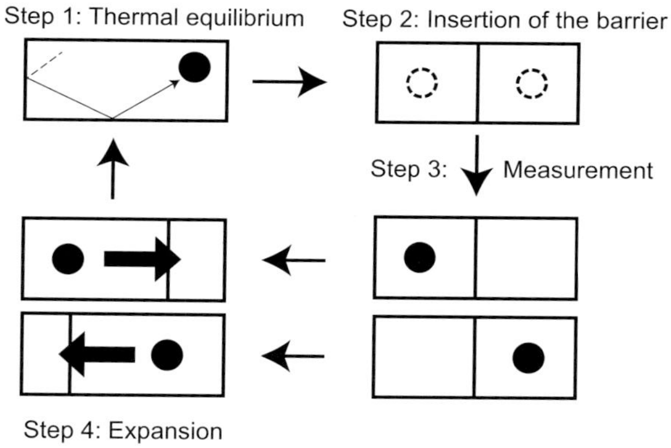

Figure 6.2 The Szilard engine. See the text for details.

Step 2: Insertion of the barrier. We insert a barrier in the middle of the box and divide it into two with equal volume $V_0/2$.
Step 3: Measurement. We perform an error-free measurement of the position of the particle to find out in which box the molecule is. Because the particle will be found to be in each box with probability $1/2$, the amount of information gained from this measurement is 1 bit. We note that 1 ($= \log_2 2$) bit of information in the binary logarithm corresponds to ln 2 nat of information in the natural logarithm.
Step 4: Feedback. If the particle is in the left (right) box, we quasi-statically expand it by moving the barrier to the rightmost (leftmost) position. By this process, we can extract work W_{ext} given by

$$W_{\mathrm{ext}} = \int_{V_0/2}^{V_0} \frac{k_{\mathrm{B}}T}{V} \mathrm{d}V = k_{\mathrm{B}}T \ln 2, \tag{6.1}$$

where we used $pV = k_{\mathrm{B}}T$ with p, V, and k_{B}, respectively, being the pressure, the volume, and the Boltzmann constant. This process corresponds to feedback control, because the direction of the expansion (i.e., left or right) depends on the measurement outcome. After this expansion, the gas returns to the initial thermal equilibrium with volume V_0.

The extracted work is proportional to the obtained information with proportionality constant $k_{\mathrm{B}}T$. This is due to the fact that the entropy of the system is effectively decreased by ln 2 via feedback control, and the decrease in entropy leads to the increase in the free energy by $k_{\mathrm{B}}T \ln 2$, which is the resource of the extracted work.

The Szilard engine *prima facie* seems to contradict the second law of thermodynamics, which dictates that one cannot extract positive work from a single heat bath with a thermodynamic cycle (Kelvin's principle). In fact, the Szilard engine is consistent with the second law, due to an additional energy cost (work) that is needed for the measurement and information erasure of the measurement device or the demon. This additional cost compensates for the excess work extracted from the engine. In his original paper, Szilard argued that there must be an entropic cost for the measurement process [47]. We stress that the work performed on the demon need not be transferred to the engine; only the information obtained by the measurement should be utilized for the feedback. This is the crucial characteristic of the information heat engine.

By utilizing the obtained information in feedback control, we can extract work from the engine without decreasing its free energy, or we can increase the engine's free energy without injecting any work to the engine directly. The resource of the work or the free energy is thermal fluctuations of the heat bath; by utilizing information via feedback, we can rectify the thermal energy of the bath and convert it into the work or the free energy. This method allows us to control the energy balance of the engine beyond the conventional thermodynamics. We shall call such a feedback-controlled heat engine as an "information heat engine." The Szilard engine works as the simplest model that illustrates the quintessence of information heat engines. Quantum versions of the Szilard engine have also been studied [12, 31, 38, 45].

6.3 Information Content in Thermodynamics

In this section, we briefly review the Shannon information and the mutual information [48, 49]. In particular, the mutual information plays a crucial role in thermodynamics of information processing.

6.3.1 Shannon Information

Let $x \in X$ be a probability variable that represents a finite set of possible events. We write as $P[x]$ the probability of event x being realized. The information content that is associated with event x is then defined as $-\ln P[x]$, which implies that the rarer the event, the more the information associated with it. The Shannon information is then given by the average of $-\ln P[x]$ over all possible events:

$$H(X) := -\sum_x P[x] \ln P[x]. \tag{6.2}$$

The Shannon information satisfies $0 \leq H(X) \leq \ln N$, where N is the number of the possible events (the size of set X). Here, the lower bound ($H(X) = 0$) is achieved if $P[x] = 1$ holds for some x; in this case, the event is indeed deterministic, while the upper bound ($H(X) = \ln N$) is achieved if $P[x] = 1/N$ for arbitrary x. In general, the Shannon information characterizes the randomness of a probability variable; the more random the variable, the greater the Shannon information. Consider, for example, a simple case in which x takes two values: $x = 0$ or $x = 1$. We set $P[0] =: p$ and $P[1] =: 1 - p$ with $0 \leq p \leq 1$. The Shannon information is then given by $H(X) = -p \ln p - (1 - p) \ln (1 - p)$, which takes the maximum value $\ln 2$ for $p = 1/2$ and the minimal value 0 for $p = 0$ or 1.

6.3.2
Mutual Information

The mutual information characterizes the correlation between two probability variables. Let $x \in X$ and $y \in Y$ be the two probability variables, and $P[x, y]$ be their joint distribution. The marginal distributions are given by $P[x] = \sum_y P[x, y]$ and $P[y] = \sum_x P[x, y]$. If the two variables are statistically independent, then $P[x, y] = P[x]P[y]$. Otherwise, they are correlated. If the two variables are perfectly correlated, the joint distribution satisfies

$$P[x, y] = \delta(x, f(y))P[x] = \delta(x, f(y))P[y], \tag{6.3}$$

where $\delta(\cdot, \cdot)$ is the Kronecker's delta and $f(\cdot)$ is a bijective function on Y. For example, if $P[0, 1] = P[1, 0] = 1/2$ and $P[0, 0] = P[1, 1] = 0$ with $X = \{0, 1\}$ and $Y = \{0, 1\}$, the two variables are perfectly correlated with $f(0) = 1$ and $f(1) = 0$. If $f(\cdot)$ is the identity function that satisfies $f(y) = y$ for any y, Eq. (6.3) reduces to $P[x, y] = \delta(x, y)P[x] = \delta(x, y)P[y]$.

The conditional probability of x for given y is given by $P[x|y] = P[x, y]/P[y]$. If the two probability variables are statistically independent, the conditional probability reduces to $P[x|y] = P[x]$. This implies that we cannot obtain any information about x from knowledge of y. On the other hand, in the case of the perfect correlation (6.3), we obtain $P[x|y] = \delta(x, f(y))$. This means that we can precisely estimate x from y by $x = f(y)$.

We next introduce the joint Shannon information and the conditional Shannon information. The Shannon information for the joint probability $P[x, y]$ is given by

$$H(X, Y) := -\sum_{xy} P[x, y] \ln P[x, y]. \tag{6.4}$$

On the other hand, the Shannon information for the conditional probability $P[x|y]$, where x is the relevant probability variable, is given by

$$H(X|y) := -\sum_{x} P[x|y] \ln P[x|y]. \tag{6.5}$$

By averaging $H(X|y)$ over y, we define the conditional Shannon information:

$$H(X|Y) := \sum_{y} P[y]H(X|y) = -\sum_{xy} P[x, y] \ln P[x|y]. \tag{6.6}$$

The conditional Shannon information satisfies the following properties:

$$H(X|Y) = H(X, Y) - H(Y), \quad H(Y|X) = H(X, Y) - H(X). \tag{6.7}$$

By definition, $H(X|y) \geq 0$ and $H(X|Y) \geq 0$. Hence,

$$H(X, Y) \geq H(Y), \quad H(X, Y) \geq H(X), \tag{6.8}$$

which implies that the randomness decreases if only one of the two variables is concerned.

The mutual information is defined by

$$I(X:Y) := H(X) + H(Y) - H(X,Y), \tag{6.9}$$

or equivalently,

$$I(X:Y) = \sum_{xy} P[x,y] \ln \frac{P[x,y]}{P[x]P[y]}. \tag{6.10}$$

As shown below, the mutual information satisfies

$$0 \le I(X:Y) \le H(X), \quad 0 \le I(X:Y) \le H(Y). \tag{6.11}$$

Here, $I(X:Y) = 0$ is achieved if X and Y are statistically independent, that is, $P[x,y] = P[x]P[y]$. On the other hand, $I(X:Y) = H(X)$ is achieved if $H(X|Y) = 0$, or equivalently, if $H(X|y) = 0$ for any y. This condition is equivalent to the condition that, for any y, there exists a single x such that $P[x|y] = 1$, which implies that we can estimate x from y with certainty. Similarly, $I(X:Y) = H(Y)$ is achieved if $H(Y|X) = 0$. In particular, if the correlation between x and y is perfect such that Eq. (6.3) holds, $I(X:Y) = H(X) = H(Y)$. In general, the mutual information describes the correlation between two probability variables; the more strongly the x and y correlated, the larger the $I(X:Y)$.

The proof of inequalities (Eq. (6.11)) goes as follows. Because $\ln(t^{-1}) \ge 1 - t$ for $t > 0$, where the equality is achieved if and only if $t = 1$, we obtain

$$-\sum_x P[x,y] \ln \frac{P[x]P[y]}{P[x,y]} \ge \sum_x P[x,y]\left(1 - \frac{P[x]P[y]}{P[x,y]}\right) = 0, \tag{6.12}$$

which implies $I(X:Y) \ge 0$. On the other hand, Eq. (6.9) leads to

$$I(X:Y) = H(X) - H(X|Y) = H(Y) - H(Y|X), \tag{6.13}$$

which implies $I(X:Y) \le H(X)$ and $I(X:Y) \le H(Y)$.

We note that Eqs. (6.7),(6.9) and (6.13) can be illustrated by using a Venn diagram shown in Figure 6.3 [48]. This diagram is useful to memorize the relationship among $H(X|Y)$, $H(Y|X)$, and $I(X:Y)$.

The mutual information can be used to characterize the effective information that can be obtained by measurements. Let us consider a situation in which x is a phase space point of a physical system and y is an outcome that is obtained from a

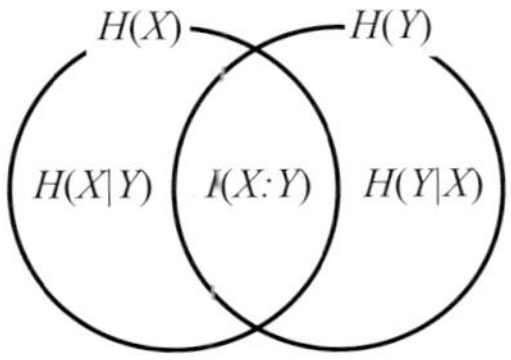

Figure 6.3 A Venn diagram [48] illustrating the relationship among different information contents. The entire region represents the joint Shannon information $H(X,Y)$.

measurement on the system. In the case of the Szilard engine, x specifies the location of the particle ("left" or "right") and y is the measurement outcome. If the measurement is error-free as we assumed in Section 6.2, $x = y$ is always satisfied and the correlation between the two variables is perfect. In this case, $I(X : Y) = H(X) = H(Y)$ holds and, therefore, the obtained information can be characterized by the Shannon information as is the argument in Section 6.2. In general, there exist measurement errors, and the obtained information by the measurement needs to be characterized by the mutual information. The less the amount of the measurement error, the more the mutual information.

We next discuss the cases in which the probability variables take continuous values. The probability distributions such as $P[x, y]$ and $P[x]$ should then be interpreted as probability densities, where the corresponding probabilities are given by $P[x, y]\mathrm{d}x\mathrm{d}y$ and $P[x]\mathrm{d}x$ with $\mathrm{d}x\mathrm{d}y$ and $\mathrm{d}x$ being the integral elements. The Shannon information of x can be formally defined as

$$H(X) := -\int \mathrm{d}x P[x] \ln P[x]. \tag{6.14}$$

However, Eq. (6.14) is not invariant under the transformation of the variable. In fact, if we change x to x' such that $P[x]\mathrm{d}x = P[x']\mathrm{d}x'$, Eq. (6.14) is given by

$$H(X) = -\int \mathrm{d}x' P[x'] \ln P[x'] - \int \mathrm{d}x' P[x'] \ln \left| \frac{\mathrm{d}x'}{\mathrm{d}x} \right|. \tag{6.15}$$

Thus, the Shannon information is not uniquely defined for the case of continuous variables. Only when we fix some probability variable, we can give the Shannon information a unique meaning. On the other hand, the mutual information is defined as

$$I(X : Y) := \int \mathrm{d}x\mathrm{d}y P[x, y] \ln \frac{P[x, y]}{P[x]P[y]}, \tag{6.16}$$

which is invariant under the transformation of the variables. In this sense, the mutual information is uniquely defined for the cases of continuous variables, regardless of the choices of probability variables.

6.3.3 Examples

We now discuss two typical examples of probability variables: discrete and continuous variables.

Example 6.1: (Binary channel)

We consider a binary channel with which at most 1 bit of information is sent from variable x to y (Figure 6.4a). Let $x = 0, 1$ be the sender's bit and $y = 0, 1$ be the receiver's bit. We regard this binary channel as a model of a measurement in which x describes the state of the measured system and y describes the measurement outcome. We assume that the error in the

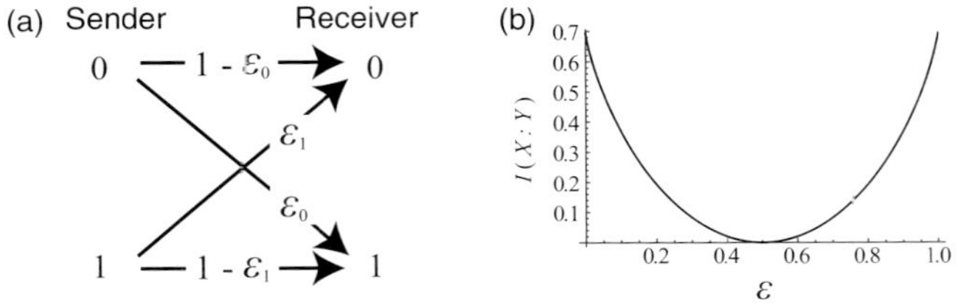

Figure 6.4 (a) A binary channel with error rates ε_0 and ε_1. (b) Mutual information $I(X:Y)$ versus error rate ε for a binary symmetric channel with $\varepsilon_0 = \varepsilon_1 =: \varepsilon$ and $p = 1/2$, which gives $I(X:Y) = \ln 2 + \varepsilon \ln \varepsilon + (1-\varepsilon)\ln(1-\varepsilon)$.

communication (or the measurement error) is characterized by

$$\begin{aligned} P[x=0|y=0] &= 1-\varepsilon_0, \quad P[x=0|y=1] = \varepsilon_0, \\ P[x=1|y=0] &= \varepsilon_1, \quad P[x=1|y=1] = 1-\varepsilon_1, \end{aligned} \tag{6.17}$$

where ε_0 and ε_1 are the error rates for $x=0$ and $x=1$, respectively. The crucial assumption here is that the error property is characterized only by a pair $(\varepsilon_0, \varepsilon_1)$, which is independent of the probability distribution of x. If $\varepsilon_0 = \varepsilon_1 =: \varepsilon$, this model is called a binary symmetric channel.

Let $P[x=0] =: p$ and $P[x=1] =: 1-p$ be the probability distribution of x. The joint distribution of x and y is then given by $P[x=0, y=0] = p(1-\varepsilon_0)$, $P[x=0, y=1] = p\varepsilon_0$, $P[x=1, y=0] = (1-p)\varepsilon_1$, $P[x=1, y=1] = (1-p)(1-\varepsilon_1)$, and the distribution of y is given by $P[y=0] = p(1-\varepsilon_0) + (1-p)\varepsilon_1 =: q$, $P[y=1] = p\varepsilon_0 + (1-p)(1-\varepsilon_1) =: 1-q$. By definition, we can show that the mutual information is given by

$$I(X:Y) = H(Y) - pH(\varepsilon_0) - (1-p)H(\varepsilon_1), \tag{6.18}$$

where $H(Y) := -q\ln q - (1-q)\ln(1-q)$ is the Shannon information for Y, and we defined $H(\varepsilon_i) := -\varepsilon_i \ln \varepsilon_i - (1-\varepsilon_i)\ln(1-\varepsilon_i)$ for $i=0$ and 1. From Eq. (6.18), $I(X:Y) = H(Y)$ holds for $(\varepsilon_0, \varepsilon_1) = (0,0)$, $(0,1)$, $(1,0)$, or $(1,1)$. For a binary symmetric channel, Eq. (6.18) reduces to $I(X:Y) = H(Y) - H(\varepsilon)$. Figure 6.4b shows $I(X:Y)$ versus ε for the case of $p = 1/2$. The mutual information takes the maximum value if $\varepsilon = 0$ or $\varepsilon = 1$. In this case, we can precisely estimate x from y. We note that, if $\varepsilon = 1$, we can just relabel "0" and "1" of y such that $x = y$ holds.

Example 6.2: (Gaussian channel)

We next consider a Gaussian channel with continuous variables x and y. Let x be the sender's variable or the "signal" and y be the receiver's variable or the "outcome." We can also interpret that x describes the phase space point of a physical system such as the position of a Brownian particle, and that y describes the outcome of the measurement on the system. We assume that the error is characterized by a Gaussian noise:

$$P[y|x] = \frac{1}{\sqrt{2\pi N}} \exp\left(-\frac{(y-x)^2}{2N}\right), \tag{6.19}$$

where N is the variance of the noise. For simplicity, we also assume that the probability density of x is also Gaussian:

$$P[x] = \frac{1}{\sqrt{2\pi S}} \exp\left(-\frac{x^2}{2S}\right), \tag{6.20}$$

where S is the variance of x. The joint probability density of x and y is then given by

$$P[x, y] = P[y|x]P[x] = \frac{1}{\sqrt{4\pi^2 SN}} \exp\left(-\frac{x^2}{2S} - \frac{(y-x)^2}{2N}\right). \tag{6.21}$$

On the other hand, the probability density of y is given by

$$P[y] = \frac{1}{\sqrt{2\pi(S+N)}} \exp\left(-\frac{y^2}{2(S+N)}\right), \tag{6.22}$$

which implies that the variance of the outcome y is enhanced by factor $1 + N/S$ compared to the original variance S of the signal x. We can also calculate the conditional probability density as

$$P[x|y] = \frac{P[x, y]}{P[y]} = \frac{1}{\sqrt{2\pi SN/(S+N)}} \exp\left(-\frac{S+N}{2SN}\left(x - \frac{S}{S+N}y\right)^2\right), \tag{6.23}$$

which implies that the variance of the conditional distribution of x is suppressed compared to the original variance S by a factor of $1 + S/N$.

We can straightforwardly calculate the mutual information as

$$I(X:Y) = \frac{1}{2} \ln\left(1 + \frac{S}{N}\right), \tag{6.24}$$

which is determined by the signal-to-noise ratio S/N alone. In the limit of $S/N \to 0$, where the noise dominates the signal, the mutual information vanishes. On the other hand, in the limit of $S/N \to \infty$ where the noise is negligible compared to the signal, the mutual information diverges, as the variable is continuous.

6.4 Second Law of Thermodynamics with Feedback Control

In this section, we discuss a universal upper bound of the work that can be extracted from information heat engines such as the Szilard engine. Starting from a general argument for isothermal processes in Section 6.4.1, we discuss two models with which the universal bound is achieved in Sections 6.4.2 and 6.4.3. Moreover, we will discuss an experimental result that demonstrates an information heat engine in Section 6.4.4. We will also discuss the Carnot efficiency with two heat baths in Section 6.4.5.

6.4.1 General Bound

In the conventional thermodynamics, we extract a work from a heat engine by changing external parameters such as the volume of a gas or the frequency of an optical tweezer. In addition to such parameter changes, we perform measurements and feedback control for the case of information heat engines.

Suppose that we have a thermodynamic engine that is in contact with a single heat bath at temperature T. We perform a measurement on a thermodynamic engine and obtain I of mutual information. After that, we extract a positive amount of work by changing external parameters. The crucial point here is that the protocol of changing the external parameters can depend on the measurement outcome via feedback control.

In the case of the Szilard engine, we obtain ln 2 nat of information, and extract $k_B T \ln 2$ of work. How much work can we extract in principle under the condition that we have I of mutual information about the system? The answer to this fundamental question is given by the following inequality [22, 26]:

$$W_{\text{ext}} \leq -\Delta F + k_B T I, \tag{6.25}$$

where W_{ext} is the average of the work that is extracted from the engine and ΔF is the free energy difference of the engine between the initial and final states. Inequality (6.25) has been proved for both quantum and classical regimes [22, 26]. However, the mutual information in (6.25) needs to be replaced by a quantum extension of the mutual information [22, 50–52] for quantum cases. We will prove inequality (6.25) for classical cases by invoking the detailed fluctuation theorem in Section 6.5.

Inequality (6.25) states that we can extract an excess work up to $k_B T I$ if we utilize I of mutual information obtained by the measurement. In the conventional thermodynamics, the upper bound of the extractable work is bounded only by the free energy difference ΔF, which is determined by the initial and final values of external parameters. For information heat engines, the mutual information is also needed to determine the upper bound of the extractable work. In this sense, inequality (6.25) is a generalization of the second law of thermodynamics for feedback-controlled processes, in which thermodynamic variables (W and ΔF) and the information content (I) are treated on an equal footing.

The equality in (6.25) is achieved with the "best" protocol, which means that the process is quasi-static and the postfeedback state is independent of the measurement outcome, that is, we utilize all the obtained information. This condition is achieved by the Szilard engine, as already discussed. Some models that achieve the equality in (6.25) have been proposed [23, 34, 35, 46]. Two of them [35, 46] are discussed in Sections 6.4.2 and 6.4.3. The Szilard engine, which gives $W = k_B T \ln 2$, $\Delta F = 0$, and $I = \ln 2$, achieves the upper bound of the extractable work, and its special role in information heat engines parallels that of the Carnot cycle in conventional thermodynamics.

6.4.2 Generalized Szilard Engine

We discuss a generalization of the Szilard engine with measurement errors and imperfect feedback [46], with which the equality in (6.25) is achieved. The control protocol is described as follows (Figure 6.5):

Step 1: Initial state. A single-particle gas is in thermal equilibrium, which is in contact with a single heat bath at temperature T.
Step 2: Insertion of the barrier. We insert a barrier to the box and divide it into two with the volume ratio being $p : 1 - p$.
Step 3: Measurement. We perform a measurement to find out in which box the particle is. The possible outcomes are "left" and "right," which we respectively denote as "0" and "1." The measurement can then be modeled by the binary channel with error rates ε_0 and ε_1 (see Section 6.3), where x $(= 0, 1)$ specifies the location of the particle and y $(= 0, 1)$ shows the measurement outcome.
Step 4: Feedback. We quasi-statically move the barrier depending on the measurement outcome. If the outcome is "left" ($y = 0$), we move the barrier, so that the final ratio of the volumes of the two boxes is given by $v_0 : 1 - v_0$ $(0 \leq v_0 \leq 1)$. If the outcome is "right" ($y = 1$), we move the barrier, so that the final ratio of the volumes of the two boxes is given by $1 - v_1 : v_1$ $(0 \leq v_1 \leq 1)$. We note that the feedback protocol is specified by (v_0, v_1).

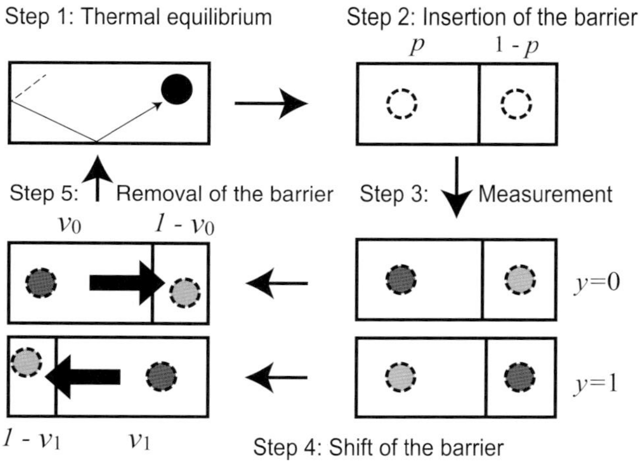

Figure 6.5 Generalized Szilard engine. See the text for details.

Step 5: Removal of the barrier. We remove the barrier and the system returns to the initial thermal equilibrium by a free expansion.

We now calculate the amount of the work that is extracted in step 4. By using the equation of states $pV = k_B T$, we find that the extracted work is $k_B T \ln [\nu_0/p]$ for $(x, y) = (0, 0)$, $k_B T \ln [(1 - \nu_1)/p]$ for $(0, 1)$, $k_B T \ln[(1 - \nu_0)/(1 - p)]$ for $(1, 0)$, and $k_B T \ln [\nu_1/(1 - p)]$ for $(1, 1)$. Therefore, the average work is given by

$$\frac{W_{\text{ext}}}{k_B T} = p(1 - \varepsilon_0) \ln \frac{\nu_0}{p} + p\varepsilon_0 \ln \frac{1 - \nu_1}{p} + (1 - p)\varepsilon_1 \ln \frac{1 - \nu_0}{1 - p} + (1 - p)(1 - \varepsilon_1) \ln \frac{\nu_1}{1 - p}. \tag{6.26}$$

The mutual information obtained by the measurement is given by Eq. (6.18). The upper bound of inequality (6.25) is not necessarily achieved with a general feedback protocol (ν_0, ν_1). We then maximize W_{ext} in terms of ν_0 and ν_1. The optimal feedback protocol with the maximum work is determined by equations $\partial W_{\text{ext}}/\partial \nu_0 = 0$ and $\partial W_{\text{ext}}/\partial \nu_1 = 0$, which lead to $\nu_0 = p(1 - \varepsilon_0)/q = P[x = 0|y = 0]$ and $\nu_1 = (1 - p)(1 - \varepsilon_1)/(1 - q) = P[x = 1|y = 1]$. Therefore, we obtain the maximum work as

$$W_{\text{ext}} = k_B T I, \tag{6.27}$$

which achieves the upper bound of the generalized second law (Eq. (6.25)).

6.4.3
Overdamped Langevin System

We next discuss a feedback protocol for an overdamped Langevin system, which also achieves the upper bound of inequality (6.25) as shown in Ref. [35]. We consider a Brownian particle with a harmonic potential, which obeys the following overdamped Langevin equation:

$$\eta \frac{dx}{dt} = -\lambda_1(x - \lambda_2) + \xi(t), \tag{6.28}$$

where η is a friction constant and $\xi(t)$ is a Gaussian white noise satisfying $\langle \xi(t)\xi(t') \rangle = 2\eta k_B T \delta(t - t')$ with $\delta(\cdot)$ being the delta function. The harmonic potential can be controlled through two external parameters (λ_1, λ_2) such that

$$V(x, \lambda_1, \lambda_2) = \frac{\lambda_1}{2}(x - \lambda_2)^2, \tag{6.29}$$

where λ_1 and λ_2 respectively describe the spring constant and the center of the potential. We consider the following feedback protocol (Figure 6.6):

Step 1: Initial state. The particle is initially in thermal equilibrium with initial external parameters $\lambda_1(0) =: k$ and $\lambda_2(0) = 0$.

Step 2: Measurement. We measure the position of the particle and obtain outcome y. The measurement error is assumed to be Gaussian that is given by Eq. (6.19), where S is the variance of x in the initial equilibrium state (i.e.,

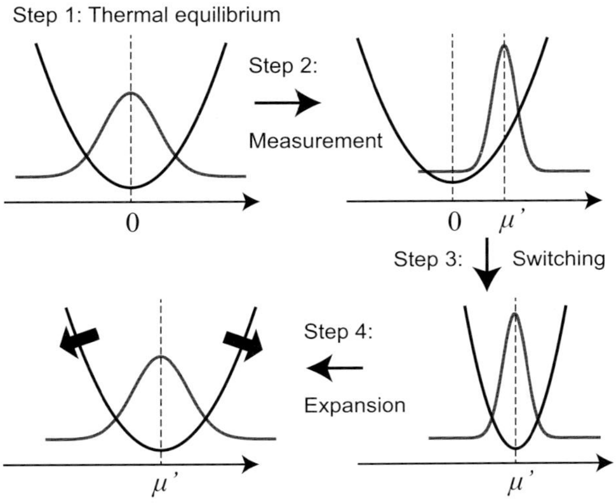

Figure 6.6 Feedback control on a Langevin system with a harmonic potential. See the text for details.

$S = k_B T/k$), and N is the variance of the noise in the measurement (Eq. (6.19)). Immediately after the measurement, the conditional probability is given by Eq. (6.23). The obtained mutual information I is given by Eq. (6.24).

Step 3: Feedback. Immediately after the measurement, we instantaneously change λ_1 from k to $k' := (1 + S/N)k$ and λ_2 from 0 to $\mu_y := Sy/(S + N)$. By this change, the conditional distribution (6.23) becomes thermally equilibrated with new parameters $(\lambda_1, \lambda_2) = (k', \mu_y)$.

Step 4: Work extraction. We quasi-statically expand the potential by changing λ_1 from k' to k thereby extracting the work. The system then gets thermally equilibrated with parameters $(\lambda_1, \lambda_2) = (k, \mu_y)$.

We now calculate the work that can be extracted from this engine. Let $P[x, t|y]$ be the probability distribution of x at time t under the condition of y. The average of the work for step 3 is given by

$$W_{\text{ext}}^{(3)} = \int \left[V(x, k, 0) - V(x, k', \mu_y) \right] P[x, 0|y] P[y] \mathrm{d}x\mathrm{d}y = 0, \tag{6.30}$$

where $P[x, 0|y]$ is given by $P[x|y]$ in Eq. (6.23) and $P[y]$ is given by Eq. (6.22). The work for step 4 is given by

$$\begin{aligned} W_{\text{ext}}^{(4)} &= -\int_0^{\tau} \mathrm{d}t \frac{\mathrm{d}\lambda_1(t)}{\mathrm{d}t} \int \mathrm{d}x\mathrm{d}y P[x, t|y] P[y] \frac{\partial V}{\partial \lambda_1}(x, \lambda_1(t), \lambda_2(t)) \\ &= -\frac{1}{2} \int_{k'}^{k} \mathrm{d}\lambda_1 \int \mathrm{d}x\mathrm{d}y P[x, t|y] P[y] (x - \mu_y)^2 = -\frac{1}{2} \int_{k'}^{k} \mathrm{d}\lambda_1 \frac{k_B T}{\lambda_1} \\ &= \frac{k_B T}{2} \ln \frac{k'}{k} = \frac{k_B T}{2} \ln \left(1 + \frac{S}{N} \right), \end{aligned} \tag{6.31}$$

where we used the fact that the expansion is quasi-static. Comparing Eqs. (6.30) and (6.31) with mutual information (6.24), we obtain

$$W_{\text{ext}} := W_{\text{ext}}^{(3)} + W_{\text{ext}}^{(4)} = k_B T I, \tag{6.32}$$

which achieves the upper bound of inequality (6.25).

6.4.4
Experimental Demonstration: Feedback-Controlled Ratchet

We next discuss a recent experiment that realized an information heat engine by using a real-time feedback control on a colloidal particle in water at room temperature [30].

In the experiment, a colloidal particle with diameter 300 nm is attached to the cover glass and another particle is attached to the first one (Figure 6.7a). The second particle then moves around the first as a rotating Brownian particle that we observe and control. An AC electric field is applied with four electrodes, and the particle undergoes an effective potential, as illustrated in Figure 6.7b. We note that the potential can take two configurations depending on the phases of the electric field. Each configuration consists of a spatially periodic potential and a constant slope. The slope is created by a constant torque around the circle along which the particle rotates. This potential is like spiral stairs. The depth of the periodic potential is about $3k_B T$ and the gradient of the slope per angle 2π is about $k_B T$.

If the periodic potential without the slope was asymmetric and the two potential configurations were periodically switched, the particle would be transported in one

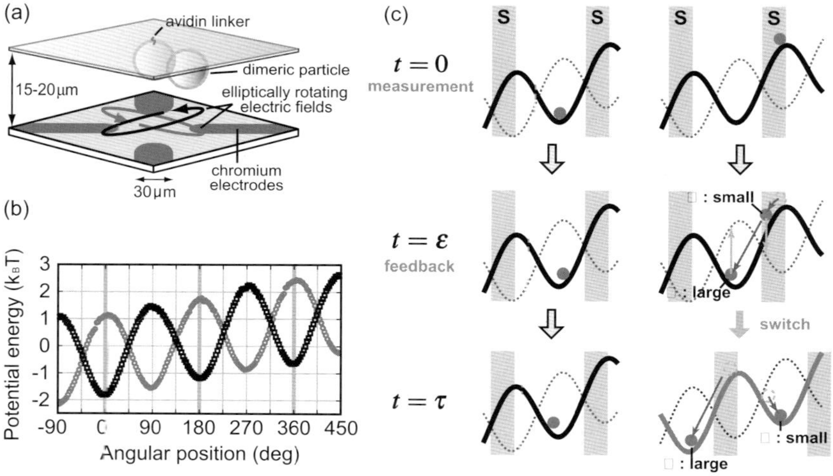

Figure 6.7 Experimental setup of the feedback-controlled ratchet. (a) A rotating Brownian particle. (b) Two possible configurations of the potential that can be switched into each other. (c) Feedback protocol. Only if the particle is found in region "S," the potential is switched. (Reproduced with permission from Ref. [30].)

direction as a flashing ratchet [53–57]. In the present setup, however, the periodic potential without the slope is symmetric and is not switched periodically, but switched in a manner that depends on the measured position of the particle via feedback control [17, 20, 25, 55]. Such a feedback-controlled ratchet has been experimentally realized [20]. In the present experiment, the work and the free energy were measured precisely for quantitatively comparing the experimental results with the theoretical bound (6.25). It has been pointed out [55, 58] that the feedback-controlled ratchet, as well as the flashing ratchet, can be a model of biological molecular motors [59].

The feedback protocol in the experiment [30] was done as follows (Figure 6.7c). The position of the particle was probed every 40 ms by a microscope, a camera, and an image analyzer. Only if the particle was found in the switching region described by "S" in Figure 6.7c, the potential configuration was switched after a short delay time ε. By this switching, when the particle reached the hilltop, the potential is inverted, so that the peak of the potential changed into the bottom of the valley, and therefore the particle is transported to the right direction. Without the switching, the particle would be more likely to go back to the left valley. This position-dependent switching via feedback control induces the reduction of the entropy in a manner analogous to the feedback control in the Szilard engine. By performing this protocol many times, the particle is expected to be transported to the right direction by climbing up the potential slope.

Figure 6.8a shows typical trajectories of the particle. If the feedback delay ε is sufficiently shorter than the relaxation time of the particle in each well ($\simeq 10$ ms), the particle climbs up the potential. If the feedback delay is longer than the

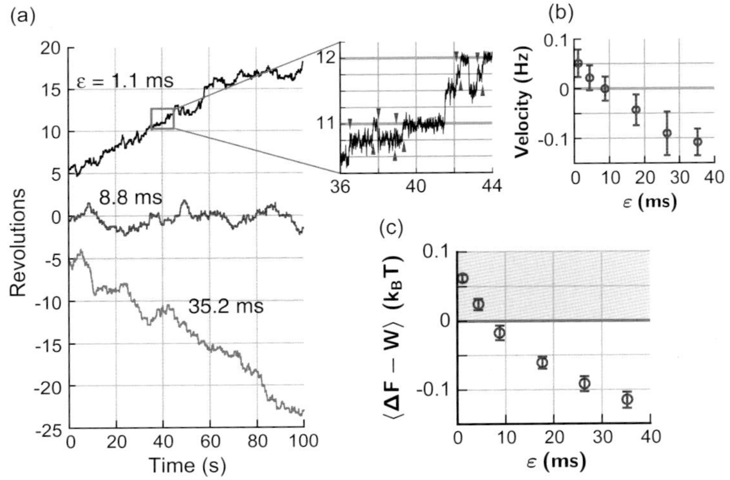

Figure 6.8 Experimental results on the feedback-controlled ratchet. (a) Typical trajectories of the particle with feedback delays $\varepsilon = 1.1$, 8.8, and 35.2 ms. (b) The averaged velocity of the particle versus the feedback delay. (c) Energy balance of feedback control. The shaded region is prohibited by the conventional second law of thermodynamics, and can only be achieved by feedback control. (Reproduced with permission from Ref. [30].)

relaxation time, the feedback does not work and the particle moves down the potential in agreement with the conventional second law of thermodynamics. Figure 6.8c shows the averaged velocity of the particle versus ε, which implies that the shorter the feedback delay, the faster the average velocity.

Figure 6.8c shows the energy balance of this engine. The shaded region is prohibited by the conventional second law of thermodynamics $\langle \Delta F - W \rangle \leq 0$, where ΔF is the free energy difference corresponding to the height of the potential, W is the work performed on the particle during the switching, and $\langle \cdots \rangle$ represents the ensemble average over all trajectories. By using information via feedback, however, the shaded region is indeed achieved if ε is sufficiently small. The resource of the excess free energy gain is thermal fluctuations of the heat bath, which are rectified by feedback control. This is an experimental realization of an information heat engine.

For the case of $\varepsilon = 1.1$ ms, $\langle \Delta F - W \rangle = 0.062 k_B T$. On the other hand, the obtained information is given by $I = 0.22$, which can be calculated from the histogram of the measurement outcomes by assuming that the measurement is error-free. By comparing these experimental data with the theoretical bound (Eq. (6.25)), the efficiency of this information heat engine is determined to be $\langle \Delta F - W \rangle / k_B T I = 0.062/0.22 \simeq 0.28$. The reason why the efficiency is less than unity is two-fold: (i) the switching is not quasi-static but instantaneous, and (ii) the obtained information is not utilized if the particle is found outside the switching region.

We note that various experiments that are analogous to Maxwell's demon have been performed with, for example, a granular gas [60, 61], supramolecules [62], and ultracold atoms [63]; however, these examples do not explicitly involve the measurement and feedback, as the controlled system and the demon constitute an autonomous system in these experiments. Such autonomous versions of Maxwell's demon have also been theoretically studied [64–69]. In contrast, in the experiment of Ref. [30], the demon (the camera and the computer) is separated from the controlled system (the colloidal particle) as in the case for the Szilard engine. We also note that an information heat engine similar to that in Ref. [30] has been proposed for an electron pump system in Ref. [39].

6.4.5
Carnot Efficiency with Two Heat Baths

We next consider the case in which there are two heat baths and the process is a thermodynamic cycle. Without feedback control, the heat efficiency is bounded by the Carnot bound. If we perform measurements and feedback control on this system, the extractable work is bounded from above by [22]

$$W_{\text{ext}} \leq \left(1 - \frac{T_L}{T_H}\right) Q_H + k_B T_L I, \tag{6.33}$$

where T_H and T_L are the temperatures of the hot and cold heat baths, respectively, and Q_H is the heat that is absorbed by the engine from the hot heat bath. The proof of inequality (6.33) will be given in Section 6.5. The last term on the right-hand side (RHS) of (6.33) describes the effect of feedback. We note that the coefficient of the last term is given by the temperature of the cold bath.

The equality in (6.33) is achieved in the following example. We consider a single-particle gas in a box, and quasi-statically control it as in the case of the usual Carnot cycle. We then perform the Szilard engine-type operation consisting of measurement and feedback on the system while it is in contact with the cold bath. In this case, we have $k_B T_L \ln 2$ of excess work, and Q_H remains unchanged. Therefore, the equality in (6.33) is achieved with $I = \ln 2$.

We can also achieve the equality in (6.33) if we perform the Szilard engine-type operation while the engine is in contact with the hot heat bath. In this case, we can extract $k_B T_H I$ of excess work, and Q_H is increased by $k_B T_H I$. Therefore, we again obtain the equality in (6.33) with $I = \ln 2$.

6.5 Nonequilibrium Equalities with Feedback Control

Since the late 1990s, a number of universal equalities have been found for nonequilibrium processes [70–78], and they have been shown to reproduce the second law of thermodynamics and the fluctuation–dissipation theorem. The fluctuation theorem and the Jarzynski equality are two prime examples of the nonequilibrium equalities. In this section, we generalize the nonequilibrium equalities to situations in which a thermodynamic system is subject to measurements and feedback control in line with Refs [26, 29, 46]. As corollaries, we derive inequalities (6.25) and (6.33).

6.5.1 Preliminaries

First of all, we review the nonequilibrium equalities without feedback control. We consider a stochastic thermodynamic system in contact with heat bath(s) with inverse temperatures $\beta_m (m = 1, 2, \ldots)$. Let x be the phase space point of the system. The system is controlled through external parameters λ, which describe, for instance, the volume of a gas or the frequency of an optical tweezer. Even when the initial state of the system is in thermal equilibrium, the system can be driven far from equilibrium by changing the external parameters. We consider such a stochastic dynamics of the system from time 0 to τ. The state of the system stochastically evolves as $x(t)$ under a deterministic protocol of the external parameters denoted collectively as $\lambda(t)$.

Let $X_\tau := \{x(t)\}_{0 \le t \le \tau}$ be the trajectory of the phase space point and $\Lambda_\tau := \{\lambda(t)\}_{0 \le t \le \tau}$ be that of the external parameters. The heat that is absorbed by the system from heat bath m is a trajectory-dependent quantity, which we write as $Q_m[X_\tau|\Lambda_\tau]$. The work that is performed on the system is also trajectory dependent and is denoted as $W[X_\tau|\Lambda_\tau]$. The first law of thermodynamics is then given by

$$H[x(\tau)|\lambda(\tau)] - H[x(0)|\lambda(0)] = W[X_\tau|\Lambda_\tau] + \sum_m Q_m[X_\tau|\Lambda_\tau], \tag{6.34}$$

where $H[x|\lambda]$ is the Hamiltonian with external parameters λ, and the work can be written as

$$W[X_\tau|\Lambda_\tau] = \int_0^\tau \frac{\partial H}{\partial \lambda}[x(t)|\lambda(t)] \frac{\mathrm{d}\lambda(t)}{\mathrm{d}t} \mathrm{d}t. \tag{6.35}$$

Let $P[X_\tau|\Lambda_\tau]$ be the probability density of trajectory X_τ with control protocol Λ_τ. It can be decomposed as $P[X_\tau|\Lambda_\tau] = P[X_\tau|x(0), \Lambda_\tau]P_f[x(0)]$, where $P[X_\tau|x(0), \Lambda_\tau]$ is the probability density under the condition that the initial state is $x(0)$ and $P_\mathrm{f}[x(0)]$ is the initial distribution of the forward process.

We next introduce backward processes of the system. Let x^* be the time reversal of x. For example, if $x = (\mathbf{r}, \mathbf{p})$ with $\mathbf{r}$ being the position and $\mathbf{p}$ being the momentum, then $x^* = (\mathbf{r}, -\mathbf{p})$. Similarly, we denote the time reversal of λ as λ^*. For example, if λ is the magnetic field, then $\lambda^* = -\lambda$. Let $X_\tau^\dagger$ be the time-reversed trajectory of X_τ defined as $X_\tau^\dagger := \{x^*(\tau - t)\}_{0 \le t \le \tau}$. We also write the time reversal of control protocol Λ_τ as $\Lambda_\tau^\dagger := \{\lambda^*(\tau - t)\}_{0 \le t \le \tau}$, and write as $P[X_\tau^\dagger|\Lambda_\tau^\dagger]$ the probability density of the time-reversed trajectory with the time-reversed control protocol. We can decompose $P[X_\tau^\dagger|\Lambda_\tau^\dagger]$ as $P[X_\tau^\dagger|\Lambda_\tau^\dagger] = P[X_\tau^\dagger|x^\dagger(0), \Lambda_\tau]P_\mathrm{b}[x^\dagger(0)]$, where $P[X_\tau^\dagger|x^\dagger(0), \Lambda_\tau^\dagger]$ is the probability density under the condition that the initial state of the backward process is $x^\dagger(0)$, and $P_\mathrm{b}[x^\dagger(0)]$ is the initial distribution of the backward process. We stress that the initial distribution of the backward process $P_\mathrm{b}[x^\dagger(0)]$ can be set independent of the final distribution of the forward process. In experiments, we can initialize the system before we start a backward process so that its initial distribution can be chosen independent of the forward process.

The detailed fluctuation theorem (the transient fluctuation theorem) is given by [71–74]

$$\frac{P[X_\tau^\dagger|x^\dagger(0), \Lambda_\tau^\dagger]}{P[X_\tau|x(0), \Lambda_\tau]} = \mathrm{e}^{\sum_m \beta_m Q_m[X_\tau|\Lambda_\tau]}. \tag{6.36}$$

Defining the entropy production as

$$\sigma[X_\tau|\Lambda_\tau] := \ln P_\mathrm{f}[x(0)] - \ln P_\mathrm{b}[x^\dagger(0)] - \sum_m Q_m[X_\tau|\Lambda_\tau], \tag{6.37}$$

we obtain

$$\frac{P[X_\tau^\dagger|\Lambda_\tau^\dagger]}{P[X_\tau|\Lambda_\tau]} = \mathrm{e}^{-\sigma[X_\tau|\Lambda_\tau]}. \tag{6.38}$$

By taking the ensemble average of Eq. (6.38), we have

$$\int \mathrm{d}X_\tau P[X_\tau|\Lambda_\tau]\mathrm{e}^{-\sigma[X_\tau|\Lambda_\tau]} = \int \mathrm{d}X_\tau P[X_\tau|\Lambda_\tau]\frac{P[X_\tau^\dagger|\Lambda_\tau^\dagger]}{P[X_\tau|\Lambda_\tau]} = \int \mathrm{d}X_\tau^\dagger P[X_\tau^\dagger|\Lambda_\tau^\dagger] = 1, \tag{6.39}$$

where we used $\mathrm{d}X_\tau = \mathrm{d}X_\tau^\dagger$. Therefore, we obtain the integral fluctuation theorem

$$\langle \mathrm{e}^{-\sigma} \rangle = 1. \tag{6.40}$$

By using the concavity of the exponential function, we find from Eq. (6.40) that

$$\langle \sigma \rangle \geq 0, \tag{6.41}$$

which is an expression of the second law of thermodynamics. In the following, we discuss the physical meanings of entropy production σ for typical situations. The equality in (6.41) is achieved if $P[X_\tau^\dagger|\Lambda_\tau^\dagger] = P[X_\tau|\Lambda_\tau]$ holds for any X_τ, which implies the reversibility of the process.

We first consider isothermal processes. In this case, we choose the initial distributions of the forward and backward processes as

$$\begin{aligned} P_{\rm f}[x(0)] &= \exp\left(\beta(F[\lambda(0)] - H[x(0)|\lambda(0)])\right), \\ P_{\rm b}[x^\dagger(0)] &= \exp\left(\beta(F[\lambda^\dagger(0)] - H[x^\dagger(0)|\lambda^\dagger(0)])\right), \end{aligned} \tag{6.42}$$

where $F[\lambda] := -k_{\rm B}T \ln \int {\rm d}x {\rm e}^{-\beta H[x|\lambda]}$ is the free energy of the system. We assume that the Hamiltonian has the time-reversal symmetry:

$$H[x|\lambda] = H[x^*|\lambda^*], \tag{6.43}$$

and therefore the canonical distribution satisfies

$$P_{\rm b}[x^\dagger(0)] = \exp\left(\beta(F[\lambda(\tau)] - H[x(\tau)|\lambda(\tau)])\right). \tag{6.44}$$

The entropy production then reduces to

$$\sigma[X_\tau] = \beta(W[X_\tau] - \Delta F), \tag{6.45}$$

where $\Delta F := F[\lambda(\tau)] - F[\lambda(0)]$. Thus, the integral fluctuation theorem (6.40) leads to the Jarzynski equality [70]

$$\langle {\rm e}^{-\beta W} \rangle = {\rm e}^{-\beta \Delta F}, \tag{6.46}$$

and inequality (6.41) gives the second law of thermodynamics:

$$\langle W \rangle \geq \Delta F, \tag{6.47}$$

where $W_{\rm ext} := -\langle W \rangle$ is the work that is extracted from the system.

We next consider a case with multiheat baths, and assume that the initial distributions of the forward and the backward processes are given by the canonical distributions as in (6.42) with a reference inverse temperature β. In practice, β can be taken as one of β_m's, which can be realized if the system is initially attached only to that particular heat bath. We then have

$$\sigma[X_\tau] = \beta(\Delta E[X_\tau] - \Delta F) - \sum_m \beta_m Q_m[X_\tau], \tag{6.48}$$

where $\Delta E[X_\tau] := H[x(\tau)|\lambda(\tau)] - H[x(0)|\lambda(0)]$ is the difference of the internal energy of the system. Inequality (6.41) leads to

$$\sum_m \beta_m \langle Q_m \rangle \leq \beta(\langle \Delta E \rangle - \Delta F). \tag{6.49}$$

In particular, if the process is a cycle such that $\Delta F = 0$ and $\langle \Delta E \rangle = 0$ hold, inequality (6.49) reduces to

$$\sum_m \beta_m \langle \mathcal{Q}_m \rangle \leq 0, \tag{6.50}$$

which is the Clausius inequality. If there are two heat baths with temperatures T_{H} and T_{L}, (6.50) gives the Carnot bound:

$$-\langle W \rangle \leq \left(1 - \frac{T_{\mathrm{L}}}{T_{\mathrm{H}}}\right) \langle Q_{\mathrm{H}} \rangle, \tag{6.51}$$

where $\langle Q_{\mathrm{H}} \rangle$ is the average of the heat that is absorbed by the engine from the hot heat bath.

6.5.2
Measurement and Feedback

We now formulate measurements and feedback on the thermodynamic system [26, 29, 46]. We perform measurements at time t_k $(k = 1, 2, \ldots, M)$ with $0 \leq t_1 < t_2 < \cdots < t_M < \tau$. Let $y(t_k)$ be the measurement outcome at time t_k. For simplicity, we assume that the measurement is instantaneous; the measurement error of $y(t_k)$ can be characterized only by the conditional probability $P_{\mathrm{c}}[y(t_k)|x(t_k)]$, which implies that only $y(t_k)$ has the information about $x(t_k)$. (Note, however, that this assumption can be relaxed [46].) We write the sequence of the measurement outcomes as $Y_\tau := (y(t_1), y(t_2), \ldots, y(t_M))$, and write

$$P_{\mathrm{c}}[Y_\tau | X_\tau] := \prod_k P_{\mathrm{c}}[y(t_k)|x(t_k)]. \tag{6.52}$$

We then introduce the following quantity:

$$I_{\mathrm{c}}[X_\tau : Y_\tau] := \ln \frac{P_{\mathrm{c}}[Y_\tau | X_\tau]}{P[Y_\tau]}, \tag{6.53}$$

which can be interpreted as a stochastic version of the mutual information. The ensemble average of Eq. (6.53) gives the mutual information obtained by the measurements as

$$\langle I_{\mathrm{c}} \rangle = \int \mathrm{d}X_\tau \mathrm{d}Y_\tau P[X_\tau, Y_\tau] \ln \frac{P_{\mathrm{c}}[Y_\tau | X_\tau]}{P[Y_\tau]}. \tag{6.54}$$

We note that $\langle I_{\mathrm{c}} \rangle$ describes the correlation between X_τ and Y_τ that is induced only by measurements, and not by feedback control. The suffix "c" represents this property of I_{c}. We then identify $\langle I_{\mathrm{c}} \rangle$ with I in the foregoing arguments. See Ref. [46] for details. We note that $\langle I_{\mathrm{c}} \rangle$ has been discussed and referred to as the transfer entropy in Ref. [79]. The following equality also holds by definition:

$$\langle \mathrm{e}^{-I_{\mathrm{c}}} \rangle = 1. \tag{6.55}$$

We next consider feedback control by using the obtained outcomes. The control protocol after time t_k can depend on the outcome $y(t_k)$ in the presence of feedback control. We write this dependence as $\Lambda_\tau(Y_\tau)$. We can show that the joint probability of (X_τ, Y_τ) is given by

$$P[X_\tau, Y_\tau] = P_c[Y_\tau|X_\tau]P[X_\tau|\Lambda(Y_\tau)], \tag{6.56}$$

which satisfies the normalization condition $\int dX_\tau dY_\tau P[X_\tau, Y_\tau] = 1$. Equality (6.56) is proved in Appendix 6.A (see also Ref. [46]). The probability of obtaining outcome Y_τ in the forward process is given by the marginal distribution as $P[Y_\tau] = \int dX_\tau P[X_\tau, Y_\tau]$, and the conditional probability is given by $P[X_\tau|Y_\tau] = P[X_\tau, Y_\tau]/P[Y_\tau]$. In the presence of measurement and feedback, the ensemble average is taken over all trajectories and all outcomes; for an arbitrary stochastic quantity $A[X_\tau, Y_\tau]$, its ensemble average is given by

$$\langle A\rangle = \int dX_\tau dY_\tau P[X_\tau, Y_\tau]A[X_\tau, Y_\tau]. \tag{6.57}$$

The detailed fluctuation theorem for a given Y_τ can be written as

$$\frac{P[X_\tau^\dagger|\Lambda_\tau(Y_\tau)^\dagger]}{P[X_\tau|\Lambda_\tau(Y_\tau)]} = e^{-\sigma[X_\tau|\Lambda_\tau(Y_\tau)]}. \tag{6.58}$$

We note that Eq. (6.58) is valid in the presence of feedback control because the detailed fluctuation theorem is satisfied once a control protocol is fixed. Equality (6.58) provides the basis for the derivations of the formulas in the following section.

6.5.3 Nonequilibrium Equalities with Mutual Information

In this section, we generalize the nonequilibrium equalities by incorporating the mutual information. First of all, from Eq. (6.53), we have

$$\frac{P[Y_\tau]}{P_c[Y_\tau|X_\tau]} = e^{-I_c}. \tag{6.59}$$

By multiplying both sides of this equality by those of Eq. (6.58), we obtain

$$\frac{P[X_\tau^\dagger|\Lambda_\tau(Y_\tau)^\dagger]P[Y_\tau]}{P[X_\tau, Y_\tau]} = e^{-\sigma[X_\tau|\Lambda_\tau(Y_\tau)]-I_c[X_\tau:Y_\tau]}. \tag{6.60}$$

To measure $P[X_\tau^\dagger|\Lambda_\tau(Y_\tau)^\dagger]P[Y_\tau]$, we follow the backward process corresponding to each forward outcome and count the occurrences of the time-reversed trajectories. By taking the ensemble average of both sides of Eq. (6.60) with formula (6.57), we obtain a generalized integral fluctuation theorem with feedback control:

$$\langle e^{-\sigma-I_c}\rangle = 1. \tag{6.61}$$

Using the concavity of the exponential function, Eq. (6.61) leads to

$$\langle\sigma\rangle \geq -\langle I_c\rangle. \tag{6.62}$$

Inequality (6.62) is a generalized second law of thermodynamics, which states that the entropy production can be decreased by feedback control, and that the lower bound of the entropy production is given by the mutual information $\langle I_c \rangle$. As shown below, inequalities (6.25) and (6.33) in Section 6.4 are special cases of inequality (6.62). The equality in (6.62) is achieved if $P[X_\tau^\dagger | \Lambda_\tau(Y_\tau)^\dagger] P[Y_\tau] = P[X_\tau, Y_\tau]$ holds for any X_τ and Y_τ, which implies the reversibility with feedback control, as discussed in Ref. [34].

The generalized integral fluctuation theorem of the form (6.61) was first shown in Ref. [26] for a single measurement, and Eq. (6.60) was obtained in Refs [29, 46] for multiple measurements. These results have also been generalized to the optimal control process with continuous measurement and the Kalman filter in Ref. [28].

A generalized fluctuation theorem was also obtained in Ref. [19], which is similar to Eq. (6.61). In Ref. [19], feedback control is performed based on information about the continuously monitored velocity of a Langevin system. The result of Ref. [19] includes a quantity that describes the decrease in the entropy by continuous feedback control, instead of the mutual information obtained by the continuous measurement.

We note that the Hatano–Sasa equality for transitions between nonequilibrium steady states [80–84] has also been generalized to classical stochastic systems in the presence of feedback control [42–44], where the entropy production σ in Eq. (6.61) is replaced by an excess part of the entropy production.

We now consider isothermal processes with a single heat bath, in which the entropy production is given by Eq. (6.45). Equality (6.61) then reduces to a generalized Jarzynski equality:

$$\langle e^{\beta(\Delta F - W) - I_c} \rangle = 1, \tag{6.63}$$

and inequality (6.62) reduces to

$$\langle \Delta F - W \rangle \geq k_B T \langle I_c \rangle, \tag{6.64}$$

which implies inequality (6.25) with identifications $W_{ext} = -\langle W \rangle$, $\Delta F = \langle \Delta F \rangle$, and $I = \langle I_c \rangle$.

We next consider the cases in which there are two heat baths and the process is a cycle, in which the entropy production is given by the ensemble average of Eq. (6.48) with $\langle \Delta E \rangle = \langle \Delta F \rangle = 0$. The generalized second law (6.62) then leads to

$$\beta_H \langle Q_H \rangle + \beta_L \langle Q_L \rangle \leq \langle I_c \rangle, \tag{6.65}$$

which can be rewritten as

$$-\langle W \rangle \leq \left(1 - \frac{T_L}{T_H}\right) \langle Q_H \rangle + k_B T_L \langle I_c \rangle. \tag{6.66}$$

By identifying $Q_H = \langle Q_H \rangle$, inequality (6.66) implies inequality (6.33).

6.5.4 Nonequilibrium Equalities with Efficacy Parameter

In this section, we discuss another generalization of nonequilibrium equalities. We define the time reversal of outcomes Y_τ as $Y_\tau^\dagger := (y(\tau - t_M)^*, \ldots, y(\tau - t_2)^*, y(\tau - t_1)^*)$, where y^* is the time reversal of y, and introduce the probability that we obtain outcome $Y_\tau^\dagger$ with control protocol $\Lambda_\tau(Y_\tau)^\dagger$, which is given by

$$P[Y_\tau^\dagger|\Lambda_\tau(Y_\tau)^\dagger] = \int dX_\tau^\dagger P_c[Y_\tau^\dagger|X_\tau^\dagger] P[X_\tau^\dagger|\Lambda_\tau(Y_\tau)^\dagger]. \tag{6.67}$$

We stress that no feedback control is performed in the backward processes. We then assume that the measurement error has the time-reversal symmetry:

$$P_c[Y_\tau^\dagger|X_\tau^\dagger] = P_c[Y_\tau|X_\tau]. \tag{6.68}$$

This assumption is satisfied if $P_c[y(t_k)|x(t_k)] = P_c[y(\tau - t_k)^*|x(\tau - t_k)^*]$ holds for $k = 1, 2, \ldots, M$. By using Eq. (6.67) and assumption (6.68), we can show that

$$\frac{P[Y_\tau^\dagger|\Lambda_\tau(Y_\tau)^\dagger]}{P[Y_\tau]} = \langle e^{-\sigma} \rangle_{Y_\tau}, \tag{6.69}$$

where $\langle \cdots \rangle_{Y_\tau}$ denotes the conditional average with condition Y_τ such that

$$\langle e^{-\sigma} \rangle_{Y_\tau} := \int dX_\tau e^{-\sigma[X_\tau|\Lambda_\tau(Y_\tau)]} P[X_\tau|Y_\tau]. \tag{6.70}$$

Equality (6.69) has been shown for Hamiltonian systems [75] and stochastic systems [26, 46]. By noting that

$$\langle e^{-\sigma} \rangle = \int dY_\tau P[Y_\tau] \langle e^{-\sigma} \rangle_{Y_\tau}, \tag{6.71}$$

we obtain yet another generalization of the integral fluctuation theorem [26, 46]:

$$\langle e^{-\sigma} \rangle = \gamma, \tag{6.72}$$

where

$$\gamma = \int dY_\tau P[Y_\tau^\dagger|\Lambda_\tau(Y_\tau)^\dagger] \tag{6.73}$$

is the sum of the probabilities that we obtain the time-reversed outcomes with a time-reversed protocol. For the cases of isothermal processes, Eq. (6.72) reduces to

$$\langle e^{\beta(\Delta F - W)} \rangle = \gamma. \tag{6.74}$$

We note that γ characterizes the efficacy of feedback control. The more efficient the feedback protocol, the larger the amount of γ. Without feedback control, $P[Y_\tau^\dagger|\Lambda_\tau^\dagger]$ reduces to a single unconditional probability distribution, and we therefore obtain

$$\gamma = \int dY_\tau P[Y_\tau^\dagger|\Lambda_\tau^\dagger] = 1, \tag{6.75}$$

which reproduces the integral fluctuation theorem (6.40) without feedback. We note that the maximum value of γ is the number of the possible outcomes of Y_τ.

We illustrate the efficacy parameter γ for the case of the Szilard engine that is described in Section 6.2. The backward control protocol of the Szilard engine is as follows (Figure 6.9) [26]:

Step 1: Initial state. The single-particle gas is initially in thermal equilibrium.
Step 2: Compression of the box. In accordance with the measurement outcome in the forward process, which is "0" (= "left") or "1" (= "right"), we quasi-statically compress the box by moving the wall in the box to the center. By this compression, the volume of the box becomes half.
Step 3: Measurement. We measure the position of the particle to find out in which box the particle is. The outcome of this backward measurement is "0" (= "left") or "1" (= "right") with unit probability corresponding to forward outcome "0" or "1," respectively.
Step 4: We remove the barrier at the center of the box, and the engine returns to the initial state by a free expansion.

In these backward processes, the measurement outcomes in step 2 satisfy $P[0|\Lambda_\tau(0)^\dagger] = 1$ and $P[1|\Lambda_\tau(1)^\dagger] = 1$, and therefore we obtain $\gamma = P[0|\Lambda_\tau(0)^\dagger] + P[1|\Lambda_\tau(1)^\dagger] = 2$, which gives the maximum value of γ for situations in which the number of possible outcomes is two. On the other hand, since $W = -k_B T \ln 2$ and $\Delta F = 0$ in the absence of fluctuations, the generalized Jarzynski equality (6.74) is satisfied as $\langle e^{\beta(\Delta F - W)} \rangle = 2 = \gamma$.

The generalized Jarzynski equality (6.74) has been experimentally verified in the experiment described in Section 6.4.4 by measuring $\Delta F - W$ and γ separately in the forward and backward experiments, respectively [30]. Equalities (6.63) and (6.74) have been obtained in Hamiltonian systems [37]. Equality (6.74) has also been generalized to quantum systems [32, 43].

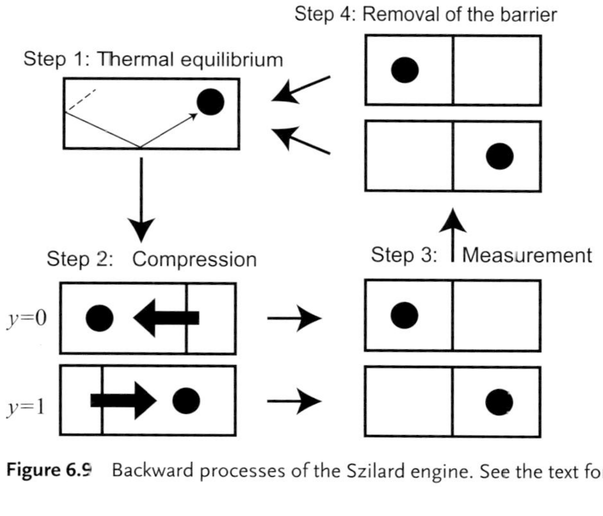

Figure 6.9 Backward processes of the Szilard engine. See the text for details.

While Eq. (6.61) only includes the obtained mutual information and does not describe how we utilize the information via feedback, Eq. (6.72) includes the term of feedback efficacy that depends on the feedback protocol. To quantitatively discuss the relationship between mutual information I_c and efficacy parameter γ, we define $C[A] := -\ln\langle e^{-A}\rangle$. By noting Eq. (6.55), we obtain

$$C[\sigma] + C[I_c] - C[\sigma + I_c] = -\ln \gamma. \tag{6.76}$$

If the joint distribution of σ and I_c is Gaussian, Eq. (6.76) reduces to

$$\langle \sigma I_c\rangle - \langle \sigma\rangle\langle I_c\rangle = -\ln \gamma. \tag{6.77}$$

Equalities (6.76) and (6.77) imply that the more efficiently we use the obtained information to decrease the entropy production by feedback control, the larger the γ. In fact, if γ is large, the left-hand sides of Eqs. (6.76) and (6.77) are both small, which means that the obtained information I_c has a large negative correlation with σ. Without feedback control, $\gamma = 1$ holds and therefore I_c is not correlated with σ. In this sense, γ characterizes the efficacy of feedback control.

6.6
Thermodynamic Energy Cost for Measurement and Information Erasure

So far, we have discussed the energy balance of information heat engines controlled by the demon. In this section, we discuss the energy cost that is needed for the demon itself, which has been a subject of active discussion [2–4, 85–102].

Suppose that the demon has a memory that can store the outcome obtained by a measurement. If the outcome is binary, the memory can be modeled by a system with a binary potential (Figure 6.10). Before the measurement, the memory is in the initial standard state 0. The memory then interacts with a measured system such as the Szilard engine and stores the measurement outcome. Figure 6.10 illustrates a case with a binary outcome. Let p_k be the probability of obtaining outcome k. After the measurement, the memory is detached from the measured system and returns to the initial standard state, which is the erasure of the obtained information. The central question in this section is how much work is needed for the demon during the measurement and the information erasure.

Let F_k^{M} be the free energy of the memory under the condition that the outcome is "k." During the measurement process, the free energy of the memory is changed on average by $\Delta F^{\mathrm{M}} := \sum_k p_k F_k^{\mathrm{M}} - F_0^{\mathrm{M}}$, where F_0^{M} is the free energy of the initial standard state. If F_k^{M}'s are the same for all k's including $k = 0$ (i.e., the memory's potential is symmetric), $\Delta F^{\mathrm{M}} = 0$ holds for every $\{p_k\}$. It has been shown [102] that the averaged work $W_{\mathrm{meas}}^{\mathrm{M}}$ that is performed on the memory during the measurement is bounded as

$$W_{\mathrm{meas}}^{\mathrm{M}} \geq \Delta F^{\mathrm{M}} - k_{\mathrm{B}} T H + k_{\mathrm{B}} T I, \tag{6.78}$$

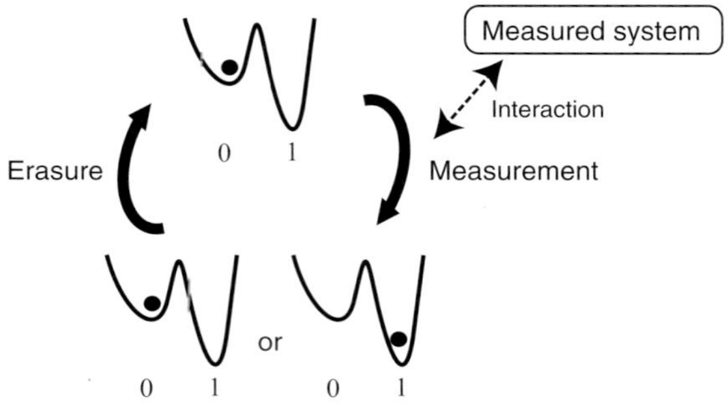

Figure 6.10 A schematic of the measurement and erasure of information for the case of an asymmetric binary memory. Although the memory is in the standard state with unit probability before the measurement, it stores the measurement outcome in accordance with the state of the measured system. The measurement and erasure processes are time reversal with each other except for the fact that the memory establishes a correlation with the measured system during the measurement process.

where $H := -\sum_k p_k \ln p_k$ is the Shannon information of the outcomes and I is the mutual information obtained by the measurement. For the special case with $\Delta F^{\mathrm{M}} = 0$ and $H = I$, the RHS of inequality (6.78) reduces to zero.

On the other hand, during the information erasure, the change of the free energy of the memory is given by $-\Delta F^{\mathrm{M}}$. The averaged work $W^{\mathrm{M}}_{\mathrm{eras}}$ that is needed for the erasure process is bounded as [102]

$$W^{\mathrm{M}}_{\mathrm{eras}} \geq -\Delta F^{\mathrm{M}} + k_{\mathrm{B}} T H. \tag{6.79}$$

If ΔF^{M} vanishes, inequality (6.79) reduces to

$$W^{\mathrm{M}}_{\mathrm{eras}} \geq k_{\mathrm{B}} T H, \tag{6.80}$$

which is known as the Landauer principle [86]. The additional term $-\Delta F^{\mathrm{M}}$ on the RHS of (6.79) arises from the asymmetry of the memory. By summing up inequalities (6.78) and (6.79), we obtain the fundamental inequality:

$$W^{\mathrm{M}}_{\mathrm{meas}} + W^{\mathrm{M}}_{\mathrm{eras}} \geq k_{\mathrm{B}} T I, \tag{6.81}$$

which implies that the work needed for the demon is bounded only by the mutual information if we take into account both the measurement and erasure processes.

We stress that while inequality (6.79) is a generalized Landauer principle for the information erasure, inequality (6.81) is completely different from the Landauer principle. In fact, while the lower bound of the Landauer principle is given by the Shannon information that characterizes the randomness of the measurement outcomes, the lower bound of (6.81) is given by the mutual information that characterizes the correlation between the measured system and the measurement outcome. Moreover, both terms on the RHS of (6.79) are exactly canceled by the first and second terms on the RHS of (6.78). The reason for the cancellation lies in the fact that the dynamics of the memory during the erasure process is the time reversal of the measurement process, except for the fact that the memory interacts with the measured system and establishes a correlation (or equivalently, gains information) only in the measurement process (Figure 6.10). The additional cost for the establishment of the correlation is given by the last term on the RHS of (6.78), which also appears in the RHS of (6.81). Therefore, the mutual information term in inequality (6.81) is induced by the measurement process.

Historically, there has been a lot of discussions [2–4] on as to what compensates for the additional work of $k_B T \ln 2$, which can be extracted from the Szilard engine. Szilard considered that an entropic cost must be needed for the measurement process [47]. Brillouin [85] argued that we need the work greater than $k_B T \ln 2$ for the measurement process, based on a specific model of measurement. Later, by explicitly constructing a model of the memory that does not require any work for the measurement, Bennett argued that, based on the Landauer principle (6.80), we always need the work of at least $k_B T \ln 2$ for the erasure process [87, 94]. The key observation here is that the erasure process is logically irreversible, while the measurement process can be logically reversible. In fact, if we assume that the Shannon information of the measurement outcome equals the thermodynamic entropy of the memory, the logically irreversible erasure should be accompanied by a reduction in the thermodynamic entropy of the memory, which implies that $k_B T \ln 2$ of heat should be transferred to the heat bath and, therefore, the same amount of the work is needed.

However, the argument by Landauer and Bennett is valid only for symmetric memories with $\Delta F^{\mathrm{M}} = 0$. As discussed in Refs [98–102], the Shannon information does not equal the thermodynamic entropy of the memory in general. If the memory is asymmetric as illustrated in Figure 6.10, the lower bound of the energy cost needed for the information erasure is not given by (6.80), and the Landauer principle needs to be generalized to inequality (6.79) for asymmetric memories. We note that the Landauer principle can also be violated for symmetric memories in the quantum regime due to the initial correlation between the memory and the heat bath [96, 97]. A more detailed historical review about the Landauer principle is given in Ref. [4].

As a consequence, the lower bound of the individual energy cost for measurement or erasure processes can be made arbitrarily small for asymmetric memories, while their sum (6.81) is bounded from below by $k_B TI$ that originates from the measurement process. The total work given on the left-hand side of (6.81) then compensates for $k_B TI$ of additional work in (6.25) that is extracted from an information heat engine by the demon. This compensation confirms the consistency between the demon and the second law of thermodynamics; we cannot extract any positive amount of work by a cycle from the total system consisting of the engine and the memory of the demon.

Nevertheless, feedback control is still useful for manipulating small thermodynamic systems. In fact, as discussed in Section 6.2, feedback control enables us to increase the engine's free energy without injecting energy to the engine directly. In other words, the work (6.81) needed for the demon is not necessarily transferred to the engine, which can be energetically separated from the demon. Therefore, by using information heat engines, we can control thermodynamic systems beyond the energy balance that is imposed by the conventional thermodynamics.

6.7 Conclusions

In this chapter, we have discussed a generalized thermodynamics that can be applied to feedback-controlled systems, which we call information heat engines. The Szilard engine described in Section 6.2 is the simplest model of information heat engines. Based on the information theory reviewed in Section 6.3, we have formulated a generalized second law involving the term of the mutual information in Section 6.4. The generalized second law gives an upper bound of the work that can be extracted from a heat bath with the assistance of feedback control. We also discussed some typical examples of information heat engines, including a recent experimental result [30]. In Section 6.5, we discussed nonequilibrium equalities with feedback control, and derived the generalized second law discussed in Section 6.4. We also discussed the energy cost that is needed for the measurement and the information erasure in Section 6.6.

Inequalities (6.25), (6.78), (6.79), and (6.81) are the generalizations of the second law of thermodynamics, giving the fundamental bounds of the work needed for information processing. In fact, if we set the information contents to be zero (i.e., $I = H = 0$) in these inequalities, all of them reduce to the conventional second law of thermodynamics. In this sense, these inequalities constitute the second law of "information thermodynamics," which is a generalized thermodynamics for information processing.

While the studies of information and thermodynamics have a long history, recent developments of nonequilibrium physics and nanotechnologies have shed new light on classical problems from the modern viewpoint. Thermodynamics of information processing opens a fruitful research arena that enables us to quantitatively analyze the energy costs of the feedback control and information processing in small thermodynamic systems. Possible applications of this new research field include designing and controlling nanomachines [103] and nanodevices.

Appendix 6.A: Proof of Eq. (6.56)

In this appendix, we prove Eq. (6.56). We introduce notations $X_{t_{k-1}<t\leq t_k} := \{x(t)\}_{t_{k-1}<t\leq t_k}$, $X_{t_k} := \{x(t)\}_{0\leq t\leq t_k}$, $\Lambda_{t_k} := \{\lambda(t)\}_{0\leq t\leq t_k}$, and $Y_{t_k} := (y(t_1), \ldots, y(t_k))$. The joint probability of (X_τ, Y_τ) is given by

$$\begin{aligned} P[X_\tau, Y_\tau] &= P[X_{t_M<t\leq\tau}|X_{t_M}, \Lambda_\tau(Y_\tau)] \\ &\cdot\prod_{k=1}^{M} P[y(t_k)|x(t_k)]P[X_{t_{k-1}<t\leq t_k}|X_{t_{k-1}}, \Lambda_{t_k}(Y_{t_{k-1}})] \cdot P_{\mathrm{f}}[x(0)], \end{aligned} \tag{6.A1}$$

where we set $t_0 := 0$. We note that Λ_{t_k} depends only on $Y_{t_{k-1}}$ due to the causality. We also note that

$$P[X_\tau|\Lambda(Y_\tau)] = P[X_{t_k<t\le\tau}|X_{t_k}, \Lambda_\tau(Y_\tau)] \prod_{k=1}^{M} P[X_{t_{k-1}<t\le t_k}|X_{t_{k-1}}, \Lambda_{t_k}(Y_{t_{k-1}})] P_{\mathrm{f}}[x(0)]. \tag{6.A2}$$

By combining Eqs. (6.52),(6.A1) and (6.A2), we obtain Eq. (6.56). We can confirm that the joint probability satisfies the normalization condition $\int dX_\tau dY_\tau P[X_\tau, Y_\tau] = 1$ by integrating Eq. (6.A1) on the order of $X_{t_M<t\le\tau} \to y(t_M) \to X_{t_{M-1}<t\le t_M} \to y(t_{M-1}) \to \cdots \to y(t_1) \to X_{0<t\le t_1} \to x(0)$ due to the causality.

Acknowledgments

The authors are grateful to Christopher Jarzynski for valuable comments and to Shoichi Toyabe, Eiro Muneyuki, and Masaki Sano for providing us the experimental data discussed in Section 6.4.4. This work was supported by KAKENHI 22340114, a grant-in-aid for Scientific Research on Innovation Areas "Topological Quantum Phenomena" (KAKENHI 22103005), a Global COE Program "the Physical Sciences Frontier," the Photon Frontier Network Program, and the Grant-in-Aid for Research Activity Start-up (KAKENHI 11025807), from MEXT of Japan.

References

1 Maxwell, J.C. (1871) *Theory of Heat*, Appleton, London.
2 Leff, H.S. and Rex, A.F. (eds) (2003) *Maxwell's Demon 2: Entropy, Classical and Quantum Information, Computing*, Princeton University Press, New Jersey.
3 Maruyama, K., Nori, F., and Vedral, V. (2009) *Rev. Mod. Phys.*, **81**, 1.
4 Maroney, O.J.E. (2009) *Information Processing and Thermodynamic Entropy* (ed. E.N. Zalta), The Stanford Encyclopedia of Philosophy.
5 Doyle, J.C., Francis, B.A., and Tannenbaum, A.R. (1992) *Feedback Control Theory*, Macmillan, New York.
6 Åstrom, K.J. and Murray, R.M. (2008) *Feedback Systems: An Introduction for Scientists and Engineers*, Princeton University Press.
7 Lloyd, S. (1989) *Phys. Rev. A*, **39**, 5378.
8 Caves, C.M. (1990) *Phys. Rev. Lett.*, **64**, 2111.
9 Lloyd, S. (1997) *Phys. Rev. A*, **56**, 3374.
10 Nielsen, M.A., Caves, C.M., Schumacher, B., and Barnum, H. (1998) *Proc. R. Soc. Lond. A*, **454**, 277.
11 Touchette, H. and Lloyd, S. (2000) *Phys. Rev. Lett.*, **84**, 1156.
12 Zurek, W.H. (2003) arXiv:quant-ph/0301076.
13 Kieu, T.D. (2004) *Phys. Rev. Lett.*, **93**, 140403.
14 Allahverdyan, A.E., Balian, R., and Nieuwenhuizen, Th.M. (2004) *J. Mod. Optics*, **51**, 2703.
15 Touchette, H. and Lloyd, S. (2004) *Physica A*, **331**, 140.
16 Quan, H.T., Wang, Y.D., Liu, Y-x., Sun, C. P., and Nori, F. (2006) *Phys. Rev. Lett.*, **97**, 180402.
17 Cao, F.J., Dinis, L., and Parrondo, J.M.R. (2004) *Phys. Rev. Lett.*, **93**, 040603.
18 Kim, K.H. and Qian, H. (2004) *Phys. Rev. Lett.*, **93**, 120602.
19 Kim, K.H. and Qian, H. (2007) *Phys. Rev. E*, **75**, 022102.

20 Lopez, B.J., Kuwada, N.J., Craig, E.M., Long, B.R., and Linke, H. (2008) *Phys. Rev. Lett.*, **101**, 220601.
21 Allahverdyan, A.E. and Saakian, D.B. (2008) *Europhys. Lett.*, **81**, 30003.
22 Sagawa, T. and Ueda, M. (2008) *Phys. Rev. Lett.*, **100**, 080403.
23 Jacobs, K. (2009) *Phys. Rev. A*, **80**, 012322.
24 Cao, F.J. and Feito, M. (2009) *Phys. Rev. E*, **79**, 041118.
25 Cao, F.J., Feito, M., and Touchette, H. (2009) *Physica A*, **388**, 113.
26 Sagawa, T. and Ueda, M. (2010) *Phys. Rev. Lett.*, **104**, 090602.
27 Ponmurugan, M. (2010) *Phys. Rev. E*, **82**, 031129.
28 Fujitani, Y. and Suzuki, H. (2010) *J. Phys. Soc. Jpn.*, **79**, 104003.
29 Horowitz, J.M. and Vaikuntanathan, S. (2010) *Phys. Rev. E*, **82**, 061120.
30 Toyabe, S., Sagawa, T., Ueda, M., Muneyuki, E., and Sano, M. (2010) *Nat. Phys.*, **6**, 988.
31 Kim, S.W., Sagawa, T., De Liberato, S., and Ueda, M. (2011) *Phys. Rev. Lett.*, **106**, 070401.
32 Morikuni, Y. and Tasaki, H. (2011) *J. Stat. Phys.*, **143**, 1.
33 Ito, S. and Sano, M. (2011) *Phys. Rev. E*, **84**, 021123.
34 Horowitz, J.M. and Parrondo, J.M.R. (2011) *Europhys. Lett.*, **95**, 10005.
35 Abreu, D. and Seifert, U. (2011) *Europhys. Lett.*, **94**, 10001.
36 Vaikuntanathan, S. and Jarzynski, C. (2011) *Phys. Rev. E*, **83**, 061120.
37 Sagawa, T. (2011) *J. Phys. Conf. Ser.*, **297**, 012015.
38 Dong, H., Xu, D.Z., Cai, C.Y., and Sun, C.P. (2011) *Phys. Rev. E*, **83**, 061108.
39 Averin, D.V., Möttönen, M., and Pekola, J.P. (2011) *Phys. Rev. B*, **84**, 245448.
40 Horowitz, J.M. and Parrondo, J.M.R. (2011) *New J. Phys.*, **13**, 123019.
41 Granger, L. and Kantz, H. (2011) *Phys. Rev. E*, **84**, 061110.
42 Abreu, D. and Seifert, U. (2012) *Phys. Rev. Lett.*, **108**, 030601.
43 Lahiri, S., Rana, S., and Jayannavar, A.M. (2012) *J. Phys. A*, **45**, 065002.
44 Deffner, S. and Lutz, E. (2012) arXiv:1201.3888.
45 Lu, Y. and Long, G.L. (2012) *Phys. Rev. E*, **85**, 011125.
46 Sagawa, T. and Ueda, M. (2012) *Phys. Rev. E*, **85**, 021104.
47 Szilard, L. (1929) *Z. Phys.*, **53**, 840.
48 Cover, T.M. and Thomas, J.A. (1991) *Elements of Information Theory*, John Wiley & Sons, Inc., New York.
49 Shannon, C. (1948) *AT&T Tech. J.*, **27**, 379–423. and 623–656.
50 Groenewold, H.J. (1971) *Int. J. Theor. Phys.*, **4**, 327.
51 Ozawa, M. (1986) *J. Math. Phys.*, **27**, 759.
52 Buscemi, F., Hayashi, M., and Horodecki, M. (2008) *Phys. Rev. Lett.*, **100**, 210504.
53 Vale, R.D. and Oosawa, F. (1990) *Adv. Biophys.*, **26**, 97.
54 Julicher, F., Ajdari, A., and Prost, J. (1997) *Rev. Mod. Phys.*, **69**, 1269.
55 Parrondo, J.M.R. and De Cisneros, B.J. (2002) *Appl. Phys. A*, **75**, 179.
56 Reimann, P. (2002) *Phys. Rep.*, **361**, 57.
57 Hänggi, P. and Marchesoni, F. (2009) *Rev. Mod. Phys.*, **81**, 387.
58 Bier, M. (2007) *Biosystems*, **88**, 201.
59 Schliwa, M. and Woehlke, G. (2003) *Nature*, **422**, 759.
60 Schlichting, H.J. and Nordmeier, V. (1996) *Math. Naturwiss. Unterr.*, **49**, 323.
61 van der Weele, K., van der Meer, D., Versluis, M., and Lohse, D. (2001) *Europhys. Lett.*, **53**, 328.
62 Serreli, V., Lee, C.-F., Kay, E.R., and Leigh, D.A. (2007) *Nature*, **445**, 523.
63 Price, G.N., Bannerman, S.T., Viering, K., Narevicius, E., and Raizen, M.G. (2008) *Phys. Rev. Lett.*, **100**, 093004.
64 Millonas, M.M. (1995) *Phys. Rev. Lett.*, **74**, 10.
65 Jayannavar, A.M. (1996) *Phys. Rev. E*, **53**, 2957.
66 Eggers, J. (1999) *Phys. Rev. Lett.*, **83**, 5322.
67 Brey, J.J., Moreno, F., Garcia-Rojo, R., and Ruiz-Montero, M.J. (2001) *Phys. Rev. E*, **65**, 011305.
68 Van den Broeck, C., Meurs, P., and Kawai, R. (2005) *New J. Phys.*, **7**, 10.
69 Ruschhaupt, A., Muga, J.G., and Raize, M.G. (2006) *J. Phys. B*, **39**, 3833.
70 Jarzynski, C. (1997) *Phys. Rev. Lett.*, **78**, 2690.

71 Crooks, G.E. (1998) *J. Stat. Phys.*, **90**, 1481.

72 Crooks, G.E. (1999) *Phys. Rev. E*, **60**, 2721.

73 Jarzynski, C. (2000) *J. Stat. Phys.*, **98**, 77.

74 Seifert, U. (2005) *Phys. Rev. Lett.*, **95**, 040602.

75 Kawai, R., Parrondo, J.M.R., and Van den Broeck, C. (2007) *Phys. Rev. Lett.*, **98**, 080602.

76 Bustamante, C., Liphardt, J., and Ritort, F. (2005) *Phys. Today*, **58**, 43.

77 Liphardt, J. *et al.* (2002) *Science*, **296**, 1832.

78 Collin, D. *et al.* (2005) *Nature*, **437**, 231.

79 Schreiber, T. (2000) *Phys. Rev. Lett.*, **85**, 461.

80 Hatano, T. and Sasa, S.-I. (2001) *Phys. Rev. Lett.*, **86**, 3463.

81 Speck, T. and Seifert, U. (2005) *J. Phys. A*, **38**, L581.

82 Sasa, S.-I. and Tasaki, H. (2006) *J. Stat. Phys.*, **125**, 125.

83 Esposito, M., Harbola, U., and Mukamel, S. (2007) *Phys. Rev. E*, **76**, 031132.

84 Esposito, M. and Van den Broeck, C. (2010) *Phys. Rev. Lett.*, **104**, 090601.

85 Brillouin, L. (1951) *J. Appl. Phys.*, **22**, 334.

86 Landauer, R. (1961) *IBM J. Res. Dev.*, **5**, 183.

87 Bennett, C.H. (1982) *Int. J. Theor. Phys.*, **21**, 905.

88 Zurek, W.H. (1989) *Nature*, **341**, 119.

89 Zurek, W.H. (1989) *Phys. Rev. A*, **40**, 4731.

90 Shizume, K. (1995) *Phys. Rev. E*, **52**, 3495.

91 Landauer, R. (1996) *Science*, **272**, 1914.

92 Matsueda, H., Goto, E., and Loe, K.-F. (1997) *RIMS Kôkyûroku*, **1013**, 187.

93 Piechocinska, B. (2000) *Phys. Rev. A*, **61**, 062314.

94 Bennett, C.H. (2003) *Stud. Hist. Philos. Mod. Phys.*, **34**, 501.

95 Dillenschneider, R. and Lutz, E. (2009) *Phys. Rev. Lett.*, **102**, 210601.

96 Allahverdyan, A.E. and Nieuwenhuizen, T.M. (2001) *Phys. Rev. E*, **64**, 0561171.

97 Horhammer, C. and Buttner, H. (2008) *J. Stat. Phys.*, **133**, 1161.

98 Barkeshli, M.M. (2005) arXiv:cond-mat/0504323.

99 Norton, J.D. (2005) *Stud. Hist. Philos. Mod. Phys.*, **36**, 375.

100 Maroney, O.J.E. (2009) *Phys. Rev. E*, **79**, 031105.

101 Turgut, S. (2009) *Phys. Rev. E*, **79**, 041102.

102 Sagawa, T. and Ueda, M. (2009) *Phys. Rev. Lett.*, **102**, 250602; (2011) **106**, 189901(E).

103 Kay, E.R., Leigh, D.A., and Zerbetto, F. (2007) *Angew. Chem.*, **46**, 72.

7
Time-Reversal Symmetry Relations for Currents in Quantum and Stochastic Nonequilibrium Systems

Pierre Gaspard

7.1
Introduction

Microreversibility is the symmetry under time reversal of the equations of motion for the microscopic particles composing matter. This symmetry is a property of the electromagnetic interaction.[1] The motion is ruled by Newton's equations for the positions and velocities of the particles in classical mechanics and by Schrödinger's equation for the wave function in quantum mechanics. Classical mechanics emerges out of quantum mechanics in the limit where de Broglie's wavelength becomes smaller than the spatial scale of features in the energy landscape where each particle moves. A particle of mass m and velocity v has its de Broglie wavelength given by

$$\lambda = \frac{h}{mv}, \tag{7.1}$$

where h is the Planck's constant. Hence, the wavelength is shorter for heavier particles in systems at higher temperatures. At room temperature, the wavelength of nuclei is smaller than the typical interatomic distances. This is not the case for electrons that maintain the interatomic chemical bonds by the spatial extension of their quantum wave function. Consequently, the motion of nuclei is essentially classical, while electrons behave quantum mechanically. At low temperatures, de Broglie's wavelengths are larger. For instance, in semiconducting nanodevices at sub-Kelvin electronic temperatures, the nuclei are nearly frozen, while the electronic waves propagate ballistically in artificial circuits of a few hundred nanometers [1–3].

Newton's and Schrödinger's equations are in common time-reversal symmetric and, moreover, deterministic in the sense that the time evolution they rule is uniquely determined by initial conditions belonging either to the classical phase space of positions and velocities or to the Hilbert space of wave functions. In both

1) For high-energy systems involving all the forces of nature including the weak interaction, the corresponding symmetry is the combination of time reversal, space inversion, and charge conjugation.

Nonequilibrium Statistical Physics of Small Systems: Fluctuation Relations and Beyond, First Edition.
Edited by Rainer Klages, Wolfram Just, and Christopher Jarzynski.

schemes, the initial conditions themselves are not uniquely determined once for all, but free to take arbitrary values depending on the particular experiment considered. In this regard, the equations of motion describe all the possible histories that a system may follow, among which there exists the unique history actually followed by the system under specific conditions. Besides, most of the histories do not coincide with their time reversal in typical classical or quantum systems. For instance, the trajectory of a free particle with a nonvanishing velocity is physically distinct from its time reversal in the phase space of classical states:

$$(\mathbf{r}_t, \mathbf{v}_t) = (\mathbf{v}_0 t + \mathbf{r}_0, \mathbf{v}_0) \neq (\mathbf{r}'_t, \mathbf{v}'_t) = (-\mathbf{v}_0 t + \mathbf{r}_0, -\mathbf{v}_0). \tag{7.2}$$

If the first trajectory comes from the star Antares toward the constellation of Taurus on the opposite side of the celestial sphere, the second would travel in the opposite direction following a reversed history [4]. A similar result holds for the wave packet of a free quantum particle. For such systems, the selection of initial conditions breaks the time-reversal symmetry. Indeed, the solutions of an equation may have a lower symmetry than the equation itself. This is well known in condensed matter physics with the phenomenon of spontaneous symmetry breaking. For example, in ferromagnetism, the orientation of magnetization is determined by fluctuations at the initial stage of the process, that is, by the initial conditions.

These considerations extend to statistical mechanics where the time evolution concerns statistical ensembles described, in classical mechanics, by a probability density ϱ evolving in phase space according to Liouville's equation:

$$\partial_t \varrho = \{H, \varrho\}_{\mathrm{cl}}, \tag{7.3}$$

where H is the Hamiltonian and $\{\cdot,\cdot\}_{\mathrm{cl}}$ is the Poisson bracket, or, in quantum mechanics, by a density operator $\hat{\varrho}$ evolving according to von Neumann's equation:

$$\partial_t \hat{\varrho} = \frac{1}{i\hbar}\left[\hat{H}, \hat{\varrho}\right], \tag{7.4}$$

where $[\cdot,\cdot]$ is the commutator, $i = \sqrt{-1}$, and $\hbar = h/(2\pi)$ [5]. Both equations are symmetric under time reversal, but they may admit solutions that correspond to histories differing from their time reversals.

However, thermodynamic equilibria are described by Gibbsian statistical ensembles such as the canonical ensemble defined by the probability distribution:

$$\varrho = \frac{1}{Z}\exp(-\beta H), \quad \text{with} \quad \beta = \frac{1}{k_{\mathrm{B}} T}, \tag{7.5}$$

for the temperature T and where k_{B} is the Boltzmann's constant. This distribution is a stationary solution of Liouville's or von Neumann's equation and it is time-reversal symmetric. As a consequence, the principle of detailed balancing holds according to which opposite fluctuations are equiprobable. The entropy production vanishes and there is no energy dissipation on average in systems at equilibrium. We note that the thermodynamic equilibrium is a state that is stationary from a statistical viewpoint, but dynamical from a mechanical viewpoint.

During the last decades, remarkable advances have been achieved in the understanding of the statistical properties of nonequilibrium systems. These results find their origins in the study of large deviation properties of chaotic dynamical systems, for which methods have been developed to characterize randomness in time (instead of space or phase space as done in equilibrium statistical mechanics) [6]. On the basis of dynamical systems theory, relationships have been established between dynamical large deviation quantities and transport properties such as diffusion, viscosity, and electric or heat conductivities [7–17]. In this context, the further consideration of time reversal led to the discovery of different types of symmetry relations depending on the nonequilibrium regime. Indeed, systems may evolve out of equilibrium under different conditions:

- In isolated systems, the solution of Liouville's equation is not stationary and evolves in time if the initial distribution ϱ_0 differs from an equilibrium one. If the dynamics is mixing, the statistical distribution ϱ_t weakly converges toward a final equilibrium distribution after nonequilibrium transients. Transient states with exponential decay exist in the forward or backward time evolutions, which are mapped onto each other by time reversal, while the equilibrium state is symmetric [13–15].
- In systems controlled by time-dependent external forces, the statistical distribution ϱ_t remains out of equilibrium since the relaxation toward equilibrium is not possible. For instance, the nonequilibrium work is a fluctuating quantity of interest in small systems such as single molecules subjected to the time-dependent forces of optical tweezers or atomic force microscopy [18–23].
- In open systems in contact with several reservoirs at different temperatures or chemical potentials, a nonequilibrium steady state is reached after some relaxation time. Such states are described by stationary statistical distributions that are no longer symmetric under time reversal contrary to the equilibrium canonical state (Eq. (7.5)). Indeed, out of equilibrium, mean fluxes of energy or matter are flowing on average across the open system and this directionality breaks the time-reversal symmetry. In small systems or on small scales, the energy or particle currents are fluctuating quantities described by the nonequilibrium statistical distribution.

Because the underlying microscopic dynamics is reversible, the fluctuations of nonequilibrium quantities obey remarkable relationships, which are valid not only close to equilibrium but also arbitrarily far from equilibrium [20–34]. Several kinds of such relationships have been obtained for transitory or stationary nonequilibrium situations, for classical or quantum systems, and for Markovian or non-Markovian stochastic processes.

Moreover, further time-reversal symmetry relations have also been obtained for the probabilities of the histories or paths followed by a system under stroboscopic observations [35–42]. It turns out that the time asymmetry of nonequilibrium statistical distributions implies that typical histories are more probable than their corresponding time reversal. In this regard, dynamical order manifests itself away from equilibrium [39]. Furthermore, the breaking of detailed balancing between forward and reversed histories is directly related to the thermodynamic entropy production, as established for classical, stochastic, and quantum systems [35–42].

The purpose of the present chapter is to give an overview of the time-reversal symmetry relations established for nonequilibrium quantum systems [42–62].[2),3)] Recently, important experimental and theoretical work has been devoted to transport in small quantum systems and, in particular, to electronic transport in quantum point contacts (QPCs), quantum dots (QDs), and coherent quantum conductors [1–3]. In these systems, single-electron transfers can be observed experimentally and subjected to full counting statistics, allowing the experimental test of the time-reversal symmetry relations and their predictions. Here, the aim is to provide a comprehensive overview of the results obtained till now on this topic and to discuss some of the open issues.

This chapter is organized as follows. In Section 7.2, functional symmetry relations are established for quantum systems driven by time-dependent external forces in a constant magnetic field, which allows us to derive the Kubo formulas and the Casimir–Onsager reciprocity relations for the linear response properties. In Section 7.3, general methods are presented to describe open quantum systems in contact with several reservoirs and transitory fluctuation theorems are obtained. In Section 7.4, an appropriate long-time limit is taken in order to reach a nonequilibrium steady state. In this limit, the stationary fluctuation theorem is obtained from the transitory one for all the currents flowing across the open system. In Section 7.5, the current fluctuation theorem is used to obtain the response properties. The Casimir–Onsager reciprocity relations and fluctuation–dissipation relations are generalized from linear to nonlinear response properties. In Section 7.6, these results are obtained for independent electrons in quantum point contacts or quantum dots. The generating function of full counting statistics is computed using Klich formula and the connection is established with the Levitov–Lesovik formula in the Landauer–Büttiker scattering approach. The equivalence with the Keldysh approach is also discussed. In Section 7.7, the current fluctuation theorem is derived in the master equation approach for the corresponding stochastic process. The second law of thermodynamics is deduced from the stationary fluctuation theorem for the currents and results on the statistics of histories are presented. In Section 7.8, the general theory is applied to electronic transport in quantum dots, quantum point contacts, and coherent quantum conductors. Conclusions are drawn in Section 7.9.

7.2
Functional Symmetry Relations and Response Theory

We consider a quantum system described by the Hamiltonian operator $\hat{H}(t;\mathcal{B})$, which depends on the time t and the external magnetic field $\mathcal{B}$. The

2) Kurchan, J. cond-mat/0007360.
3) Tasaki, H. cond-mat/0009244.

time-reversal operator $\hat{\Theta}$ reverses the magnetic field but otherwise leaves invariant the Hamiltonian:

$$\hat{\Theta}\hat{H}(t;\mathcal{B})\hat{\Theta}^{-1} = \hat{H}(t;-\mathcal{B}). \tag{7.6}$$

The reason is that the time-reversal symmetry of the electromagnetic interaction concerns the whole physical system, including the external electric currents generating the magnetic field. The magnetic field changes its sign because the external currents are reversed under time reversal. At the level of the system itself, the symmetry should thus be implemented by the combination of the internal operator $\hat{\Theta}$ and the reversal of the background magnetic field.

Since the system is driven by time-dependent forces, a comparison should be carried out between some protocol, called the forward protocol and the reversed protocol [21–23].

In the *forward protocol*, the system starts in the equilibrium distribution:

$$\hat{\varrho}(0;\mathcal{B}) = \frac{\mathrm{e}^{-\beta\hat{H}(0;\mathcal{B})}}{Z(0)} \tag{7.7}$$

at the inverse temperature $\beta = (k_{\mathrm{B}}T)^{-1}$ and the free energy $F(0) = -k_{\mathrm{B}}T\ln Z(0)$ with $Z(0) = \mathrm{tr}\,\mathrm{e}^{-\beta\hat{H}(0;\mathcal{B})}$. The system evolves from the initial time $t = 0$ until the final time $t = \mathcal{T}$ under the unitary operator that is the solution of Schrödinger's equation:

$$i\hbar\frac{\partial}{\partial t}\hat{U}_{\mathrm{F}}(t;\mathcal{B}) = \hat{H}(t;\mathcal{B})\,\hat{U}_{\mathrm{F}}(t;\mathcal{B}), \tag{7.8}$$

with the initial condition $\hat{U}_{\mathrm{F}}(0;\mathcal{B}) = 1$. The average of an observable $\hat{A}$ is given by

$$\langle\hat{A}_{\mathrm{F}}(t)\rangle = \mathrm{tr}\,\hat{\varrho}(0)\hat{A}_{\mathrm{F}}(t), \tag{7.9}$$

where

$$\hat{A}_{\mathrm{F}}(t) = \hat{U}_{\mathrm{F}}^{\dagger}(t)\hat{A}\hat{U}_{\mathrm{F}}(t) \tag{7.10}$$

is the operator in the Heisenberg representation.

In the *reversed protocol*, the system starts in the other equilibrium distribution:

$$\hat{\varrho}(\mathcal{T};-\mathcal{B}) = \frac{\mathrm{e}^{-\beta\hat{H}(\mathcal{T};-\mathcal{B})}}{Z(\mathcal{T})} \tag{7.11}$$

of free energy $F(\mathcal{T}) = -k_{\mathrm{B}}T\ln Z(\mathcal{T})$ with $Z(\mathcal{T}) = \mathrm{tr}\,\mathrm{e}^{-\beta\hat{H}(\mathcal{T};-\mathcal{B})}$ and evolves under the reversed time evolution operator obeying

$$i\hbar\frac{\partial}{\partial t}\hat{U}_{\mathrm{R}}(t;\mathcal{B}) = \hat{H}(\mathcal{T}-t;\mathcal{B})\,\hat{U}_{\mathrm{R}}(t;\mathcal{B}), \tag{7.12}$$

with the initial condition $\hat{U}_{\mathrm{R}}(0;\mathcal{B}) = 1$. In the reversed protocol, the system thus follows the reversed external driving in the reversed magnetic field, starting at the time $t = 0$ with the Hamiltonian $\hat{H}(\mathcal{T};-\mathcal{B})$ and ending at the time $t = \mathcal{T}$ with the Hamiltonian $\hat{H}(0;-\mathcal{B})$.

We have the following:

Lemma 7.1 [48]
The forward and reversed time evolution unitary operators are related to each other according to

$$\hat{U}_{\mathrm{R}}(t;-\mathcal{B}) = \hat{\Theta}\hat{U}_{\mathrm{F}}(\mathcal{T}-t;\mathcal{B})\hat{U}_{\mathrm{F}}^{\dagger}(\mathcal{T};\mathcal{B})\hat{\Theta}^{-1}, \quad \text{with} \quad 0 \le t \le \mathcal{T}. \tag{7.13}$$

This lemma is proved by using the antiunitarity of the time-reversal operator (Eqs. (7.8) and (7.12)) obeyed by the unitary evolution operators, as well as their initial conditions [48].

Let us consider a time-independent observable $\hat{A}$ with a definite parity under time reversal, that is, such that $\hat{\Theta}\hat{A}\hat{\Theta}^{-1} = \varepsilon_A \hat{A}$ with $\varepsilon_A = \pm 1$. The lemma allows us to relate its forward and reversed Heisenberg representations according to

$$\hat{A}_{\mathrm{F}}(t) = \varepsilon_A \hat{U}_{\mathrm{F}}^{\dagger}(\mathcal{T})\hat{\Theta}^{-1}\hat{A}_{\mathrm{R}}(\mathcal{T}-t)\hat{\Theta}\hat{U}_{\mathrm{F}}(\mathcal{T}). \tag{7.14}$$

In this setting, we have the following:

Theorem 7.1 [48]
If $\lambda(t)$ denotes an arbitrary function of time, the following functional relation holds:

$$\left\langle \mathrm{e}^{\int_0^{\mathcal{T}} \mathrm{d}t \lambda(t)\hat{A}_{\mathrm{F}}(t)} \mathrm{e}^{-\beta \hat{H}_{\mathrm{F}}(\mathcal{T})} \mathrm{e}^{\beta \hat{H}(0)} \right\rangle_{\mathrm{F},\mathcal{B}} = \mathrm{e}^{-\beta \Delta F} \left\langle \mathrm{e}^{\varepsilon_A \int_0^{\mathcal{T}} \mathrm{d}t \lambda(\mathcal{T}-t)\hat{A}_{\mathrm{R}}(t)} \right\rangle_{\mathrm{R},-\mathcal{B}}, \tag{7.15}$$

where $\hat{H}_{\mathrm{F}}(\mathcal{T}) = \hat{U}_{\mathrm{F}}^{\dagger}(\mathcal{T})\hat{H}(\mathcal{T};\mathcal{B})\hat{U}_{\mathrm{F}}(\mathcal{T})$ a and $\Delta F = F(\mathcal{T}) - F(0)$ is the free-energy difference between the initial equilibrium states (7.11) and (7.7) *of the reversed and forward protocols.*

This theorem, which has been demonstrated in Ref. [48], extends the results previously obtained in Refs [63, 64] where a different approach is used, as thoroughly discussed in Ref. [53].

Moreover, the quantum Jarzynski equality is deduced in the special case where $\lambda = 0$ in Eq. (7.15):

$$\left\langle \mathrm{e}^{-\beta \hat{H}_{\mathrm{F}}(\mathcal{T})} \mathrm{e}^{\beta \hat{H}(0)} \right\rangle_{\mathrm{F},\mathcal{B}} = \mathrm{e}^{-\beta \Delta F}. \tag{7.16}$$

The quantity within the bracket on the left-hand side of Eq. (7.16) can be interpreted in terms of the work performed on the system in the quantum scheme where von Neumann measurements of the energy are carried out at the initial and final times [43, 51–53][2),3)]. In the classical limit where the operators commute, both exponential functions combine into the exponential of the classical work $W_{\mathrm{cl}} = [H_{\mathrm{F}}(\mathcal{T}) - H(0)]_{\mathrm{cl}}$ and the classical Jarzynski equality [20]:

$$\langle \mathrm{e}^{-\beta W_{\mathrm{cl}}} \rangle_{\mathrm{F},\mathcal{B}} = \mathrm{e}^{-\beta \Delta F} \tag{7.17}$$

is recovered for the nonequilibrium work performed on the system during the forward protocol. We note that ΔF does not depend on the sign of the magnetic field because the canonical equilibrium distributions are time-reversal symmetric.

Remarkably, the functional symmetry relation (7.15) unifies the work relations and the response theory in a common framework. Indeed, the Kubo formulas as well as the Casimir–Onsager reciprocity relations can also be deduced from this relation. With this aim, we assume that the system is composed of N particles of electric charges $\{e_n\}_{n=1}^N$ subjected to an external time-dependent electric field $\mathcal{E}_\nu(t)$ in the spatial direction $\nu = 1, 2$, or 3. The electric field is supposed to vanish at the initial and final times $\mathcal{E}_\nu(0) = \mathcal{E}_\nu(\mathcal{T}) = 0$ so that the free energy difference is here equal to zero, $\Delta F = 0$. The time-dependent Hamiltonian is given by

$$\hat{H}(t) = \hat{H}_0 - \mathcal{E}_\nu(t) \sum_{n=1}^{N} e_n \hat{x}_{n\nu}, \tag{7.18}$$

with the position operators $\hat{x}_{n\nu}$ of the N particles. The observable we consider is the electric current density in the spatial direction $\mu = 1, 2$, or 3:

$$\hat{A} = \frac{1}{V}\hat{J}_\mu = \frac{1}{V}\sum_{n=1}^{N} e_n \frac{\mathrm{d}\hat{x}_{n\mu}}{\mathrm{d}t}, \tag{7.19}$$

where V is the volume of the system. In order to obtain the linear response of the observable $\hat{A}$ with respect to the perturbation due to an electric field $\mathcal{E}_\nu(t)$ of small amplitude, the functional derivative of Eq. (7.15) is taken with respect to $\lambda(\mathcal{T})$ around $\lambda = 0$, which yields

$$\left\langle \hat{A}_{\mathrm{F}}(\mathcal{T}) \mathrm{e}^{-\beta \hat{H}_{\mathrm{F}}(\mathcal{T})} \mathrm{e}^{\beta \hat{H}_0} \right\rangle_{\mathrm{F},\mathcal{B}} = \langle \hat{A} \rangle_{\mathrm{eq},\mathcal{B}}. \tag{7.20}$$

Developing this expression to first order in the electric field, the mean value of the current density at the final time $t = \mathcal{T}$ is obtained as [48]

$$j_\mu(\mathcal{T}) = \langle \hat{A}_{\mathrm{F}}(\mathcal{T}) \rangle_{\mathcal{B}} = \langle \hat{A} \rangle_{\mathrm{eq},\mathcal{B}} + \int_0^{\mathcal{T}} \mathrm{d}t\, \mathcal{E}_\nu(\mathcal{T} - t) \phi_{\mu\nu}(t; \mathcal{B}) + O(\mathcal{E}_\nu^2) \tag{7.21}$$

with the response function

$$\phi_{\mu\nu}(t; \mathcal{B}) = \frac{1}{V} \int_0^\beta \mathrm{d}u \langle \hat{J}_\nu(-i\hbar u) \hat{J}_\mu(t) \rangle_{\mathrm{eq},\mathcal{B}}. \tag{7.22}$$

Since the mean value of the current is vanishing at equilibrium $\langle \hat{A} \rangle_{\mathrm{eq},\mathcal{B}} = 0$, we find the expression of Ohm's law

$$j_\mu = \sigma_{\mu\nu} \mathcal{E}_\nu \tag{7.23}$$

in the long-time limit $\mathcal{T} \to \infty$. The electric conductivity is given by

$$\sigma_{\mu\nu} = \int_0^\infty \phi_{\mu\nu}(t; \mathcal{B}) \mathrm{d}t = \sigma_{\mu\nu}(\omega = 0; \mathcal{B}) \tag{7.24}$$

in terms of Kubo's formulas for the ac conductivities [65]:

$$\sigma_{\mu\nu}(\omega; \mathcal{B}) = \frac{1}{V}\int_0^{\infty} \mathrm{d}t e^{i\omega t} \int_0^{\beta} \mathrm{d}u \langle \hat{J}_\nu(-i\hbar u)\hat{J}_\mu(t)\rangle_{\mathrm{eq},\mathcal{B}}. \tag{7.25}$$

These conductivities satisfy the Casimir–Onsager reciprocity relations [66, 67]:

$$\sigma_{\mu\nu}(\omega; \mathcal{B}) = \sigma_{\nu\mu}(\omega; -\mathcal{B}). \tag{7.26}$$

In this way, the linear response properties are recovered from the functional symmetry relation (7.15). Higher order terms in the expansion can also be obtained for the nonlinear response properties.

The functional time-reversal symmetry relation thus provides a unifying framework to study the fundamental properties of quantum systems driven out of equilibrium by time-dependent external fields.

7.3
Transitory Current Fluctuation Theorem

A related problem of interest is the transport of particles and energy across an open quantum subsystem in contact with several reservoirs at different temperatures and chemical potentials. The subsystem is coupled to the reservoirs by a time-dependent potential $\hat{V}(t)$ during some time interval $0 \leq t \leq \mathcal{T}$. The total system evolves in time under the quantum-mechanical unitary time evolution from the initial state specified by taking an equilibrium distribution in each separate part of the total system until the final time $\mathcal{T}$, as schematically depicted in Figure 7.1. Here, the general case is considered where the time-dependent potential may leave some permanent changes inside the subsystem and the reservoirs after the final time $t = \mathcal{T}$. The flow of energy and particles between the reservoirs is determined by two successive von Neumann quantum measurements in each part of the total system, the first at the initial time $t = 0$ and the second at the final time $\mathcal{T}$. In this setup, a time-reversal symmetry relation is established for the transfers of energy and particles between the parts.

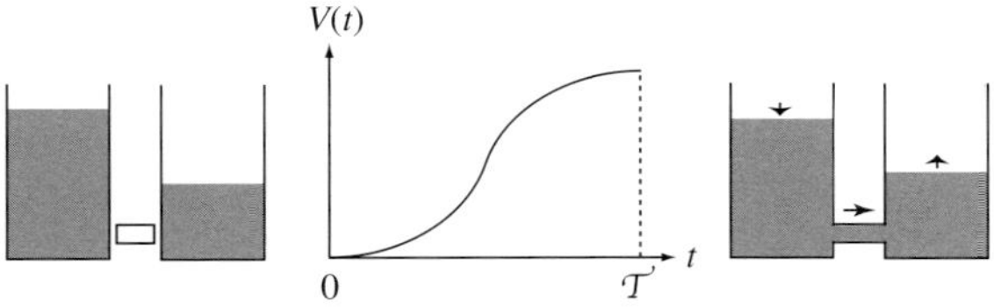

Figure 7.1 Schematic representation of the coupling of a subsystem to a pair of reservoirs by the potential $\hat{V}(t)$ during the time interval $0 \leq t \leq \mathcal{T}$, for the protocol used to deduce the *transitory* current fluctuation theorem. At the initial time $t = 0$, the potential is vanishing, $\hat{V}(0) = 0$. At the final time $t = \mathcal{T}$, the potential may be nonvanishing, but then given by a sum of operators separately acting on the quantum state spaces of the subsystem and the reservoirs.

The total Hamiltonian operator is given by

$$\hat{H}(t;\mathcal{B}) = \begin{cases} \hat{\mathcal{H}}_{\mathrm{S}} + \sum_{j=1}^{r} \hat{\mathcal{H}}_j, & \text{for} \quad t \leq 0, \\ \hat{\mathcal{H}}_{\mathrm{S}} + \sum_{j=1}^{r} \hat{\mathcal{H}}_j + \hat{V}(t), & \text{for} \quad 0 < t < \mathcal{T}, \\ \hat{\mathcal{H}}'_{\mathrm{S}} + \sum_{j=1}^{r} \hat{\mathcal{H}}'_j, & \text{for} \quad \mathcal{T} \leq t, \end{cases} \tag{7.27}$$

where $\hat{\mathcal{H}}_{\mathrm{S}}$ and $\hat{\mathcal{H}}_j$ denote the Hamiltonian operators of the isolated subsystem and the jth reservoir before the interaction is switched on. $\hat{V}(t)$ is the time-dependent interaction between the subsystem and the reservoirs. After the final time $t = \mathcal{T}$, the subsystem and the reservoirs no longer interact with each other, but the final potential $\hat{V}(\mathcal{T})$ may have added possible contributions to the Hamiltonian operators of the subsystem and the reservoirs, $\hat{\mathcal{H}}'_{\mathrm{S}}$ and $\hat{\mathcal{H}}'_j$, which could thus differ from the initial ones. Moreover, the total Hamiltonian always obeys the time-reversal symmetry (7.6).

The observables of the total system include not only the Hamiltonian operators but also the particle numbers of species $\alpha = 1, 2, \ldots, c$ in the subsystem $\hat{\mathcal{N}}_{\mathrm{S}\alpha}$ and the reservoirs $\hat{\mathcal{N}}_{j\alpha}$, with $j = 1, 2, \ldots, r$. These latter observables are also time-reversal symmetric: $\hat{\Theta}\hat{\mathcal{N}}_{j\alpha}\hat{\Theta}^{-1} = \hat{\mathcal{N}}_{j\alpha}$, with $j = \mathrm{S}, 1, 2, \ldots, r$.

The total numbers of particles are given by

$$\hat{\mathrm{N}}_\alpha = \hat{\mathcal{N}}_{\mathrm{S}\alpha} + \sum_{j=1}^{r} \hat{\mathcal{N}}_{j\alpha}, \quad \text{for} \quad \alpha = 1, 2, \ldots, c \tag{7.28}$$

and they commute with the total Hamiltonian operator because they are conserved:

$$[\hat{H}(t;\mathcal{B}), \hat{\mathrm{N}}_\alpha] = 0. \tag{7.29}$$

The system may admit further constant numbers of particles if the reservoirs are not all connected to each other. If the total system was composed of d disconnected parts between which there is no flux of particles, the particle numbers of every species would be separately conserved in each part so that the system would admit $c \times d$ independent constants of motion.

Moreover, before the initial time and after the final time, the reservoirs are decoupled from each other so that their Hamiltonian operator commutes with the particle numbers:

$$\left[\hat{\mathcal{H}}_j, \hat{\mathcal{N}}_{j'\alpha}\right] = 0, \quad \text{for} \quad t < 0 \quad \text{and} \quad \left[\hat{\mathcal{H}}'_j, \hat{\mathcal{N}}_{j'\alpha}\right] = 0, \quad \text{for} \quad \mathcal{T} < t, \tag{7.30}$$

for every $j, j' = \mathrm{S}, 1, 2, \ldots, r$ and $\alpha = 1, 2, \ldots, c$.

In nonequilibrium systems, the currents are determined by the differences of temperature and chemical potentials between the reservoirs. Since the reservoirs are large, the initial temperature and chemical potentials of the subsystem are not relevant if the subsystem is small enough. In this regard, we can simplify the formulation of the problem by regrouping the subsystem with one of the reservoirs, for instance, the first one, and redefine the Hamiltonian and particle number operators as follows:

$$\begin{aligned}
&\hat{H}_1 = \hat{\mathcal{H}}_S + \hat{\mathcal{H}}_1, \quad \hat{H}_1' = \hat{\mathcal{H}}_S' + \hat{\mathcal{H}}_1', \quad \hat{N}_{1\alpha} = \hat{\mathcal{N}}_{S\alpha} + \hat{\mathcal{N}}_{1\alpha}, \quad \text{for} \quad j = 1, \\
&\hat{H}_j = \hat{\mathcal{H}}_j, \qquad\qquad \hat{H}_j' = \hat{\mathcal{H}}_j', \qquad\qquad \hat{N}_{j\alpha} = \hat{\mathcal{N}}_{j\alpha}, \qquad\qquad \text{for} \quad j = 2, \ldots, r.
\end{aligned} \tag{7.31}$$

In this case, the total Hamiltonian given by Eq. (7.27) can be rewritten as

$$\hat{H}(t; \mathcal{B}) = \begin{cases} \sum_{j=1}^{r} \hat{H}_j, & \text{for} \quad t \leq 0, \\ \sum_{j=1}^{r} \hat{H}_j + \hat{V}(t), & \text{for} \quad 0 < t < \mathcal{T}, \\ \sum_{j=1}^{r} \hat{H}_j', & \text{for} \quad \mathcal{T} \leq t, \end{cases} \tag{7.32}$$

and the total particle numbers of species $\alpha = 1, 2, \ldots, c$ as

$$\hat{N}_\alpha = \sum_{j=1}^{r} \hat{N}_{j\alpha}. \tag{7.33}$$

As in previous Section 7.2, a comparison is made between the outcomes of some forward and reversed protocols.

The *forward protocol* starts with the total system in the grand canonical equilibrium state of the decoupled parts at the different inverse temperatures $\beta_j = (k_B T_j)^{-1}$ and chemical potentials $\mu_{j\alpha}$:

$$\hat{\varrho}(0; \mathcal{B}) = \prod_j \Xi_j(\mathcal{B})^{-1} e^{-\beta_j [\hat{H}_j(\mathcal{B}) - \sum_\alpha \mu_{j\alpha} \hat{N}_{j\alpha}]} = \prod_j e^{-\beta_j [\hat{H}_j(\mathcal{B}) - \sum_\alpha \mu_{j\alpha} \hat{N}_{j\alpha} - \Phi_j(\mathcal{B})]}, \tag{7.34}$$

where $\Phi_j(\mathcal{B}) = -k_B T_j \ln \Xi_j(\mathcal{B})$ denotes the grand canonical thermodynamic potential of the jth part in the initial equilibrium state. We note that the grand canonical potential is even in the magnetic field if the corresponding Hamiltonian has the time-reversal symmetry (7.6). The system evolves from the initial time $t = 0$ until the final time $t = \mathcal{T}$ under the forward time evolution unitary operator obeying Eq. (7.8) with the initial condition $\hat{U}_F(0; \mathcal{B}) = 1$.

In order to determine the fluxes of energy and particles, quantum measurements are carried out on the reservoirs before and after the unitary time evolution.

An *initial quantum measurement* is performed that prepares the system in the eigenstate $|\Psi_k\rangle$ of the energy and particle number operators:

$$t \leq 0: \quad \hat{H}_j |\Psi_k\rangle = \varepsilon_{jk} |\Psi_k\rangle, \tag{7.35}$$

$$\hat{N}_{j\alpha} |\Psi_k\rangle = \nu_{j\alpha k} |\Psi_k\rangle. \tag{7.36}$$

After the time interval $0 < t < \mathcal{T}$, a *final quantum measurement* is performed in which the system is observed in the eigenstate $|\Psi'_l\rangle$ of the final energy and particle number operators:

$$\mathcal{T} \leq t : \hat{H}'_j|\Psi'_l\rangle = \varepsilon'_{jl}|\Psi'_l\rangle, \tag{7.37}$$

$$\hat{N}_{ja}|\Psi'_l\rangle = \nu'_{jal}|\Psi'_l\rangle. \tag{7.38}$$

We note that semi-infinite time intervals are available to perform the initial and final quantum measurements of well-defined eigenvalues. Accordingly, this scheme based on two quantum measurements provides a systematic way to measure the energies and the numbers of particles transferred between the reservoirs during the time interval $0 < t < \mathcal{T}$ of their mutual interaction. During the forward protocol, the energy and the particle numbers are observed to vary in the jth part of the system by the amounts

$$\Delta\varepsilon_j = \varepsilon'_{jl} - \varepsilon_{jk}, \tag{7.39}$$

$$\Delta\nu_{ja} = \nu'_{jal} - \nu_{jak}. \tag{7.40}$$

The probability distribution function to observe these variations is defined as

$$P_{\mathrm{F}}(\Delta\varepsilon_j, \Delta\nu_{ja}; \mathcal{B}) \equiv \sum_{kl} \prod_j \delta\left[\Delta\varepsilon_j - (\varepsilon'_{jl} - \varepsilon_{jk})\right] \prod_{ja} \delta\left[\Delta\nu_{ja} - (\nu'_{jal} - \nu_{jak})\right] \times |\langle\Psi'_l(\mathcal{B})|\hat{U}_{\mathrm{F}}(\mathcal{T};\mathcal{B})|\Psi_k(\mathcal{B})\rangle|^2 \langle\Psi_k(\mathcal{B})|\hat{\varrho}(0;\mathcal{B})|\Psi_k(\mathcal{B})\rangle \tag{7.41}$$

in terms of Dirac distributions $\delta(\cdot)$.

In the *reversed protocol*, the system starts in the other equilibrium distribution:

$$\hat{\varrho}(\mathcal{T}; -\mathcal{B}) = \prod_j \Xi'_j(\mathcal{B})^{-1} \mathrm{e}^{-\beta_j\left[\hat{H}'_j(-\mathcal{B}) - \sum_a \mu_{ja}\hat{N}_{ja}\right]} = \prod_j \mathrm{e}^{-\beta_j\left[\hat{H}'_j(-\mathcal{B}) - \sum_a \mu_{ja}\hat{N}_{ja} - \Phi'_j(\mathcal{B})\right]} \tag{7.42}$$

at the same inverse temperatures $\beta_j = (k_{\mathrm{B}}T_j)^{-1}$ and chemical potentials μ_{ja} as in the forward protocol. Here, $\Phi'_l(\mathcal{B}) = -k_{\mathrm{B}}T_j \ln \Xi'_j(\mathcal{B})$ denotes the grand canonical thermodynamic potential of the jth part in the final equilibrium state and reversed magnetic field. The reversed time evolution unitary operator obeys Eq. (7.12) and it is related to one of the forward protocols by Eq. (7.13) with $t = \mathcal{T}$ of Lemma 7.1. Before and after the reversed protocol, quantum measurements are similarly carried out in order to determine the variations of energy and particle numbers in the reservoirs. The probability distribution function of these variations during the reversed protocol is defined by an expression similar to Eq. (7.41).

Comparing the probability distribution functions of opposite variations during the forward and reversed protocols, the following symmetry relation is obtained:

$$\frac{P_{\mathrm{F}}(\Delta\varepsilon_j, \Delta\nu_{ja}; \mathcal{B})}{P_{\mathrm{R}}(-\Delta\varepsilon_j, -\Delta\nu_{ja}; -\mathcal{B})} = \mathrm{e}^{\sum_j \beta_j(\Delta\varepsilon_j - \sum_a \mu_{ja}\Delta\nu_{ja} - \Delta\Phi_j)}, \tag{7.43}$$

with the differences $\Delta\Phi_j \equiv \Phi'_j - \Phi_j$ in the grand potentials [49, 55].

For the forward and reversed protocols, the *generating functions of the statistical moments* are defined as

$$\mathcal{G}_{\mathrm{F,R}}(\xi_j, \eta_{ja}; \mathcal{B}) \equiv \int \prod_{ja} \mathrm{d}\Delta\varepsilon_j \mathrm{d}\Delta\nu_{ja} \mathrm{e}^{-\sum_j \xi_j \Delta\varepsilon_j - \sum_a \eta_{ja} \Delta\nu_{ja}} P_{\mathrm{F,R}}(\Delta\varepsilon_j, \Delta\nu_{ja}; \mathcal{B}) \tag{7.44}$$

in terms of the so-called *counting parameters* $\{\xi_j, \eta_{ja}\}$. The statistical moments of the energy and particle number variations can be obtained from this generating function by taking derivatives with respect to these counting parameters. Now, the symmetry relation (7.43) can be equivalently expressed as

$$\mathcal{G}_{\mathrm{F}}(\xi_j, \eta_{ja}; \mathcal{B}) = \mathrm{e}^{-\sum_j \beta_j \Delta\Phi_j} \mathcal{G}_{\mathrm{R}}(\beta_j - \xi_j, -\beta_j \mu_{ja} - \eta_{ja}; -\mathcal{B}) \tag{7.45}$$

in terms of the temperatures and chemical potentials of the reservoirs. We point out that this relation is symmetric only with respect to the inverse temperatures and the chemical potentials of the reservoirs.

Such symmetry relations are useful for quantum system driven by time-dependent external forces. If the relation (7.43) is restricted to the energy variation, the fluctuating quantity is the work W performed on the system and we recover the quantum version of Crooks' fluctuation theorem [22]:

$$\frac{P_{\mathrm{F}}(W; \mathcal{B})}{P_{\mathrm{R}}(-W; -\mathcal{B})} = \mathrm{e}^{\beta(W - \Delta F)}, \tag{7.46}$$

with the free energy difference $\Delta F = F' - F$. We note that this relation implies Eq. (7.16) [53]. Similar symmetry relations may be obtained for cold atoms or molecules in rotating frames, in which case the rotation rate Ω plays the role of the magnetic field $\mathcal{B}$.

7.4
From Transitory to the Stationary Current Fluctuation Theorem

If r infinitely large reservoirs are coupled together via the subsystem of interest, a stationary state will be reached in the long-time limit [58–70]. According to thermodynamics, the whole system is at equilibrium if the temperature and the chemical potentials of the different particle species are all uniform, that is, if every reservoir shares the same temperature and chemical potentials. This is no longer the case if the reservoirs have different temperatures or chemical potentials whereupon energy and particles are exchanged between the reservoirs across the subsystem coupling them together. Therefore, energy and particle currents are induced by the so-called thermodynamic forces or affinities defined in terms of the differences of temperatures and chemical potentials with respect to some reference values [71–73].

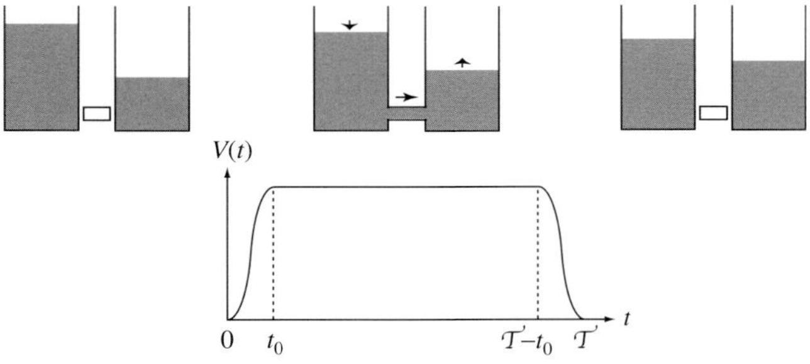

Figure 7.2 Schematic representation of the coupling of a subsystem to a pair of reservoirs by the potential $\hat{V}(t)$ during the time interval $0 \leq t \leq \mathcal{T}$, for the protocol used to obtain the *stationary* current fluctuation theorem. Here, the time-dependent potential has the symmetry $\hat{V}(t) = \hat{V}(\mathcal{T} - t)$, remaining constant during the time interval $t_0 \leq t \leq \mathcal{T} - t_0$ and vanishing at the initial and final times: $\hat{V}(0) = \hat{V}(\mathcal{T}) = 0$.

If we consider an open subsystem in contact with $r = 2$ reservoirs as depicted in Figure 7.2, the *thermodynamic forces* or *affinities* are defined as

$$A_E \equiv \beta_1 - \beta_2, \tag{7.47}$$

$$A_\alpha \equiv \beta_2\mu_{2\alpha} - \beta_1\mu_{1\alpha}, \quad \text{for} \quad \alpha = 1, 2, \ldots, c, \tag{7.48}$$

driving, respectively, the energy and particle currents from reservoir No. 2 to reservoir No. 1. The nonequilibrium conditions are thus specified by these affinities.

At the microscopic level of description, this system is described by the following Hamiltonian operator:

$$\hat{H}(t; \mathcal{B}) = \hat{H}_1 + \hat{H}_2 + \hat{V}(t), \quad \text{for} \quad 0 \leq t \leq \mathcal{T}. \tag{7.49}$$

Here, the interaction $\hat{V}(t)$ is switched on during a short time interval $0 \leq t < t_0$ at the beginning of the forward protocol with $t_0 \ll \mathcal{T}$, remains constant $\hat{V}(t) = \hat{V}_0$ during most of the protocol for $t_0 \leq t \leq \mathcal{T} - t_0$, and is switched off at the end when $\mathcal{T} - t_0 < t \leq \mathcal{T}$, as shown in Figure 7.2. Moreover, the time dependence of the interaction is assumed to be symmetric under the transformation $t \to \mathcal{T} - t$. Therefore, the forward and reversed protocols are identical except for the reversal of the magnetic field. Consequently, $\Delta\Phi_j = 0$ and the functions (7.44) of the forward and reversed protocols coincide, $\mathcal{G}_\mathrm{F} = \mathcal{G}_\mathrm{R} \equiv \mathcal{G}_\mathcal{T}$. In this case, the symmetry relation (7.45) becomes

$$\begin{aligned}\mathcal{G}_\mathcal{T}(\xi_1, \xi_2, \eta_{1\alpha}, \eta_{2\alpha}; \mathcal{B}) = \mathcal{G}_\mathcal{T}(&\beta_1 - \xi_1, \beta_2 - \xi_2,\\ &-\beta_1\mu_{1\alpha} - \eta_{1\alpha}, -\beta_2\mu_{2\alpha} - \eta_{2\alpha}; -\mathcal{B}).\end{aligned} \tag{7.50}$$

For applying to nonequilibrium steady states, the shortcoming is that this relation is not a symmetry with respect to the affinities or thermodynamic forces driving the system out of equilibrium. Accordingly, this relation does not yet concern nonequilibrium steady states associated with given affinities and further considerations are required.

We introduce the *cumulant-generating function* as

$$\mathcal{Q} \equiv \lim_{\mathcal{T}\to\infty} -\frac{1}{\mathcal{T}} \ln \mathcal{G}_{\mathcal{T}}. \tag{7.51}$$

According to Eq. (7.44), the statistical averages of the energy and the particle number variations are given by

$$\left.\frac{\partial \mathcal{Q}}{\partial \xi_j}\right|_0 = \lim_{\mathcal{T}\to\infty} \frac{\langle \Delta\varepsilon_j \rangle}{\mathcal{T}} \quad \text{and} \quad \left.\frac{\partial \mathcal{Q}}{\partial \eta_{ja}}\right|_0 = \lim_{\mathcal{T}\to\infty} \frac{\langle \Delta\nu_{ja} \rangle}{\mathcal{T}}, \tag{7.52}$$

while further differentiations give the cumulants. Besides, the short time interval t_0 is assumed to be constant in the long-time limit $\mathcal{T} \to \infty$. Remarkably, it is possible to prove the following:

Proposition 7.1 [49]
If the long-time limit (7.51) exists, the cumulant-generating function depends only on the differences $\xi = \xi_1 - \xi_2$ and $\eta_a = \eta_{1a} - \eta_{2a}$ between the counting parameters associated with the reservoirs:

$$\mathcal{Q}(\xi, \eta_a; \mathcal{B}) = \lim_{\mathcal{T}\to\infty} -\frac{1}{\mathcal{T}} \ln \mathcal{G}_{\mathcal{T}}\left(\xi_0 + \frac{\xi}{2}, \xi_0 - \frac{\xi}{2}, \eta_{0a} + \frac{\eta_a}{2}, \eta_{0a} - \frac{\eta_a}{2}; \mathcal{B}\right), \quad \text{for any} \quad \{\xi_0, \eta_{0a}\} \tag{7.53}$$

and the cumulant-generating function of the energy and particle currents is given by

$$\mathcal{Q}(\xi, \eta_a; \mathcal{B}) = \lim_{t\to\infty} -\frac{1}{t} \ln \tilde{\mathcal{G}}_t(\xi, \eta_a; \mathcal{B}), \tag{7.54}$$

where

$$\tilde{\mathcal{G}}_t(\xi, \eta_a; \mathcal{B}) = \text{tr}\, \hat{\varrho}_0\, e^{i\hat{H}t}\, e^{-(\xi/2)(\hat{H}_1 - \hat{H}_2) - \sum_a \frac{\eta_a}{2}(\hat{N}_{1a} - \hat{N}_{2a})}\, e^{-i\hat{H}t}\, e^{(\xi/2)(\hat{H}_1 - \hat{H}_2) + \sum_a \frac{\eta_a}{2}(\hat{N}_{1a} - \hat{N}_{2a})} \tag{7.55}$$

with $t = \mathcal{T} - 2t_0$, $\hat{H} = \hat{H}_1 + \hat{H}_2 + \hat{V}_0$, *and $\hat{\varrho}_0$ is the initial grand canonical distribution (7.34), which fixes the temperatures and the chemical potentials of the reservoirs.*

This proposition is proved by bounding together the functions (7.50) and (7.55) according to

$$L\tilde{\mathcal{G}}_t \le \mathcal{G}_{\mathcal{T}} \le K\tilde{\mathcal{G}}_t, \tag{7.56}$$

where $\tilde{\mathcal{G}}_t = \tilde{\mathcal{G}}_t(\xi_1 - \xi_2, \eta_{1a} - \eta_{2a}; \mathcal{B})$ and the two factors L and K are independent of the time interval $\mathcal{T}$ [49]. Therefore, the contributions of the factors L and K disappear in the long-time limit $\mathcal{T} \to \infty$. Moreover, $\lim_{\mathcal{T}\to\infty}(t/\mathcal{T}) = 1$ since $t = \mathcal{T} - 2t_0$ and t_0 is constant. For these reasons, the results (7.53) and (7.54) are obtained. Q.E.D.

The inequalities (7.56) are established in several steps [49]. The first two steps consist in removing the contributions from the initial and final lapses of time t_0, during which the interaction is switched on or off. Indeed, these contributions become negligible in the long-time limit. The third step has the effect of letting appear the differences $\xi_1 - \xi_2$ and $\eta_{1\alpha} - \eta_{2\alpha}$ between the counting parameters of the two reservoirs, which is the main result of the proposition. Accordingly, Eqs. (7.54) and (7.55) genuinely concern the fluxes of energy and particles transferred between the reservoirs.

Now, the symmetry relation (7.50) implies the following:

Current fluctuation theorem [49]

As the consequence of the time-reversal symmetry $\hat{\Theta}\hat{H}(\mathcal{B})\hat{\Theta}^{-1} = \hat{H}(-\mathcal{B})$, the cumulant-generating function (7.54) of the energy and particle currents satisfies the symmetry relation:

$$\mathcal{Q}(\xi, \eta_\alpha; \mathcal{B}) = \mathcal{Q}(A_E - \xi, A_\alpha - \eta_\alpha; -\mathcal{B}), \tag{7.57}$$

with respect to the thermal and chemical affinities (7.47) and (7.48).

We note that the function (7.55) takes the unit value if $\xi = \eta_\alpha = 0$ so that the cumulant-generating function vanishes with the counting parameters:

$$\mathcal{Q}(0, 0; \mathcal{B}) = \mathcal{Q}(A_E, A_\alpha; -\mathcal{B}) = 0. \tag{7.58}$$

The current fluctuation theorem (7.57) can be extended to the case of systems with more than two reservoirs [49]. If all the r reservoirs are coupled together, the total energy as well as the total numbers of particles are conserved in the transport process so that there are $(c + 1)$ constants of motion. The grand canonical equilibrium state of each reservoir is specified by one temperature and c chemical potentials. One of the r reservoirs can be taken as a reference with respect to which the affinities are defined. Consequently, the nonequilibrium steady states are specified by $p = (c + 1)(r - 1)$ different affinities and so many independent currents may flow across the open subsystem in between the reservoirs.

Some systems may be composed of several separated circuits between which no current is flowing. This is the case, for instance, in quantum dots monitored by a secondary circuit with a quantum point contact [56–58]. Both circuits are coupled only by the Coulomb interaction so that there is no electron transfer between them. In such systems, there are more independent affinities and currents. If a system composed of r reservoirs is partitioned into d disconnected circuits, each containing at least two reservoirs, a reference reservoir should be taken in each separate circuit. Therefore, $(r - d)$ reservoirs are controlling the $(c + 1)$ possible currents. The nonequilibrium steady states are thus specified by $p = (c + 1)(r - d)$ different affinities and so many independent currents can flow in such systems.

7.5 Current Fluctuation Theorem and Response Theory

If we collect together all the independent affinities and counting parameters as

$$\mathbf{A} = \{A_{jE}, A_{j\alpha}\} \tag{7.59}$$

$$\boldsymbol{\lambda} = \{\xi_j - \xi_k, \eta_{j\alpha} - \eta_{k\alpha}\}, \tag{7.60}$$

where k stands for the indices of the d reference reservoirs, the current fluctuation theorem can be expressed as

$$\mathcal{Q}(\boldsymbol{\lambda}, \mathbf{A}; \mathcal{B}) = \mathcal{Q}(\mathbf{A} - \boldsymbol{\lambda}, \mathbf{A}; -\mathcal{B}), \tag{7.61}$$

where we have introduced explicitly the dependence of the generating function on the affinities $\mathbf{A}$ that specify the nonequilibrium steady state of the open system. The maximum number of independent affinities and counting parameters is equal to $p = (c+1)(r-d)$. If the total system is isothermal, the number of independent affinities and counting parameters is equal to $p = c \times (r-d)$ where c is the number of different particle species.

All the statistical cumulants of the energy and particles transferred between the reservoirs are obtained by taking successive derivatives with respect to the counting parameters and setting afterward all these parameters equal to zero. The mean values of the currents are given by the first derivatives, the diffusivities or second cumulants by the second derivatives, and similarly for the third or higher cumulants:

$$J_\alpha(\mathbf{A}; \mathcal{B}) \equiv \frac{\partial \mathcal{Q}}{\partial \lambda_\alpha}(\mathbf{0}, \mathbf{A}; \mathcal{B}), \tag{7.62}$$

$$D_{\alpha\beta}(\mathbf{A}; \mathcal{B}) \equiv -\frac{1}{2}\frac{\partial^2 \mathcal{Q}}{\partial \lambda_\alpha \partial \lambda_\beta}(\mathbf{0}, \mathbf{A}; \mathcal{B}), \tag{7.63}$$

$$C_{\alpha\beta\gamma}(\mathbf{A}; \mathcal{B}) \equiv \frac{\partial^3 \mathcal{Q}}{\partial \lambda_\alpha \partial \lambda_\beta \partial \lambda_\gamma}(\mathbf{0}, \mathbf{A}; \mathcal{B}), \tag{7.64}$$

$$\vdots$$

with $\alpha, \beta, \ldots = 1, 2, \ldots, p$. All these cumulants characterize the full counting statistics of the coupled fluctuating currents in the nonequilibrium steady state associated with the affinities $\mathbf{A}$ in the external magnetic field $\mathcal{B}$.

At equilibrium, the mean currents vanish with the affinities. Therefore, we may expect that the mean currents can be expanded in powers of the affinities close to equilibrium:

$$J_\alpha = \sum_\beta L_{\alpha,\beta} A_\beta + \frac{1}{2}\sum_{\beta,\gamma} M_{\alpha,\beta\gamma} A_\beta A_\gamma + \cdots, \tag{7.65}$$

which defines the response coefficients:

$$L_{\alpha,\beta}(\mathcal{B}) \equiv \frac{\partial^2 \mathcal{Q}}{\partial \lambda_\alpha \partial A_\beta}(\mathbf{0}, \mathbf{0}; \mathcal{B}), \tag{7.66}$$

$$M_{\alpha,\beta\gamma}(\mathcal{B}) \equiv \frac{\partial^3 \mathcal{Q}}{\partial \lambda_\alpha \partial A_\beta \partial A_\gamma}(\mathbf{0}, \mathbf{0}; \mathcal{B}), \tag{7.67}$$

$$\vdots$$

We observe that the cumulants are given by successive derivatives with respect to the counting parameters. On the other hand, the response coefficients are given by one derivative with respect to a counting parameter and further successive derivatives with respect to the affinities. The remarkable result is that differentiating the symmetry relation (7.61) with respect to the affinities leads to derivatives with respect to the counting parameters. In this way, the current fluctuation theorem implies fundamental relationships between the cumulants and the response coefficients.

Considering all the second derivatives of the generating function with respect to the counting parameters and the affinities, it is possible to deduce the Casimir–Onsager reciprocity relations:

$$L_{\alpha,\beta}(\mathcal{B}) = L_{\beta,\alpha}(-\mathcal{B}), \tag{7.68}$$

as well as the identities

$$L_{\alpha,\beta}(\mathcal{B}) + L_{\beta,\alpha}(\mathcal{B}) = 2D_{\alpha\beta}(\mathbf{0};\mathcal{B}), \tag{7.69}$$

which relate the linear response coefficients to the diffusivities at equilibrium where $\mathbf{A} = \mathbf{0}$.

In the absence of magnetic field $\mathcal{B} = 0$, we recover the Onsager reciprocity relations:

$$L_{\alpha,\beta}(0) = L_{\beta,\alpha}(0), \tag{7.70}$$

as well as the expression of the fluctuation–dissipation theorem according to which the linear response coefficients are equal to the diffusivities of the currents around equilibrium:

$$L_{\alpha,\beta}(0) = D_{\alpha\beta}(\mathbf{0};0). \tag{7.71}$$

Going to the third derivatives of the symmetry relation (7.61), the third cumulants (7.64) characterize a magnetic asymmetry at equilibrium where $\mathbf{A} = 0$ and they are related to the responses of the diffusivities with respect to the affinities according to

$$C_{\alpha\beta\gamma}(\mathbf{0};\mathcal{B}) = -C_{\alpha\beta\gamma}(\mathbf{0};-\mathcal{B}) = 2\frac{\partial D_{\alpha\beta}}{\partial A_\gamma}(\mathbf{0};\mathcal{B}) - 2\frac{\partial D_{\alpha\beta}}{\partial A_\gamma}(\mathbf{0};-\mathcal{B}). \tag{7.72}$$

Moreover, the following relations hold for the nonlinear response coefficients at second order in the perturbations with respect to equilibrium [49]:

$$M_{\alpha,\beta\gamma}(\mathcal{B}) + M_{\alpha,\beta\gamma}(-\mathcal{B}) = 2\frac{\partial D_{\alpha\beta}}{\partial A_\gamma}(\mathbf{0};\mathcal{B}) + 2\frac{\partial D_{\alpha\gamma}}{\partial A_\beta}(\mathbf{0};-\mathcal{B}), \tag{7.73}$$

$$M_{\alpha,\beta\gamma}(\mathcal{B}) + M_{\beta,\gamma\alpha}(\mathcal{B}) + M_{\gamma,\alpha\beta}(\mathcal{B}) = 2\left(\frac{\partial D_{\beta\gamma}}{\partial A_\alpha} + \frac{\partial D_{\gamma\alpha}}{\partial A_\beta} + \frac{\partial D_{\alpha\beta}}{\partial A_\gamma} - \frac{1}{2}C_{\alpha\beta\gamma}\right)_{\mathbf{A}=\mathbf{0};\mathcal{B}}. \tag{7.74}$$

In the absence of magnetic field $\mathcal{B} = 0$, the magnetic asymmetry disappears because the third cumulants vanish, $C_{\alpha\beta\gamma}(\mathbf{0};0) = 0$, and the response coefficients are given in terms of the diffusivities as follows [74–76]:

$$M_{\alpha,\beta\gamma}(0) = \left(\frac{\partial D_{\alpha\beta}}{\partial A_\gamma} + \frac{\partial D_{\alpha\gamma}}{\partial A_\beta}\right)_{\mathbf{A}=\mathbf{0};\mathcal{B}=0}. \tag{7.75}$$

These relations are the generalizations of Eq. (7.71) to nonlinear response coefficients.

Such relations, which can also be extended to higher orders [49, 74–76], find their origin in the microreversibility expressed by the current fluctuation theorem (7.61).

7.6
Case of Independent Particles

In order to illustrate the previous results, let us consider systems with independent fermionic particles ruled by the Hamiltonian:

$$\hat{H} = \sum_{\sigma=\pm} \int d\mathbf{r} \hat{\psi}_\sigma^\dagger(\mathbf{r}) \hat{h} \hat{\psi}_\sigma(\mathbf{r}), \tag{7.76}$$

written in terms of the anticommuting field operators $\hat{\psi}_\sigma(\mathbf{r})$ and the one-particle Hamiltonian operator:

$$\hat{h} = -\frac{\hbar^2}{2m}\nabla^2 + u(\mathbf{r}), \quad \text{with} \quad \mathbf{r} = (x, y, z). \tag{7.77}$$

The confining potential $u(\mathbf{r})$ is minimum in the conducting region of the circuit and presents a barrier at the borders of this region. Such confining potentials can be fabricated by nanolithography on semiconducting heterojunctions and various shapes can be given to the circuit [2]. If the temperature is on the order of a few Kelvins or less, only the electrons around the Fermi energy are transported. The de Broglie wavelength can be larger than the interwall distance so that the circuit behaves as an electronic waveguide.

A quantum point contact (QPC) is a circuit with a barrier separating two semi-infinite waveguides, which form two reservoirs (Figure 7.3a). The barrier is a bottleneck for the transport of electrons from one reservoir to the other. This bottleneck presents a saddle point in the potential energy landscape. If the Fermi energy is lower than the energy of the saddle point, the transport proceeds by quantum tunneling through the barrier. If the Fermi energy is larger, the electronic waves undergo direct scattering on the obstacle formed by the barrier. Different models can be envisaged for the potential of the QPC. At large distance from the barrier,

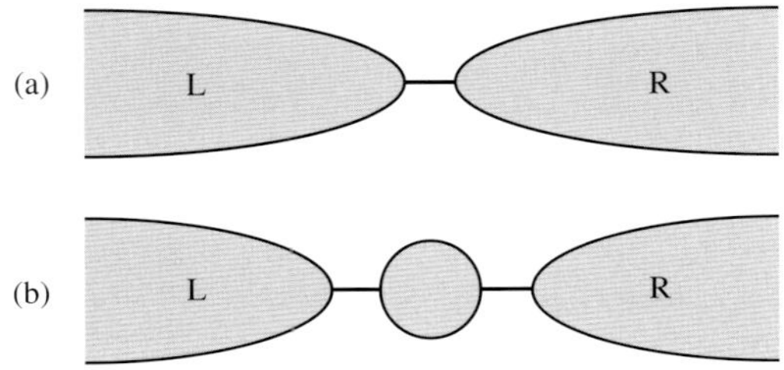

Figure 7.3 Schematic representation of different electronic circuits. (a) Quantum point contact between two reservoirs. (b) Quantum dot in tunneling contact with two reservoirs.

electrons can be assumed to propagate in a waveguide with a potential $u(\mathbf{r}) = u_\perp(y,z)$, which is invariant under translations in the x-direction and confining in the transverse y- and z-directions. The propagation modes in the infinite waveguide are given by the eigenstates of the one-electron Hamiltonian:

$$\hat{h}_0 \psi_{k_x n_y n_z}(\mathbf{r}) = \varepsilon_{k_x n_y n_z} \psi_{k_x n_y n_z}(\mathbf{r}), \tag{7.78}$$

where k_x is the wave number in the x-direction of propagation, while n_y and n_z are the quantum numbers of the modes in the transverse directions. Each of these modes is a possible channel of conduction, which opens at the energy threshold given by the minimum of the energy band $\varepsilon_{k_x n_y n_z}$ at $k_x = 0$.

The propagation modes $\psi_{k_x n_y n_z}(\mathbf{r})$ are scattered by the barrier. If the potential can be assumed to be the sum of the transverse and barrier potentials, $u(\mathbf{r}) = u_{\|}(x) + u_\perp(y,z)$, the problem is unidimensional for the passage of the barrier. If the incoming wave function is a plane wave in the positive x-direction, the scattering generates reflected and transmitted waves:

$$\psi_k(x) = \begin{cases} \dfrac{1}{\sqrt{2\pi}}\left(\mathrm{e}^{ikx} + r_k \mathrm{e}^{-ikx}\right), & \text{for} \quad x < 0, \\ \dfrac{1}{\sqrt{2\pi}} t_k \mathrm{e}^{ikx}, & \text{for} \quad x > 0, \end{cases} \tag{7.79}$$

where r_k and t_k denote the reflection and transmission amplitudes, while $k = k_x$ is the wave number. At the energy $\varepsilon = \varepsilon(k)$, the outgoing waves are related to the incoming waves by the unitary scattering matrix according to

$$\Psi_{\text{out}} = \hat{S}(\varepsilon)\Psi_{\text{in}}, \tag{7.80}$$

where Ψ_{in} and Ψ_{out} denote vectors with two components. The first component is the wave amplitude in the left-hand reservoir and the second in the right-hand reservoir. Therefore, the unitary scattering matrix of the one-dimensional barrier is given by

$$\hat{S}(\varepsilon) = \begin{pmatrix} r_k & t_k \\ t_k & r_k \end{pmatrix}, \quad \text{such that} \quad \hat{S}^\dagger(\varepsilon)\hat{S}(\varepsilon) = 1. \tag{7.81}$$

The unitarity of the scattering matrix expresses the conservation of probability and, in particular, the fact that the transmission probability $T(\varepsilon) = |t_k|^2$ and the reflection probability $R(\varepsilon) = |r_k|^2$ add to the unit value: $T(\varepsilon) + R(\varepsilon) = 1$. Here, the scattering matrix is 2×2 for every transverse mode because the potential defines a problem that is separable into the longitudinal wave and the transverse modes. This would not be the case if the potential was no longer the sum of two potentials in perpendicular directions. In more complicated circuits such as billiards, it is known that the scattering matrix is infinite and couples together the different transverse modes [77, 78].

In the case of separable potentials, every mode is scattered independent of the other ones and the scattering is thus characterized by the transmission probability $T(\varepsilon) = |t_k|^2$, which depends on the wave number $k = k_x$ or, equivalently, on the corresponding energy $\varepsilon = \varepsilon(k)$ of the electron in the longitudinal direction. A well-known example is the inverted parabolic barrier $u(x) = u_0 - m\lambda^2 x^2/2$, for which the transmission probability is given by [3, 79]

$$T(\varepsilon) = \left[1 + \exp\left(-2\pi \frac{\varepsilon - u_0}{\hbar\lambda}\right)\right]^{-1}. \tag{7.82}$$

The transmission probability converges to the unit value at high energy, well above the height of the barrier. For energies lower than the height of the barrier, the transmission proceeds by tunneling. The broader the barrier, the smaller the transmission probability. Since the inverted parabolic potential is unbounded from below, the transmission probability (7.82) remains positive for $\varepsilon \leq 0$ and vanishes only in the limit $\varepsilon \to -\infty$. For potentials that are vanishing at large distances, the transmission probability is strictly equal to zero if $\varepsilon \leq 0$.

A further example of electronic circuit is composed of a quantum dot between two reservoirs, as depicted in Figure 7.3b. A quantum dot is an artificial atom in which several electrons are confined to quasi-discrete energy levels [2, 3]. In general, there exist several electronic energy levels, but the transmission can be observed around a single electronic level. If the tunneling to the reservoirs is small enough, the energy levels are observed as resonances in the transmission probability. For quantum circuits such as classically chaotic billiards, the scattering resonances may form an irregular spectrum in the complex plane of the wave number or the energy and semiclassical methods become useful [77, 78].

An important remark is that the Hamiltonian operator (7.76) can also be written in the following form:

$$\hat{H} = \hat{H}_{\mathrm{L}} + \hat{H}_{\mathrm{R}} + \hat{V}_0, \tag{7.83}$$

where $\hat{H}_{\mathrm{L}}$ and $\hat{H}_{\mathrm{R}}$ are the Hamiltonian operators of the left- and right-hand reservoirs, while $\hat{V}_0$ is the interaction between the reservoirs, this interaction being at the origin of the scattering.

For these systems, we consider the function (7.55) in the absence of magnetic field and for one species of particles, namely, the electrons. This function has the following form:

$$\tilde{\mathcal{G}}_t(\xi, \eta) = \mathrm{tr}\ \hat{\varrho}_0 \mathrm{e}^{-\hat{A}_t} \mathrm{e}^{\hat{A}_0} \tag{7.84}$$

with

$$\hat{A} = \frac{\xi}{2}(\hat{H}_{\mathrm{R}} - \hat{H}_{\mathrm{L}}) + \frac{\eta}{2}(\hat{N}_{\mathrm{R}} - \hat{N}_{\mathrm{L}}) \tag{7.85}$$

and

$$\hat{A}_t = \mathrm{e}^{i\hat{H}t} \hat{A} \mathrm{e}^{-i\hat{H}t}, \tag{7.86}$$

where $\hat{H}_1 = \hat{H}_{\mathrm{R}}$ and $\hat{H}_2 = \hat{H}_{\mathrm{L}}$, in order to define the affinities (7.47) and (7.48) corresponding to currents from the left-hand reservoir to the right-hand one.

According to Eq. (7.34), the initial density operator is given by

$$\hat{\varrho}_0 = \frac{\mathrm{e}^{-\hat{B}}}{\mathrm{tr} \mathrm{e}^{-\hat{B}}}, \quad \text{with} \quad \hat{B} = \beta_{\mathrm{L}}(\hat{H}_L - \mu_L \hat{N}_{\mathrm{L}}) + \beta_{\mathrm{R}}(\hat{H}_R - \mu_{\mathrm{R}} \hat{N}_{\mathrm{R}}), \tag{7.87}$$

in terms of inverse temperatures, chemical potentials, and particle number operators, $\hat{N}_{\rm L}$ and $\hat{N}_{\rm R}$, for both reservoirs. Since $[\hat{A}, \hat{B}] = 0$, the function (7.84) can be written as

$$\tilde{\mathcal{G}}_t(\xi, \eta) = \frac{\mathrm{tr}\,\mathrm{e}^{-\hat{A}_t}\mathrm{e}^{\hat{A}-\hat{B}}}{\mathrm{tr}\,\mathrm{e}^{-\hat{B}}}. \tag{7.88}$$

In systems with independent particles, many-particle operators such as the Hamiltonian and particle number operators are of the following form:

$$\hat{X} = \sum_{kl} x_{kl}\hat{c}_k^{\dagger}\hat{c}_l = \Gamma(\hat{x}), \tag{7.89}$$

where $\hat{x} = (x_{kl})$ denotes the corresponding one-particle operator, while $\hat{c}_k^{\dagger}$ and $\hat{c}_k$ are the anticommuting creation–annihilation operators in the one-particle state k. The correspondence between one-particle and many-particle operators is thus established by the mapping Γ defined by Eq. (7.89). In this framework, Klich has shown that the trace of products of exponential functions of many-particle operators can be expressed as appropriate determinants involving the one-particle operators [80, 81]. For fermions, Klich's formula reads

$$\mathrm{tr}\,\mathrm{e}^{\hat{X}}\mathrm{e}^{\hat{Y}} = \mathrm{tr}\,\mathrm{e}^{\Gamma(\hat{x})}\mathrm{e}^{\Gamma(\hat{y})} = \det\left(1 + \mathrm{e}^{\hat{x}}\mathrm{e}^{\hat{y}}\right). \tag{7.90}$$

Applying Klich's formula to the function (7.88), this latter becomes

$$\tilde{\mathcal{G}}_t(\xi, \eta) = \det\left[\left(1 + \mathrm{e}^{-\hat{b}}\right)^{-1}\left(1 + \mathrm{e}^{-\hat{a}_t}\mathrm{e}^{\hat{a}-\hat{b}}\right)\right] = \det\left[1 + \hat{f}\left(\mathrm{e}^{-\hat{a}_t}\,\mathrm{e}^{\hat{a}} - 1\right)\right] \tag{7.91}$$

with the operator

$$\hat{f} = \frac{1}{\mathrm{e}^{\hat{b}} + 1} = \frac{1}{\mathrm{e}^{\beta_{\rm L}(\hat{h}_{\rm L}-\mu_{\rm L}\hat{n}_{\rm L})+\beta_{\rm R}(\hat{h}_{\rm R}-\mu_{\rm R}\hat{n}_{\rm R})} + 1} \tag{7.92}$$

of the Fermi–Dirac distributions in the reservoirs.

In the long-time limit, the unitary scattering operator is defined as

$$\hat{S} = \lim_{t\to\infty} \mathrm{e}^{i\hat{h}_0 t/2}\,\mathrm{e}^{-i\hat{h}t}\mathrm{e}^{i\hat{h}_0 t/2} \tag{7.93}$$

in terms of the full Hamiltonian $\hat{h}$ and the noninteracting Hamiltonian $\hat{h}_0 = \hat{h}_{\rm L} + \hat{h}_{\rm R}$, which commutes with the operators $\hat{a}$ and $\hat{b}$. Since

$$\hat{a}_t = \mathrm{e}^{i\hat{h}t}\hat{a}\mathrm{e}^{-i\hat{h}t}, \tag{7.94}$$

we have

$$\mathrm{e}^{-\hat{a}_t} \simeq \mathrm{e}^{i\hat{h}_0 t/2}\hat{S}^{\dagger}\mathrm{e}^{-\hat{a}}\,\hat{S}\mathrm{e}^{-i\hat{h}_0 t/2} \tag{7.95}$$

in the long-time limit. Using the commutativity of $\hat{h}_0$ with $\hat{a}$ and $\hat{f}$, the function (7.91) can be written as

$$\tilde{\mathcal{G}}_t(\xi, \eta) = \det\left[1 + \hat{f}\left(\hat{S}^{\dagger}\mathrm{e}^{-\hat{a}}\,\hat{S}\,\mathrm{e}^{\hat{a}} - 1\right)\right]. \tag{7.96}$$

For this formula to make sense, the determinant should be taken over an appropriate discrete set of electronic states forming a quasi-continuum in the long-time limit. Indeed, the quantity on the left-hand side of Eq. (7.96) concerns the whole many-particle system, although the expression on the right-hand side concerns single independent electrons flowing one by one across the system. Every one-particle operator can be decomposed on the eigenstates of the noninteracting Hamiltonian $\hat{h}_0 = \hat{h}_{\mathrm{L}} + \hat{h}_{\mathrm{R}}$. This is the case in particular for the scattering operator:

$$\hat{S} = \int d\varepsilon \, \hat{S}(\varepsilon)\delta(\varepsilon - \hat{h}_0), \tag{7.97}$$

where $\hat{S}(\varepsilon)$ is the 2×2 scattering matrix (7.81) acting on the two wave amplitudes with opposite wave numbers $\pm k$ at the given energy $\varepsilon = \varepsilon(k)$. Accordingly, the function (7.96) becomes a product over all the relevant single-electron states $\{\varepsilon, \sigma\}$ of the determinants of the corresponding operators at this particular energy:

$$\mathcal{G}_t(\xi, \eta) = \prod_{\varepsilon,\sigma} \det \left\{ 1 + \hat{f}(\varepsilon) \left[\hat{S}^\dagger(\varepsilon) e^{-\hat{a}(\varepsilon)} \hat{S}(\varepsilon) e^{\hat{a}(\varepsilon)} - 1 \right] \right\}, \tag{7.98}$$

where the matrix containing the Fermi–Dirac distributions of the left- and right-hand reservoirs is given by

$$\hat{f}(\varepsilon) = \begin{pmatrix} f_{\mathrm{L}}(\varepsilon) & 0 \\ 0 & f_{\mathrm{R}}(\varepsilon) \end{pmatrix}, \quad \text{with} \quad f_{\mathrm{L,R}}(\varepsilon) = \frac{1}{e^{\beta_{\mathrm{L,R}}(\varepsilon - \mu_{\mathrm{L,R}})} + 1} \tag{7.99}$$

and

$$e^{\hat{a}(\varepsilon)} = \begin{pmatrix} e^{-(\varepsilon\xi+\eta)/2} & 0 \\ 0 & e^{+(\varepsilon\xi+\eta)/2} \end{pmatrix}. \tag{7.100}$$

In order to define the cumulant-generating function with Eq. (7.54), the logarithm is taken of the function (7.98). The logarithm of the product over the relevant states $\{\varepsilon, \sigma\}$ gives a sum of logarithms. For a process lasting over the lapse of time t, the spacing between the relevant energies $\{\varepsilon\}$ is equal to $\Delta\varepsilon = 2\pi\hbar/t$. Since these relevant energies form a quasi-continuum, the sum is replaced by an integral in the long-time limit. Moreover, the spin orientation takes the two values $\sigma = \pm$ for electrons. For spin one-half particles such as electrons, we thus find that

$$\lim_{t\to\infty} \frac{1}{t} \sum_{\varepsilon,\sigma} (\cdot) = 2_s \int \frac{d\varepsilon}{2\pi\hbar} (\cdot), \tag{7.101}$$

where 2_s denotes the factor 2 due to the electron spin [3].

After evaluating the determinant, the cumulant-generating function (7.54) is finally given by

$$\mathcal{Q}(\xi, \eta) = -2_s \int \frac{d\varepsilon}{2\pi\hbar} \ln \left\{ 1 + T(\varepsilon) \left[f_{\mathrm{L}}(1 - f_{\mathrm{R}})(e^{-\varepsilon\xi-\eta} - 1) + f_{\mathrm{R}}(1 - f_{\mathrm{L}})(e^{\varepsilon\xi+\eta} - 1) \right] \right\} \tag{7.102}$$

for independent electrons. Therefore, we have recovered the Levitov–Lesovik formula for the full counting statistics of electron quantum transport [82, 83]. The

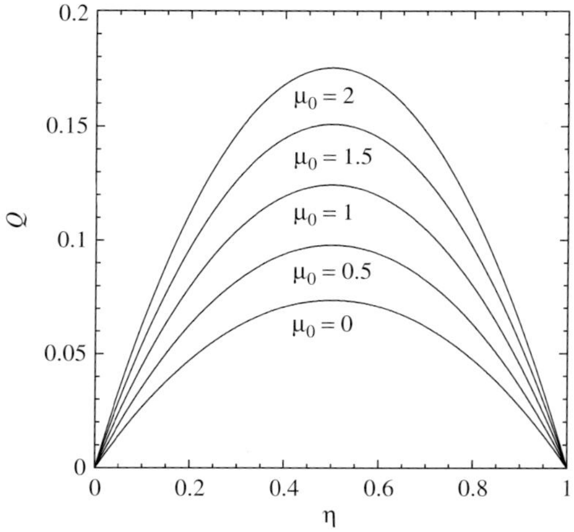

Figure 7.4 The cumulant-generating function (7.102) of the particle current versus the counting parameter η for the inverted parabolic barrier $u(x) = u_0 - m\lambda^2 x^2/2$ with the transmission probability (7.82). The temperature difference and thus the thermal affinity (7.104) are vanishing. In this regard, the generating function is considered at $\xi = 0$. The chemical potentials of the left- and right-hand reservoirs are given by $\mu_{L,R} = \mu_0 \pm eV/2$ with $eV = 1$ and $\mu_0 = 0, 0.5, 1, 1.5, 2$. The other parameter values are $u_0 = 1$, $\hbar\lambda = 3$, and $\beta = (k_B T)^{-1} = 1$. Planck's constant is taken as $h = 2_s$. The plot confirms the symmetry $\eta \to A_N - \eta$ of the fluctuation theorem with respect to the chemical affinity $A_N = \beta(\mu_L - \mu_R) = \beta eV = 1$.

symmetry of the fluctuation theorem is satisfied:

$$\mathcal{Q}(\xi, \eta) = \mathcal{Q}(A_E - \xi, A_N - \eta) \tag{7.103}$$

with respect to the thermal and chemical affinities:

$$A_E = \beta_R - \beta_L, \tag{7.104}$$

$$A_N = \beta_L \mu_L - \beta_R \mu_R, \tag{7.105}$$

defined by Eqs. (7.47) and (7.48) if the reservoir No. 1 (respectively, No. 2) is identified with the right-hand (respectively, left-hand) reservoir.

The cumulant-generating function (7.102) with $\xi = A_E = 0$ is depicted in Figure 7.4 for the inverted parabolic barrier with the transmission probability (7.82). We see that the generating function has indeed the symmetry $\eta \to A_N - \eta$ of Eq. (7.103). The generating function is depicted for several values of the mean chemical potential $\mu_0 = (\mu_L + \mu_R)/2$ of both reservoirs. We observe that the generating function increases with the mean chemical potential μ_0 because the transmission probability (7.82) does so.

The generating function given by the Levitov–Lesovik formula can be interpreted in terms of an exclusion process ruled by a trinomial distribution [84]. Suppose that at every relevant energy, three types of random events may happen for an electron: it may move from the left- to the right-hand reservoir with the probability p_+, in the other direction with the probability p_-, and stay in some reservoir with the probability $p_0 = 1 - p_+ - p_-$. The probability that N_0 electrons do not move, N_+ electrons move to the right-hand reservoir, and N_- move to the left-hand reservoir during the time interval t is given by the trinomial distribution:

$$\tilde{P}_{N_0 N_+ N_-} = \frac{N!}{N_0! N_+! N_-!} p_0^{N_0} p_+^{N_+} p_-^{N_-}, \quad \text{with} \quad N = N_0 + N_+ + N_-. \tag{7.106}$$

For this random process, the generating function is defined as

$$\tilde{Q}(\eta) = \lim_{t\to\infty} -\frac{1}{t} \ln \langle e^{-\eta \Delta N} \rangle_t, \tag{7.107}$$

where $\Delta N = N_+ - N_-$ is the total number of electrons that have been transported from the left- to the right-hand reservoir during the time interval t. Taking the average over the trinomial probability distribution, we find that

$$\tilde{Q}(\eta) = -r \ln \left[1 + p_+(e^{-\eta} - 1) + p_-(e^{\eta} - 1)\right], \tag{7.108}$$

where $r = \lim_{t\to\infty}(N/t)$ is the attempt frequency, that is, the mean number of attempted electron transfers per unit time [3]. We observe that the special form of the generating function (7.102) is indeed obtained at every relevant energy if we take $p_+ = T(\varepsilon) f_L (1 - f_R)$ and $p_- = T(\varepsilon) f_R (1 - f_L)$, as it should. The generating function of the trinomial process has the symmetry

$$\tilde{Q}(\eta) = \tilde{Q}(A_N - \eta), \quad \text{with} \quad A_N = \ln \frac{p_+}{p_-}. \tag{7.109}$$

This provides a stochastic interpretation for the quantum transport process described by the Levitov–Lesovik formula (7.102).

Many known results for quantum transport can be deduced in this framework. In particular, the mean currents are given by

$$J_N = \frac{\partial Q}{\partial \eta}(0,0) = 2_s \int \frac{d\varepsilon}{2\pi\hbar} T(\varepsilon)(f_L - f_R), \tag{7.110}$$

$$J_E = \frac{\partial Q}{\partial \xi}(0,0) = 2_s \int \frac{d\varepsilon}{2\pi\hbar} \varepsilon T(\varepsilon)(f_L - f_R), \tag{7.111}$$

which are the well-known Landauer–Büttiker formula for the particle current and the equivalent formula for the energy current [85]. On the other hand, the noise is characterized by the diffusivities:

$$D_{NN} = -\frac{1}{2}\frac{\partial^2 Q}{\partial \eta^2}(0,0) = D_0, \quad D_{EN} = -\frac{1}{2}\frac{\partial^2 Q}{\partial \eta \partial \xi}(0,0) = D_1,$$
$$D_{EE} = -\frac{1}{2}\frac{\partial^2 Q}{\partial \xi^2}(0,0) = D_2 \tag{7.112}$$

with

$$D_m = \int \frac{d\varepsilon}{2\pi\hbar} \varepsilon^m T(\varepsilon) \left[f_L(1 - f_R) + f_R(1 - f_L) - T(\varepsilon)(f_L - f_R)^2\right], \quad \text{for} \quad m = 0, 1, 2, \tag{7.113}$$

as reported in Refs [3, 86, 87]. The power spectrum of the electric noise is defined as

$$S(\omega) = \int dt\, e^{i\omega t} \langle \Delta\hat{I}(t)\Delta\hat{I}(0) + \Delta\hat{I}(0)\Delta\hat{I}(t)\rangle, \tag{7.114}$$

with $\Delta\hat{I}(t) = \hat{I}(t) - \langle\hat{I}(t)\rangle$ [87]. Since $\hat{I} = e\hat{J}_N$, the zero-frequency limit of the noise power is thus related to the corresponding diffusivity by

$$S(\omega = 0) = 4e^2 D_{NN}. \tag{7.115}$$

Close to equilibrium, the mean currents can be expanded in powers of the affinities (7.104) and (7.105). However, alternative expansions can be considered in terms of the potential difference V and the temperature difference ΔT between the left- and right-hand reservoirs at the temperatures $T_{L,R} = T_0 \pm \Delta T/2$ and chemical potentials $\mu_{L,R} = \mu_0 \pm eV/2$. If we introduce the electric and heat currents as

$$I = eJ_N, \tag{7.116}$$

$$J_Q = J_E - \mu_0 J_N, \tag{7.117}$$

with the electric charge e, the alternative expansions around equilibrium read as follows:

$$I \simeq GV + L_1 \frac{\Delta T}{T_0}, \tag{7.118}$$

$$J_Q \simeq L_1 V + L_2 \frac{\Delta T}{T_0}, \tag{7.119}$$

if higher order terms are neglected. In this linear approximation, $G = L_0$ is the electric conductance, L_1 is the thermoelectric coupling coefficient, and L_2 is the thermal conductance, which is given by [85]

$$L_m = 2_s e^{2-m} \int \frac{d\varepsilon}{2\pi\hbar} (\varepsilon - \mu_0)^m T(\varepsilon) \left(-\frac{\partial f}{\partial \varepsilon}\right)_0, \quad \text{with} \quad m = 0, 1, 2. \tag{7.120}$$

We note that the Onsager reciprocity relation is satisfied.

As Eq. (7.82) shows, the transmission probability tends to the unit value for energies higher than the barrier in every open channel corresponding to a propagation mode. In this limit where $T(\varepsilon) = 1$, the electric conductance given by Eq. (7.120) with $m = 0$ takes the universal value:

$$G_q = \frac{2_s e^2}{2\pi\hbar} = \frac{2e^2}{h}. \tag{7.121}$$

If several channels are open at the Fermi energy ε_F, the zero-temperature conductance is given by

$$G = \frac{2e^2}{h} \sum_n T_n(\varepsilon_F), \tag{7.122}$$

where $T_n(\varepsilon)$ is the transmission probability of the channel corresponding to the mode $n = (n_y, n_z)$ in Eq. (7.78). As the Fermi energy increases, the channels open successively and their transmission probability tends to the unit value, which explains the phenomenon of conductance quantization [1–3, 87, 88].

In summary, the results obtained in the previous sections allow us to deduce systematically many well-known results on quantum transport such as the Levitov–Lesovik and the Landauer–Büttiker formulas in the scattering approach, which is appropriate to treat coherent quantum transport.

Similar considerations also apply to boson transport [46, 50] as well as to many other quantum transport processes in physics or chemistry where the theory of full counting statistics extends and complements standard scattering theory and reaction rate theory.

We note that these quantum transport properties can also be deduced in the so-called Keldysh formalism, which has several conceptual advantages [2, 3, 89]. On the one hand, it formulates transport in terms of quantum fields, allowing in principle to treat the spatial dependence of local properties such as densities or current densities. On the other hand, the Keldysh formalism provides the systematic introduction of all the moments of the quantum fields, starting from the nonequilibrium Green's functions, which is suitable for perturbative calculations. Several nonequilibrium Green's functions are introduced corresponding to different time contours, which come with the doubling of the space of quantum states from pure states to statistical mixtures described by density operators. An equivalent theory is the Liouville space formalism developed by Mukamel and coworkers [55, 90, 91], which starts from the quantum Liouville superoperator, that is, the generator of von Neumann's equation (7.4). Another equivalent theory is provided by the formalism of thermofield dynamics, which also takes into account the need to double the state space for the description of nonequilibrium systems [92]. The latter formalism has recently been applied to study electron transport in molecular junctions [93, 94].

7.7 Time-Reversal Symmetry Relations in the Master Equation Approach

7.7.1 Current Fluctuation Theorem for Stochastic Processes

The master equation approach allows us to establish connections with the theory of stochastic processes. The first step of this approach consists in identifying the relevant coarse-grained states of the system or subsystem of interest. The second step usually proceeds with a perturbative calculation with respect to some small coupling parameter λ, which typically controls the timescale separation between the slow time evolution of the coarse-grained states and the fast dynamics of the other degrees of freedom. The Hamiltonian may have the form

$$\hat{H} = \hat{H}_0 + \lambda \hat{V}, \tag{7.123}$$

where the unperturbed Hamiltonian $\hat{H}_0$ leaves invariant the coarse-grained states and the perturbation $\lambda \hat{V}$ causes the interaction between the coarse-grained states and the other degrees of freedom. Without a clear separation of timescales, the time evolution of the coarse-grained states continues to keep the memory of the past and is non-Markovian. In quantum-mechanical systems, non-Markovian behavior can manifest itself as a slippage of initial conditions on the timescale of the fast degrees of freedom in the early time evolution of a subsystem in contact with a reservoir. The slippage of initial conditions allows the density operator to remain positive definite [95, 96]. Thereafter, the density operator of the subsystem follows a time evolution that is essentially Markovian on the long timescale characterizing the interaction of the coarse-grained states with the fast degrees of freedom. We note that although the subsystem density operator evolves slowly, the individual realizations present quantum jumps between the coarse-grained states. These jumps occur over the short timescale of the fast degrees of freedom and the dwell times between the jumps are on the order of the long timescale. Indeed, these dwell times are inversely proportional to the transition rates, which are on the order of λ^2 according to the second-order perturbation theory.

In transport problems where the subsystem of interest is coupled to several reservoirs, it is important to specify the coarse-grained states not only of the subsystem but also of the reservoirs, in order to proceed with the counting of particles and energy transferred between the reservoirs. This corresponds to taking two successive quantum measurements, as done in Section 7.3, and this is sometimes referred to as unraveling the master equation [97–99]. Consequently, the master equation rules the density operator of the subsystem conditioned to the numbers of particles and energy that have been transferred between the reservoirs since the initial time.

Many experimental observations are not sensitive to the quantum coherences described by the off-diagonal elements of the density operator. Under such circumstances, the knowledge of the diagonal elements, that is, the probabilities of the coarse-grained states, is enough for the description of the observations. In these cases, the time evolution of the probabilities is ruled by a Markovian master equation. For these stochastic processes, a current fluctuation theorem has been proved using graph analysis [75, 100]. This theorem is expressed by the symmetry relation (7.61) in terms of the cumulant-generating function of the fluctuating currents. Another proof of this theorem is based on the symmetry of the master equation modified to include the counting parameters, as explained here below.

We consider electron transport through an isothermal open subsystem in contact with r reservoirs. If the circuit is composed of d disconnected subcircuits, the numbers of electrons in these subcircuits are as many conserved quantities. We suppose that each subcircuit connects at least two reservoirs so that $r \geq 2d$. In this system, the nonequilibrium conditions are imposed with $p = r - d$ differences of chemical potentials, which define the independent affinities $A_j = \beta(\mu_j - \mu_k)$ with $j = 1, 2, \ldots, p$ and $k = 1, 2, \ldots, d$. These affinities may drive as many independent currents. We denote by $\mathbf{n} = \{n_j\}_{j=1}^{p}$ the random numbers of electrons transferred from the driving reservoirs $j = 1, 2, \ldots, p$ to the reference reservoirs $k = 1, 2, \ldots, d$. The probabilities $\mathbf{p}(\mathbf{n}) = \{p_\sigma(\mathbf{n})\}$ that the subsystem has evolved to some coarse-grained state σ while $\mathbf{n}$ electrons have been transferred during the time interval t

are ruled by the master equation:

$$\partial_t \mathbf{p}_t(\mathbf{n}) = \hat{\mathsf{L}} \cdot \mathbf{p}_t(\mathbf{n}), \tag{7.124}$$

with the operator

$$\hat{\mathsf{L}} = \sum_\varrho \left[\mathsf{L}_\varrho^{(+)} \mathrm{e}^{-(\partial/\partial n_\varrho)} + \mathsf{L}_\varrho^{(0)} + \mathsf{L}_\varrho^{(-)} \mathrm{e}^{+(\partial/\partial n_\varrho)} \right], \tag{7.125}$$

where $\mathsf{L}_\varrho^{(\pm)}$ is the matrix with the rates of the transitions $n_\varrho \to n_\varrho \pm 1$, while $\mathsf{L}_\varrho^{(0)}$ is the matrix with the rates of the other transitions and all the loss rates. Examples of such transport processes in quantum dots have been studied, for instance, in Refs [101, 102]. The cumulant-generating function is defined as

$$\mathcal{Q}(\boldsymbol{\eta}) = \lim_{t\to\infty} -\frac{1}{t} \ln \langle \mathrm{e}^{-\boldsymbol{\eta}\cdot n} \rangle_t, \tag{7.126}$$

where the statistical average is carried out over the probability distribution $\mathbf{p}_t(\mathbf{n})$. As a consequence, the generating function turns out to be given by the eigenvalue problem:

$$\mathsf{L}(\boldsymbol{\eta}) \cdot \mathbf{v} = -\mathcal{Q}(\boldsymbol{\eta}) \mathbf{v} \tag{7.127}$$

for the modified matricial operator

$$\mathsf{L}(\boldsymbol{\eta}) = \mathrm{e}^{-\boldsymbol{\eta}\cdot n} \hat{\mathsf{L}} \mathrm{e}^{+\boldsymbol{\eta}\cdot n} = \sum_\varrho \left[\mathsf{L}_\varrho^{(+)} \mathrm{e}^{-\eta_\varrho} + \mathsf{L}_\varrho^{(0)} + \mathsf{L}_\varrho^{(-)} \mathrm{e}^{+\eta_\varrho} \right]. \tag{7.128}$$

Such matricial operators may obey the following symmetry relation:

$$\mathsf{M}^{-1} \cdot \mathsf{L}(\boldsymbol{\eta}) \cdot \mathsf{M} = \mathsf{L}(\mathbf{A} - \boldsymbol{\eta})^T, \tag{7.129}$$

where the superscript T denotes the matricial transpose and M is the matrix of the thermal distribution for the subsystem at equilibrium with the d reference reservoirs [27, 102]. If this symmetry relation holds for the operator (7.128), this property extends to all its eigenvalues and, in particular, to the leading eigenvalue that gives the cumulant-generating function. In this way, the current fluctuation theorem is demonstrated for the subsystem in the nonequilibrium steady state specified by the affinities $\mathbf{A}$:

$$\mathcal{Q}(\boldsymbol{\eta}) = \mathcal{Q}(\mathbf{A} - \boldsymbol{\eta}). \tag{7.130}$$

Accordingly, the results of Section 7.5 apply here also for the response coefficients.

An equivalent expression of the current fluctuation theorem is given in terms of the probability $P_t(\mathbf{n})$ that $\mathbf{n}$ electrons have been transferred during the time interval t between the reservoirs under the stationary conditions $\mathbf{A}$. Equation (7.130) implies that opposite fluctuations of these numbers have probabilities obeying the symmetry relation:

$$\frac{P_t(\mathbf{n})}{P_t(-\mathbf{n})} \simeq \mathrm{e}^{\mathbf{A}\cdot\mathbf{n}}, \quad \text{for} \quad t \to \infty. \tag{7.131}$$

At equilibrium where the affinities are vanishing, $\mathbf{A} = 0$, the probabilities of opposite fluctuations are equal so that we recover the principle of detailed balancing. In contrast, a directionality manifests itself out of equilibrium where the nonvanishing affinities introduce a bias between the probabilities. Soon, one of the probabilities becomes dominant as time increases and the currents tend to flow in one direction so that the mean currents are no longer vanishing:

$$\mathbf{J} = \frac{\partial Q}{\partial \boldsymbol{\eta}}(\mathbf{0}) = \lim_{t\to\infty} \frac{\langle \mathbf{n} \rangle_t}{t}. \tag{7.132}$$

7.7.2
Thermodynamic Entropy Production

A consequence of the current fluctuation theorem is the nonnegativity of the entropy production, which is known to be equal to the sum of the affinities multiplied by the mean currents [66, 71–73]. By Jensen's inequality $\langle e^{-X} \rangle \geq e^{-\langle X \rangle}$ [103], we find that

$$\left\langle e^{-\mathbf{A}\cdot\mathbf{n}} \right\rangle_t \geq e^{-\langle \mathbf{A}\cdot\mathbf{n} \rangle_t}. \tag{7.133}$$

Consequently, the entropy production is nonnegative because

$$\frac{1}{k_B}\frac{d_i S}{dt} = \mathbf{A}\cdot\mathbf{J} = \lim_{t\to\infty} \frac{1}{t}\mathbf{A}\cdot\langle \mathbf{n} \rangle_t \geq \lim_{t\to\infty} -\frac{1}{t}\ln\left\langle e^{-\mathbf{A}\cdot\mathbf{n}} \right\rangle_t = \mathcal{Q}(\mathbf{A}) = \mathcal{Q}(\mathbf{0}) = 0. \tag{7.134}$$

The last equality is the consequence of the definition (7.126) for the cumulant-generating function at $\boldsymbol{\eta} = 0$. The second law of thermodynamics can thus be deduced from the current fluctuation theorem.

The entropy production vanishes at equilibrium and is positive out of equilibrium where energy is dissipated. In order to drive a particular current with the other currents, energy should thus be supplied and the second law constitutes a limit to the efficiency of energy transduction. Powering the process γ by the other ones requires that the corresponding mean current is opposite to its associated affinity: $A_\gamma J_\gamma < 0$. In this case, a thermodynamic efficiency can be defined and the second law implies that it cannot reach values larger than unity:

$$0 \leq \eta_\gamma \equiv -\frac{A_\gamma J_\gamma}{\sum_{\alpha\neq\gamma} A_\alpha J_\alpha} \leq 1. \tag{7.135}$$

Such thermodynamic efficiencies have been considered for molecular motors [104] as well as for mass separation by effusion [105].

7.7.3
Case of Effusion Processes

The current fluctuation theorem has also been established for effusion processes [105, 106]. Effusion processes are the classical analogues of electron quantum transport through a constriction, such as the quantum point contact of Figure 7.3a. In effusion, two gases at different pressures and temperatures are separated by a wall with a pore that is smaller than the mean free path so that the flow of particles across the pore is essentially ballistic. The master equation of effusion has been known in classical kinetic theory for a century [107, 108]. The particles of energy $\varepsilon = m\mathbf{v}^2/2$ are crossing a pore of cross-sectional area σ with the rate:

$$W_j(\varepsilon) = \frac{\sigma n_j}{\sqrt{2\pi m}} \beta_j^{3/2} \varepsilon \mathrm{e}^{-\beta_j \varepsilon}, \quad \text{for} \quad j = \mathrm{L}, \mathrm{R}, \tag{7.136}$$

if they come from the left- or the right-hand reservoir at the inverse temperature $\beta_j = (k_\mathrm{B} T_j)^{-1}$ and the particle density n_j related to the pressure by $P_j = n_j k_\mathrm{B} T_j$. These rates satisfy the condition

$$\frac{W_\mathrm{L}(\varepsilon)}{W_\mathrm{R}(\varepsilon)} = \mathrm{e}^{\varepsilon A_E + A_N} \tag{7.137}$$

in terms of the affinities (7.104) and (7.105). In effusion, the cumulant-generating function of the energy and particle currents is given by

$$\mathcal{Q}(\xi, \eta) = \int_0^\infty \mathrm{d}\varepsilon \left[W_L(\varepsilon)\left(1 - \mathrm{e}^{-\varepsilon\xi - \eta}\right) + W_\mathrm{R}(\varepsilon)\left(1 - \mathrm{e}^{\varepsilon\xi + \eta}\right) \right] \tag{7.138}$$

in terms of the transition rates (7.136) [105, 106]. By their property (7.137), the generating function obeys the same symmetry relation (7.103) as in the quantum case.

The generating functions (7.102) and (7.138) of the quantum and classical processes can be compared. At a given energy ε for $\xi = 0$, the generating function has the form (7.108) for the quantum process. In the limit where $p_\pm \ll 1$, it becomes

$$\tilde{\mathcal{Q}}(\eta) = r p_+ (1 - \mathrm{e}^{-\eta}) + r p_- (1 - \mathrm{e}^{\eta}), \tag{7.139}$$

which is the classical form appearing in Eq. (7.138) with the identification $r p_\pm = W_{\mathrm{L,R}}(\varepsilon)$. The expression (7.139) is characteristic of the combination of two independent Poisson processes, that is, the particles can be transferred in both directions.

7.7.4
Statistics of Histories and Time Reversal

Another time-reversal symmetry relationship concerns the statistical properties of the histories or paths followed by a system under stroboscopic observations at some sampling time Δt. These observations generate sequences of coarse-grained states such as

$$\boldsymbol{\omega} = \omega_1 \omega_2 \cdots \omega_n \tag{7.140}$$

corresponding to the successive times $t_j = j\Delta t$ with $j = 1, 2, \ldots, n$. This history or path has a certain probability $P(\boldsymbol{\omega})$ to happen if the system is in the stationary state corresponding to the affinities $\mathbf{A}$. Because of the randomness of the molecular fluctuations, these path probabilities typically decrease exponentially at some rate h that characterizes the temporal disorder in the process. This characterization applies to stochastic processes as well as chaotic dynamical systems, for which the temporal disorder h is called the Kolmogorov–Sinai or dynamical entropy [6, 13, 14, 109].

In nonequilibrium stationary states, the time-reversed path

$$\boldsymbol{\omega}^{\mathrm{R}} = \omega_n \cdots \omega_2 \omega_1 \tag{7.141}$$

is expected to happen with a different probability decreasing at the different rate h^{R} now characterizing the temporal disorder of the time-reversed paths [36]. The remarkable result is that the difference between the disorders of the time-reversed and typical paths is equal to the thermodynamic entropy production [36]:

$$\frac{1}{k_{\mathrm{B}}} \frac{\mathrm{d}_i S}{\mathrm{d}t} = h^{\mathrm{R}} - h \geq 0. \tag{7.142}$$

The second law of thermodynamics is satisfied because this difference is known in mathematics to be a relative entropy, which is always nonnegative [103]. At equilibrium, detailed balancing holds so that every history and its time reversal are equiprobable, their temporal disorders are equal $h = h^{\mathrm{R}}$, and the entropy production vanishes. This is no longer the case away from equilibrium where the typical paths are more probable than their time reversals. Consequently, the time-reversal symmetry is broken at the level of the statistical description in terms of the probability distribution of the nonequilibrium stationary state. In this regard, the entropy production is a measure of the time asymmetry in the temporal disorders of the typical histories and their time reversals. As a corollary of the second law, the typical histories are more ordered in time than their corresponding time reversals in the sense that $h < h^{\mathrm{R}}$ in nonequilibrium stationary states [39].

In the case of effusion processes, the temporal disorders of the histories and their time reversals are characterized by

$$h = \left(\ln \frac{\mathrm{e}}{\Delta\varepsilon\Delta t}\right) \int_0^\infty \mathrm{d}\varepsilon [W_{\mathrm{L}}(\varepsilon) + W_{\mathrm{R}}(\varepsilon)] - \int_0^\infty \mathrm{d}\varepsilon [W_{\mathrm{L}}(\varepsilon) \ln W_{\mathrm{L}}(\varepsilon) + W_{\mathrm{R}}(\varepsilon) \ln W_{\mathrm{R}}(\varepsilon)] + O(\Delta t), \tag{7.143}$$

$$h^{\mathrm{R}} = \left(\ln \frac{\mathrm{e}}{\Delta\varepsilon\Delta t}\right) \int_0^\infty \mathrm{d}\varepsilon [W_{\mathrm{L}}(\varepsilon) + W_{\mathrm{R}}(\varepsilon)] - \int_0^\infty \mathrm{d}\varepsilon [W_{\mathrm{L}}(\varepsilon) \ln W_{\mathrm{R}}(\varepsilon) + W_{\mathrm{R}}(\varepsilon) \ln W_{\mathrm{L}}(\varepsilon)] + O(\Delta t), \tag{7.144}$$

in terms of the transition rates (7.136), the sampling time Δt, and the coarse-graining in energy $\Delta\varepsilon$. The first term of these quantities is a feature of stochastic processes that are continuous in time and energy. For such processes, randomness manifests itself on arbitrarily small timescales Δt and energy scales $\Delta\varepsilon$ so that the temporal disorders increase as $\Delta\varepsilon\Delta t \to 0$.

The difference between these two quantities is equal to the entropy production because

$$h^{\mathrm{R}} - h = \int_0^\infty \mathrm{d}\varepsilon [W_{\mathrm{L}}(\varepsilon) - W_{\mathrm{R}}(\varepsilon)] \ln \frac{W_{\mathrm{L}}(\varepsilon)}{W_{\mathrm{R}}(\varepsilon)} = A_E J_E + A_N J_N = \frac{1}{k_{\mathrm{B}}} \frac{\mathrm{d}_i S}{\mathrm{d}t}, \tag{7.145}$$

in the limit where $\Delta\varepsilon\Delta t \to 0$.

The validity of the formula (7.142) has also been verified in experiments where the nonequilibrium constraints are imposed by fixing the currents instead of the affinities [40, 41].

Similar considerations have been developed for quantum systems [42, 110]. We note that quantum mechanics naturally limits the randomness on the scale where $\Delta\varepsilon\Delta t = 2\pi\hbar$, as explained in Ref. [110].

7.8
Transport in Electronic Circuits

Several types of electronic circuits are considered such as single-electron transistors, quantum dots with resonant levels, or quantum dots capacitively coupled with a quantum point contact [2, 3, 87].

7.8.1
Quantum Dot with One Resonant Level

A quantum dot is a kind of artificial atom with quantized electronic levels. Because of the tunneling to the reservoirs, these levels are broadened into resonances characterized by a lifetime. We suppose that the transport process involves a single resonant level, which may thus be either empty or occupied. In this case, the process is ruled by the master equation (7.124) with a modified operator (7.128) given by

$$\mathsf{L}(\eta) = \begin{pmatrix} -a_{\mathrm{L}} - a_{\mathrm{R}} & b_{\mathrm{L}}\mathrm{e}^{+\eta} + b_{\mathrm{R}} \\ a_{\mathrm{L}}\mathrm{e}^{-\eta} + a_{\mathrm{R}} & -b_{\mathrm{L}} - b_{\mathrm{R}} \end{pmatrix} \tag{7.146}$$

in terms of the following charging and discharging rates:

$$a_\varrho = \Gamma_\varrho f_\varrho \quad \text{and} \quad b_\varrho = \Gamma_\varrho(1 - f_\varrho), \quad \text{for} \quad \varrho = \mathrm{L, R}. \tag{7.147}$$

The rate constants Γ_ϱ are proportional to the square of the interaction parameters between the quantum dot and the reservoir ϱ, as well as to the density of states of the reservoir ϱ. The Fermi–Dirac distributions are given by

$$f_\varrho = \frac{1}{1 + \mathrm{e}^{\beta(\varepsilon_0 + \Delta U_0 - \mu_\varrho)}}, \quad \text{for} \quad \varrho = \mathrm{L, R}, \tag{7.148}$$

where ε_0 is the bare energy of the dot level, ΔU_0 is the electrostatic charging energy, and $\mu_\varrho = eV_\varrho$ are the chemical potentials of the reservoirs [2, 3, 111].

The cumulant-generating function is given by the eigenvalue problem (7.127) for the matrix (7.146) as

$$Q(\eta) = \frac{1}{2}\left[a_{\mathrm{L}} + a_{\mathrm{R}} + b_{\mathrm{L}} + b_{\mathrm{R}} - \sqrt{(a_{\mathrm{L}} + a_{\mathrm{R}} - b_{\mathrm{L}} - b_{\mathrm{R}})^2 + 4(a_{\mathrm{L}}\mathrm{e}^{-\eta} + a_{\mathrm{R}})(b_{\mathrm{L}}\mathrm{e}^{+\eta} + b_{\mathrm{R}})}\right]. \tag{7.149}$$

This function obeys the symmetry relation $\mathcal{Q}(\eta) = \mathcal{Q}(A - \eta)$ whereupon the current fluctuation theorem (7.131) is satisfied with the affinity [75]:

$$A = \ln \frac{a_L b_R}{a_R b_L} = \beta(\mu_L - \mu_R) = \beta e V. \tag{7.150}$$

7.8.2 Capacitively Coupled Circuits

In order to perform the full counting statistics of electron transport in quantum dots (QDs), the main circuit should be monitored by a secondary circuit that is made of a quantum point contact (QPC) [56–58]. Both circuits are capacitively coupled so that the current in the QPC is sensitive to the occupancy of the QDs. This latter modulates the QPC current by Coulomb repulsion. In this way, the quantum jumps in the occupancy of the QDs can be experimentally observed. Typically, the electric current in the QPC is seven or eight orders of magnitude larger than the current in the circuit with the QDs (Table 7.1). From the viewpoint of quantum measurement, the QPC plays the role of a measuring device and the QDs of the small observed quantum system. Because of the large ratio between their currents, the measuring device is essentially in a classical regime. Although the ratio of the noise to the current is negligible in this regime, the noise gets larger as the current increases. As a consequence, the QDs themselves are subjected to an important noise, which affects their

Table 7.1 Electronic temperature T, mean electric currents I_α, voltages V_α, dissipated powers $\Pi_\alpha = V_\alpha I_\alpha$, mean electron currents $J_\alpha = I_\alpha/|e|$, and affinities $A_\alpha = |e|V_\alpha/(k_B T)$ in the quantum dot (α = QD) and the quantum point contact (α = QPC), as well as the ratio of QPC to QD currents, for the experiments reported in Ref. [56] with a single QD and in Ref. [58] with a double QD.

	Single QD [56]	Double QD [58]
T	350 mK	130 mK
I_{QD}	1.3×10^{-16} A	6.5×10^{-17} A
V_{QD}	2.7 mV	0.3 mV
Π_{QD}	3.4×10^{-19} W	2.0×10^{-20} W
J_{QD}	792 Hz	406 Hz
A_{QD}	90	27
I_{QPC}	4.5×10^{-9} A	1.2×10^{-8} A
V_{QPC}	0.5 mV	0.8 mV
Π_{QPC}	2.3×10^{-12} W	9.6×10^{-12} W
J_{QPC}	2.8×10^{10} Hz	7.5×10^{10} Hz
A_{QPC}	17	71
$\frac{I_{QPC}}{I_{QD}}$	3.5×10^{7}	1.8×10^{8}

According to Eq. (7.134), the thermodynamic entropy production in the whole system is equal to $d_i S/dt = k_B \sum_\alpha A_\alpha J_\alpha = \sum_\alpha \Pi_\alpha / T$, which is dominated by the energy dissipation in the QPC under these experimental conditions.

kinetics. In particular, the charging and discharging rates of the QDs are modified by the presence of the large current in the QPC. Therefore, the Coulomb interaction between both circuits also causes a back action of the QPC onto the QDs.

Several types of circuits have been considered. In Ref. [56], a single QD with a strong bias voltage is monitored by a QPC. The QD is successively empty or occupied so that the QPC current is observed to jump between both corresponding values. This setup allows to carry out the full counting statistics of the occupancy of the QD. Since the bias voltage is large, $\mu_{1L} \gg \mu_{1R}$, the transition rates of the backward fluctuations can be assumed to be negligible. Under these conditions, the statistics of the fluctuating current can be inferred from the statistics of the occupancy. However, the backward fluctuations of the current are also negligible in this fully irreversible regime. Therefore, the current fluctuation theorem cannot be tested for lack of backward fluctuations.

In order to overcome such limitations, a setup with two QDs has been considered in Ref. [58]. The two QDs are positioned at different distances from the QPC so that different Coulomb interactions U and U' exist between the QPC and both QDs, as shown in Figure 7.5b. If an electron travels from the left- to the right-hand QD, the Coulomb repulsion is successively larger and smaller or vice versa for the reversed motion of an electron. Therefore, this setup is suitable for bidirectional counting statistics. Remarkably, the symmetry relation $P_t(n) = P_t(-n) \exp(\tilde{A}_{\rm QD} n)$ is observed, but, depending on the experimental conditions [58–60], the effective affinity $\tilde{A}_{\rm QD}$ may significantly differ from the thermodynamic affinity $A_{\rm QD} = |e| V_{\rm QD}/(k_{\rm B} T)$ driving the current in the QD. In Refs [58, 59], the gate voltages of the QDs are tuned so that transitions occur between charge states where the QDs are singly or doubly occupied and the effective affinity $\tilde{A}_{\rm QD} \simeq 3$ differs by one order of magnitude from the thermodynamic affinity $A_{\rm QD} = 27$ (Table 7.1). In Ref. [60], the experiment has been carried out closer to equilibrium with an optimized sample design and gate voltages tuned so that transitions occur between charge states where the QDs are empty or singly occupied. For the bias voltage $V_{\rm QD} = 20\,\mu{\rm V}$ and the electronic temperature $T = 330\,{\rm mK}$, the thermodynamic

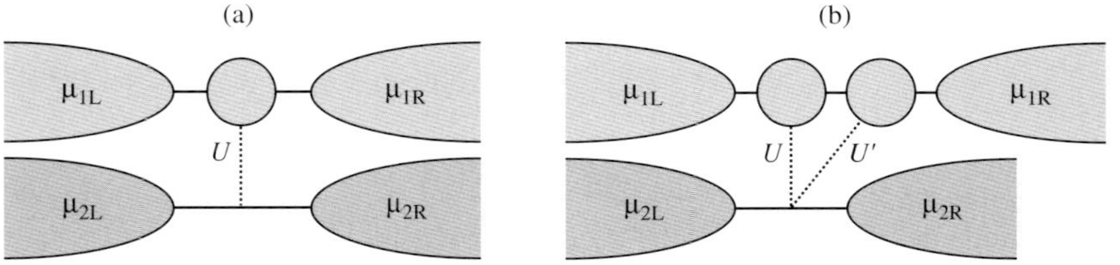

Figure 7.5 (a) Schematic representation of a single quantum dot (QD) monitored by a quantum point contact (QPC) to which it is capacitively coupled by the Coulomb interaction U, as in the circuit of Ref. [56]. (b) Schematic representation of two QDs monitored by a QPC asymmetrically positioned with respect to the QDs in order to get different capacitive couplings of the QPC to the two QDs, as in the circuits of Refs [58–60]. The Coulomb repulsion U is larger than U', which allows the bidirectional counting of electrons in the circuit with the two QDs.

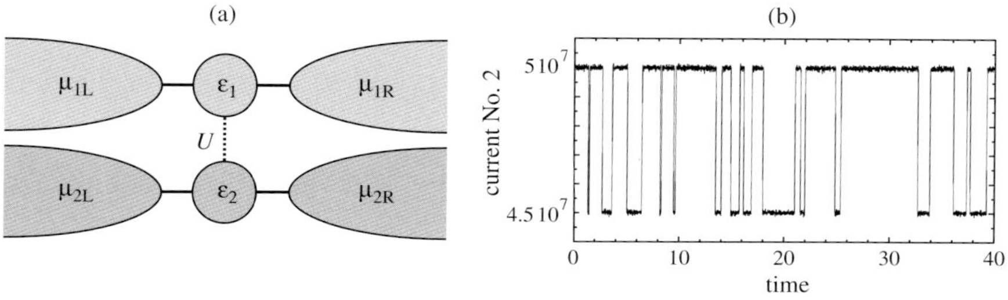

Figure 7.6 (a) Schematic representation of two quantum dots in parallel. Each quantum dot is coupled to two reservoirs of electrons. Moreover, both quantum dots influence each other by the Coulomb electrostatic interaction U. (b) Simulation with Gillespie's algorithm [113] of the detector current in circuit No. 2 measuring the occupancy of the QD No. 1. The parameter values are given by $\beta\varepsilon_1 = 0$, $\beta\varepsilon_2 = 35$, $\beta U = 32.8$, $\beta\mu_{1L} = 25$, $\beta\mu_{1R} = 0$, $\Gamma_{1L} = \Gamma_{1R} = \bar{\Gamma}_{1L} = \bar{\Gamma}_{1R} = 1$, $\beta\mu_{2L} = 70$, $\beta\mu_{2R} = 0$, and $\Gamma_{2L} = \Gamma_{2R} = \bar{\Gamma}_{2L} = \bar{\Gamma}_{2R} = 10^8$. The affinities of both circuits are $A_1 = 25$ and $A_2 = 70$. The effective affinity of the circuit No. 1 is $\tilde{A}_1 = 1.17$. The mean value of the QD current is $J_1 \simeq 0.17$ electrons per unit time. The mean value of the secondary current is $J_2 \simeq 4.8 \times 10^7$ electrons per unit time. The QD is empty (respectively, occupied) when the secondary current takes the value of 5×10^7 (respectively, 4.5×10^7). Adapted from Ref. [102].

affinity is equal to $A_{\rm QD} = 0.7$, while the effective affinity is observed to be $\tilde{A}_{\rm QD} = 0.5$. The differences observed in the experiments reported in Refs [58–60] arise because the QPC may induce back action effects onto the dynamics of the QDs under some conditions, while the limited bandwidth of the charge detection also affects the counting statistics.

In order to understand the back action effects, the capacitive coupling between the QDs and the QPC should be taken into account. For this purpose, the simple model of Figure 7.6a with two capacitively coupled QDs in two parallel circuits may be considered [101, 102, 112]. The system is composed of $d = 2$ disconnected circuits and $r = 4$ reservoirs, so that the nonequilibrium steady states are specified with $p = r - d = 2$ affinities:

$$A_1 = \beta(\mu_{1L} - \mu_{1R}) \quad \text{and} \quad A_2 = \beta(\mu_{2L} - \mu_{2R}). \tag{7.151}$$

The system can be described in terms of the probabilities $p_{\nu_1\nu_2}(n_1, n_2)$ that each quantum dot ($\alpha = 1$ or 2) is occupied by $\nu_\alpha = 0$ or 1 electron, while n_α electrons have been transferred from the left- to the right-hand reservoirs in the circuit α during the time interval t. The stochastic process is ruled by the master equation (7.124) with the operator (7.125). For $\alpha = 1, 2$ and $\varrho = \mathrm{L, R}$, the charging and discharging rates are given either by Eq. (7.152) or by Eq. (7.153):

$$a_{\alpha\varrho} = \Gamma_{\alpha\varrho} f_{\alpha\varrho} \quad \text{and} \quad b_{\alpha\varrho} = \Gamma_{\alpha\varrho}(1 - f_{\alpha\varrho}), \quad \text{with} \quad f_{\alpha\varrho} = \frac{1}{e^{\beta(\varepsilon_\alpha - \mu_{\alpha\varrho})} + 1}, \tag{7.152}$$

$$\bar{a}_{\alpha\varrho} = \bar{\Gamma}_{\alpha\varrho} \bar{f}_{\alpha\varrho} \quad \text{and} \quad \bar{b}_{\alpha\varrho} = \bar{\Gamma}_{\alpha\varrho}(1 - \bar{f}_{\alpha\varrho}), \quad \text{with} \quad \bar{f}_{\alpha\varrho} = \frac{1}{e^{\beta(\varepsilon_\alpha + U - \mu_{\alpha\varrho})} + 1}, \tag{7.153}$$

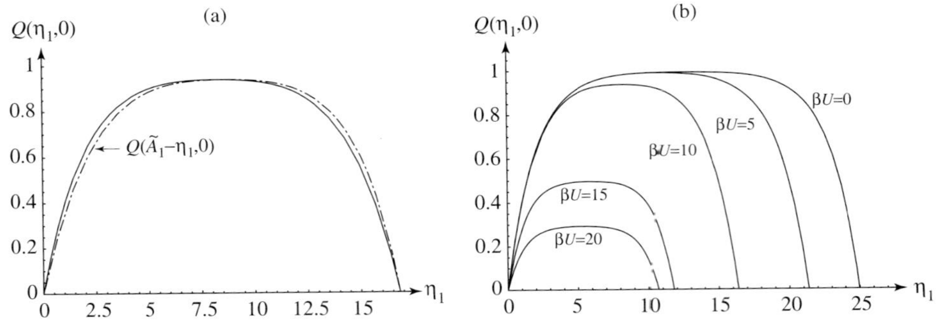

Figure 7.7 The cumulant-generating function $Q(\eta_1, \eta_2 = 0)$ for the circuit of Figure 7.6a versus the counting parameter η_1. (a) Together with the symmetric function with respect to the effective affinity $\tilde{A}_1 = 16.8356$ (dotted–dashed line) for $\beta U = 10$ and $\Gamma_{2L} = \Gamma_{2R} = \bar{\Gamma}_{2L} = \bar{\Gamma}_{2R} = 2$. (b) For different values of the electrostatic coupling parameter βU and $\Gamma_2 \equiv \Gamma_{2L} = \Gamma_{2R} = \bar{\Gamma}_{2L} = \bar{\Gamma}_{2R} = 100$. In both cases, the other parameters take the following values: $\beta\varepsilon_1 = 10$, $\beta\varepsilon_2 = 35$, $\beta\mu_{1L} = 25$, $\beta\mu_{1R} = 0$, $\beta\mu_{2L} = 70$, $\beta\mu_{2R} = 0$, and $\Gamma_1 \equiv \Gamma_{1L} = \Gamma_{1R} = \bar{\Gamma}_{1L} = \bar{\Gamma}_{1R} = 1$. In these plots, the effective affinity is given by the second root of the function at $\eta_1 = \tilde{A}_1 \neq 0$. The thermodynamic affinity of the secondary circuit is $A_2 = \beta(\mu_{2L} - \mu_{2R}) = 70$. Adapted from Ref. [102].

if the other quantum dot is either empty or occupied. The effect of the tunneling barrier capacitances is neglected in this simplified model [101, 102]. We note that their effect has been analyzed in Ref. [112].

Now, the cumulant-generating function (7.126) is obtained by solving the eigenvalue problem (7.127) and it obeys the current fluctuation theorem (7.130) for the two currents flowing across the parallel circuits:

$$Q(\eta_1, \eta_2) = Q(A_1 - \eta_1, A_2 - \eta_2), \tag{7.154}$$

with respect to the two affinities (7.151). However, this result does not imply that the symmetry relation holds for the single current flowing in one of the circuits. The generating function of this single current is given by $Q(\eta_1, 0)$ and the symmetry does not hold in general, $Q(\eta_1, 0) \neq Q(A_1 - \eta_1, 0)$, as shown in the counterexample of Figure 7.7a.

Nevertheless, under some circumstances, the single-current fluctuation theorem

$$Q(\eta_1, 0) = Q(\tilde{A}_1 - \eta_1, 0) \tag{7.155}$$

holds with respect to an effective affinity $\tilde{A}_1$, which can be calculated.

This symmetry holds with the effective affinity equal to the thermodynamic affinity, $\tilde{A}_1 = A_1$, in a few cases: (1) if the other circuit is at equilibrium, that is, if $A_2 = 0$; (2) if there is no Coulomb interaction between both circuits, $U = 0$, in which case the two circuits are independent of each other; and (3) in the limit where the Coulomb interaction is very large, $U = \infty$. Indeed, the probability that both quantum dots are occupied is vanishing in this limit, which reduces the Markov process to three states and implies the symmetry (7.155) [102].

The single-current fluctuation theorem (7.155) also holds in the limit where the rate constants of one circuit are much larger than that in the other circuit:

$$\Gamma_{2\varrho}, \bar{\Gamma}_{2\varrho} \gg \Gamma_{1\varrho}, \bar{\Gamma}_{1\varrho}, \quad \text{for} \quad \varrho = \text{L, R}, \tag{7.156}$$

as in the experiments of Refs [56, 58] where the ratio between the QPC and QD currents is many orders of magnitude larger than unity (Table 7.1). In the system of Figure 7.6a, the fast QD in the secondary circuit rapidly jumps between its two states $\nu_2 = 0$ and $\nu_2 = 1$ so that the charging and discharging rates of the slow QD are effectively averaged over both states. The conditional probabilities $P_{\nu_2|\nu_1}$ that the second QD has the occupancy ν_2 provided that the first QD is in the state ν_1 are given by

$$P_{0|0} = \frac{b_{2\text{L}} + b_{2\text{R}}}{a_{2\text{L}} + a_{2\text{R}} + b_{2\text{L}} + b_{2\text{R}}}, \qquad P_{1|0} = \frac{a_{2\text{L}} + a_{2\text{R}}}{a_{2\text{L}} + a_{2\text{R}} + b_{2\text{L}} + b_{2\text{R}}}, \tag{7.157}$$

$$P_{0|1} = \frac{\bar{b}_{2\text{L}} + \bar{b}_{2\text{R}}}{\bar{a}_{2\text{L}} + \bar{a}_{2\text{R}} + \bar{b}_{2\text{L}} + \bar{b}_{2\text{R}}}, \qquad P_{1|1} = \frac{\bar{a}_{2\text{L}} + \bar{a}_{2\text{R}}}{\bar{a}_{2\text{L}} + \bar{a}_{2\text{R}} + \bar{b}_{2\text{L}} + \bar{b}_{2\text{R}}}. \tag{7.158}$$

Therefore, the effective charging and discharging rates of the first QD are given by the following averages:

$$\tilde{a}_\varrho = a_{1\varrho} P_{0|0} + \bar{a}_{1\varrho} P_{1|0} \quad \text{and} \quad \tilde{b}_\varrho = b_{1\varrho} P_{0|1} + \bar{b}_{1\varrho} P_{1|1}, \quad \text{with} \quad \varrho = \text{L, R}. \tag{7.159}$$

Under the conditions (7.156), the Markovian process is similar to the one ruled by the operator (7.146) but with these effective charging and discharging rates. Accordingly, the cumulant-generating function has the form (7.149) with the rates (7.159) so that the single-current fluctuation theorem (7.155) is indeed satisfied with the effective affinity:

$$\tilde{A}_1 \equiv \ln \frac{\tilde{a}_\text{L} \tilde{b}_\text{R}}{\tilde{a}_\text{R} \tilde{b}_\text{L}}. \tag{7.160}$$

Figure 7.7b shows that not only the single-current generating function $\mathcal{Q}(\eta_1, 0)$ is already nearly symmetric if $\Gamma_2/\Gamma_1 = 100$ but also the symmetry holds with respect to an effective affinity $\tilde{A}_1$ that varies with the electrostatic interaction βU. If the Coulomb interaction vanishes, $\beta U = 0$, both circuits are decoupled so that the thermodynamic affinity $A_1 = 25$ is recovered. However, the effective affinity decreases if the electrostatic interaction increases till $\beta U = 20$, as seen in Figure 7.7b. For larger interaction βU, the effective affinity (7.160) recovers its thermodynamic value in the limit $U \to \infty$, as already explained [102].

The conclusion is that the secondary circuit monitoring the quantum jump dynamics in quantum dots may have important back action effects on the full counting statistics. In general, the current fluctuation theorem holds at the fundamental level of description for all the currents interacting in the system. In this regard, the secondary circuit performing the measurement of the quantum state in the first circuit is part of the total system and cannot be separated in general. Nevertheless, the symmetry relation of the fluctuation theorem is still observed to hold for a single fluctuating current, but with an effective affinity that may differ from the thermodynamic value due to the back action effect of the monitoring circuit onto the observed circuit. The effective affinity (7.160) may drop down to a value more than one order of magnitude lower than the thermodynamic affinity [102]. Similar conclusions have been reached with related approaches where the averages (7.159) of the transition rates have been carried out in the framework of the $P(E)$ theory [59, 114, 115].

A general remark is that quantum measurement implies the dissipation of energy. In order to observe the quantum jumps of the QD with a short enough time resolution, the current and the dissipated power in the monitoring circuit should be higher than that in the QD. If the QD state is monitored with a sampling time Δt, the secondary circuit playing the role of the detector should have a current $J_2 \gtrsim (\Delta t)^{-1}$. Hence, the dissipated power should be bounded by $\Pi_2 = k_B T A_2 J_2 \gtrsim k_B T A_2 (\Delta t)^{-1}$. The higher the time resolution, the higher the dissipation rate. Therefore, the experimental conditions are variously influenced by the bandwidth of charge detection.

7.8.3 Coherent Quantum Conductor

The current fluctuation theorem also applies to transport in coherent quantum conductors such as Aharonov–Bohm rings in a magnetic field [1–3]. We suppose that the circuit is connected to two reservoirs so that there is a single affinity. In this case, the results of Section 7.5 are the following. The particle current is expanded in powers of the affinity according to Eq. (7.65):

$$J(A; \mathcal{B}) = L(\mathcal{B})A + \frac{1}{2}M(\mathcal{B})A^2 + \cdots . \tag{7.161}$$

Similar expansions can be introduced for the diffusivity (Eq. (7.63)) and the third cumulant (Eq. (7.64)):

$$D(A; \mathcal{B}) = D_0(\mathcal{B}) + D_1(\mathcal{B})A + \cdots , \tag{7.162}$$

$$C(A; \mathcal{B}) = C_0(\mathcal{B}) + \cdots . \tag{7.163}$$

As shown in Section 7.5, all these coefficients are interconnected by the following relations. For linear response and the second cumulant, Eqs. (7.68) and (7.69) give

$$L(\mathcal{B}) = L(-\mathcal{B}) = D_0(\mathcal{B}). \tag{7.164}$$

For the third cumulant and the second response coefficient, Eqs. (7.72)–(7.74) give

$$C_0(\mathcal{B}) = -C_0(-\mathcal{B}) = 2[D_1(\mathcal{B}) - D_1(-\mathcal{B})], \tag{7.165}$$

$$M(\mathcal{B}) + M(-\mathcal{B}) = 2[D_1(\mathcal{B}) + D_1(-\mathcal{B})], \tag{7.166}$$

$$M(\mathcal{B}) = 2D_1(\mathcal{B}) - \frac{1}{3}C_0(\mathcal{B}). \tag{7.167}$$

Similar relations hold at higher orders [49].

Comparing with usual quantities, we note that the electric current is equal to $I(V;\mathcal{B}) = eJ(A;\mathcal{B})$ and the affinity A is related to the voltage V by

$$A = \frac{eV}{k_B T}. \tag{7.168}$$

Moreover, the noise power (7.114) at zero frequency $S = S(\omega = 0)$ is proportional to the diffusivity according to Eq. (7.115). It is standard to expand the electric current and the noise power as follows:

$$I(V;\mathcal{B}) = G_1(\mathcal{B})V + \frac{1}{2}G_2(\mathcal{B})V^2 + \cdots, \tag{7.169}$$

$$S(V;\mathcal{B}) = S_0(\mathcal{B}) + S_1(\mathcal{B})V + \cdots. \tag{7.170}$$

Comparing with the expansions (7.161) and (7.162), we get

$$e^2 L = k_B T G_1, \quad e^3 M = (k_B T)^2 G_2, \quad \ldots, \tag{7.171}$$

$$4e^2 D_0 = S_0, \quad 4e^3 D_1 = k_B T S_1, \quad \ldots. \tag{7.172}$$

Accordingly, we recover the Johnson–Nyquist and Casimir–Onsager relations [87]:

$$S_0(\mathcal{B}) = 4k_B T G_1(\mathcal{B}) \quad \text{and} \quad G_1(\mathcal{B}) = G_1(-\mathcal{B}). \tag{7.173}$$

For the nonlinear response properties, the following relations can be deduced:

$$S_1(\mathcal{B}) + S_1(-\mathcal{B}) = 2k_B T[G_2(\mathcal{B}) + G_2(-\mathcal{B})], \tag{7.174}$$

$$S_1(\mathcal{B}) - S_1(-\mathcal{B}) = 6k_B T[G_2(\mathcal{B}) - G_2(-\mathcal{B})], \tag{7.175}$$

at the level of the third cumulant $C_0(\mathcal{B})$, as well as higher order relations.

The expression (7.174) is only the consequence of the second equality in Eq. (7.58), $\mathcal{Q}(A, A;\mathcal{B}) = 0$, which has been called the "global detailed balancing condition" and which is weaker than the current fluctuation theorem itself [116]. The expression (7.175) is the consequence of the current fluctuation theorem and thus of the underlying microreversibility in the sense that $\hat{\Theta}\hat{H}(\mathcal{B})\hat{\Theta}^{-1} = \hat{H}(-\mathcal{B})$. We note that the quantity (7.175) is proportional to the magnetic asymmetry given by the third cumulant (7.165).

The symmetry relations (7.174) and (7.175) have been tested in recent experiments on a circuit with an Aharonov–Bohm ring in an external magnetic field [61, 62]. The circuit is fabricated by local oxidation on a GaAs/AlGaAs 2DEG and the ring has a diameter of about 500 nm. The electronic temperature $T = 125$ mK is determined with the Johnson–Nyquist fluctuation–dissipation relation. The current is observed as a function of the applied voltage V, the gate voltage V_g, and the external magnetic field $\mathcal{B}$. The noise power is measured with the cross-correlation technique between the signals from two sets of resonant circuit and amplifier, which are external to the circuit. Beyond linear response, the coefficients S_1 and G_2 have been measured and they have complex dependences on the magnetic field and the gate voltage. Remarkably, the proportionalities predicted by Eqs. (7.174) and (7.175) are confirmed. The proportionality constants are observed to be 12.00 ± 1.96 instead of 2 in Eq. (7.174) and 9.66 ± 1.32 instead of 6 in Eq. (7.175) [62]. The antisymmetric relation (7.175) is in better agreement with theory than the symmetric relation (7.174). The reason for the considerable deviation in the case of the symmetric relation (7.174) is not explained in Refs [61, 62]. The amplifiers used to measure the noise power may have back action effects.

The fundamental questions associated with these issues deserve further experiments to understand the origins of the quantitative deviations and to test predictions at the higher orders if the concerned quantities are experimentally accessible.

7.9 Conclusions

This chapter has been devoted to recent advances in the nonequilibrium statistical mechanics of quantum systems. For more than a century, statistical mechanics has offered the conceptual framework to formulate the various statistical aspects of mechanics. These aspects play a prominent role in quantum mechanics, which is an initial condition theory as well as classical mechanics. In such theories, the nonreproducibility of random phenomena is naturally explained by the arbitrariness of initial conditions in due respect to the principle of causality. This is the case in open systems under nonequilibrium conditions sustaining currents of energy and particles between reservoirs at different temperatures and chemical potentials. After transients, the system evolves toward a nonequilibrium steady state that describes the different properties such as the mean currents and their statistical cumulants. The set of all these cumulants characterize the full counting statistics and thus the large deviation properties of the fluctuating currents, which have been the focus of recent advances in nonequilibrium statistical mechanics.

These advances find their origins in earlier work on chaotic dynamical systems modeling transport properties such as diffusion, viscosity, or heat conductivity [7–17]. In this context, large deviation properties have been studied for physical quantities fluctuating in time. Previously, the large deviation properties were mainly considered in equilibrium statistical mechanics to infer thermodynamics from the spatial fluctuations of observables in the large size limit. In dynamical systems theory, large deviation properties have been introduced in close relation with the concepts of Lyapunov exponent, Kolmogorov–Sinai entropy per unit time, and fractal dimensions [6]. Furthermore, relationships have been established between these quantities characterizing dynamical chaos and the transport properties [7–17]. In this context, the consideration of time-reversal symmetry for nonequilibrium steady states has opened new perspectives for a fundamental understanding of nonequilibrium systems at small scales where the microscopic degrees of freedom manifest themselves as fluctuations. Several types of time-reversal symmetry relations have been established [20–38]. On the one hand, so-called fluctuation theorems have been proved, which compare the probabilities of opposite fluctuations for nonequilibrium quantities of interest. On the other hand, the thermodynamic entropy production has been shown to result from the breaking of time-reversal symmetry at the statistical level of description. These advances have turned out to play a fundamental role in the understanding of many nonequilibrium nanosystems such as molecular machines.

More recently, time-reversal symmetry relations have been extended to quantum systems and, in particular, to electron quantum transport in semiconducting nanodevices [42–55][2),3)]. Several approaches have been followed.

Functional relations have been established that provide a unifying framework to deduce many results such as the Casimir–Onsager reciprocity relations, the Kubo formula of linear response theory, the quantum version of Jarzynski's nonequilibrium work equality, as well as their generalization beyond linear response or to novel physical contexts [48].

Moreover, a theoretical approach has been developed for the transport of energy and particles across a driven open quantum system between reservoirs under nonequilibrium conditions. In quantum mechanics, the issue of measurement is always subtle because of the nonseparability of a system into its different parts. Here, quantum measurements at the initial and final times are required in order to determine the amounts of energy and particles transferred between the reservoirs during the lapse of time when the system is driven by time-dependent forces. Using the time-reversal symmetry of the microscopic dynamics or its extension in the presence of an external magnetic field, symmetry relations have been obtained that are quantum versions of the transitory current fluctuation theorem.

In the long-time limit, this theorem leads to the stationary current fluctuation theorem. An essential aspect is that, for nonequilibrium stationary conditions, the symmetry relation should hold with respect to the affinities or thermodynamic forces due to the *differences* of temperatures or chemical potentials between the reservoirs. Remarkably, a proposition guarantees that the cumulant-generating function of the fluctuating currents indeed depends only on the differences between the counting parameters associated with the reservoirs between which the thermodynamic forces are maintained [49]. Thanks to this proposition enunciated in Section 7.4, the stationary current fluctuation theorem can be established for all the currents flowing across open quantum systems.

Close to equilibrium, known results of linear response theory such as the Casimir–Onsager reciprocity relations and fluctuation–dissipation formulas can be deduced from the stationary current fluctuation theorem. The fact is that this theorem also holds arbitrarily far away from equilibrium, allowing the generalization of these results to nonlinear response properties [49, 74–76].

These advances concern in particular electron quantum transport in mesoscopic semiconducting circuits. At low temperatures, electrons have ballistic motion on the size of such circuits, which form electronic waveguides. Many transport properties of mesoscopic conductors can thus be understood by supposing that the electrons are independent of each other and undergo scattering in the effective potential formed by the semiconducting circuit. In this approximation where electron–electron and electron–phonon interactions are neglected, the cumulant-generating function of the aforementioned theorem can be expressed with the Levitov–Lesovik formula in terms of the scattering matrix of the independent electrons on the obstacles of the circuit. Therefore, this generating function has the time-reversal symmetry of the fluctuation theorem with respect to the affinities driving the currents across the circuit. This fundamental result has many consequences, in particular, concerning the full counting statistics of electrons and the relationships between conductance and noise power beyond linear response. Recently, these consequences have been investigated experimentally [59–62]. Since the full counting statistics is performed by the capacitive coupling to a secondary circuit, back action effects should be taken into account to interpret the experimental results. Accordingly, the symmetry relation for the single observed current holds with respect to an effective affinity that differs from the thermodynamic affinity because of the noise coming from the secondary circuit [59, 102]. In this context, new opportunities arise for a better understanding of the intriguing fact that quantum jumps can be observed continuously in time [96–98]. For electronic conduction in an external magnetic field, the current fluctuation theorem predicts relations beyond linear response between conductance and noise power [47, 49]. Experimental observations on Aharonov–Bohm rings designed with GaAs semiconductor are in semiquantitative agreement with the theoretical predictions, possibly because of back action from the noise measurement [61, 62].

In conclusion, the advent of time-reversal symmetry relations among the dynamical large deviation properties has opened new perspectives in our fundamental understanding of nonequilibrium quantum systems. These relations can be extended to many nonequilibrium systems beyond electron quantum transport. In particular, they may apply to fermion and boson quantum transport in the physics of cold atoms or molecules or in quantum optics. The properties that can be deduced from the symmetry relations notably concern the properties of energy transduction when several currents are coupled together in thermoelectric, photoelectric, or mass separation devices. In a broader perspective, time-reversal symmetry relations are also considered in relativistic and gravitational systems [117, 118]. Similar relations can be envisaged for other discrete symmetries or their combinations, for example, in the context of quantum field theory [119].

Acknowledgments

The author thanks Professors Markus Büttiker, Michele Campisi, Peter Hänggi, and Keiji Saito for helpful discussions and comments. This research has been financially supported by the F.R.S.-FNRS, the "Communauté française de Belgique" (contract "Actions de Recherche Concertées" No. 04/09-312), and the Belgian Federal Government (IAP project "NOSY").

References

1 Imry, Y. (1997) *Introduction to Mesoscopic Physics*, Oxford University Press, New York.

2 Ferry, D.K., Goodnick, S.M., and Bird, J. (2009) *Transport in Nanostructures*, 2nd

edn., Cambridge University Press, Cambridge, UK.
3 Nazarov, Y.V. and Blanter, Y.M. (2009) *Quantum Transport*, Cambridge University Press, Cambridge, UK.
4 Gaspard, P. (2006) *Physica A*, **369**, 201.
5 Balescu, R. (1975) *Equilibrium and Nonequilibrium Statistical Mechanics*, John Wiley & Sons, Inc., New York.
6 Eckmann, J.-P. and Ruelle, D. (1985) *Rev. Mod. Phys.*, **57**, 617.
7 Posch, H.A. and Hoover, W.G. (1988) *Phys. Rev. A*, **38**, 473.
8 Gaspard, P. and Nicolis, G. (1990) *Phys. Rev. Lett.*, **65**, 1693.
9 Gaspard, P. and Baras, F. (1995) *Phys. Rev. E*, **51**, 5332.
10 Dorfman, J.R. and Gaspard, P. (1995) *Phys. Rev. E*, **51**, 28.
11 Gaspard, P. and Dorfman, J.R. (1995) *Phys. Rev. E*, **52**, 3525.
12 van Beijeren, H., Dorfman, J.R., Posch, H.A., and Dellago, Ch. (1997) *Phys. Rev. E*, **56**, 5272.
13 Gaspard, P. (1998) *Chaos, Scattering and Statistical Mechanics*, Cambridge University Press, Cambridge, UK.
14 Dorfman, J.R. (1999) *An Introduction to Chaos in Nonequilibrium Statistical Mechanics*, Cambridge University Press, Cambridge, UK.
15 Gaspard, P., Claus, I., Gilbert, T., and Dorfman, J.R. (2001) *Phys. Rev. Lett.*, **86**, 1506.
16 Klages, R. (2007) *Microscopic Chaos, Fractals and Transport in Nonequilibrium Statistical Mechanics*, World Scientific, Singapore.
17 Evans, D.J. and Morriss, G.P. (2008) *Statistical Mechanics of Nonequilibrium Liquids*, 2nd edn, Cambridge University Press, Cambridge, UK.
18 Collin, D., Ritort, F., Jarzynski, C., Smith, S.B., Tinoco, I., Jr., and Bustamante, C. (2005) *Nature*, **437**, 231.
19 Lussis, P., Svaldo-Lanero, T., Bertocco, A., Fustin, C.-A., Leigh, D.A., and Duwez, A.-S. (2011) *Nat. Nanotechnol.*, **6**, 553.
20 Jarzynski, C. (1997) *Phys. Rev. Lett.*, **78**, 2690.
21 Crooks, G.E. (1998) *J. Stat. Phys.*, **90**, 1481.
22 Crooks, G.E. (1999) *Phys. Rev. E*, **60**, 2721.
23 Jarzynski, C. (2000) *J. Stat. Phys.*, **98**, 77.
24 Evans, D.J., Cohen, E.G.D., and Morriss, G.P. (1993) *Phys. Rev. Lett.*, **71**, 2401.
25 Gallavotti, G. and Cohen, E.G.D. (1995) *Phys. Rev. Lett.*, **74**, 2694.
26 Gallavotti, G. (1996) *Phys. Rev. Lett.*, **77**, 4334.
27 Kurchan, J. (1998) *J. Phys. A*, **31**, 3719.
28 Lebowitz, J.L. and Spohn, H. (1999) *J. Stat. Phys.*, **95**, 333.
29 Maes, C. (1999) *J. Stat. Phys.*, **95**, 367.
30 Evans, D.J. and Searles, D.J. (2002) *Adv. Phys.*, **51**, 1529.
31 van Zon, R. and Cohen, E.G.D. (2003) *Phys. Rev. E*, **67**, 046102.
32 Altland, A., De Martino, A., Egger, R., and Narozhny, B. (2010) *Phys. Rev. B*, **82**, 115323.
33 Mallick, K., Moshe, M., and Orland, H. (2011) *J. Phys. A*, **44**, 095002.
34 Jarzynski, C. (2011) *Annu. Rev. Condens. Matter Phys.*, **2**, 329.
35 Maes, C. and Netočný, K. (2003) *J. Stat. Phys.*, **110**, 269.
36 Gaspard, P. (2004) *J. Stat. Phys.*, **117**, 599.
37 Gaspard, P. (2005) *New J. Phys.*, **7**, 77.
38 Kawai, R., Parrondo, J.M.R., and Van den Broeck, C. (2007) *Phys. Rev. Lett.*, **98**, 080602.
39 Gaspard, P. (2007) *C. R. Phys.*, **8**, 598.
40 Andrieux, D., Gaspard, P., Ciliberto, S., Garnier, N., Joubaud, S., and Petrosyan, A. (2007) *Phys. Rev. Lett.*, **98**, 150601.
41 Andrieux, D., Gaspard, P., Ciliberto, S., Garnier, N., Joubaud, S., and Petrosyan, A. (2008) *J. Stat. Mech.*, P01002.
42 Callen, I., De Roeck, W., Jacob, T., Maes, C., and Netočný, K. (2004) *Physica D*, **187**, 383.
43 Mukamel, S. (2003) *Phys. Rev. Lett.*, **90**, 170604.
44 Tobiska, J. and Nazarov, Yu.V. (2005) *Phys. Rev. B*, **72**, 235328.
45 Esposito, M., Harbola, U., and Mukamel, S. (2007) *Phys. Rev. B*, **75**, 155316.
46 Harbola, U., Esposito, M., and Mukamel, S. (2007) *Phys. Rev. B*, **76**, 085408.
47 Saito, K. and Utsumi, Y. (2008) *Phys. Rev. B*, **78**, 115429.
48 Andrieux, D. and Gaspard, P. (2008) *Phys. Rev. Lett.*, **100**, 230404.

49 Andrieux, D., Gaspard, P., Monnai, T., and Tasaki, S. (2009) *New J. Phys.*, **11**, 043014 (Erratum, *ibid.*, p. 109802).
50 Saito, K. and Dhar, A. (2007) *Phys. Rev. Lett.*, **99**, 180601.
51 Talkner, P. and Hänggi, P. (2007) *J. Phys. A*, **40**, F569.
52 Talkner, P., Lutz, E., and Hänggi, P. (2007) *Phys. Rev. E*, **75**, 050102.
53 Campisi, M., Hänggi, P., and Talkner, P. (2011) *Rev. Mod. Phys.*, **83**, 771 (Erratum, *ibid.*, p. 1653).
54 Gelin, M.F. and Kosov, D.S. (2008) *Phys. Rev. E*, **78**, 011116.
55 Esposito, M., Harbola, U., and Mukamel, S. (2009) *Rev. Mod. Phys.*, **81**, 1665.
56 Gustavsson, S., Leturcq, R., Simovič, B., Schleser, R., Ihn, T., Studerus, P., Ensslin, K., Driscoll, D.C., and Gossard, A.C. (2006) *Phys. Rev. Lett.*, **96**, 076605.
57 Gustavsson, S., Leturcq, R., Studer, M., Shorubalko, I., Ihn, T., Ensslin, K., Driscoll, D.C., and Gossard, A.C. (2009) *Surf. Sci. Rep.*, **64**, 191.
58 Fujisawa, T., Hayashi, T., Tomita, R., and Hirayama, Y. (2006) *Science*, **312**, 1634.
59 Utsumi, Y., Golubev, D.S., Marthaler, M., Saito, K., Fujisawa, T., and Schön, G. (2010) *Phys. Rev. B*, **81**, 125331.
60 Küng, B., Rössler, C., Beck, M., Marthaler, M., Golubev, D.S., Utsumi, Y., Ihn, T., and Ensslin, K. (2012) *Phys. Rev. X*, **2**, 011001.
61 Nakamura, S., Yamauchi, Y., Hashisaka, M., Chida, K., Kobayashi, K., Ono, T., Leturcq, R., Ensslin, K., Saito, K., Utsumi, Y., and Gossard, A.C. (2010) *Phys. Rev. Lett.*, **104**, 080602.
62 Nakamura, S., Yamauchi, Y., Hashisaka, M., Chida, K., Kobayashi, K., Ono, T., Leturcq, R., Ensslin, K., Saito, K., Utsumi, Y., and Gossard, A.C. (2011) *Phys. Rev. B*, **83**, 155431.
63 Bochkov, G.N. and Kuzovlev, Yu.E. (1977) *Sov. Phys.*, **45**, 125.
64 Stratonovich, R.L. (1994) *Nonlinear Nonequilibrium Thermodynamics II: Advanced Theory*, Springer, Berlin.
65 Kubo, R. (1957) *J. Phys. Soc. Jpn.*, **12**, 570.
66 Onsager, L. (1931) *Phys. Rev.*, **37**, 405.
67 Casimir, H.B.G. (1945) *Rev. Mod. Phys.*, **17**, 343.
68 Tasaki, S. (2001) *Chaos Soliton Fract.*, **12**, 2657.
69 Tasaki, S. and Matsui, T. (2003) *Fundamental Aspects of Quantum Physics* (eds L. Accardi and S. Tasaki), World Scientific, New Jersey, pp. 100–119.
70 Tasaki, S. and Takahashi, J. (2006) *Prog. Theor. Phys. Suppl.*, **165**, 57.
71 De Donder, T. and Van Rysselberghe, P. (1936) *Affinity*, Stanford University Press, Menlo Park, CA.
72 Prigogine, I. (1967) *Introduction to Thermodynamics of Irreversible Processes*, John Wiley & Sons, Inc., New York.
73 Callen, H.B. (1985) *Thermodynamics and an Introduction to Thermostatistics*, John Wiley & Sons, Inc., NewYork.
74 Andrieux, D. and Gaspard, P. (2004) *J. Chem. Phys.*, **121**, 6167.
75 Andrieux, D. and Gaspard, P. (2006) *J. Stat. Mech.*, P01011.
76 Andrieux, D. and Gaspard, P. (2007) *J. Stat. Mech.*, P02006.
77 Gaspard, P., Alonso, D., Okuda, T., and Nakamura, K. (1994) *Phys. Rev. E*, **50**, 2591.
78 Nakamura, K. and Harayama, T. (2004) *Quantum Chaos and Quantum Dots*, Oxford University Press, Oxford.
79 Büttiker, M. (1990) *Phys. Rev. B*, **41**, 7906.
80 Klich, I. (2003) *Quantum Noise in Mesoscopic Physics* (ed. Y.V. Nazarov), Kluwer, Dordrecht.
81 Avron, J.E., Bachmann, S., Graff, G.M., and Klich, I. (2008) *Commun. Math. Phys.*, **280**, 807.
82 Levitov, L.S. and Lesovik, G.B. (1993) *JETP Lett.*, **58**, 230.
83 Levitov, L.S., Lee, H.W., and Lesovik, G. B. (1996) *J. Math. Phys.*, **37**, 4845.
84 Roche, P.-E., Derrida, B., and Douçot, B. (2005) *Eur. Phys. J. B*, **43**, 529.
85 Sivan, U. and Imry, Y. (1986) *Phys. Rev. B*, **33**, 551.
86 Takahashi, J. and Tasaki, S. (2005) *J. Phys. Soc. Jpn Suppl.*, **74**, 261.
87 Blanter, Y.M. and Büttiker, M. (2000) *Phys. Rep.*, **336**, 1.
88 van Wees, B.J., van Houten, H., Beenakker, C.W.J., Williamson, J.G., Kouwenhoven, L.P., van der Marel, D., and Foxon, C.T. (1988) *Phys. Rev. Lett.*, **60**, 848.

89 Rammer, J. (2007) *Quantum Field Theory of Non-Equilibrium States*, Cambridge University Press, Cambridge, UK.

90 Mukamel, S. (1995) *Principles of Nonlinear Optical Spectroscopy*, Oxford University Press, Oxford.

91 Harbola, U. and Mukamel, S. (2008) *Phys. Rep.*, **465**, 191.

92 Umezawa, H., Matsumoto, H., and Tachiki, M. (1982) *Thermo Field Dynamics and Condensed States*, North-Holland, Amsterdam.

93 Kosov, D.S. (2009) *J. Chem. Phys.*, **131**, 171102.

94 Dzhioev, A.A. and Kosov, D.S. (2011) *J. Chem. Phys.*, **134**, 044121.

95 Gaspard, P. and Nagaoka, M. (1999) *J. Chem. Phys.*, **111**, 5668.

96 Gaspard, P. and Nagaoka, M. (1999) *J. Chem. Phys.*, **111**, 5676.

97 Carmichael, H. (1993) *An Open Systems Approach to Quantum Optics*, Springer, Berlin.

98 Goan, H.-S., Milburn, G.J., Wiseman, H. M., and Sun, H.B. (2001) *Phys. Rev. B*, **63**, 125326.

99 Dereziński, J., De Roeck, W., and Maes, C. (2008) *J. Stat. Phys.*, **131**, 341.

100 Andrieux, D. and Gaspard, P. (2007) *J. Stat. Phys.*, **127**, 107.

101 Schaller, G., Kiesslich, G., and Brandes, T. (2010) *Phys. Rev. B*, **82**, 041303.

102 Bulnes Cuetara, G., Esposito, M., and Gaspard, P. (2011) *Phys. Rev. B*, **84**, 165114.

103 Cover, T.M. and Thomas, J.A. (2006) *Elements of Information Theory*, 2nd edn, John Wiley & Sons, Inc., Hoboken.

104 Gerritsma, E. and Gaspard, P. (2010) *Biophys. Rev. Lett.*, **5**, 163.

105 Gaspard, P. and Andrieux, D. (2011) *J. Stat. Mech.*, P03024.

106 Cleuren, B., Van den Broeck, C., and Kawai, R. (2006) *Phys. Rev. E*, **74**, 021117.

107 Knudsen, M. (1909) *Ann. Phys.*, **28**, 999.

108 Present, R.D. (1958) *Kinetic Theory of Gases*, McGraw-Hill, New York.

109 Gaspard, P. and Wang, X.-J. (1993) *Phys. Rep.*, **235**, 291.

110 Gaspard, P. (1994) *Prog. Theor. Phys. Suppl.*, **116**, 369.

111 Sánchez, R. and Büttiker, M. (2011) *Phys. Rev. B*, **83**, 085428.

112 Sánchez, R., López, R., Sánchez, D., and Büttiker, M. (2010) *Phys. Rev. Lett.*, **104**, 076801.

113 Gillespie, D.T. (1976) *J. Comput. Phys.*, **22**, 403.

114 Golubev, D.S., Utsumi, Y., Marthaler, M., and Schön, G. (2011) *Phys. Rev. B*, **84**, 075323.

115 Ingold, G.-L. and Nazarov, Yu.V. (1992) *Single Charge Tunneling, NATO Advanced Studies Institute, Series B: Physics*, Vol. 294 (eds H. Grabert and M. Devoret), Plenum, New York, pp. 21–107.

116 Förster, H. and Büttiker, M. (2008) *Phys. Rev. Lett.*, **101**, 136805.

117 Fingerle, A. (2007) *C.R. Phys.*, **8**, 696.

118 Iso, S. and Okazawa, S. (2011) *Nucl. Phys. B*, **851**, 380.

119 Calzetta, E.A. and Hu, B.-L.B. (2008) *Nonequilibrium Quantum Field Theory*, Cambridge University Press, Cambridge, UK.

8
Anomalous Fluctuation Relations

Rainer Klages, Aleksei V. Chechkin, and Peter Dieterich

8.1
Introduction

With *fluctuation relations* (FRs) we denote a set of symmetry relations describing large deviation properties (see Chapter 11) of the probability distribution functions (PDFs) of statistical physical observables far from equilibrium. First forms of one subset of them, often referred to as *fluctuation theorems*, emerged from generalizing fluctuation–dissipation relations to nonlinear stochastic processes [1, 2]. They were then discovered as generalizations of the second law of thermodynamics for thermostated dynamical systems, that is, systems interacting with thermal reservoirs, in nonequilibrium steady states [3–6] (see Chapters 2 and 3 for this deterministic approach). Another subset, so-called *work relations*, generalizes the relation between work and free energy, known from equilibrium thermodynamics, to nonequilibrium situations [7, 8] (see Chapters 1 and 5 for this line of research). These two fundamental classes were later on amended and generalized by a variety of other FRs from which they can partially be derived as special cases [9–12], as has already been discussed starting from Chapter 1 up to Chapter 7. Research performed over the past 10 years has shown that FRs hold for a great variety of systems, thus featuring one of the rare statistical physical principles that is valid even very far from equilibrium: see summaries in Refs [13–18] for stochastic processes, Refs [19–24] for deterministic dynamics, and Refs [25, 26] for quantum systems. Many of these relations have meanwhile been verified in experiments on small systems, that is, systems on molecular scales featuring only a limited number of relevant degrees of freedom [27–32] (cf. Chapters 4–6).

The term *anomalous* in the title of this chapter refers to *anomalous dynamics*, which are loosely speaking processes that do not obey the laws of conventional statistical physics and thermodynamics [33–35] (see, for example, Chapter 10 for anomalous deviations from Fourier's law of heat conduction in small systems). Paradigmatic examples are diffusion processes where the long-time mean square displacement does not grow linearly in time; that is, $\langle x^2 \rangle \sim t^\alpha$, where the angle brackets denote an ensemble average, does not increase with $\alpha = 1$ as expected for Brownian motion but either *subdiffusively* with $\alpha < 1$ or *superdiffusively* with $\alpha > 1$ [36–38]. After

Nonequilibrium Statistical Physics of Small Systems: Fluctuation Relations and Beyond, First Edition.
Edited by Rainer Klages, Wolfram Just, and Christopher Jarzynski.

pioneering work on amorphous semiconductors [39], anomalous transport phenomena have more recently been observed in a wide variety of complex systems, such as plasmas [40], nanopores [41], epidemic spreading [42], biological cell migration [43], and glassy materials [44], to mention a few [45, 46]. This raises the question to which extent conventional FRs are valid for anomalous dynamics. Theoretical results for generalized Langevin equations [47–50], Lévy flights [51, 52], and continuous-time random walk models [53] as well as computer simulations for glassy dynamics [54] showed both validity and violations of the various types of conventional FRs referred to above, depending on the specific type of anomalous dynamics considered and the nonequilibrium conditions that have been applied [55].

The purpose of this chapter is to outline how the two different fields of FRs and anomalous dynamics can be cross-linked in order to explore to which extent conventional forms of FRs are valid for anomalous dynamics. With the term *anomalous fluctuation relations* we refer to deviations from conventional forms of FRs as they have been discussed in the previous chapters, which are due to anomalous dynamics. Here we focus on generic types of stochastic anomalous dynamics by only checking *transient fluctuation relations* (TFRs), which describe the approach from a given initial distribution toward a (non)equilibrium steady state. Section 8.2 motivates the latter type of FRs by introducing simple scaling relations, as they are partially used later on in this chapter. As a warm-up, we then first derive the conventional TFR for the trivial case of Brownian motion of a particle moving under a constant external force modeled by standard Langevin dynamics. Section 8.3 introduces three generic types of stochastic anomalous dynamics: long-time correlated Gaussian stochastic processes, Lévy flights, and time-fractional kinetics. We check these three stochastic models for the existence of conventional TFRs under the simple nonequilibrium condition of a constant external force. Section 8.4 introduces a system exhibiting anomalous dynamics that is experimentally accessible, which is biological cell migration. We then outline how an anomalous transient fluctuation relation might be verified for cells migrating under chemical gradients. We summarize our results in Section 8.5 by highlighting an intimate connection between the validity of conventional TFRs and the validity of fluctuation–dissipation relations.

8.2 Transient Fluctuation Relations

8.2.1 Motivation

Consider a particle system evolving from some initial state at time $t = 0$ into a nonequilibrium steady state for $t \to \infty$. A famous example that has been investigated experimentally [27] (cf. also Chapter 5) is a colloidal particle immersed into water and confined by an optical harmonic trap (see Figure 8.1). The trap is first at rest but then dragged through water with a constant velocity v^*. Another paradigmatic example, whose nonequilibrium fluctuations have been much studied by molecular dynamics

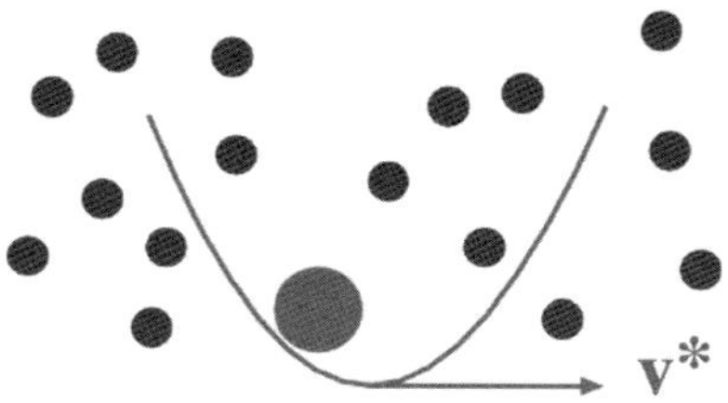

Figure 8.1 Sketch of a colloidal particle confined within a harmonic trap that is dragged through water with a constant velocity v^* (cf. the experiment by Wang *et al.* [27]).

computer simulations [3], is an interacting many-particle fluid under a shear force, which starts in thermal equilibrium by evolving into a nonequilibrium steady state [4].

The key for obtaining FRs in such systems is to obtain the PDF $\rho(\xi_t)$ of suitably defined dimensionless entropy production ξ_t over trajectory segments of time length t. The goal is to quantify the asymmetry between positive and negative entropy production in $\rho(\xi_t)$ for different times t since, as we will demonstrate in a moment, this relation is intimately related to the second law of thermodynamics. For a very large class of systems, and under rather general conditions, it was shown that the following equation holds [13, 14, 20, 22, 23]:

$$\ln \frac{\rho(\xi_t)}{\rho(-\xi_t)} = \xi_t. \tag{8.1}$$

Given that here we consider the transient evolution of a system from an initial into a steady state, this formula became known as the *transient fluctuation relation*. We may call the left-hand side the fluctuation ratio. Relations exhibiting this functional form have first been proposed in the seminal work by Evans, Cohen, and Morriss [3], although in the different situation of considering nonequilibrium steady states. Such a steady-state relation was proved a few years later by Gallavotti and Cohen for deterministic dynamical systems, based on the so-called chaotic hypothesis [5, 6]. The idea to consider such relations for transient dynamics was first put forward by Evans and Searles [4].

Figure 8.2 displays the temporal evolution of the PDF for entropy production in such a situation and may be compared to Figure 4.12 for analogous results

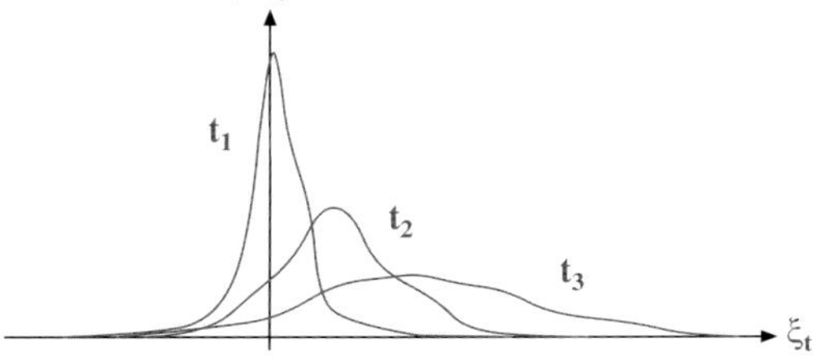

Figure 8.2 Illustration of the dynamics of the probability density function for entropy production $\rho(\xi_t)$ for different times $t_1 < t_2 < t_3$.

extracted from experimental measurements. The asymmetry of the evolving distribution, formalized by the fluctuation relation (Eq. (8.1)), is in line with the second law of thermodynamics. This easily follows from Eq. (8.1) by noting that

$$\rho(\xi_t) = \rho(-\xi_t)\exp(\xi_t) \geq \rho(-\xi_t), \tag{8.2}$$

where ξ_t is taken to be positive or zero. Integration from zero to infinity over both sides of this inequality after multiplication with ξ_t and defining the ensemble average over the given PDF as $\langle\cdot\rangle = \int_{-\infty}^{\infty} d\xi_t\, \rho(\xi_t)\ldots$ yields

$$\langle\xi_t\rangle \geq 0. \tag{8.3}$$

8.2.2 Scaling

By using FRs one is typically interested in assessing large deviation properties of the PDF of entropy production. That is, one wishes to sample the tails of the distributions for large times, and not so much the short-time dynamics, or the center of the distribution. For this purpose, it is useful to introduce suitably scaled variables that enable us to eliminate the drift associated with the positive average entropy production (Eq. (8.3)). The first option is to look at the PDF $\rho(\tilde{\xi}_t)$ of the scaled variable [52]

$$\tilde{\xi}_t = \frac{\xi_t}{\langle\xi_t\rangle}, \tag{8.4}$$

as illustrated in Figure 8.3. By definition, the PDF is now centered at $\langle\tilde{\xi}_t\rangle = 1$; hence, we have eliminated any contribution to the left-hand side of Eq. (8.1) that comes from the drift, by purely focusing on the asymmetric shape of the distribution.

Another way of scaling was used by Gallavotti and Cohen [5, 6] by employing the scaled time average

$$\hat{\xi}_t = \frac{\xi_t}{t\langle\xi_t\rangle}, \tag{8.5}$$

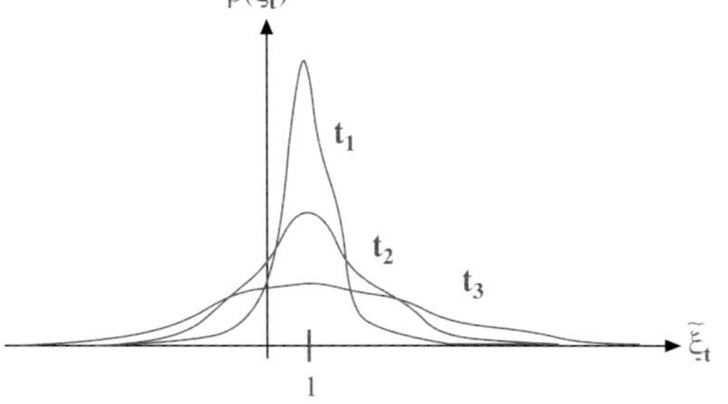

Figure 8.3 Illustration of the dynamics of the probability density function for entropy production $\rho(\tilde{\xi}_t)$ for different times $t_1 < t_2 < t_3$ by using the scaled variable (Eq. (8.4)).

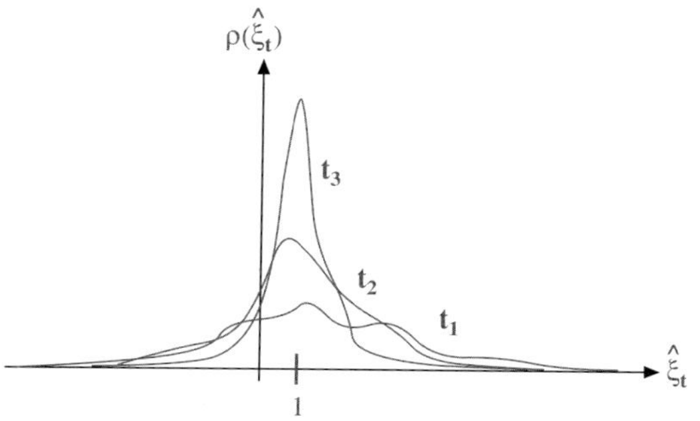

Figure 8.4 Illustration of the dynamics of the probability density function for entropy production $\rho(\hat{\xi}_t)$ for different times $t_1 < t_2 < t_3$ by using the scaled time average (Eq. (8.5)).

yielding the PDF for entropy production displayed in Figure 8.4. With this scaling, and for ergodic systems, clearly

$$\rho(\hat{\xi}_t) \to \delta(1 - \hat{\xi}_t) \quad (t \to \infty), \tag{8.6}$$

with

$$\frac{\xi_t}{t} \to \langle \xi_t \rangle \geq 0 \quad (t \to \infty), \tag{8.7}$$

thus illustrating the relation between FRs and the second law again.

8.2.3
Transient Fluctuation Relation for Ordinary Langevin Dynamics

As a preparation for what follows, we may first check the TFR for the ordinary overdamped *Langevin equation* [56]

$$\dot{x} = F + \zeta(t), \tag{8.8}$$

with a constant external force given by F and Gaussian white noise $\zeta(t)$. Note that for sake of simplicity, here we set all the other constants that are not relevant within this specific context equal to 1. For Langevin dynamics with a constant force, the entropy production ξ_t defined by the heat, or equivalently the dissipative work, is simply equal to the mechanical work [57]

$$W_t = Fx(t). \tag{8.9}$$

It follows that the PDF for entropy production, which here is identical to the one for the mechanical work, is trivially related to the PDF of the position x of the Langevin particle via

$$\rho(W_t) = F^{-1}\varrho(x, t). \tag{8.10}$$

This is very convenient, since it implies that all that remains to be done in order to check the TFR (Eq. (8.1)) is to solve the Fokker–Planck equation for the position PDF $\varrho(x, t)$ for a given initial condition. Here and in the following, we choose $x(0) = 0$; that is, in terms of position PDFs, we start with a delta distribution at $x = 0$. Note that for ordinary Langevin dynamics in a given potential, typically the equilibrium density is taken as the initial density [57, 58]. However, since in the following we will consider dynamics that may not exhibit a simple equilibrium state, without loss of generality here we make a different choice.

For the ordinary Langevin dynamics (Eq. (8.8)) modeling a linear Gaussian stochastic process, the position PDF is Gaussian exhibiting normal diffusion [56, 59] (cf. also Chapter 1),

$$\varrho(x, t) = \frac{1}{\sqrt{2\pi\sigma_{x,0}^2}} \exp\left(-\frac{(x - \langle x \rangle)^2}{2\sigma_{x,0}^2}\right). \tag{8.11}$$

With the subscript zero we denote ensemble averages in case of zero external field. By using the PDF scaling (Eq. (8.10)) and plugging this result into the TFR (Eq. (8.1)), we easily derive that the TFR for the work W_t holds if

$$\langle W_t \rangle = \frac{\sigma_{W_t,0}^2}{2}, \tag{8.12}$$

which is nothing else than an example of the *fluctuation–dissipation relation of the first kind* (FDR1) [56, 60] (cf. also Chapter 9). We thus arrive at the seemingly trivial but nevertheless important result that for this simple Gaussian stochastic process, the validity of FDR1 (Eq. (8.12)) implies the validity of the work TFR (Eq. (8.1)). For a full analysis of FRs of ordinary Langevin dynamics, we refer to Refs [57, 58].

Probably inspired by the experiment of Ref. [27], typically Langevin dynamics in a harmonic potential moving with a constant velocity has been studied in the literature [48–50, 61] (cf. Figure 8.1). Note that in this slightly more complicated case, the (total) work is not equal to the heat [57]. While for the work one recovers the TFR in its conventional form (Eq. (8.1)) in analogy to the calculation above, surprisingly the TFR for heat looks different for large enough fluctuations. This is due to the system being affected by the singularity of the harmonic potential, as has nicely been elucidated by van Zon and Cohen [58]. A similar effect has been reported by Harris *et al.* for a different type of stochastic dynamics, the asymmetric zero-range process [62]. For deterministic dynamics involving Nosé–Hoover thermostats, analogous consequences for the validity of the Gallavotti–Cohen FR have been discussed in Ref. [63]. See Ref. [13] for a brief review about the general mechanism underlying this type of violation of conventional forms of TFRs.

In the following, we check for yet another source of deviations from the conventional TFR (Eq. (8.1)) than the one induced by singular potentials. We explore the validity of work TFRs if one makes the underlying microscopic dynamics more complicated by modeling dynamical correlations or using non-Gaussian PDFs. In order to illustrate the main ideas along these lines, it suffices to consider a nonequilibrium situation simply generated by a constant external force.

8.3
Transient Work Fluctuation Relations for Anomalous Dynamics

Our goal is to check the TFR (Eq. (8.1)) for three generic types of stochastic processes modeling anomalous diffusion [46]: (1) *Gaussian stochastic processes*, (2) *Lévy flights*, and (3) *time-fractional kinetics*. We model all these dynamics by generalized Langevin equations. This section reports results from Ref. [55], which may be consulted for further details.

8.3.1
Gaussian Stochastic Processes

The first type we consider are Gaussian stochastic processes defined by the overdamped generalized Langevin equation

$$\int_0^t \mathrm{d}t' \, \dot{x}(t')\gamma(t-t') = F + \zeta(t), \tag{8.13}$$

with Gaussian noise $\zeta(t)$ and friction that is modeled with a memory kernel $\gamma(t)$. By using this equation, a stochastic process can be defined that exhibits normal statistics but with anomalous memory properties in the form of non-Markovian long-time correlated Gaussian noise. Equations of this type can be traced back at least to work by Mori and Kubo around 1965 (see Ref. [60] and references cited therein). They form a class of standard models generating anomalous diffusion that has been widely investigated (see, for example, Refs [56, 64, 65]). FRs for this type of dynamics have more recently been analyzed in Refs [47–50]. Examples of applications for this type of stochastic modeling are given by generalized elastic models [66], polymer dynamics [67], and biological cell migration [43].

We now split this class into two specific cases.

8.3.1.1 Correlated Internal Gaussian Noise

The first case corresponds to *internal* Gaussian noise, in the sense that we require the system to exhibit the *fluctuation–dissipation relation of the second kind* (FDR2) [56, 60]

$$\langle \zeta(t)\zeta(t')\rangle \sim \gamma(t-t'), \tag{8.14}$$

again by neglecting all constants that are not relevant for the main point we wish to make here. We now consider the specific case that both the noise and the friction are correlated by a simple power law,

$$\gamma(t) \sim t^{-\beta}, \quad 0 < \beta < 1. \tag{8.15}$$

Because of the linearity of the generalized Langevin equation (Eq. (8.13)), the position PDF must be Gaussian (Eq. (8.11)), and by the scaling of Eq. (8.10) we have $\rho(W_t) \sim \varrho(x,t)$. It thus remains to solve Eq. (8.13) for mean and variance, which can be done in Laplace space [55] yielding *subdiffusion*,

$$\sigma^2_{x,F} \sim t^\beta, \tag{8.16}$$

by preserving the FDR1 (Eq. (8.12)). Here and in the following, we denote ensemble averages in case of a nonzero external field with the subscript F. For Gaussian stochastic processes, we have seen in Section 8.2.3 that the conventional work TFR follows from FDR1. Hence, for the above power law correlated internal Gaussian noise, we recover the conventional work TFR (Eq. (8.1)).

8.3.1.2 Correlated External Gaussian Noise

As a second case, we consider the overdamped generalized Langevin equation

$$\dot{x} = F + \zeta(t), \tag{8.17}$$

which represents a special case of Eq. (8.13) with a memory kernel modeled by a delta function. Again we use correlated Gaussian noise defined by the power law

$$\langle \zeta(t)\zeta(t')\rangle \sim |t - t'|^{-\beta}, \quad 0 < \beta < 1, \tag{8.18}$$

which one may call *external*, because in this case we do not postulate the existence of FDR2. The position PDF is again Gaussian, and as before $\rho(W_t) \sim \varrho(x,t)$. However, by solving the Langevin equation along the same lines as in the previous case, here one obtains *superdiffusion* by breaking FDR1,

$$\langle W_t\rangle \sim t, \quad \sigma^2_{W_t,F} \sim t^{2-\beta}. \tag{8.19}$$

Calculating the fluctuation ratio, that is, the left-hand side of Eq. (8.1), from these results yields the *anomalous work TFR*

$$\ln \frac{\rho(W_t)}{\rho(-W_t)} = C_\beta t^{\beta-1} W_t, \quad 0 < \beta < 1, \tag{8.20}$$

where C_β is a constant that depends on physical parameters [55]. Comparing this equation with the conventional form of the TFR (Eq. (8.1)), one observes that the fluctuation ratio is still linear in W_t, thus exhibiting the exponential large deviation form [52] (cf. Chapter 11). However, there are two important deviations: (1) the slope of the fluctuation ratio as a function of W_t is not equal to 1 anymore, and in particular (2) it decreases with time. We may thus classify Eq. (8.20) as a *weak violation of the conventional TFR*.

We remark that for driven glassy systems FRs have already been obtained displaying slopes that are not equal to 1. Within this context, it has been suggested to capture these deviations from 1 by introducing the concept of an "effective temperature" [61, 68, 69]. As far as the time dependence of the coefficient is concerned, such behavior has recently been observed in computer simulations of a paradigmatic two-dimensional lattice gas model generating glassy dynamics [54]. Figure 8.5 shows the fluctuation ratio as a function of the entropy production at different

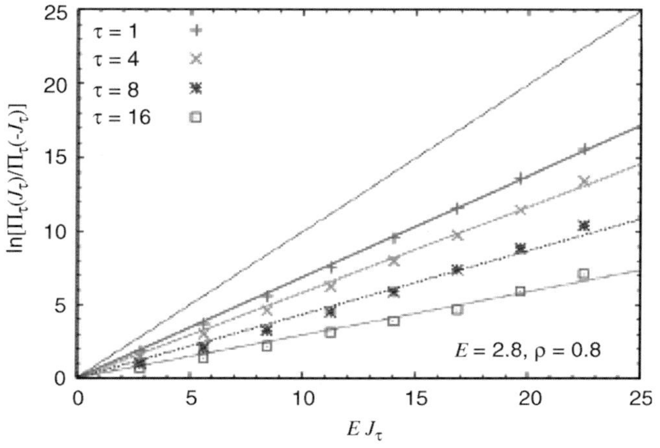

Figure 8.5 The fluctuation ratio $\ln(\Pi_\tau(J_\tau)/\Pi_\tau(J_\tau))$ for the entropy production $W_\tau = EJ_\tau$ with particle current J_τ and field strength E for particle density ρ at different times τ. The full line, with slope 1, displays the result of the conventional FR (Eq. (8.1)) in a nonequilibrium steady state. (Reproduced from Ref. [54].)

times τ as extracted from computer simulations of this model, where the PDF has first been relaxed into a nonequilibrium steady state. It is clearly seen that the slope decreases with time, which is in line with the prediction of the anomalous TFR (Eq. (8.20)). However, to which extent the nonequilibrium dynamics of this lattice gas model can be mapped onto the generalized Langevin equation (Eq. (8.17)) is an open question.

8.3.2
Lévy Flights

A second fundamental type of anomalous dynamics can as well be defined by the overdamped Langevin equation (Eq. (8.17)). However, this time we choose white *Lévy noise*; that is, the random variable ζ is distributed according to the PDF

$$\chi(\zeta) \sim \zeta^{-1-\alpha} \quad (\zeta \to \infty), \quad 0 \leq \alpha < 2. \tag{8.21}$$

In general, the full Lévy stable PDF is defined by its characteristic function. In this case, we are thus dealing with Markovian stochastic processes that are not Gaussian distributed generating so-called *Lévy flights*, which are due to the heavy tails of the underlying PDF. An introduction to the theory of Lévy flights can be found in Chapter 5 of Ref. [46]; the rigorous mathematical theory is presented in, for example, Ref. [70]. Lévy flights define one of the most paradigmatic models of anomalous dynamics with wide applications, for example, in fluid dynamics [71], in the foraging of biological organisms [72], and in glassy optical material [44], to highlight only a few cases.

It can be shown that the position PDF $\varrho(x,t)$ characterizing the process defined by Eqs. (8.17) and (8.21) obeys the *space-fractional* Fokker–Planck equation

$$\frac{\partial \varrho}{\partial t} = -F\frac{\partial \varrho}{\partial x} + \frac{\partial^\alpha \varrho}{\partial |x|^\alpha}, \tag{8.22}$$

where the last term is given by the *Riesz fractional derivative*, which in real space is a complicated integrodifferential operator. It is thus more convenient to represent this derivative by its Fourier transform, which takes the simple expression

$$\mathcal{F}\{\partial^\alpha \varrho/\partial|x|^\alpha\} = -|k|^\alpha \mathcal{F}\{\varrho\}. \tag{8.23}$$

Fractional derivatives provide generalizations of ordinary derivatives by reproducing them in case of integer values of the derivative parameter. Being defined by power law memory kernels, they have proven to be extremely useful in order to mathematically model anomalous dynamics. The well-developed discipline of *fractional calculus* rigorously explores the properties of these mathematical objects; for introduction to fractional derivatives, see, for example. Refs [34, 35, 46, 73]. A systematic and comprehensive mathematical exposition of fractional calculus is given in Ref. [74]. After solving Eq. (8.22) in Fourier space, the resulting position PDF needs to be converted into the work PDF by using Eq. (8.10). In this case, it is sensible to apply the scaling (Eq. (8.4)) [52], which here yields the scaled variable $\tilde{W}_t = W_t/F^2 t$. Expressing the work PDF in this variable and using the asymptotics of the Lévy stable PDF (Eq. (8.21)), we arrive at the asymptotic TFR for Lévy flights

$$\lim_{\tilde{W}_t \to \pm\infty} \frac{\rho(\tilde{W}_t)}{\rho(-\tilde{W}_t)} = 1. \tag{8.24}$$

This result has first been reported by Touchette and Cohen [51] by using a different technique for the different situation of a harmonic potential dragged with a constant velocity. Note that for $\alpha = 2$ in the above model we recover the conventional TFR (Eq. (8.1)). For $0 < \alpha < 2$, however, we obtain the surprising result that asymptotically large positive and negative fluctuations of the scaled work are equally probable for Lévy flights. The underlying work PDF is nevertheless still generically asymmetric. Note that the fluctuation ratio (Eq. (8.24)) does not display the exponential large deviation form; hence, one may denote this as a *strong violation of the conventional TFR.*

8.3.3
Time-Fractional Kinetics

The third and final fundamental type of stochastic anomalous dynamics that we consider here can be modeled by the so-called *subordinated* Langevin equation [75, 76]

$$\frac{\mathrm{d}x(u)}{\mathrm{d}u} = F + \zeta(u), \qquad \frac{\mathrm{d}t(u)}{\mathrm{d}u} = \tau(u), \tag{8.25}$$

with Gaussian white noise $\zeta(u)$ and white Lévy noise $\tau(u) > 0$ with $0 < \alpha < 1$. It can be shown that subordinated Langevin dynamics is intimately related to *continuous-time random walk theory*, which provides a generalization of ordinary random walk theory by generating nontrivial jump dynamics. The latter approach has in turn been used, for example, to understand measurements of anomalous photocurrents in copy machines [39], microsphere diffusion in the cell membrane [45], translocations of biomolecules through membrane pores [77], and even dynamics of prices in financial markets [78]. It was demonstrated that this Langevin description leads to the time-fractional Fokker–Planck equation [75, 76]

$$\frac{\partial \varrho}{\partial t} = \frac{\partial^{1-\alpha}}{\partial t^{1-\alpha}} \left[-\frac{\partial F \varrho}{\partial x} + \frac{\partial^2 \varrho}{\partial x^2} \right] \tag{8.26}$$

for $0 < \alpha < 1$ with *Riemann–Liouville fractional derivative* on the right semi-axis

$$\frac{\partial^\delta \varrho}{\partial t^\delta} = \frac{\partial}{\partial t} \left[\frac{1}{\Gamma(1-\delta)} \int_0^t \mathrm{d}t' \, \frac{\varrho(t')}{(t-t')^\delta} \right] \tag{8.27}$$

for $0 < \delta < 1$. This equation obeys a (generalized) Einstein relation for friction and diffusion coefficients (which here are both set to unity, for sake of simplicity). From Eq. (8.26), equations for the first and second moments can be derived and then solved in Laplace space. The second moment in the absence of an external force yields *subdiffusion*,

$$\sigma_{x,0}^2 \sim t^\alpha . \tag{8.28}$$

A calculation of the current $\langle x \rangle$ shows that the FDR1 (Eq. (8.12)) is preserved by this dynamics. Solving Eq. (8.26) in Laplace space and putting everything together, one recovers the conventional form of the TFR (Eq. (8.1)) for this type of dynamics. This confirms again that a distinctive role is played by FDR1 for the validity of conventional TFRs, even if the work PDFs are not Gaussian, as in this case.

We remark that analogous results are obtained by studying these three types of anomalous dynamics for the case of a particle moving in a harmonic potential that is dragged with a constant velocity [55].

8.4 Anomalous Dynamics of Biological Cell Migration

In order to illustrate the application of anomalous dynamics, and possibly of anomalous FRs, to realistic situations, in this section we discuss experiments and theory about the migration of single biological cells crawling on surfaces or in 3D matrices as examples. We first introduce the problem of cell migration by considering cells in an equilibrium situation, that is, not moving under the influence of any external gradient or field. This case is investigated by extracting results for the mean square displacement (MSD) and for the position PDFs from experimental data. We then show how the experimental results can be understood by a mathematical model in the form of a fractional Klein–Kramers equation. As far as MSD and velocity autocorrelation function are concerned, this equation bears some similarity to a generalized Langevin equation that is of the same type as the one that has been discussed in Section 8.3.1.2. We finally give an outlook to the nonequilibrium problem of cell migration under chemical gradients and describe first results obtained from experiments and data analysis. This research paves the way to eventually check the existence of anomalous work TFR in biological cell migration. The results on cell migration in equilibrium outlined in this section are based on Ref. [43].

8.4.1
Cell Migration in Equilibrium

Nearly all cells in the human body are mobile at a given time during their life cycle. Embryogenesis, wound healing, immune defense, and the formation of tumor metastases are well known phenomena that rely on cell migration [79–81]. Figure 8.6 depicts the path of a single biological cell crawling on a substrate measured in an *in vitro* experiment [43]. At first sight, the path looks like the trajectory of a Brownian particle generated, for example, by the ordinary Langevin dynamics of Eq. (8.8). On the other hand, according to Einstein's theory of Brownian motion, a Brownian particle is *passively* driven by collisions from the surrounding fluid molecules, whereas biological cells move *actively* by themselves converting chemical into kinetic energy. This raises the question whether the random-looking paths of crawling biological cells can really be understood in terms of simple Brownian motion [82, 83] or whether more advanced concepts of dynamical modeling have to be applied [84–88].

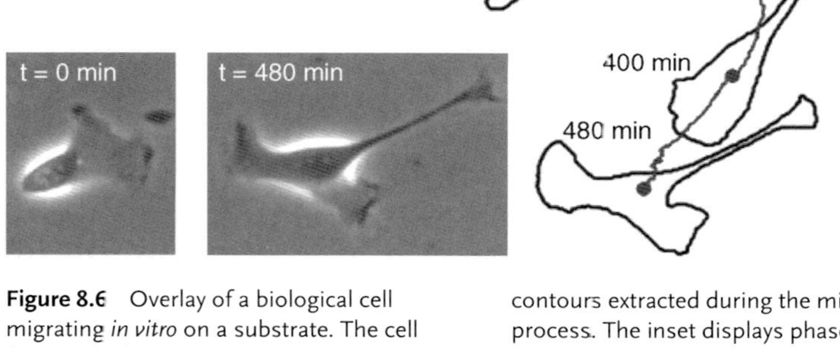

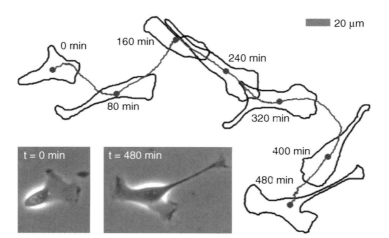

Figure 8.6 Overlay of a biological cell migrating *in vitro* on a substrate. The cell frequently changes its shape and direction during migration, as shown by several cell contours extracted during the migration process. The inset displays phase contrast images of the cell at the beginning and end of its migration process [43].

8.4.1.1 Experimental Results

The cell migration experiments that we now discuss have been performed on two types of tumor-like migrating *transformed renal epithelial Madin–Darby canine kidney (MDCK-F)* cell strains: wild-type (NHE^+) and NHE-deficient (NHE^-) cells. Here NHE^+ stands for a molecular sodium hydrogen exchanger that is either present or deficient. It can thus be checked whether this microscopic exchanger has an influence on cell migration, which is a typical question asked particularly by cell physiologists. The cell diameter is about 20–50 μm and the mean velocity of the cells is about 1 μm min^{-1}. Cells are driven by active protrusions of growing actin filaments (*lamellipodial dynamics*) and coordinated interactions with myosin motors and dynamically reorganizing cell–substrate contacts. The leading edge dynamics of a polarized cell proceeds at the order of seconds. Thirteen cells were observed for up to 1000 min. Sequences of microscopic phase contrast images were taken and segmented to obtain the cell boundaries shown in Figure 8.6 (see Ref. [43] for full details of the experiments).

According to the Langevin description of Brownian motion outlined in Section 8.2.3, Brownian motion is characterized by a MSD $\sigma_{x,0}^2(t) \sim t\,(t \to \infty)$ designating normal diffusion. Figure 8.7 shows that both types of cells behave differently: First of all, MDCK-F NHE^- cells move less efficiently than NHE^+ cells resulting in a reduced MSD for all times. As displayed in Figure 8.7a, the MSD of

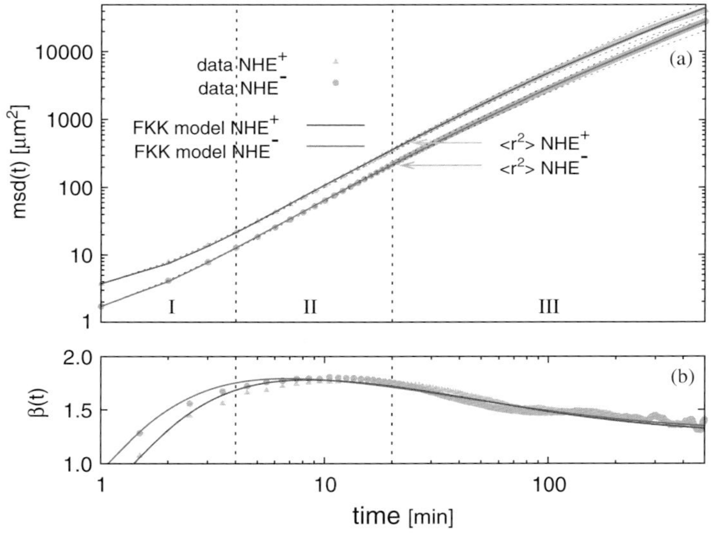

Figure 8.7 (a) Double logarithmic plot of the MSD as a function of time. Experimental data points for both cell types are shown by symbols. Different timescales are marked as phases I, II, and III, as discussed in the text. The solid lines represent fits to the MSD from the solution of our model (see Eq. (8.34)). All parameter values of the model are given in Ref. [43]. The dashed lines indicate the uncertainties of the MSD values according to Bayes data analysis. (b) Logarithmic derivative $\beta(t)$ of the MSD for both cell types as defined by Eq. (8.29).

both cell types exhibits a crossover between three different dynamical regimes. These three phases can be best identified by extracting the time-dependent exponent β of the MSD $\sigma_{x,0}^2(t) \sim t^\beta$ from the data, which can be done by using the logarithmic derivative

$$\beta(t) = \frac{d \ln \mathrm{msd}(t)}{d \ln t}. \tag{8.29}$$

The results are shown in Figure 8.7b. Phase I is characterized by an exponent $\beta(t)$ roughly below 1.8. In the subsequent intermediate phase II, the MSD reaches its strongest increase with a maximum exponent β. When the cell has approximately moved beyond a square distance larger than its own mean square radius (indicated by arrows in the figure), $\beta(t)$ gradually decreases to about 1.4. Both cell types therefore do not exhibit normal diffusion, which would be characterized by $\beta(t) \to 1$ in the long-time limit, but move anomalously, where the exponent $\beta > 1$ indicates superdiffusion.

We next study the PDF of cell positions. Since no correlations between x and y positions could be found, it suffices to restrict ourselves to one dimension. Figure 8.8a and b reveals the existence of non-Gaussian distributions at different times. The transition from a peaked distribution at short times to rather broad distributions at long times suggests again the existence of distinct dynamical processes acting on different timescales. The shape of these distributions can be quantified by calculating the *kurtosis*

$$\kappa(t) := \frac{\langle x^4(t)\rangle}{\langle x^2(t)\rangle^2}, \tag{8.30}$$

which is displayed as a function of time in Figure 8.8c. For both cell types, $\kappa(t)$ rapidly decays to a constant that is clearly below 3 in the long-time limit. A value of 3 would be the result for the spreading Gaussian distributions characterizing Brownian motion. These findings are another strong manifestation of the anomalous nature of cell migration.

8.4.1.2 Theoretical Modeling

We now present the stochastic model that we have used to reproduce the experimental data yielding the fit functions shown in the previous two figures. The model is defined by the *fractional Klein–Kramers equation* [89]

$$\frac{\partial \varrho}{\partial t} = -\frac{\partial}{\partial x}[v\varrho] + \frac{\partial^{1-\alpha}}{\partial t^{1-\alpha}}\gamma_\alpha \left[\frac{\partial}{\partial v}v + v_{\mathrm{th}}^2 \frac{\partial^2}{\partial v^2}\right]\varrho, \quad 0 < \alpha < 1. \tag{8.31}$$

Here $\varrho = \varrho(x, v, t)$ is the PDF depending on time t, position x, and velocity v in one dimension, γ_α is a friction term, and $v_{\mathrm{th}}^2 = k_B T/M$ stands for the thermal velocity squared of a particle of mass $M = 1$ at temperature T, where k_B is Boltzmann's constant. The last term in this equation models diffusion in velocity space. In contrast to Fokker–Planck equations such as Eq. (8.22), this equation features time

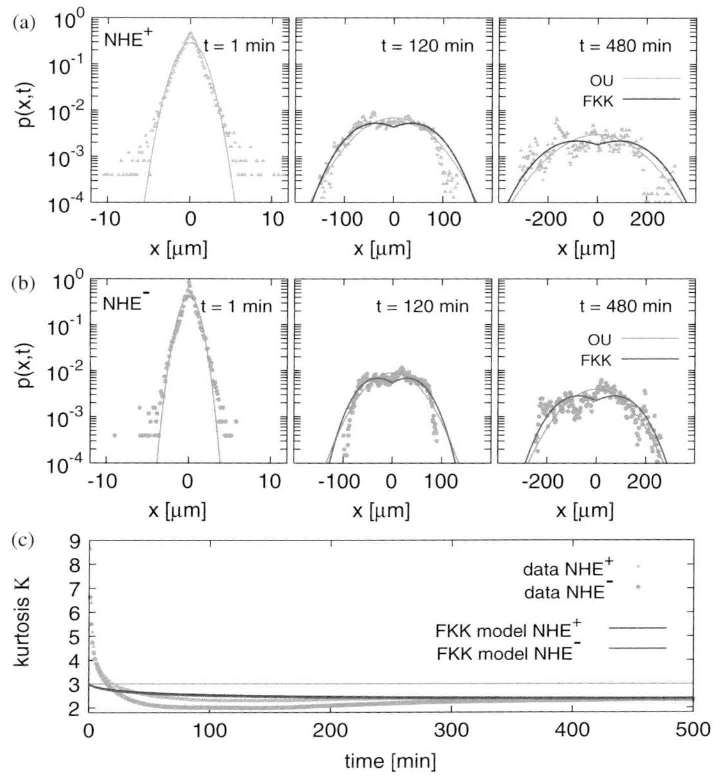

Figure 8.8 Spatiotemporal probability distributions $P(x, t)$. (a and b) Experimental data for both cell types at different times in semilogarithmic representation. The dark lines, labeled FKK, show the long-time asymptotic solutions of our model (Eq. (8.31)) with the same parameter set used for the MSD fit. The light lines, labeled OU, depict fits by the Gaussian distributions (Eq. (8.11)) representing Brownian motion. For $t = 1$ min, both $P(x, t)$ show a peaked structure clearly deviating from a Gaussian form. (c) The kurtosis $\kappa(t)$ of $P(x, t)$ (cf. Eq. (8.30)) plotted as a function of time saturates at a value different from that of Brownian motion (line at $\kappa = 3$). The other two lines represent $\kappa(t)$ obtained from the model (Eq. (8.31)) [43]. (For a color version of this figure, please see the color plate at the end of this book.)

evolution in both position and velocity space. What distinguishes this equation from an ordinary Klein–Kramers equation, the most general model of Brownian motion [59], is the presence of the Riemann–Liouville fractional derivative of order $1 - \alpha$ (Eq. (8.27)) in front of the terms in square brackets. Note that for $\alpha = 1$ the ordinary Klein–Kramers equation is recovered. The analytical solution of this equation for the MSD has been calculated in Ref. [89] to

$$\sigma_{x,0}^2(t) = 2v_{\text{th}}^2 t^2 E_{\alpha,3}(-\gamma_\alpha t^\alpha) \to 2\frac{D_\alpha t^{2-\alpha}}{\Gamma(3-\alpha)} \quad (t \to \infty), \tag{8.32}$$

with $D_\alpha = v_{\text{th}}^2/\gamma_\alpha$ and the *two-parametric* or *generalized Mittag–Leffler function* (see, for example, Chapter 4 of Ref. [46] and Refs [73, 90])

$$E_{\alpha,\beta}(z) = \sum_{k=0}^{\infty} \frac{z^k}{\Gamma(\alpha k + \beta)}, \quad \alpha, \beta > 0, \quad z \in \mathbb{C}. \tag{8.33}$$

Note that $E_{1,1}(z) = \exp(z)$; hence, $E_{\alpha,\beta}(z)$ is a generalized exponential function. We see that for long times Eq. (8.32) yields a power law, which reduces to the long-time Brownian motion result in case of $\alpha = 1$.

In view of the experimental data shown in Figure 8.7, Eq. (8.32) was amended by including the impact of random perturbations acting on very short timescales for which we take Gaussian white noise of variance η^2. This leads to [91]

$$\sigma_{x,0;\text{noise}}^2(t) = \sigma_{x,0}^2(t) + 2\eta^2. \tag{8.34}$$

The second term mimics both measurement errors and fluctuations of the cell cytoskeleton. In case of the experiments with MDCK-F cells [43], the value of η can be extracted from the experimental data and is larger than the estimated measurement error. Hence, this noise must largely be of a biological nature and may be understood as being generated by microscopic fluctuations of the lamellipodia in the experiment.

The analytical solution of Eq. (8.31) for $\varrho(x, v, t)$ is not known; however, for large friction γ_α, this equation boils down to a fractional diffusion equation for which $\varrho(x, t)$ can be calculated in terms of a Fox function [92]. The experimental data in Figures 8.7 and 8.8 were then fitted consistently by using the above solutions with the four parameters v_{th}^2, α, γ, and η^2 in Bayesian data analysis [43].

In summary, by statistical analysis of experimental data we have shown that the equilibrium migration of the biological cells under consideration is anomalous. Related anomalies have also been observed for other types of migrating cells [84–88]. These experimental results are coherently reproduced by a mathematical model in the form of a stochastic fractional equation. We now elaborate on possible physical and biological interpretations of our findings.

First of all, we remark that the solutions of Eq. (8.31) for both the MSD and the velocity autocorrelation function match precisely to the solutions of the generalized Langevin equation [65]

$$\dot{v} = -\int_0^t dt'\, \gamma(t - t')v(t') + \xi(t). \tag{8.35}$$

Here $\xi(t)$ holds for Gaussian white noise and $\gamma(t) \sim t^{-\alpha}$ for a time-dependent friction coefficient with a power law memory kernel, which alternatively could be written by using a fractional derivative [65]. For $\gamma(t) \sim \delta(t)$, the ordinary Langevin equation is recovered. Note that the position PDF generated by this equation is Gaussian in the long-time limit and thus does not match to the one of the fractional Klein–Kramers equation (Eq. (8.31)). However, alternatively one could sample from a non-Gaussian $\xi(t)$ to generate a non-Gaussian position PDF. Strictly speaking, despite equivalent MSD and velocity correlations, Eqs. (8.31) and (8.35) define different classes of anomalous stochastic processes. The precise cross-links between the Langevin description and the fractional Klein–Kramers equation are subtle [93] and to some extent still unknown. The advantage of Eq. (8.35) is that it allows more straightforwardly a possible biophysical interpretation of the origin of the observed anomalous MSD and velocity correlations, at least partially, in terms of the existence of a memory-dependent friction coefficient. The latter, in turn, might be explained by anomalous rheological properties of the cell cytoskeleton, which consists of a complex biopolymer gel [94].

Second, what could be the possible biological significance of the observed anomalous cell migration? There is an ongoing debate about whether biological organisms such as albatrosses, marine predators, and fruit flies have managed to minimize the search time for food in a way that matches to optimizing search strategies in terms of stochastic processes (see Refs [72, 95] and references cited therein). In particular, it has been argued that Lévy flights are superior to Brownian motion in order to find sparsely, randomly distributed, replenishing food sources [95]. However, it was also shown that in other situations *intermittent dynamics* is more efficient than pure Lévy motion [95]. For our cell experiment, both the experimental data and the theoretical modeling suggest that there exists a slow diffusion on short timescales, whereas the long-time motion is much faster, which resembles intermittency as discussed in Ref. [95]. Hence, the results on anomalous cell migration presented above might be biologically relevant in view of suitably optimized foraging strategies.

8.4.2
Cell Migration Under Chemical Gradients

We conclude this section with a brief outlook to cell migration under chemical gradients [96]. In new experiments conducted by Lindemann and Schwab (O. Lindemann and A. Schwab 2011, unpublished.), *murine neutrophil* cells have been exposed to concentration gradients of chemoattractants. A plot of trajectories of an ensemble of cells crawling under chemotaxis is shown in Figure 8.9.

Statistical analysis of the experimental data (P. Dieterich, 2011, private communication.) yielded a linear drift in the direction of the gradient,

$$\langle x(t)\rangle \sim t. \tag{8.36}$$

The MSD in the comoving frame, on the other hand, was found to be

$$\sigma^2_{x,F}(t) - \langle x(t)\rangle^2 \sim t^\beta, \tag{8.37}$$

with the same exponent $\beta > 1$ as obtained for the equilibrium dynamics discussed before. Consequently, FDR1 (cf. Eq. (8.12)) is broken. These results suggest that, for obtaining a stochastic model, the force-free fractional Klein–Kramers equation (Eq. (8.31)) needs to be generalized by including an external force as discussed by

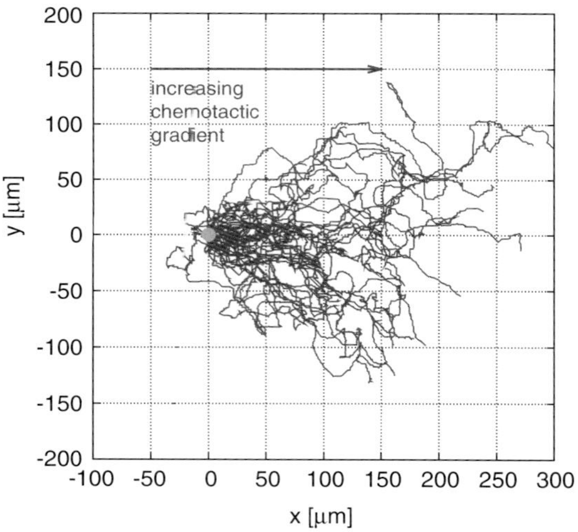

Figure 8.9 Trajectories of an ensemble of 40 murine neutrophil cells exposed to a chemical gradient of the chemoattractant fMLP that increases along the positive x-axis. Cells were observed over 30 min with a time interval of 5 s. Starting points have been transformed to the origin of the coordinate system (gray circle). It can be seen that there is an average drift of the ensemble toward the positive x-axis (O. Lindemann and A. Schwab 2011, unpublished.).

Metzler and Sokolov [97],

$$\frac{\partial \varrho}{\partial t} = -\frac{\partial}{\partial x}[v\varrho] + \frac{\partial^{1-\alpha}}{\partial t^{1-\alpha}}\gamma_\alpha \left[\frac{\partial}{\partial v}v - \frac{F}{\gamma_\alpha m}\frac{\partial}{\partial v} + v_{\mathrm{th}}^2 \frac{\partial^2}{\partial v^2}\right]\varrho. \tag{8.38}$$

Note that there exist two different ways in the literature of how to include the force F in Eq. (8.31) [89, 97]. These choices lead to different results for drift and MSD. The above results obtained from experimental data analysis clearly select the version of Ref. [97] as the adequate type of stochastic model in this case by rejecting that of Ref. [89], which, however, might well work in other situations. The (approximate) analytical solutions of Eq. (8.38) reproduce correctly drift, MSD, velocity correlations, and (for large enough friction coefficient γ_α and long enough times) the position PDFs of the measured nonequilibrium cell dynamics (P. Dieterich, 2011, private communication.).

Along these lines, one might also check for the form of the work TFR in case of cell migration. This has already been done in an experiment on the cellular slime mold *Dictyostelium discoideum*, in this case under electrotaxis [98]. By plotting the fluctuation ratio as a function of the cell positions at two different times, it was concluded that the conventional TFR (Eq. (8.1)) was confirmed by this experiment. In Figure 8.10, however, we show experimental results for the fluctuation ratio of the neutrophils of Figure 8.9 as a function of the cell positions at three different times. In complete formal analogy to Figure 8.5, the slopes clearly decrease with increasing time, which indicates a violation of the conventional TFR (Eq. (8.1)). To

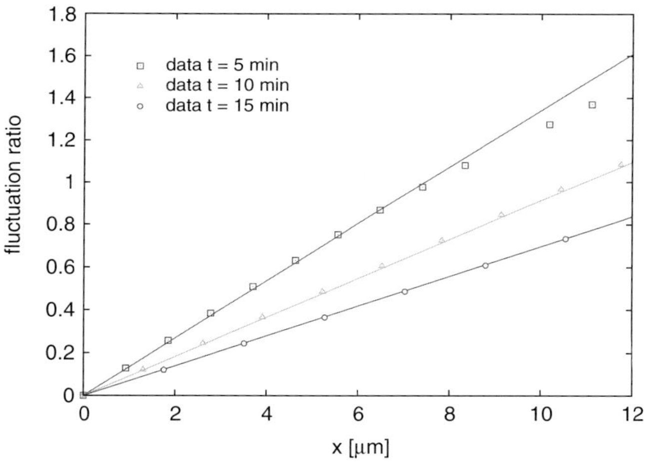

Figure 8.10 The fluctuation ratio $\ln(P(A)/P(-A))$ as a function of $A = x(\tau) - x(0)$ for $\tau = 5$ min (□), $\tau = 10$ min (Δ), and $\tau = 15$ min (∘) obtained from 90 independent trajectories of murine neutrophils moving in a chemotactic gradient of the substance fMLP, as depicted in Figure 8.9. Data show a linear increase in A; however, the reduction of the slope as a function of τ indicates deviations from the conventional TFR (Eq. (8.1)).

further explore the validity of work TFRs in cell migration experiments thus appears to be a very interesting, important open problem.

8.5 Conclusions

In this chapter, we have applied the concept of FRs as discussed in the previous chapters to anomalous stochastic processes. This cross-linking enables us to address the question whether conventional forms of FRs are valid for more complicated types of dynamics involving non-Markovian memory and non-Gaussian distributions. We have answered this question for three fundamental types of anomalous stochastic dynamics.

For *Gaussian stochastic processes with correlated noise,* the existence of FDR2 implies the existence of FDR1, and we have found that FDR1 in turn implies the existence of work TFR in conventional form. That is, analytical calculations showed that the conventional work TFR holds for internal noise. However, a weak violation of the conventional form was detected in case of external noise yielding a prefactor that is not equal to 1 and in particular depends on time. A strong violation of the conventional work TFR was derived for *space-fractional Lévy dynamics* confirming previous results from the literature. We have also found that the conventional work TFR holds for a typical example of *time-fractional dynamics.* These generic models suggest an intimate connection between FDRs and FRs in case of anomalous dynamics.

As a realistic example of anomalous dynamics, we have then discussed biological cell migration. By extracting the MSD and the position PDF from experimental data for cells crawling in an equilibrium *in vitro* situation, we found that the cells under investigation exhibited different dynamics on different timescales deviating from simple Brownian motion. For longer times, these cells moved superdiffusively. These experimental findings were reproduced by a stochastic model in the form of a fractional Klein–Kramers equation. For cells moving in nonequilibrium under chemotaxis, new data showed a breaking of FDR1 leading to a stochastic modeling in the form of a suitably extended fractional Klein–Kramers equation. Further analysis of these data indicated the existence of anomalous work TFRs.

To better understand work TFRs in biological cell migration both theoretically and experimentally remains an important open problem. However, it might also be interesting to experimentally check for anomalous work TFR in case of a particle dragged through a highly viscous gel instead of through water [27], for the fluctuations of a driven pendulum in gel [99], for granular gases exhibiting subdiffusion [100], or for glassy systems [54, 61, 69]. On the theoretical side, the basic results reported in this chapter suggest to systematically check the remaining variety of conventional fluctuation relations [9–11] for anomalous generalizations.

Acknowledgments

R.K. and A.V.C. thank the London Mathematical Society for a travel grant. Financial support by the EPSRC for a small grant within the framework of the QMUL *Bridging the Gap* initiative is gratefully acknowledged.

References

1 Bochkov, G. and Kuzovlev, Y. (1981) Nonlinear fluctuation–dissipation relations and stochastic models in nonequilibrium thermodynamics. I. Generalized fluctuation–dissipation theorem. *Physica A*, **106**, 443–479.

2 Bochkov, G. and Kuzovlev, Y. (1981) Nonlinear fluctuation–dissipation relations and stochastic models in nonequilibrium thermodynamics. II. Kinetic potential and variational principles for nonlinear irreversible processes. *Physica A*, **106**, 480–520.

3 Evans, D.J., Cohen, E.G.D., and Morriss, G.P. (1993) Probability of second law violations in shearing steady flows. *Phys. Rev. Lett.*, **71**, 2401–2404.

4 Evans, D.J. and Searles, D.J. (1994) Equilibrium microstates which generate second law violating steady states. *Phys. Rev. E*, **50**, 1645–1648.

5 Gallavotti, G. and Cohen, E.G.D. (1995) Dynamical ensembles in nonequilibrium statistical mechanics. *Phys. Rev. Lett.*, **74**, 2694–2697.

6 Gallavotti, G. and Cohen, E.G.D. (1995) Dynamical ensembles in stationary states. *J. Stat. Phys.*, **80**, 931–970.

7 Jarzynski, C. (1997) Nonequilibrium equality for free energy differences. *Phys. Rev. Lett.*, **78**, 2690–2693.

8 Jarzynski, C. (1997) Equilibrium free-energy differences from nonequilibrium measurements: a master-equation approach. *Phys. Rev. E*, **56**, 5018–5035.

9 Crooks, G. (1999) Entropy production fluctuation theorem and the nonequilibrium work relation for free

energy differences. *Phys. Rev. E*, **60**, 2721–2726.
10 Hatano, T. and Sasa, S. (2001) Steady-state thermodynamics of Langevin systems. *Phys. Rev. Lett.*, **86**, 3463–3466.
11 Seifert, U. (2005) Entropy production along a stochastic trajectory and an integral fluctuation theorem. *Phys. Rev. Lett.*, **95**, 040602/1–040602/4.
12 Sagawa, T. and Ueda, M. (2010) Generalized Jarzynski equality under nonequilibrium feedback control. *Phys. Rev. Lett.*, **104**, 090602/1–090602/4.
13 Harris, R.J. and Schütz, G.M. (2007) Fluctuation theorems for stochastic dynamics. *J. Stat. Mech.*, P07020/1–P07020/45.
14 Kurchan, J. (2007) Non-equilibrium work relations. *J. Stat. Mech.*, P07005/1–P07005/14.
15 Seifert, U. (2008) Stochastic thermodynamics: principles and perspectives. *Eur. Phys. J. B*, **64**, 423–431.
16 Jarzynski, C. (2008) Nonequilibrium work relations: foundations and applications. *Eur. Phys. J. B*, **64**, 331–340.
17 Boksenbojm, E., Wynants, B., and Jarzynski, C. (2010) Nonequilibrium thermodynamics at the microscale: work relations and the second law. *Physica A*, **389**, 4406–4417.
18 Van den Broeck, C. (2010) The many faces of the second law. *J. Stat. Mech.*, P10009/1–P10009/22.
19 Gallavotti, G. (1998) Chaotic dynamics, fluctuations, nonequilibrium ensembles. *Chaos*, **8**, 384–393.
20 Evans, D.J. and Searles, D.J. (2002) The fluctuation theorem. *Adv. Phys.*, **51** (7), 1529–1585.
21 Klages, R. (2007) *Microscopic Chaos, Fractals and Transport in Nonequilibrium Statistical Mechanics, Advanced Series in Nonlinear Dynamics*, vol. 24, World Scientific, Singapore.
22 Rondoni, L. and Mejía-Monasterio, C. (2007) Fluctuations in nonequilibrium statistical mechanics: models, mathematical theory, physical mechanisms. *Nonlinearity*, **20** (10), R1–R37.
23 Jepps, O.G. and Rondoni, L. (2010) Deterministic thermostats, theories of nonequilibrium systems and parallels with the ergodic condition. *J. Phys. A: Math. Theor.*, **43**, 133001/1–133001/42.
24 Jaksic, V., Pillet, C.A., and Rey-Bellet, L. (2011) Entropic fluctuations in statistical mechanics. I. Classical dynamical systems. *Nonlinearity*, **24** (3), 699.
25 Esposito, M., Harbola, U., and Mukamel, S. (2009) Nonequilibrium fluctuations, fluctuation theorems, and counting statistics in quantum systems. *Rev. Mod. Phys.*, **81**, 1665–1702.
26 Campisi, M., Hänggi, P., and Talkner, P. (2011) Quantum fluctuation relations: foundations and applications. *Rev. Mod. Phys.*, **83**, 771–791.
27 Wang, G.M., Sevick, E.M., Mittag, E., Searles, D.J., and Evans, D.J. (2002) Experimental demonstration of violations of the second law of thermodynamics for small systems and short time scales. *Phys. Rev. Lett.*, **89**, 050601/1–050601/4.
28 Ritort, F. (2003) Work fluctuations, transient violations of the second law and free-energy recovery methods: perspectives in theory and experiments. *Semin. Poincaré*, **2**, 195–229.
29 Bustamante, C., Liphardt, J., and Ritort, F. (2005) The nonequilibrium thermodynamics of small systems. *Phys. Today*, **58**, 43–48.
30 Ciliberto, S., Joubaud, S., and Petrosyan, A. (2010) Fluctuations in out-of-equilibrium systems: from theory to experiment. *J. Stat. Mech.*, P12003/1–P12003/27.
31 Toyabe, S., Sagawa, T., Ueda, M., Muneyuki, E., and Sano, M. (2010) Experimental demonstration of information-to-energy conversion and validation of the generalized Jarzynski equality. *Nat. Phys.*, **6**, 988–992.
32 Alemany, A., Ribezzi, M., and Ritort, F. (2011) Recent progress in fluctuation theorems and free energy recovery. *AIP Conf. Proc.*, **1332** (1), 96–110.
33 Bouchaud, J. and Georges, A. (1990) Anomalous diffusion in disordered media: statistical mechanisms, models and physical applications. *Phys. Rep.*, **195**, 127–293.
34 Metzler, R. and Klafter, J. (2000) The random walk's guide to anomalous

diffusion: a fractional dynamics approach. *Phys. Rep.*, **339**, 1–77.
35 Klafter, J. and Sokolov, I.M. (2011) *First Steps in Random Walks: From Tools to Applications*, Oxford University Press, Oxford.
36 Shlesinger, M.F., Zaslavsky, G.M., and Klafter, J. (1993) Strange kinetics. *Nature*, **363**, 31–37.
37 Klafter, J., Shlesinger, M.F., and Zumofen, G. (1996) Beyond Brownian motion. *Phys. Today*, **49**, 33–39.
38 Sokolov, I., Klafter, J., and Blumen, A. (2002) Fractional kinetics. *Phys. Today*, **55**, 48–54.
39 Scher, H. and Montroll, E. (1975) Anomalous transit-time dispersion in amorphous solids. *Phys. Rev. B*, **12**, 2455–2477.
40 Balescu, R. (2005) *Aspects of Anomalous Transport in Plasmas, Series in Plasma Physics*, vol. 18, CRC Press, London.
41 Kukla, V., Kornatowski, J., Demuth, D., Girnus, I., Pfeifer, H., Rees, L.V.C., Schunk, S., Unger, K.K., and Kärger, J. (1996) NMR studies of single-file diffusion in unidimensional channel zeolites. *Science*, **272** (5262), 702–704.
42 Brockmann1, D., Hufnagel, L., and Geisel, T. (2006) The scaling laws of human travel. *Nature*, **439**, 462–465.
43 Dieterich, P., Klages, R., Preuss, R., and Schwab, A. (2008) Anomalous dynamics of cell migration. *Proc. Natl. Acad. Sci. USA*, **105**, 459–463.
44 Barthelemy, P., Bertolotti, J., and Wiersma, D. (2008) A Lévy flight for light. *Nature*, **453**, 495–498.
45 Metzler, R. and Klafter, J. (2004) The restaurant at the end of the random walk: recent developments in the description of anomalous transport by fractional dynamics. *J. Phys. A: Math. Gen.*, **37**, R161–R168.
46 Klages, R., Radons, G., and Sokolov, I.M. (2008) *Anomalous Transport*, Wiley-VCH Verlag GmbH, Weinheim.
47 Beck, C. and Cohen, E.G.D. (2004) Superstatistical generalization of the work fluctuation theorem. *Physica A*, **344**, 393–402.
48 Ohkuma, T. and Ohta, T. (2007) Fluctuation theorems for non-linear generalized Langevin systems. *J. Stat. Mech.*, **10**, P10010/1–P10010/28.
49 Mai, T. and Dhar, A. (2007) Nonequilibrium work fluctuations for oscillators in non-Markovian baths. *Phys. Rev. E*, **75**, 061101/1–061101/7.
50 Chaudhury, S., Chatterjee, D., and Cherayil, B.J. (2008) Resolving a puzzle concerning fluctuation theorems for forced harmonic oscillators in non-Markovian heat baths. *J. Stat. Mech.*, P10006/1–P10006/13.
51 Touchette, H. and Cohen, E.G.D. (2007) Fluctuation relation for a Lévy particle. *Phys. Rev. E*, **76**, 020101(R)/1–020101(R)/4.
52 Touchette, H. and Cohen, E.G.D. (2009) Anomalous fluctuation properties. *Phys. Rev. E*, **80**, 011114/1–011114/11.
53 Esposito, M. and Lindenberg, K. (2008) Continuous-time random walk for open systems: fluctuation theorems and counting statistics. *Phys. Rev. E*, **77**, 051119/1–051119/12.
54 Sellitto, M. (2009) Fluctuation relation and heterogeneous superdiffusion in glassy transport. *Phys. Rev. E*, **80**, 011134/1–011134/5.
55 Chechkin, A.V. and Klages, R. (2009) Fluctuation relations for anomalous dynamics. *J. Stat. Mech.*, L03002/1–L03002/11.
56 Kubo, R., Toda, M., and Hashitsume, N. (1992) *Statistical Physics, Springer Series in Solid State Sciences*, 2nd edn, vol. 2, Springer, Berlin.
57 van Zon, R. and Cohen, E.G.D. (2003) Stationary and transient work-fluctuation theorems for a dragged Brownian particle. *Phys. Rev. E*, **67**, 046102/1–046102/10.
58 van Zon, R. and Cohen, E.G.D. (2003) Extension of the fluctuation theorem. *Phys. Rev. Lett.*, **91**, 110601/1–110601/4.
59 Risken, H. (1996) *The Fokker–Planck Equation*, 2nd edn, Springer, Berlin.
60 Kubo, R. (1966) The fluctuation–dissipation theorem. *Rep. Prog. Phys.*, **29**, 255–284.
61 Zamponi, F., Bonetto, F., Cugliandolo, L., and Kurchan, J. (2005) A fluctuation theorem for non-equilibrium relaxational systems driven by external forces. *J. Stat. Mech.*, P09013/1–P09013/51.

62 Harris, R.J., Rákos, A., and Schütz, G.M. (2006) Breakdown of Gallavotti–Cohen symmetry for stochastic dynamics. *Europhys. Lett.*, **75**, 227–233.

63 Evans, D.J., Searles, D.J., and Rondoni, L. (2005) Application of the Gallavotti–Cohen fluctuation relation to thermostated steady states near equilibrium. *Phys. Rev. E*, **71**, 056120/1–056120/12.

64 Porra, J.M., Wang, K.G., and Masoliver, J. (1996) Generalized Langevin equations: anomalous diffusion and probability distributions. *Phys. Rev. E*, **53**, 5872–5881.

65 Lutz, E. (2001) Fractional Langevin equation. *Phys. Rev. E*, **64**, 051106/1–051106/4.

66 Taloni, A., Chechkin, A.V., and Klafter, J. (2010) Generalized elastic model yields a fractional Langevin equation description. *Phys. Rev. Lett.*, **104**, 160602/1–160602/4.

67 Panja, D. (2010) Anomalous polymer dynamics is non-Markovian: memory effects and the generalized Langevin equation formulation. *J. Stat. Mech.*, P06011/1–P06011/34.

68 Sellitto, M. (1998) Fluctuations of entropy production in driven glasses. Preprint arXiv:q-bio.PE/0404018.

69 Zamponi, F., Ruocco, G., and Angelani, L. (2005) Generalized fluctuation relation and effective temperature in a driven fluid. *Phys. Rev. E*, **71**, 020101(R)/1–020101(R)/4.

70 Samorodnitsky, G. and Taqqu, M.S. (1994) *Stable Non-Gaussian Random Processes*, Chapman & Hall, New York.

71 Solomon, T.H., Weeks, E.R., and Swinney, H.L. (1993) Observation of anomalous diffusion and Lévy flights in a two-dimensional rotating flow. *Phys. Rev. Lett.*, **71**, 3975–3978.

72 Viswanathan, G., da Luz, M.G.E., Raposo, E.P., and Stanley, H.E. (2011) *The Physics of Foraging*, Cambridge University Press, Cambridge.

73 Podlubny, I. (1999) *Fractional Differential Equations*, Academic Press, New York.

74 Samko, S.G., Kilbas, A.A., and Marichev, O.I. (1993) *Fractional Integrals and Derivatives: Theory and Applications*, Gordon and Breach, Amsterdam.

75 Fogedby, H.C. (1994) Langevin equations for continuous time Lévy flights. *Phys. Rev. E*, **50**, 1657–1660.

76 Baule, A. and Friedrich, R. (2005) Joint probability distributions for a class of non-Markovian processes. *Phys. Rev. E*, **71**, 026101/1–026101/9.

77 Metzler, R. and Klafter, J. (2003) When translocation dynamics becomes anomalous. *Biophys. J.*, **85**, 2776–2779.

78 Scalas, E. (2006) The application of continuous-time random walks in finance and economics. *Physica A*, **362**, 225–239.

79 Lauffenburger, D.A. and Horwitz, A.F. (1996) Cell migration: a physically integrated molecular process. *Cell*, **84**, 359–369.

80 Lämmermann, T. and Sixt, M. (2009) Mechanical modes of "amoeboid" cell migration. *Curr. Opin. Cell Biol.*, **21** (5), 636–644.

81 Friedl, P. and Wolf, K. (2010) Plasticity of cell migration: a multiscale tuning model. *J. Cell Biol.*, **188**, 11–19.

82 Dunn, G.A. and Brown, A.F. (1987) A unified approach to analysing cell motility. *J. Cell Sci. Suppl.*, **8**, 81–102.

83 Stokes, C.L., Lauffenburger, D.A., and Williams, S.K. (1991) Migration of individual microvessel endothelial cells: stochastic model and parameter measurement. *J. Cell Sci.*, **99**, 419–430.

84 Hartmann, R.S., Lau, K., Chou, W., and Coates, T.D. (1994) The fundamental motor of the human neutrophil is not random: evidence for local non-Markov movement in neutrophils. *Biophys. J.*, **67**, 2535–2545.

85 Upadhyaya, A., Rieu, J.P., Glazier, J.A., and Sawada, Y. (2001) Anomalous diffusion and non-Gaussian velocity distributions of hydra cells in cellular aggregates. *Physica A*, **293**, 549–558.

86 Li, L., Norrelykke, S.F., and Cox, E.C. (2008) Persistent cell motion in the absence of external signals: a search strategy for eukaryotic cells. *PLoS ONE*, **3**, e2093/1–e2093/11.

87 Takagi, H., Sato, M.J., Yanagida, T., and Ueda, M. (2008) Functional analysis of spontaneous cell movement under different physiological conditions. *PLoS ONE*, **3**, e2648/1–e2648/8.

88 Bödeker, H.U., Beta, C., Frank, T.D., and Bodenschatz, E. (2010) Quantitative analysis of random ameboid motion. *Europhys. Lett.*, **90**, 28005/1–28005/5.

89 Barkai, E. and Silbey, R. (2000) Fractional Kramers equation. *J. Phys. Chem. B*, **104**, 3866–3874.

90 Gorenflo, R. and Mainardi, F. (1997) Differential equations of fractional order, in *Fractals and Fractional Calculus in Continuum Mechanics, CISM Courses and Lecture Notes*, vol. 378 (eds A. Carpinteri and F. Mainardi), Springer, Berlin, pp. 223–276.

91 Martin, D.S., Forstner, M.B., and Kaes, J.A. (2002) Apparent subdiffusion inherent to single particle tracking. *Biophys. J.*, **83**, 2109–2117.

92 Schneider, W.R. and Wyss, W. (1989) Fractional diffusion and wave equations. *J. Math. Phys.*, **30**, 134–144.

93 Eule, S., Friedrich, R., Jenko, F., and Kleinhans, D. (2007) Langevin approach to fractional diffusion equations including inertial effects. *J. Phys. Chem. B*, **111**, 11474–11477.

94 Semmrich, C., Storz, T., Glaser, J., Merkel, R., Bausch, A.R., and Kroy, K. (2007) Glass transition and rheological redundancy in F-actin solutions. *Proc. Natl. Acad. Sci. USA*, **104**, 20199–20203.

95 Bénichou, O., Loverdo, C., Moreau, M., and Voituriez, R. (2011) Intermittent search strategies. *Rev. Mod. Phys.*, **83** (1), 81.

96 Song, L., Nadkarni, S.M., Bödeker, H.U., Beta, C., Bae, A., Franck, C., Rappel, W.J., Loomis, W.F., and Bodenschatz, E. (2006) *Dictyostelium discoideum* chemotaxis: threshold for directed motion. *Eur. J. Cell Biol.*, **85**, 981–989.

97 Metzler, R. and Sokolov, I.M. (2002) Superdiffusive Klein–Kramers equation: normal and anomalous time evolution and Lévy walk moments. *Europhys. Lett.*, **58**, 482–488.

98 Hayashi, K. and Takagi, H. (2007) Fluctuation theorem applied to *Dictyostelium discoideum* system. *J. Phys. Soc. Jpn.*, **76**, 105001/1–105001/2.

99 Douarche, F., Joubaud, S., Garnier, N.B., Petrosyan, A., and Ciliberto, S. (2006) Work fluctuation theorems for harmonic oscillators. *Phys. Rev. Lett.*, **97**, 140603/1–140603/4.

100 Brilliantov, N.V. and Pöschel, T. (2004) *Kinetic Theory of Granular Gases*, Oxford University Press, Oxford.

Part II
Beyond Fluctuation Relations

Part II of the book embeds the topics of Part I into the wider remit of statistical mechanics and nonequilibrium dynamics. It discusses in particular anomalous transport processes, large fluctuations and dissipation on small scales, and the role of entropy and currents. The final five chapters thus establish cross-links to other fundamental theoretical concepts of small systems science by including further experimental applications to selected problems in the bio- and nanosciences.

The initial chapter by Gradenigo, Puglisi, Sarracino, Villamaina, and Vulpiani takes up the main thread of the previous part by embedding fluctuation relations into the more general context of fluctuations in statistical physics. By focusing on fluctuation–dissipation relations, the authors present an overview about how to relate entropy production and generalized response functions to fluctuation properties of systems far from equilibrium. These concepts are illustrated by applications to anomalous dynamics and to granular media.

The following contribution by Zhang, Liu, and Li widens the scope toward the study of anomalous energy fluctuations on the nanoscale. The authors summarize recent studies about heat transport in nanomaterials, such as nanotubes and nanowires, from both the theoretical and the experimental perspective. Heat transport in these structures is crucially determined by anomalous fluctuations resulting in superdiffusive behavior.

The third chapter by Touchette and Harris focuses again on fundamental aspects of statistical mechanics. The authors review how large deviation theory, a prominent topic within mathematical physics with applications to the study of extreme events, can be exploited for the investigation of fundamental problems in nonequilibrium statistical physics, such as the notion of macroscopic hydrodynamic limits or generalized nonequilibrium statistical ensembles. These concepts are illustrated by applications to nonequilibrium processes such as, for example, the analysis of current fluctuations in interacting particle systems.

Nonequilibrium Statistical Physics of Small Systems: Fluctuation Relations and Beyond, First Edition.
Edited by Rainer Klages, Wolfram Just, and Christopher Jarzynski.

The contribution by Yang and Radons adds a new facet to the previous discussion by reviewing a very recent direction of research in dynamical systems theory applied to nonequilibrium statistical mechanics. The authors use Lyapunov stability analysis in order to study the implications of microscopic chaos in liquids and glasses for macroscopic statistical behavior. They derive relevant structure functions assessing these dynamical features, which provide a more profound understanding of the emergence of hydrodynamic modes in complex and disordered dynamical systems.

The book concludes with a review by Mackowiak and Bräuchle about the state of the art of experimental investigations using single-molecule microscopy. By experimentally investigating the diffusion of individual molecules in meso- and nanostructures, they explore a particularly interesting case of fluctuations in small systems physics. Their contribution demonstrates how to link experimental research to topics of theoretical relevance by suggesting a wide range of applications in biology, medicine, and bionanotechnology.

9
Out-of-Equilibrium Generalized Fluctuation–Dissipation Relations

G. Gradenigo, A. Puglisi, A. Sarracino, D. Villamaina, and A. Vulpiani

9.1
Introduction

Surely one of the most important and general results of statistical mechanics is the existence of a relation between the spontaneous fluctuations and the response to external fields of physical observables. This result has applications both in equilibrium statistical mechanics, where it is used to relate the correlation functions to macroscopically measurable quantities such as specific heats, susceptibilities, and compressibilities, and in nonequilibrium systems, where it offers the possibility of studying the response to time-dependent external fields, by analyzing time-dependent correlations.

The idea of relating the amplitude of the dissipation to that of the fluctuations dates back to Einstein's work on Brownian motion [1]. Later, Onsager [2, 3] put forward his regression hypothesis stating that the relaxation of a macroscopic nonequilibrium perturbation follows the same laws that govern the dynamics of fluctuations in equilibrium systems. This principle is the basis of the fluctuation–dissipation relation (FDR) of Callen and Welton, and of Kubo's theory of time-dependent correlation functions [4]. This result represents a fundamental tool in nonequilibrium statistical mechanics since it allows one to predict the average response to external perturbations, without applying any perturbation.

Although the FDR theory was originally obtained for Hamiltonian systems near thermodynamic equilibrium, it has been realized that a generalized FDR holds for a vast class of systems. A renewed interest toward the theory of fluctuations has been motivated by the study of nonequilibrium processes. In 1993, Evans, Cohen, and Morriss [5] considered the fluctuations of the entropy production rate in a shearing fluid, and proposed the so-called fluctuation relation (FR). Such relation was then developed by Evans and Searles [6] and, under different conditions, by Gallavotti and Cohen [7]. Noteworthy applications of those theories are the Crooks and Jarzynski relations [8, 9].

In the recent years, there has been a great interest in the study of the relation between responses and correlation functions, also with the aim to understand the features of aging and glassy systems. While, on the one hand, a failure of the

Nonequilibrium Statistical Physics of Small Systems: Fluctuation Relations and Beyond, First Edition.
Edited by Rainer Klages, Wolfram Just, and Christopher Jarzynski.

validity of the equilibrium statistical mechanics (e.g., lack of the ergodicity) implies a violation of the fluctuation–dissipation theorem, on the other hand, as we will see in Sections 9.3, 9.5, and 9.6, the opposite is not true: a failure of the fluctuation–dissipation theorem (at least in its "naive version") does not imply the presence of glassy phenomena. It may, for instance, reside in the presence of stationary non-equilibrium currents.

9.1.1 The Relevance of Fluctuations: Few Historical Comments

Let us spend a few words on the conceptual relevance of the fluctuations in statistical physics. Even Boltzmann and Gibbs, who already knew the expression for the mean-square energy fluctuation,

$$\langle (E - \langle E \rangle)^2 \rangle = k_B T^2 C_V, \tag{9.1}$$

pointed out that fluctuations were too small to be observed in macroscopic systems: *In the molecular theory we assume that the laws of the phenomena found in nature do not essentially deviate from the limits that they would approach in the case of an infinite number of infinitely small molecules . . . [10] . . . [the fluctuations] would be in general vanishing quantities, since such experience would not be wide enough to embrace the more considerable divergences from the mean values [11].*

On the contrary, Einstein and Smoluchowski attributed a central role to the fluctuations. For instance, Einstein understood very well that the well-known equation (9.1) . . . *would yield an exact determination of the universal constant* (i.e., the Avogadro number), *if it were possible to determine the average of the square of the energy fluctuation of the system; this is, however, not possible according to our present knowledge* [12].

It is well known that the study of the fluctuations in the Brownian motion allows one to determine the Avogadro number (i.e., a quantity at microscopic level) from experimentally accessible macroscopic quantities. Therefore, we have a nonambiguous link between the microscopic and macroscopic levels; this is clear from the celebrated Einstein relation that gives the diffusion coefficient D in terms of macroscopic variables and the Avogadro number:

$$D = \lim_{t \to \infty} \frac{\langle (x(t) - x(0)^2 \rangle}{2t} = \frac{RT}{6 N_A \pi \eta a}, \tag{9.2}$$

where T and η are the temperature and dynamic viscosity of the fluid, respectively, a is the radius of the colloidal particle, $R = N_A k_B$ is the perfect gas constant, and N_A is the Avogadro number.

The theoretical work by Einstein [1] and the experiments by Perrin [13] gave a clear and conclusive evidence of the relationship between the diffusion coefficient (which is measurable at the macroscopic level) and the Avogadro number (which is related to the microscopic description). Therefore, already at equilibrium, small-scale fluctuations *do matter* in the statistical description of a certain system.

The main purpose of this chapter is to show *how* such fluctuations must be accounted for when the system is out of equilibrium. A relevant outcome will be the evidence that in order to properly describe small-scale out-of-equilibrium fluctuations some *correlations* not present at equilibrium must be taken into account.

The celebrated paper of Einstein, apart from its historical relevance for the atomic hypothesis and the development of the modern theory of stochastic processes, shows the first example of FDR. Let us write the Langevin [14] equation for the colloidal particle of mass m:

$$\frac{dV}{dt} = -\gamma V + \sqrt{\frac{2\gamma k_B T}{m}}\zeta, \tag{9.3}$$

where $\gamma = 6N_A\pi\eta a/m$ and ζ is the white noise, that is, a Gaussian stochastic process with $\langle\zeta(t)\rangle = 0$ and $\langle\zeta(t)\zeta(t')\rangle = \delta(t-t')$. An easy computation gives $D = \langle V^2\rangle\tau$. Consider now a (small) perturbating force $f(t) = F\Theta(t)$, where $\Theta(t)$ is the Heaviside step function. It is simple to determine the average response of the velocity after a long time (i.e., the drift) to such a perturbation: $\langle\delta V\rangle = F/\gamma$. Defining the mobility μ as $\langle\delta V\rangle = \mu F$, one easily obtains the celebrated Einstein relation (EFDR)

$$\mu = \frac{D}{k_B T}, \tag{9.4}$$

which gives a link between the diffusion coefficient (a property of the unperturbed system) and the mobility that measures how the system reacts to a small perturbation.

An interesting point addressed here is that out-of-equilibrium conditions can be hardly obtained from a single Langevin equation when the noise and drag terms are local in time. Preserving this locality, and hence the Markovian nature of the model, we will see that out of equilibrium a proper description of the relation between correlations and responses can be given within an enriched space of variables, where also some *local* field coupled to the variable of interest must be considered. This new field is itself a *fluctuating* object whose dynamics is ruled by another stochastic equation.

9.2 Generalized Fluctuation–Dissipation Relations

The fluctuation–response theory was originally developed in the context of equilibrium statistical mechanics of Hamiltonian systems. Sometimes this induced confusion; for example, one can find misleading claims on the limited validity of the FDR: we can cite an important review on turbulence containing the following (wrong) conclusion: *This absence of correlation between fluctuations and relaxation is reflected in the nonexistence of a fluctuation–dissipation theorem for turbulence* [15].

On the contrary, as we will discuss in the following, a generalized FDR holds under rather general hypotheses [16–18].

9.2.1 Chaos and the FDR: van Kampen's Objection

It has been argued by van Kampen [19] that the usual derivation of the FDR that relies on a first-order truncation of the time-dependent perturbation theory, for the evolution of probability density, can be severely criticized. In a nutshell, using the dynamical systems terminology, the van Kampen's argument is as follows. Given a perturbation $\delta\mathbf{x}(0)$ on the state of the system at time $t = 0$, one can write a Taylor expansion for $\delta\mathbf{x}(t)$, the difference between the perturbed trajectory and the unperturbed one:

$$\delta x_i(t) = \sum_j \frac{\partial x_i(t)}{\partial x_j(0)} \delta x_j(0) + O(|\delta\mathbf{x}(0)|^2). \tag{9.5}$$

Averaging over the initial condition, one has the mean response function:

$$R_{i,j}(t) = \left\langle \frac{\delta x_i(t)}{\delta x_j(0)} \right\rangle = \int \frac{\partial x_i(t)}{\partial x_j(0)} \varrho(\mathbf{x}(0))\mathrm{d}\mathbf{x}(0). \tag{9.6}$$

In the case of the equilibrium statistical mechanics, we have $\varrho(x) \propto \exp(-\beta H(x))$ so that, after integration by parts, one obtains

$$R_{i,j}(t) = \beta \left\langle x_i(t) \frac{\partial H(\mathbf{x}(0))}{\partial x_j(0)} \right\rangle, \tag{9.7}$$

which is nothing but the differential form of the usual FDR.

In the presence of chaos, the terms $\partial x_i(t)/\partial x_j(0)$ grow exponentially as $e^{\lambda t}$, where λ is the Lyapunov exponent; therefore, it is not possible to use the linear expansion (9.5) for a time larger than $(1/\lambda) \ln (L/|\delta\mathbf{x}(0)|)$, where L is the typical fluctuation of the variable $\mathbf{x}$. On account of that estimate, the linear response theory is expected to be valid only for extremely small and unphysical perturbations (or times). For instance, according to the argument by van Kampen, if one wants that the FDR holds up to 1 s when applied to the electrons in a conductor, then a perturbing electric field should be smaller than 10^{-20} V m^{-1}, in clear disagreement with the experience.

This result, at first glance, sounds as bad news for the statistical mechanics. However, this criticism had the merit to stimulate for a deeper understanding of the basic ingredients for the validity of the FDR relation and its validity range. On the other hand, the success of the linear theory, for the computation of the transport coefficients (e.g., electric conductivity) in terms of correlation function of the unperturbed systems, is transparent, and its validity has been, directly and indirectly, tested in a huge number of cases. Therefore, it is really difficult to believe that the FDR cannot be applied for physically relevant values of the perturbations.

Kubo suggested that the origin of this effectiveness of the linear response theory may reside in the "constructive role of chaos" because "instability [of the trajectories] instead favors the stability of distribution functions, working as the cause of the mixing" [20]. The following derivation [17] of a generalized FDR supports this intuition.

9.2.2
Generalized FDR for Stationary Systems

Let us briefly discuss a derivation, under rather general hypothesis, of a generalized FDR (also in nonequilibrium or non-Hamiltonian systems), for which the van

Kampen critique does not hold. Consider a dynamical system $\mathbf{x}(0) \to \mathbf{x}(t) = U^t\mathbf{x}(0)$ with states $\mathbf{x}$ belonging to an N-dimensional vector space. For the sake of generality, we will consider the case in which the time evolution can also be not completely deterministic (e.g., stochastic differential equations). We assume the existence of an invariant probability distribution $\varrho(\mathbf{x})$, for which some "absolute continuity" type conditions are required (see later), and the mixing character of the system (from which its ergodicity follows). Note that no assumption is made on N.

Our aim is to express the average response of a generic observable A to a perturbation, in terms of suitable correlation functions, computed according to the invariant measure of the unperturbed system. At the first step, we study the behavior of one component of $\mathbf{x}$, say x_i, when the system, described by $\varrho(\mathbf{x})$, is subjected to an initial (nonrandom) perturbation such that $\mathbf{x}(0) \to \mathbf{x}(0) + \Delta\mathbf{x}_0$.[1)] This instantaneous kick modifies the density of the system into $\varrho'(\mathbf{x})$, related to the invariant distribution by $\varrho'(\mathbf{x}) = \varrho(\mathbf{x} - \Delta\mathbf{x}_0)$. We introduce the probability of transition from $\mathbf{x}_0$ at time 0 to $\mathbf{x}$ at time t, $w(\mathbf{x}_0, 0 \to \mathbf{x}, t)$. For a deterministic system, with evolution law $\mathbf{x}(t) = U^t\mathbf{x}(0)$, the probability of transition reduces to $w(\mathbf{x}_0, 0 \to \mathbf{x}, t) = \delta(\mathbf{x} - U^t\mathbf{x}_0)$, where $\delta(\cdot)$ is the Dirac's delta. Then we can write an expression for the mean value of the variable x_i, computed with the density of the perturbed system:

$$\langle x_i(t)\rangle' = \int\int x_i\varrho'(\mathbf{x}_0)w(\mathbf{x}_0, 0 \to \mathbf{x}, t)\mathrm{d}\mathbf{x}\mathrm{d}\mathbf{x}_0. \tag{9.8}$$

The mean value of x_i during the unperturbed evolution can be written in a similar way:

$$\langle x_i(t)\rangle = \int\int x_i\varrho(\mathbf{x}_0)w(\mathbf{x}_0, 0 \to \mathbf{x}, t)\mathrm{d}\mathbf{x}\mathrm{d}\mathbf{x}_0. \tag{9.9}$$

Therefore, defining $\overline{\delta x_i} = \langle x_i\rangle' - \langle x_i\rangle$, we have

$$\overline{\delta x_i}(t) = \iint x_i \frac{\varrho(\mathbf{x}_0 - \Delta\mathbf{x}_0) - \varrho(\mathbf{x}_0)}{\varrho(\mathbf{x}_0)}\varrho(\mathbf{x}_0)w(\mathbf{x}_0, 0 \to \mathbf{x}, t)\mathrm{d}\mathbf{x}\mathrm{d}\mathbf{x}_0 = \langle x_i(t)F(\mathbf{x}_0, \Delta\mathbf{x}_0)\rangle, \tag{9.10}$$

where

$$F(\mathbf{x}_0, \Delta\mathbf{x}_0) = \left[\frac{\varrho(\mathbf{x}_0 - \Delta\mathbf{x}_0) - \varrho(\mathbf{x}_0)}{\varrho(\mathbf{x}_0)}\right]. \tag{9.11}$$

Let us note here that the mixing property of the system is required so that the decay to zero of the time correlation functions assures the switching off of the deviations from equilibrium.

1) The study of an "impulsive" perturbation is not a limitation; for example, in the linear regime from the (differential) linear response, one understands the effect of a generic perturbation.

For an infinitesimal perturbation $\delta x(0) = (\delta x_1(0) \cdots \delta x_N(0))$, if $\varrho(\mathbf{x})$ is non-vanishing and differentiable, the function in (9.11) can be expanded to first order and one obtains

$$\overline{\delta x_i}(t) = -\sum_j \left\langle x_i(t) \frac{\partial \ln \varrho(\mathbf{x})}{\partial x_j}\bigg|_{t=0} \right\rangle \delta x_j(0) \equiv \sum_j R_{i,j}(t)\delta x_j(0), \quad (9.12)$$

which defines the linear response

$$R_{i,j}(t) = -\left\langle x_i(t) \frac{\partial \ln \varrho(\mathbf{x})}{\partial x_j}\bigg|_{t=0} \right\rangle \quad (9.13)$$

of the variable x_i with respect to a perturbation of x_j.

We note that in the above derivation of the FDR relation we never used any approximation on the evolution of $\delta\mathbf{x}(t)$. Starting with the exact expression (9.9) for the response, only a linearization on the initial time perturbed density is needed, and this implies nothing but the smallness of the initial perturbation. We have to stress again that, from the evolution of the trajectories difference, one can define the leading Lyapunov exponent λ by considering the absolute values of $\delta\mathbf{x}(t)$: at small $|\delta\mathbf{x}(0)|$ and large enough t, one has

$$\langle \ln|\delta\mathbf{x}(t)|\rangle \simeq \ln|\delta\mathbf{x}(0)| + \lambda t. \quad (9.14)$$

On the other hand, in the FDR issue one deals with averages of quantities with sign, such as $\overline{\delta\mathbf{x}(t)}$. This apparently marginal difference is very important and it is at the basis of the possibility to derive the FDR relation avoiding the van Kampen's objection.

9.2.3
Remarks on the Invariant Measure

At this point, one could object that in a chaotic deterministic dissipative system the above machinery cannot be applied, because the invariant measure is not smooth at all.[2)] In chaotic dissipative systems, the invariant measure is singular; however, the previous derivation of the FDR relation is still valid if one considers perturbations along the expanding directions. A general response function has two contributions, corresponding, respectively, to the expanding (unstable) and the contracting (stable) directions of the dynamics. The first contribution can be associated with some correlation function of the dynamics on the attractor (i.e., the unperturbed system). On the contrary, this is not true for the second contribution (from the contracting directions); this part of the response is very difficult to extract numerically [21]. Nevertheless, a small amount of noise, which is always present in a physical

2) Typically, the invariant measure of a chaotic attractor has a multifractal character and its Renyi dimensions d_q are not constant.

system, smoothens the $\varrho(\mathbf{x})$ and the FDR relation can be derived. We recall that this "beneficial" noise has the important role of selecting the natural measure, and, in the numerical experiments, it is provided by the round-off errors of the computer. We stress that the assumption on the smoothness of the invariant measure allows one to avoid subtle technical difficulties.

In Hamiltonian systems, taking the canonical ensemble as the equilibrium distribution, one has $\ln \varrho = -\beta H(\mathbf{Q}, \mathbf{P}) + \text{constant}$. Recalling Eq. (9.13), if we indicate there by x_i the component q_k of the position vector $\mathbf{Q}$ and by x_j the corresponding component p_k of the momentum $\mathbf{P}$, from the Hamilton's equations ($\mathrm{d}q_k/\mathrm{d}t = \partial H/\partial p_k$) one has the differential form of the usual FDR [4, 20]

$$\frac{\overline{\delta q_k}(t)}{\delta p_k(0)} = \beta \left\langle q_k(t) \frac{\mathrm{d}q_k(0)}{\mathrm{d}t} \right\rangle = -\beta \frac{\mathrm{d}}{\mathrm{d}t} \langle q_k(t) q_k(0) \rangle . \tag{9.15}$$

When the stationary state is described by a multivariate Gaussian distribution,

$$\ln \varrho(\mathbf{x}) = -\frac{1}{2} \sum_{i,j} \alpha_{ij} x_i x_j + \text{constant}, \tag{9.16}$$

with $\{\alpha_{ij}\}$ a positive matrix, and the elements of the linear response matrix can be written in terms of the usual correlation functions:

$$R_{i,j}(t) = \sum_k \alpha_{i,k} \langle x_i(t) x_k(0) \rangle . \tag{9.17}$$

The above result is nothing but the Onsager regression originally obtained for linear Langevin equations. It is interesting to note that there are important nonlinear systems with a Gaussian invariant distribution, for example, the inviscid hydrodynamics [22, 23], where the Liouville theorem holds, and a quadratic invariant exists.

In non-Hamiltonian systems, where usually the shape of $\varrho(\mathbf{x})$ is not known, relation (9.13) does not give very detailed quantitative information. However, from it one can see the existence of a connection between the mean response function $R_{i,j}$ and a suitable correlation function, computed in the nonperturbed systems:

$$\langle x_i(t) f_j(\mathbf{x}(0)) \rangle, \quad \text{with} \quad f_j(\mathbf{x}) = -\frac{\partial \ln \varrho}{\partial x_j}, \tag{9.18}$$

where, in the general case, the function f_j is unknown.

Let us stress that in spite of the technical difficulty for the determination of the function f_j, which depends on the invariant measure, a FDR always holds in mixing systems whose invariant measure is "smooth enough." We note that the nature of the statistical steady state (either equilibrium or nonequilibrium) has no role at all for the validity of the FDR.

We close this section noting that, as clear even in the case of Gaussian variables, the knowledge of a marginal distribution

$$p_i(x_i) = \int \varrho(x_1, x_2, \ldots) \prod_{j \neq i} \mathrm{d}x_j \tag{9.19}$$

is not enough for the computation of the auto response:

$$R_{i,i}(t) \neq -\left\langle x_i(t) \frac{\partial \ln p_i(x_i)}{\partial x_i}\bigg|_{t=0} \right\rangle. \tag{9.20}$$

This observation is particularly relevant for what follows. Consider that the description of our system has been restricted to a single *slow* degree of freedom, for instance, the coordinate of a colloidal particle ruled by the Langevin equation. The above discussion has indeed made clear that there can be other fluctuating variables coupled to the one we are interested in, and it is not correct to project them out by the marginalization in Eq. (9.19). Conversely, a stationary probability distribution with new variables coupled to the colloidal particle position must be taken into account. In the examples discussed in the last two sections, these additional fluctuating variables turn out to be local fields coupled with the probe particle.

9.2.4
Generalized FDR for Markovian Systems

In the previous section, we have discussed a formula useful to write an explicit expression for the response to an external perturbation at *stationarity* in a case where the invariant measure is known. The possibility to work out explicitly a response formula is indeed quite general. In particular, for Markov processes a general FDR has been derived [24–26], and also experimentally verified [27, 28] (see Chapter 4), which also holds for nonstationary, aging processes, even in the absence of detailed balance. Here we report a derivation that strictly follows the one given in Ref. [24]. An interesting outcome for the purpose of the whole discussion will be the evidence that the response function includes quite generally the correlation of the field of interest with a local field.

Let us briefly recall that a Markov process is univocally defined by the initial distribution probability $\varrho(\mathbf{x}, 0)$ and the transition rates $W(\mathbf{x}' \to \mathbf{x})$ from state $\mathbf{x}'$ to state $\mathbf{x}$, with normalization

$$\sum_{\mathbf{x}} W(\mathbf{x}' \to \mathbf{x}) = 0. \tag{9.21}$$

The two-time conditional probability is then approximated by

$$P(\mathbf{x}, t + \Delta t|\mathbf{x}', t) = \delta_{\mathbf{x},\mathbf{x}'} + W(\mathbf{x}' \to \mathbf{x})\Delta t + \mathcal{O}(\Delta t^2), \tag{9.22}$$

where Δt is a small time interval. The average of a given observable $A(\mathbf{x})$ at time t reads

$$\langle A(t)\rangle = \sum_{\mathbf{x},\mathbf{x}'} A(\mathbf{x}) P(\mathbf{x}, t|\mathbf{x}', t')\varrho(\mathbf{x}', t'). \tag{9.23}$$

Now we discuss a perturbation to our system in the form of a time-dependent external field $h(s)$, which couples to the potential $V(\mathbf{x})$ and changes the energy of the system from $\mathcal{H}(\mathbf{x})$ to $\mathcal{H}(\mathbf{x}) - h(s)V(\mathbf{x})$. The dynamics in the presence of the perturbation is described by a new Markov process defined through the perturbed transition rates $W^h(\mathbf{x}|\mathbf{x}')$. There is not a univocal prescription constraining the choice of a particular form of W^h. Here we focus on the FDR following from the particular choice that obeys the so-called *local detailed balance* [29]. In this case, to linear order in h, the perturbed transition rates read, for $\mathbf{x} \neq \mathbf{x}'$,

$$W^h(\mathbf{x}' \to \mathbf{x}) = W(\mathbf{x}' \to \mathbf{x})\left\{1 - \frac{\beta h}{2}[V(\mathbf{x}) - V(\mathbf{x}')] + M(\mathbf{x}, \mathbf{x}')\right\}, \tag{9.24}$$

where β is the inverse temperature and $M(\mathbf{x}, \mathbf{x}')$ is an arbitrary function of order βh, symmetric with respect to the exchange of its arguments. The diagonal elements are obtained imposing the normalization condition (9.21). Notice that once the local detailed balance is imposed, there remains a further degree of arbitrariness in the choice of the function M. The dependence of the FDR on the form of such function is studied in Refs [24, 26]. Here we focus on the particular case $M = 0$.

Let us consider an impulsive perturbation turned on at time s for the time interval Δt. The linear response function of the observable A is, at $t > s + \Delta t$, with $t - s \gg \Delta t$,

$$R(t, s) = \lim_{\Delta t \to 0} \frac{1}{\Delta t} \frac{\delta \langle A(t) \rangle_h}{\delta h(s)}\bigg|_{h=0}, \tag{9.25}$$

where $\langle \cdot \rangle_h$ denotes an average on the perturbed dynamics. The derivative with respect to h is written in terms of the conditional probabilities as follows:

$$\frac{\delta \langle A(t) \rangle_h}{\delta h(s)}\bigg|_{h=0} = \sum_{\mathbf{x},\mathbf{x}',\mathbf{x}''} A(\mathbf{x}) P(\mathbf{x}, t|\mathbf{x}', s + \Delta t) \frac{\delta P^h(\mathbf{x}', s + \Delta t|\mathbf{x}'', s)}{\delta h}\bigg|_{h=0} P(\mathbf{x}'', s). \tag{9.26}$$

In order to derive a FDR relating the response function to correlation functions computed in the unperturbed dynamics, the derivative with respect to the external field on the right-hand side of Eq. (9.26) has to be worked out explicitly. To do this, we notice that, enforcing the normalization condition (9.21), the transition rates can be separated in diagonal and off-diagonal contributions, namely,

$$W^h(\mathbf{x}' \to \mathbf{x}) = -\delta_{\mathbf{x},\mathbf{x}'} \sum_{\mathbf{x}'' \neq \mathbf{x}'} W^h(\mathbf{x}' \to \mathbf{x}'') + \left(1 - \delta_{\mathbf{x},\mathbf{x}'}\right) W^h(\mathbf{x}' \to \mathbf{x}). \tag{9.27}$$

Then, using Eqs. (9.22),(9.24) and (9.27), one obtains

$$\begin{aligned} \frac{1}{\Delta t} \frac{\delta P^h(\mathbf{x}, s + \Delta t|\mathbf{x}', s)}{\delta h}\bigg|_{h=0} &= \frac{\beta}{2}\Bigg\{\delta_{\mathbf{x},\mathbf{x}'} \sum_{\mathbf{x}'' \neq \mathbf{x}'} W(\mathbf{x}''|\mathbf{x}')[V(\mathbf{x}') - V(\mathbf{x}'')] + (1 - \delta_{\mathbf{x},\mathbf{x}'}) \\ &\quad \times W(\mathbf{x}' \to \mathbf{x})[V(\mathbf{x}) - V(\mathbf{x}')]\Bigg\}. \end{aligned} \tag{9.28}$$

Such separation allows us to single out two different terms. Indeed, substituting Eq. (9.28) into Eq. (9.26), one obtains

$$\left.\frac{\delta\langle A(t)\rangle_h}{\delta h(s)}\right|_{h=0} = \frac{\beta}{2}\sum_{\mathbf{x},\mathbf{x}',\mathbf{x}''} A(\mathbf{x})P(\mathbf{x},t|\mathbf{x}',s+\Delta t)\{\Delta t W(\mathbf{x}'\to\mathbf{x}'')[V(\mathbf{x}')-V(\mathbf{x}'')]$$
$$\times P(\mathbf{x}',s)+\Delta t W(\mathbf{x}''\to\mathbf{x}')[V(\mathbf{x}')-V(\mathbf{x}'')]P(\mathbf{x}'',s)\}. \tag{9.29}$$

Then, exploiting the time translation invariance of the conditional probability, namely, $P(\mathbf{x},t+\Delta t|\mathbf{x}',t) = P(\mathbf{x},t|\mathbf{x}',t-\Delta t)$, in the first line of the above equation, and using Eq. (9.22) in the second one, in the limit $\Delta t \to 0$ one obtains the response function

$$R(t,s) = \frac{\beta}{2}\left[\frac{\partial\langle A(t)V(s)\rangle}{\partial s} - \langle A(t)B(s)\rangle\right], \tag{9.30}$$

where

$$B(s) \equiv B[\mathbf{x}(s)] = \sum_{x''}\{V(\mathbf{x}'') - V[\mathbf{x}(s)]\}W[\mathbf{x}(s)\to\mathbf{x}''] \tag{9.31}$$

is an observable quantity, namely, it depends only on the state of the system at a given time. The relation (9.30) is known since a long time in the context of overdamped Langevin equation for continuous variables [30].

For instance, in the case of the overdamped Langevin dynamics of a colloidal particle diffusing in a space-dependent potential $U(x)$.

$$\dot{x}(t) = -\frac{\partial U(x)}{\partial x} + \sqrt{2T}\zeta(t), \tag{9.32}$$

with $\zeta(t)$ a white noise with zero mean and unit variance, one can derive a response formula, with respect to a perturbing force F, analogous to Eq. (9.30) that reads as

$$\frac{\delta\langle x(t)\rangle_F}{\delta F(s)} = \frac{\beta}{2}\left[\frac{\partial\langle x(t)x(s)\rangle}{\partial s} - \langle x(t)B[x(s)]\rangle\right], \tag{9.33}$$

with $B[x(s)] = -\partial U/\partial x|_{x(s)}$. At equilibrium, it can be easily proved that $\langle x(t)B[x(s)]\rangle = -\partial\langle x(t)x(s)\rangle/\partial s$, recovering the standard FDT formula. Differently, out of equilibrium the contribution coming from the local field $B[x(s)]$ must be explicitly taken into account, as will be shown in examples in the next sections.

9.3
Random Walk on a Comb Lattice

9.3.1
Anomalous Diffusion and FDR

As discussed above for Brownian motion, in the absence of external forcing one has, for large times $t \to \infty$,

$$\langle x(t)\rangle = 0, \qquad \langle x^2(t)\rangle \simeq 2Dt, \tag{9.34}$$

where x is the position of the Brownian particle and the average is taken over the unperturbed dynamic. Once a small constant external force F is applied, one has a linear drift

$$\overline{\delta x}(t) = \langle x(t)\rangle_F - \langle x(t)\rangle \simeq \mu F t, \tag{9.35}$$

where $\langle\cdot\rangle_F$ indicates the average on the perturbed system, and μ is the mobility of the colloidal particle. It is remarkable that $\langle x^2(t)\rangle$ is proportional to $\overline{\delta x}(t)$ at any time:

$$\frac{\langle x^2(t)\rangle}{\overline{\delta x}(t)} = \frac{2}{\beta F}, \tag{9.36}$$

and the Einstein relation (see Eq. (9.4)) holds.

On the other hand, it is now well established that beyond the standard diffusion, as in (9.34), one can have systems with anomalous diffusion (see, for instance, Refs [31–36] (see Chapter 8)), that is,

$$\langle x^2(t)\rangle \sim t^{2\nu} \quad \text{with} \quad \nu \neq 1/2. \tag{9.37}$$

Formally, this corresponds to having $D = \infty$ if $\nu > 1/2$ (superdiffusion) and $D = 0$ if $\nu < 1/2$ (subdiffusion). In the following, we will limit the study to the case $\nu < 1/2$.

It is quite natural to wonder if (and how) the FDR changes in the presence of anomalous diffusion, that is, if instead of (9.34), Eq. (9.37) holds. In some systems, it has been shown that (9.36) holds even in the subdiffusive case. This has been explicitly proved in systems described by a fractional Fokker–Planck equation [37] (see also Refs [38, 39]). In addition, there is clear analytical [40] and numerical [41] evidence that (9.36) is valid for the elastic single file, that is, a gas of hard rods on a ring with elastic collisions, driven by an external thermostat, which exhibits subdiffusive behavior, $\langle x^2\rangle \sim t^{1/2}$ [42].

Here we discuss the validity of the fluctuation–dissipation relation in the form (9.36) for a system with anomalous diffusion that is not fully described by a fractional Fokker–Planck equation. In particular, we will investigate the relevance of the anomalous diffusion, the presence of nonequilibrium conditions, and the (possible) role of finite size. The example discussed here is the diffusion of a particle on a comb lattice. The dynamics of the model is defined by transition rates and therefore a straightforward application of the generalized FDR introduced in Section 9.2.4, Eq. (9.30), is possible.

9.3.2
Transition Rates of the Model

The comb lattice is a discrete structure consisting of an infinite linear chain (backbone), the sites of which are connected with other linear chains (teeth) of length L [43]. We denote by $x \in (-\infty, \infty)$ the position of the particle performing the random walk along the backbone and with $y \in [-L, L]$ that along a tooth. The transition probabilities from state $\mathbf{x} \equiv (x, y)$ to $\mathbf{x}' \equiv (x', y')$ are

$$\begin{aligned} W^d[(x,0) \to (x \pm 1, 0)] &= 1/4 \pm d, \\ W^d[(x,0) \to (x, \pm 1)] &= 1/4, \\ W^d[(x,y) \to (x, y \pm 1)] &= 1/2 \quad \text{for} \quad y \neq 0, \pm L. \end{aligned} \tag{9.38}$$

On the boundaries of each tooth, $y = \pm L$, the particle is reflected with probability 1. The case $L = \infty$ is obtained in numerical simulations by letting the y coordinate increase without boundaries. Here we consider a discrete time process and, of course, the normalization $\sum_{(x',y')} W^d[(x,y) \to (x',y')] = 1$ holds. The parameter $d \in [0, 1/4]$ allows us to consider also the case where a constant external field is applied along the x-axis, producing a nonzero drift of the particle. A state with a nonzero drift can be considered as a perturbed state (in that case we denote the perturbing field by ε), or it can be itself the stationary state where a further perturbation can be added changing $d \to d + \varepsilon$.

9.3.3 Anomalous Dynamics

Let us start by considering as a reference state the case $d = 0$. For finite teeth length $L < \infty$, we have numerical evidence of a dynamical crossover, at a time t^*, from a subdiffusive to a simple diffusive asymptotic behavior (see Figure 9.1)

$$\langle x^2(t) \rangle_0 \simeq \begin{cases} C t^{1/2}, & t < t^*(L), \\ 2D(L)t, & t > t^*(L), \end{cases}$$

where C is a constant and $D(L)$ is an effective diffusion coefficient depending on L. The symbol $\langle \cdot \rangle_0$ denotes an average over different realizations of the dynamics (9.38)

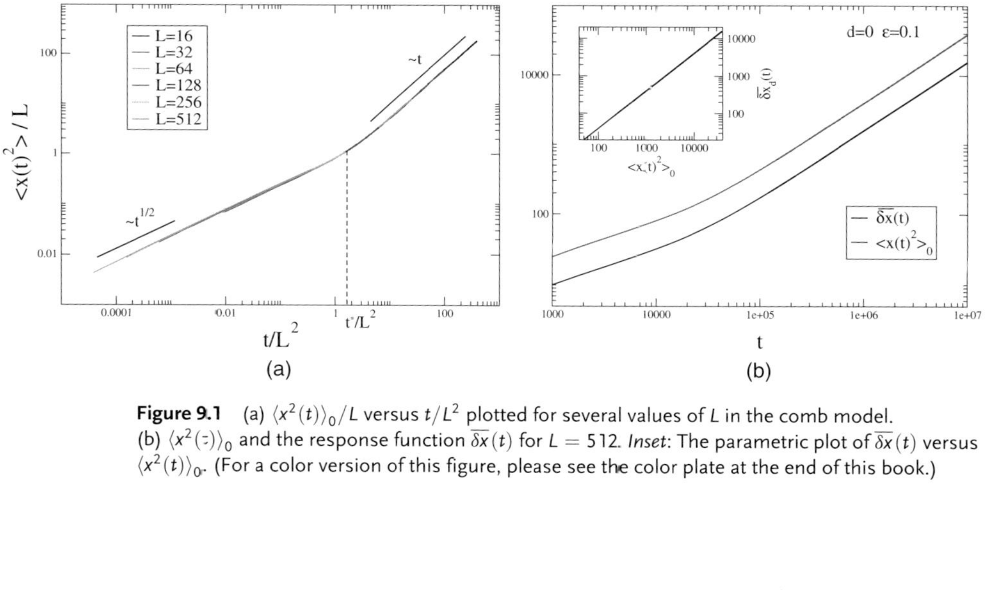

Figure 9.1 (a) $\langle x^2(t) \rangle_0 / L$ versus t/L^2 plotted for several values of L in the comb model. (b) $\langle x^2(t) \rangle_0$ and the response function $\overline{\delta x}(t)$ for $L = 512$. *Inset*: The parametric plot of $\overline{\delta x}(t)$ versus $\langle x^2(t) \rangle_0$. (For a color version of this figure, please see the color plate at the end of this book.)

with $d = 0$ and initial condition $x(0) = y(0) = 0$. We find $t^*(L) \sim L^2$ and $D(L) \sim 1/L$. In Figure 9.1a, we plot $\langle x^2(t)\rangle_0/L$ as a function of t/L^2 for several values of L, showing an excellent data collapse.

In the limit of infinite teeth, $L \to \infty$, $D \to 0$, and $t^* \to \infty$ and the system shows a pure subdiffusive behavior [44]

$$\langle x^2(t)\rangle_0 \sim t^{1/2}. \tag{9.39}$$

In this case, the probability distribution function behaves as

$$P_0(x,t) \sim t^{-1/4}\, \mathrm{e}^{-c(|x|/t^{1/4})^{4/3}}, \tag{9.40}$$

where c is a constant, in agreement with an argument *à la Flory* [31]. The behavior (9.40) also holds in the case of finite L, provided that $t < t^*$. For larger times, a simple Gaussian distribution is observed.

In the comb model with infinite teeth, the FDR in its standard form is fulfilled; namely, if we apply a constant perturbation ε pulling the particles along the 1D lattice, one has numerical evidence that

$$\langle x^2(t)\rangle_0 \simeq C\overline{\delta x}(t) \sim t^{1/2}, \tag{9.41}$$

where $\overline{\delta x}$ is the average in the presence of the perturbation ε. Moreover, the proportionality between $\langle x^2(t)\rangle_0$ and $\overline{\delta x}(t)$ is fulfilled also with $L < \infty$, where both the mean square displacement (MSD) and the drift with an applied force exhibit the same crossover from subdiffusive, $\sim t^{1/2}$, to diffusive behavior, $\sim t$ (see Figure 9.1b). Therefore, we can say that the FDR is somehow "blind" to the dynamical crossover experienced by the system. When the perturbation is applied to a state without any current, the proportionality between response and correlation holds despite anomalous transport phenomena.

9.3.4 Application of the Generalized FDR

Our aim here is to show that, differently from that depicted above about the zero current situation, within a state with a nonzero drift the emergence of a dynamical crossover is connected to the breaking of the Einstein FDR (9.4). Indeed, the MSD in the presence of a nonzero current, even with $L = \infty$, shows a dynamical crossover

$$\langle x^2(t)\rangle_d \sim a\, t^{1/2} + b\, t, \tag{9.42}$$

$$\langle [x(t) - \langle x(t)\rangle_d]^2\rangle_d \sim a' t^{1/2} + b' t, \tag{9.43}$$

where a, b, a', and b' are constants, whereas

$$\overline{\delta x}_d(t) \sim t^{1/2}, \tag{9.44}$$

with $\overline{\delta x}_d(t) = \langle x(t)\rangle_{d+\varepsilon} - \langle x(t)\rangle_d$: hence, at large times the Einstein relation breaks down (see Figure 9.2). Notice that the proportionality between response and fluctuations cannot be recovered by simply replacing $\langle x^2(t)\rangle_d$ with $\langle x^2(t)\rangle_d - \langle x(t)\rangle_d^2$, as it

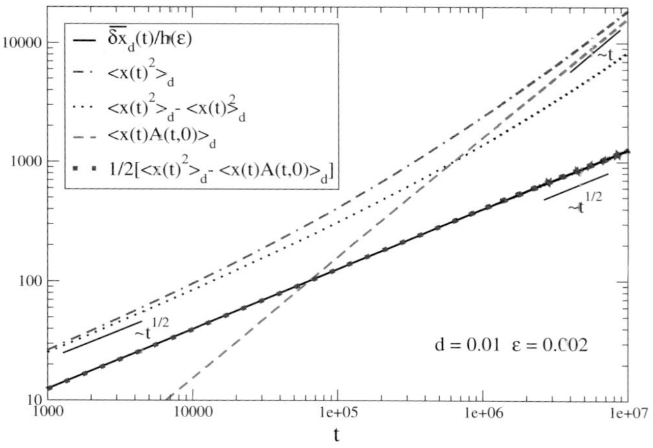

Figure 9.2 Response function (solid line), MSD (dash-dotted line), and second cumulant (dotted line) measured in the comb model with $L = \infty$, field $d = 0.01$, and perturbation $\varepsilon = 0.002$. The correlation involving the quantity B (dashed line) yields the right correction to recover the full response function (line with "■" symbols), in agreement with the FDR (9.47).

happens for Gaussian processes (see discussion below). The constants a' and b' can be computed analytically in the case $L = \infty$ (E. Barkai, private communication).

The first moment $\langle x(t)\rangle_{d+\varepsilon}$ of $P_{d+\varepsilon}(x,t)$ is not proportional to the second cumulant of $P_d(x,t)$, namely, $\langle x^2(t)\rangle_d - \langle x(t)\rangle_d^2$. In order to point out a relation between such quantities, we need a generalized fluctuation–dissipation relation.

According to the definition (9.38), one has for the backbone

$$\begin{aligned} W^{d+\varepsilon}[(x,y)\to(x',y')] &= W^d[(x,y)\to(x',y')]\left(1+\frac{\varepsilon(x'-x)}{W^0+d(x'-x)}\right) \\ &\simeq W^d\,\mathrm{e}^{(\varepsilon/W^0)(x'-x)}, \end{aligned} \tag{9.45}$$

where $W^0 = 1/4$, and the last expression holds under the condition $d/W^0 \ll 1$. The above expression can be rewritten in the form of a *local detailed balance* (Eq. (9.24)), with $V(x) = x$ and $\beta = 1$, yielding, for $(x,y) \neq (x',y')$,

$$W^{d+\varepsilon}[(x,y)\to(x',y')] = W^d[(x,y)\to(x',y')]\mathrm{e}^{(h(\varepsilon)/2)(x'-x)}, \tag{9.46}$$

where $h(\varepsilon) = 2\varepsilon/W^0$. Using Eqs. (9.30) and (9.31), we obtain

$$\frac{\overline{\delta\mathcal{O}}_d}{h(\varepsilon)} = \frac{\langle\mathcal{O}(t)\rangle_{d+\varepsilon}-\langle\mathcal{O}(t)\rangle_d}{h(\varepsilon)} = \frac{1}{2}\left[\langle\mathcal{O}(t)x(t)\rangle_d - \langle\mathcal{O}(t)x(0)\rangle_d - \langle\mathcal{O}(t)A(t,0)\rangle_d\right], \tag{9.47}$$

where $\mathcal{O}$ is a generic observable, and $A(t,0) = \sum_{t'=0}^{t} B(t')$, with, in this case,

$$B[(x,y)] = \sum_{(x',y')} (x'-x)\,W^d[(x,y)\to(x',y')]. \tag{9.48}$$

Recalling the definitions (9.38), from the above equation we have $B[(x,y)] = 2d\delta_{y,0}$ and therefore the sum on B has an intuitive meaning: it counts the time spent by the particle on the x-axis. The results described in Section 9.3.3 can then be read in the light of the fluctuation–dissipation relation (9.47):

i) Putting $\mathcal{O}(t) = x(t)$, in the case without drift, that is, $d = 0$, one has $B = 0$ and, recalling the choice of the initial condition $x(0) = 0$,

$$\frac{\overline{\delta x}}{h(\varepsilon)} = \frac{\langle x(t)\rangle_\varepsilon - \langle x(t)\rangle_0}{h(\varepsilon)} = \frac{1}{2}\langle x^2(t)\rangle_0. \tag{9.49}$$

This explains the observed behavior (9.41) even in the anomalous regime and predicts the correct proportionality factor, $\overline{\delta x}(t) = \varepsilon/W^0\langle x^2(t)\rangle_0$.

ii) Putting $\mathcal{O}(t) = x(t)$, in the case with $d \neq 0$, one has

$$\frac{\overline{\delta x}_d}{h(\varepsilon)} = \frac{1}{2}\left[\langle x^2(t)\rangle_d - \langle x(t)A(t,0)\rangle_d\right]. \tag{9.50}$$

This explains the observed behaviors (9.42) and (9.44): the leading behavior at large times of $\langle x^2(t)\rangle_d \sim t$ turns out to be exactly canceled by the term $\langle x(t)A(t,0)\rangle_d$, so that the relation between response and unperturbed correlation functions is recovered (see Figure 9.2).

iii) As discussed above, it is not enough to substitute $\langle x^2(t)\rangle_d$ with $\langle x^2(t)\rangle_d - \langle x(t)\rangle_d^2$ to recover the proportionality with $\overline{\delta x}_d(t)$ when the process is not Gaussian. This can be explained in the following manner. By making use of the second-order out-of-equilibrium FDR derived by Lippiello *et al.* in Refs [45–47], which is needed due to the vanishing of the first-order term for symmetry, we can explicitly evaluate

$$\langle x^2(t)\rangle_d = \langle x^2(t)\rangle_0 + h^2(d)\frac{1}{2}\left[\frac{1}{4}\langle x^4(t)\rangle_0 + \frac{1}{4}\langle x^2(t)A^{(2)}(t,0)\rangle_0\right], \tag{9.51}$$

where $A^{(2)}(t,0) = \sum_{t'=0}^{t} B^{(2)}(t')$ with $B^{(2)} = -\sum_{x'}(x'-x)^2 W[(x,y) \to (x',y')] = -1/2\delta_{y,0}$. Then, recalling Eq. (9.49), we obtain

$$\langle x^2(t)\rangle_d - \langle x(t)\rangle_d^2 = \langle x^2(t)\rangle_0 + h^2(d)\left[\frac{1}{8}\langle x^4(t)\rangle_0 + \frac{1}{8}\langle x^2(t)A^{(2)}(t,0)\rangle_0 - \frac{1}{4}\langle x^2(t)\rangle_0^2\right]. \tag{9.52}$$

Numerical simulations show that the term in the square brackets grows like t yielding a scaling behavior with time consistent with Eq. (9.43). On the other hand, in the case of the simple random walk, one has $B^{(2)} = -1$ and $A^{(2)}(t,0) = -t$ and then

$$\langle x^2(t)\rangle_d - \langle x(t)\rangle_d^2 = \langle x^2(t)\rangle_0 + h^2(d)\left[\frac{1}{8}\langle x^4(t)\rangle_0 - \frac{1}{8}t\langle x^2(t)\rangle_0 - \frac{1}{4}\langle x^2(t)\rangle_0^2\right]. \tag{9.53}$$

Since in the Gaussian case $\langle x^4(t)\rangle_0 = 3\langle x^2(t)\rangle_0^2$ and $\langle x^2(t)\rangle_0 = t$, the term in the square brackets vanishes identically and that explains why, in the presence of a drift, the second cumulant grows *exactly* as the second moment with no drift.

9.4
Entropy Production

Nonequilibrium regimes are always characterized by some sort of current flowing across the systems. In the following sections, we will suggest, showing examples, some connections between such nonequilibrium currents and the new degrees of freedom entering the fluctuation–dissipation relation, while in this section we introduce a general symmetry relation of the probability distribution of such currents, the fluctuation relation.

In general, when one deals with nonequilibrium dynamics, very few results independent of the details of the model are available. Actually, in the past decade, a group of relations known as "fluctuation relations" have captured the interest of the scientific community, especially for the vast range of applicability. Initially, a numerical evidence given by Evans and collaborators showed a particular symmetry in the distribution of an observable of a molecular fluid under shear. In a second moment, such a symmetry has been proved as a theorem by Gallavotti and Cohen, under quite general hypothesis. The interested reader can see, among others, Refs [48] (see Chapter 1) (see Chapter 3). For a review of large deviation theory applied to nonequilibrium systems, (see Chapter 11).

According to the point of view adopted here, we will focus on systems in which some randomness is present. Thanks to this assumption, it is possible to skip several technical problems and some forms of fluctuation theorems for stochastic systems can be used. Among others, we are going to use the Lebowitz–Spohn functional introduced for Markovian dynamics. In order to fix ideas, let us consider a one-dimensional process discrete in time (the generalization to the other cases is straightforward) and let us identify a trajectory in a time interval $[0, t]$:

$$\Omega_t = \{x(0), x(1), \ldots, x(t)\}. \tag{9.54}$$

Clearly, because of the stochastic nature of the dynamics, it is possible to associate with the trajectories a probability $P(\Omega_t)$; using the Markovian nature of the process, one has

$$P(\Omega_t) = p(x(0)) \prod_{n=1}^{t} W(x(n-1) \to x(n)). \tag{9.55}$$

Analogously, one can consider the time-reversed trajectory, that is, $\overline{\Omega}_t = x(t), x(t-1), \ldots, x(0)$ with its probability $P(\overline{\Omega}_t)$. For clarity, here we are considering variables that are *even* in time. This is not the most general case: for example, when one deals also with velocities one must take into account the different parity of the variables. Then the Lebowitz–Spohn functional is defined as

$$\Sigma_t = \log \frac{P(\Omega_t)}{P(\overline{\Omega}_t)}. \tag{9.56}$$

Note that this quantity, in general, is not related to a specific thermodynamic observable. However, in what follows, we will call it "entropy production" according to the recent literature. If the dynamics satisfies detailed balance conditions, one has that $P(\Omega_t) = P(\overline{\Omega}_t)$ and then Σ_t is identically equal to zero. On the contrary, when the detailed balance condition is broken, its average is strictly positive and does increase in time:

$$\langle \Sigma_t \rangle \sim t \quad \text{for } t \text{ large}, \tag{9.57}$$

where the angle brackets mean an average on the space of trajectories in the stationary ensemble. In this sense, one of the main features of this quantity is that it captures the "nonequilibrium" nature of the system.

Let us discuss with a pedagogical example how the entropy production is related to nonequilibrium currents. Consider a Markov process where the perturbation of an external force F inducing nonequilibrium currents enters in the transition rates according to *local detailed balance* condition

$$\frac{W_F(x \to x')}{W_F(x' \to x)} = \frac{W_0(x \to x')}{W_0(x' \to x)} e^{2\beta F j(x \to x')}, \tag{9.58}$$

with $j(x \to x')$ the current associated with the transition $x \to x'$, which obeys the symmetry property $j(x \to x') = -j(x' \to x)$. According to the definition of entropy production (9.56), one finds, for large times,

$$\frac{\Sigma_t}{t} = 2\beta F \frac{1}{t} \sum_{n=1}^{t} J(x(n-1) \to x(n)) = 2\beta F J(t), \tag{9.59}$$

where $J(t)$ is the average current over a time window of duration t. The fluctuation theorem is a symmetry property of the probability distribution of the variable $y = \Sigma_t / t$, which reads

$$\frac{P(y)}{P(-y)} = e^y \Rightarrow \frac{P(2\beta F J(t))}{P(-2\beta F J(t))} = e^{2\beta F J(t)}. \tag{9.60}$$

Namely, the fluctuation theorem describes a symmetry in the fluctuations of currents. Also, for large times we can assume a large deviation hypothesis $P(y) \sim e^{-tS(y)}$, with $S(y)$ a Cramer function. For small fluctuations around the mean value of y, the Cramer function can be approximated to $S(y) = S(2\beta F J) \simeq \beta^2 F^2 (J - \bar{J})^2 / \sigma_J^2$, where σ_J is the variance. The fluctuation relation reads as $S(y) - S(-y) = y$; in the Gaussian limit (y close to $\bar{y}$), the previous constraint can be easily demonstrated to be equivalent to $\bar{J}/F = \beta \sigma_J^2$, which is nothing but the standard fluctuation–dissipation relation. Therefore, the fluctuation relation, which in some simple situation can be directly related to the fluctuation–dissipation relation, is a more general symmetry expected to be satisfied by the fluctuating entropy production. For a more general discussion of the link between the Lebowitz–Spohn entropy production and currents, see Refs [49, 50]. The remarkable fact appearing in Eq. (9.60) is that it does not contain any free parameter, and so, in this sense, is model independent.

9.5 Langevin Processes without Detailed Balance

Sometimes the properties of statistical systems are investigated studying the diffusional behavior of probe particles that customarily have larger mass and size compared with the constituents of the environment. This is, for instance, the case of Brownian motion, in which the erratic motion of flower dust within water molecules was considered. The central point is that such probe particle is always coupled to a *small* portion of the whole system, so that small-scale fluctuations always do matter. We are going to study the feedback of local fluctuations of the fluid surrounding the probe on the dynamics of the probe itself when the environment is out of equilibrium. The dynamics of a single probe particle in contact with an equilibrium thermal bath is well described by a single stochastic differential equation:

$$M\dot{V}(t) = -\int_{-\infty}^{t} \mathrm{d}t'\,\Gamma(t-t')V(t') + \mathcal{E}(t'), \tag{9.61}$$

where $\mathcal{E}(t)$ is a Gaussian white noise with zero mean and variance $\langle \mathcal{E}(t)\mathcal{E}(t')\rangle = \Gamma(|t-t'|)$ (FDT of the second kind). The equilibrium condition is always guaranteed by the proportionality between memory kernel and noise autocorrelation. The simplest way to account for the coupling between our probe particle and *nonequilibrium* fluctuations of the environments is by breaking the FDT of the second kind.

In the rest of the section, we will discuss the following points:

- How nonequilibrium arises in a multidimensional linear Langevin model?
- How the coupling between the variable of interest and others does matter in the response formula?
- What form is taken by the fluctuation-response relation in such a system?

9.5.1 Markovian Linear System

The coupling between the probe and the portion of fluid surrounding it, Eq. (9.61), in certain regimes may be effectively described in terms of two coupled Langevin equations: one variable is the velocity of our tagged particle and the other is a local force field. Such a simplification can be realized in not too dense cases, where the memory kernel Γ has a single characteristic finite time: a more detailed discussion can be found in Refs [51, 52] and a specific example, in the context of granular systems, is given in Section 9.6. The system can then be put in the following form [53]:

$$\begin{aligned} M\dot{V} &= -\Gamma(V-U) + \sqrt{2\Gamma T_1}\phi_1, \\ M'\dot{U} &= -\Gamma' U - \Gamma V + \sqrt{2\Gamma' T_2}\phi_2, \end{aligned} \tag{9.62}$$

where M and M' are masses, Γ and Γ' are sort of viscosities, and T_1 and T_2 are two different energy scales.

Model (9.62), in a more compact form, reads

$$\frac{\mathrm{d}\mathbf{X}}{\mathrm{d}t} = -A\mathbf{X} + \phi, \tag{9.63}$$

where $\mathbf{X} \equiv (X_1, X_2)$, $\phi \equiv (\phi_1, \phi_2)$ are two-dimensional vectors, A is a real 2×2 matrix, $\phi(t)$ a Gaussian process, with zero mean and covariance matrix:

$$\langle \phi_i(t')\phi_j(t)\rangle = 2\delta(t-t')D_{ij}, \tag{9.64}$$

and

$$A = \begin{pmatrix} \dfrac{\Gamma}{M} & -\dfrac{\Gamma}{M} \\[2ex] \dfrac{\Gamma}{M'} & \dfrac{\Gamma'}{M'} \end{pmatrix}, \qquad D = \begin{pmatrix} \dfrac{\Gamma T_1}{M^2} & 0 \\[2ex] 0 & \dfrac{\Gamma' T_2}{(M')^2} \end{pmatrix}. \tag{9.65}$$

The stability conditions on the dynamical matrix that guarantee the existence of a stationary state are $\mathrm{Tr}(A) > 0$ and $\det(A) > 0$, which are clearly fulfilled in the present case. The invariant measure of the steady state is represented by a 2-variate Gaussian distribution:

$$\varrho(\mathbf{X}) = N \exp\left(-\frac{1}{2}\mathbf{X}\sigma^{-1}\mathbf{X}\right), \tag{9.66}$$

where N is a normalization coefficient and the matrix of covariances σ is obtained by solving

$$D = A\sigma + \sigma A^T, \tag{9.67}$$

which yields

$$\sigma = \begin{pmatrix} \dfrac{T}{M} + \Theta\Delta T & \Theta\Delta T \\[2ex] \Theta\Delta T & \dfrac{T_1}{M'} + \dfrac{\Gamma}{\Gamma'}\Theta\Delta T \end{pmatrix}, \tag{9.68}$$

where $\Theta = \Gamma\Gamma'/(\Gamma+\Gamma')(M'\Gamma + M\Gamma')$ and $\Delta T = T_1 - T_2$. It is now clear that when $T_1 = T_2$ the two variables are uncorrelated. This is the fingerprint of an equilibrium condition, as we shall see in the following sections.

9.5.2 Fluctuation–Response Relation

We can now explicitly study the fluctuation and response properties of the system since the dynamics is linear. First, the correlation matrix $C_{ij}(t,t') = \langle X_i(t)X_j(t')\rangle$ in the stationary state is time-translational invariant; that is, it depends only on the time difference $t-t'$. Then, using the equation of motion, it is easy to verify that $\dot{C}(t) = -AC(t)$, with initial condition given by the covariance matrix, that is, $C(0) = \sigma$. The corresponding solution (in the matrix form) is $C(t) = \mathrm{e}^{-At}\sigma$. Note that, in general, σ and A do not commute. Moreover, considering the Gaussian shape of the steady-state distribution function (9.66), it is possible to calculate the response function of the system, by a straightforward application of (9.17):

$$\mathbf{R}(t) \equiv \frac{\overline{\delta X_i(t)}}{\delta X_j(0)} = \sum_{i,j} \sigma_{ij}^{-1} \langle X_i(t) X_j(0) \rangle = \mathbf{C}(t)\sigma^{-1}. \tag{9.69}$$

In the case of our interest, by considering a perturbation applied to the variable V, one obtains

$$\frac{\overline{\delta V(t)}}{\delta V(0)} = \sigma_{11}^{-1} \langle V(t) V(0) \rangle + \sigma_{21}^{-1} \langle V(t) U(0) \rangle. \tag{9.70}$$

As it appears clear from Eq. (9.70), the response of the variable V to an external perturbation in general is not proportional to the unperturbed autocorrelation. It is possible to observe the following scenario: in the case $T_1 = T_2 \equiv T$, that is, $\Delta T = 0$, σ is diagonal with $\sigma_{11} = T/M$ and $\sigma_{22} = T/M'$, independently of the values of other parameters and a direct proportionality between C_{VV} and R_{VV} is obtained. This is not the only case for this to happen: also for $\Gamma \ll \Gamma'$, or $\Gamma \gg \Gamma'$, σ_{12} goes to zero and a direct proportionality between response and autocorrelation of velocity is recovered. More in general, when $T_2 \neq T_1$, the coupling term σ_{12} differs from zero and a "violation" of the Einstein relation emerges, triggered by the coupling between different degrees of freedom.

The situation is clarified by Figure 9.3, where the Einstein relation is violated and the generalized FDR holds: response $R_{VV}(t)$, when plotted against $C_{VV}(t)$, shows a nonlinear (and nonmonotonic) graph. Anyway, a simple linear plot is restored when the response is plotted against the linear combination of correlations indicated by formula (9.70). In this case, it is evident that the "violation" cannot be

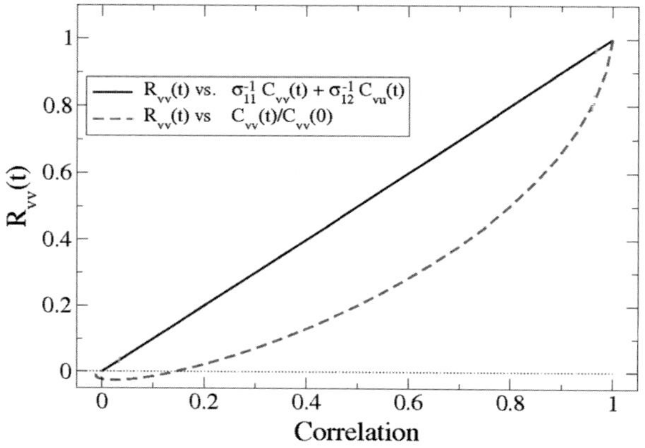

Figure 9.3 Free particle with viscosity and memory, whose dynamics is given by Eq. (9.62): here we show the parametric plot of the velocity response to an impulsive perturbation at time 0, versus two different correlations. The Einstein relation, which is not satisfied, would correspond to a linear shape with slope 1 for the dashed line. The solid line shows that the generalized response relation holds.

interpreted by means of any effective temperature [54]: on the contrary, it is a consequence of having "missed" the coupling between variables V and U, which gives an additive contribution to the response of V.

The above consideration can be easily generalized to the case where several auxiliary fields are present: let us suppose that there are $N-1$ variables at temperature T_1 (the velocity V and $N-2$ auxiliary variables U_i) and one at temperature T_2. In this case, it is possible to show that the off-diagonal terms in σ are proportional to $(T_1 - T_2)$ and a qualitative similar analysis can be performed.

It is useful to stress the role of Markovianity, which is relevant for a correct prediction of the response. In fact, the marginal probability distribution of velocity $P_m(V)$ can be computed straightforwardly and has always a Gaussian shape. By that, one could be tempted to conclude, inserting $P_m(V)$ inside, that proportionality between response and correlation holds also if $T_2 \neq T_1$. This conclusion is wrong, as shown in Eq. (9.20), because the process is Markovian only if both the variable V and the "hidden" variable U are considered.

We conclude by mentioning a sentence of Onsager: *How do you know you have taken enough variables, for it to be Markovian?* [55]. We have seen in this section how a correct response–fluctuation analysis is intimately related to this important *caveat*.

9.5.3
Entropy Production

Due to the linearity of the model (9.63), the explicit calculation of the entropy production according to the Lebowitz–Spohn definition is quite straightforward. There is a one-to-one mapping between the trajectories of the system and the realizations of the noise that allows to write the probability of each trajectory as a path integral over all the possible realizations of noise:

$$P(\{\mathbf{X}(s)\}_0^t) = \int \mathcal{D}\phi P(\phi)\delta(\dot{\mathbf{X}} \to +A\mathbf{X} - \phi). \tag{9.71}$$

Due to the Gaussianity of $P(\phi)$, the above integral is easily evaluated yielding the well-known *Onsager–Machlup* expression for the path probabilities:

$$P(\{\mathbf{X}(s)\}_0^t) \sim \mathrm{e}^{-(\dot{\mathbf{X}}+A\mathbf{X})D^{-1}(\dot{\mathbf{X}}+A\mathbf{X})}. \tag{9.72}$$

In our model, it is possible to compute the entropy production functional of a single trajectory. Let us consider a general trajectory $\{\mathbf{X}(s)\}_0^t$ and its time reversal $\{\bar{\mathbf{X}}(s)\}_0^t$. Lebowitz and Spohn defined a fluctuating entropy production functional W_t as follows [50]:

$$\Sigma'_t = \log \frac{p[\mathbf{X}(0)]P(\{\mathbf{X}(s)\}_0^t)}{p[\mathbf{X}(t)]P(\{\bar{\mathbf{X}}(s)\}_0^t)} = W_t + b_t, \tag{9.73}$$

with

$$b_t = \log\{fp[\mathbf{X}(0)]\} - \log\{fp[(\mathbf{X})]\}, \tag{9.74}$$

where $p[\mathbf{X}(0)]$ is the stationary distribution, that is, a bivariate Gaussian with covariance given by Eq. (9.67). Lebowitz and Spohn have shown that the average (over the steady ensemble) of W_t, if detailed balance is not satisfied, increases linearly with time, while the term b_t, usually known as "border term," is usually negligible for large times, unless particular conditions of "singularity" occur [56–58].

For simplicity of the notation, let us define

$$\begin{pmatrix} F_1(\mathbf{X}) \\ F_2(\mathbf{X}) \end{pmatrix} \equiv -A\mathbf{X}.$$

In order to write down an explicit expression for the entropy production, it is necessary to establish parity of variables under time reversal (i.e., positions are even and velocities are odd under time inversion transformation). Let us assume that under time reversal $\overline{X_i} = \varepsilon_i X_i$, where ε_i can be $+1$ or -1, using also $\varepsilon\mathbf{X} \equiv (\varepsilon_1 X_1, \varepsilon_2 X_2)$. Then it is possible to define

$$F_i^{\mathrm{rev}}(\mathbf{X}) = \frac{1}{2}[F_i(\mathbf{X}) - \varepsilon_i F_i(\varepsilon\mathbf{X})] = -\varepsilon_i F_i^{\mathrm{rev}}(\varepsilon\mathbf{X}), \tag{9.75}$$

$$F_i^{\mathrm{ir}}(\mathbf{X}) = \frac{1}{2}[F_i(\mathbf{X}) + \varepsilon_i F_i(\varepsilon\mathbf{X})] = \varepsilon_i F_i^{\mathrm{ir}}(\varepsilon\mathbf{X}). \tag{9.76}$$

Given this notation [59], it is possible to write down a compact form for the entropy production[3] by simply substituting Eq. (9.72) into (9.73) and obtaining

$$\Sigma_t = \sum_k D_{kk}^{-1} \int_0^t \mathrm{d}s\, F_k^{\mathrm{ir}}[\dot{X}_k - F_k^{\mathrm{rev}}], \tag{9.77}$$

where the sum is over k such that $D_{kk} \neq 0$. Formula (9.77) is valid also in the presence of nonlinear terms and with several variables [52].

For the model system in Eq. (9.62), it appears clear that the variable V, being velocity, is odd under the change of time, while variable U, which represents the local force field (divided by Γ), is even. According to the above formula, the entropy production reads as

$$\Sigma_t \simeq \Gamma\left(\frac{1}{T_2} - \frac{1}{T_1}\right) \int_0^t V(t')U(t')\mathrm{d}t'. \tag{9.78}$$

Note that, from (9.78), one recovers the equilibrium case, that is, $T_1 = T_2$, for which there is no entropy production. Remarkably, the equality of the two temperatures is the same equilibrium condition given by the FDT analysis.

9.6
Granular Intruder

Models of granular fluids [60] are an interesting framework where the issues discussed in the previous sections can be addressed. Due to dissipative interactions

3) Note that if the correlation matrix of the noise is not diagonal, this formula is slightly different.

among the microscopic constituents, energy is not conserved and external sources are necessary in order to maintain a nontrivial stationary state: time reversal invariance is broken, and consequently, properties such as the equilibrium fluctuation–dissipation relation (EFDR) do not hold. In recent years, a rather complete theory for fluctuations in granular systems has been developed for the dilute limit, in good agreement with numerical simulations [61, 62]. However, a general understanding of dense cases is still lacking. A common approach is the so-called Enskog correction [61, 63], which renormalizes the collision frequency to take into account the breakdown of molecular chaos due to high density. In cooling regimes, the Enskog theory may describe strong nonequilibrium effects, due to the explicit cooling time dependence [64]. However, it cannot describe dynamical effects in stationary regimes, such as violations of the Einstein relation [65, 66]. Indeed, as discussed before, violations of Einstein relation comes from having neglected *coupled* degrees of freedom that are expected to be independent at equilibrium. The Enskog approximation is not sufficient to describe the effect of such a coupling, because it does not break factorization of velocities:

$$\varrho(x_1,\dots x_N, v_1,\dots v_N) = \varrho_x(x_1,\dots,x_N)\prod_{i=1}^{N} p_1(v_i), \tag{9.79}$$

with $p_1(v)$ the single-particle velocity distribution. Such an approximation, which predicts the validity of Einstein relation, is revealed to be wrong already at moderate densities in granular fluids, as discussed in the following.

9.6.1 Model

A good benchmark for new nonequilibrium theories is provided by the dynamics of a massive tracer interacting with a gas of smaller granular particles, both coupled to an external bath. In particular, taking as a reference point the dilute limit, where the system has a closed analytical description [67], it is shown that more dense configurations are well described by a generalized Langevin equation (GLE) with an exponential memory kernel of the form given in Eq. (9.61), at least as a first approximation capable of describing violations of EFDR and other nonequilibrium properties [53]. Here, the main features are as follows:

- the decay of correlation and response functions is not a simple exponential and shows backscattering [68, 69];
- the EFDRs [70, 71] of the first and the second kind do not hold.

In the model described here, detailed balance is not satisfied in general, nonequilibrium effects can be taken into account, and the correct behavior of correlation and response functions is predicted. Furthermore, the model has a remarkable property: it can be mapped onto the two-variable Markovian system discussed in the previous section, that is, two coupled Langevin equations, as in Eq. (9.62). The dilute limit is then naturally recovered by putting to zero the coupling constant between the original variable and the auxiliary one. The auxiliary variable can be identified in the local velocity field spontaneously appearing in the surrounding fluid. This allows us to measure the fluctuating entropy production (see Eq. (9.56)) [72] and fairly verify the fluctuation relation (9.60) [50, 71, 73], a remarkable result, if considered the interest of the community [74] and compared with unsuccessful past attempts [75, 76].

The model considered here [53] is the following: an "intruder" disk of mass $m_0 = M$ and radius R, moving in a gas of N granular disks with mass $m_i = m$ $(i > 0)$ and radius r, in a two-dimensional box of area $A = L^2$. We denote by $n = N/A$ the number density of the gas and by ϕ the occupied volume fraction, that is, $\phi = \pi(Nr^2 + R^2)/A$, and we denote by $\boldsymbol{V}$ (or $\boldsymbol{v}_0$) and $\boldsymbol{v}$ (or $\boldsymbol{v}_i$ with $i > 0$) the velocity vectors of the tracer and the gas particles, respectively. Interactions among the particles are hard-core binary instantaneous inelastic collisions, such that particle i, after a collision with particle j, comes out with a velocity

$$\boldsymbol{v}_{i'} = \boldsymbol{v}_i - (1+\alpha)\frac{m_j}{m_i + m_j}\left[(\boldsymbol{v}_i - \boldsymbol{v}_j)\cdot\hat{\boldsymbol{n}}\right]\hat{\boldsymbol{n}}, \tag{9.80}$$

where $\hat{\boldsymbol{n}}$ is the unit vector joining the particles' centers of mass and $\alpha \in [0, 1]$ is the restitution coefficient ($\alpha = 1$ is the elastic case). The mean free path of the intruder is proportional to $l_0 = 1/((r + R)n)$ and we denote by τ_c its mean collision time. Two kinetic temperatures can be introduced for the two species: the gas granular temperature $T_{\mathrm{g}} = m\langle \boldsymbol{v}^2\rangle/2$ and the tracer temperature $T_{\mathrm{tr}} = M\langle \boldsymbol{V}^2\rangle/2$.

In order to maintain a granular medium in a fluidized state, an external energy source is coupled to each particle in the form of a thermal bath [77–79] (hereafter, exploiting isotropy, we consider only one component of the velocities):

$$m_i\dot{v}_i(t) = -\gamma_{\mathrm{b}} v_i(t) + f_i(t) + \xi_{\mathrm{b}}(t). \tag{9.81}$$

Here $f_i(t)$ is the force taking into account the collisions of particle i with other particles, and $\xi_{\mathrm{b}}(t)$ is a white noise (different for all particles), with $\langle \xi_{\mathrm{b}}(t)\rangle = 0$ and $\langle \xi_{\mathrm{b}}(t)\xi_{\mathrm{b}}(t')\rangle = 2T_{\mathrm{b}}\gamma_{\mathrm{b}}\delta(t - t')$. The effect of the external energy source balances the energy lost in the collisions and a stationary state is attained with $m_i\langle v_i^2\rangle \leq T_{\mathrm{b}}$.

At low packing fractions, $\phi < 0.1$, and in the large mass limit, $m/M \ll 1$, using the Enskog approximation it has been shown [67] that the dynamics of the intruder is described by a linear Langevin equation:

$$M\dot{V} = -\Gamma_{\mathrm{E}} V + \mathcal{E}_{\mathrm{E}}, \tag{9.82}$$

where $\mathcal{E}_{\mathrm{E}}$ is a white noise such that

$$\langle \mathcal{E}_i(t)\mathcal{E}_j(t')\rangle = 2\left[\gamma_{\mathrm{b}} T_{\mathrm{b}} + \gamma_{\mathrm{g}}^{\mathrm{E}}\left(\frac{1+\alpha}{2}T_{\mathrm{g}}\right)\right]\delta_{ij}\delta(t - t') \tag{9.83}$$

and

$$\Gamma_{\mathrm{E}} = \gamma_{\mathrm{b}} + \gamma_{\mathrm{g}}^{\mathrm{E}}, \quad \text{with} \quad \gamma_{\mathrm{g}}^{\mathrm{E}} = \frac{g_2(r+R)}{l_0}\sqrt{2\pi m T_{\mathrm{g}}}(1+\alpha), \tag{9.84}$$

where $g_2(r+R)$ is the pair correlation function for a gas particle and the intruder at contact. In this limit, the velocity autocorrelation function shows a simple exponential decay, with characteristic time M/Γ_E. Time reversal and the EFDR, which are very weakly modified for uniform dilute granular gases [65, 80, 81], become perfectly satisfied for a massive intruder. The temperature of the tracer is computed as $T_{tr}^E = [\gamma_b T_b + \gamma_g^E((1+\alpha)/2)T_g]/\Gamma_E$. For a general study of a Langevin equation with "two temperatures" but a single timescale (which is always at equilibrium), see also Ref. [82].

9.6.2
Dense Case: Double Langevin with Two Temperatures

As the packing fraction is increased, the Enskog approximation is less and less effective in predicting memory effects and the dynamical properties. In particular, velocity autocorrelation $C(t) = \langle V(t)V(0)\rangle/\langle V^2\rangle$ and linear response function $R(t) = \overline{\delta V(t)/\delta V(0)}$ (i.e., the mean response at time t to an impulsive perturbation applied at time 0) show an exponential decay modulated in amplitude by oscillating functions [69]. Moreover, violations of the EFDR $C(t) = R(t)$ (Einstein relation) are observed for $\alpha < 1$ [41, 66].

Molecular dynamics simulations [53] of the system have been performed, giving access to $C(t)$ and $R(t)$, for several different values of the parameters α and ϕ (see Figure 9.4).

Notice that the Enskog approximation [61, 67] cannot predict the observed functional forms, because it only modifies by a constant factor the collision frequency.

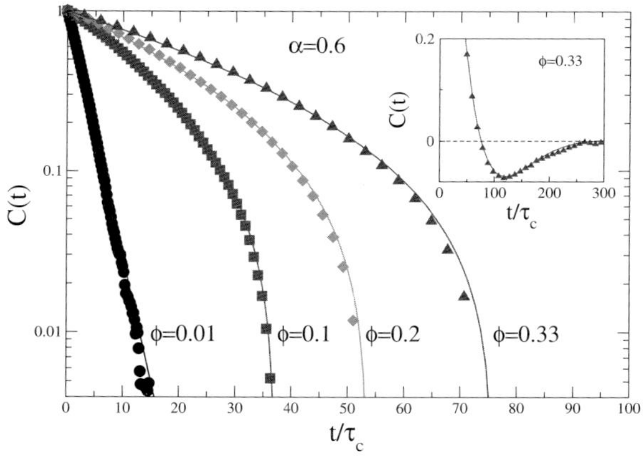

Figure 9.4 Semi-log plot of $C(t)$ (symbols) for different values of $\phi = 0.01, 0.1, 0.2$, and 0.33 at $\alpha = 0.6$. Times are rescaled by the mean collision time τ_c. The other parameters are fixed: $N = 2500$, $m = 1$, $M = 25$, $r = 0.005$, $R = 0.025$, $T_b = 1$, and $\gamma_b = 200$. Times are rescaled by the mean collision times τ_c, as measured in the different cases. Numerical data are averaged over $\sim 10^5$ realizations. Continuous lines are the best fits obtained with Eq. (9.88). *Inset*: $C(t)$ and the best fit in linear scale for $\phi = 0.33$ and $\alpha = 0.6$.

In order to describe the full phenomenology, a model with more than one characteristic time is needed. The proposed model is a Langevin equation with a single exponential memory kernel [53, 83, 84], as in Eq. (9.61):

$$M\dot{V}(t) = -\int_{-\infty}^{t} dt'\, \Gamma(t-t')V(t') + \mathcal{E}'(t). \tag{9.85}$$

In this case,

$$\Gamma(t) = 2\gamma_0\delta(t) + \gamma_1/\tau_1\, \mathrm{e}^{-t/\tau_1} \tag{9.86}$$

and $\mathcal{E}'(t) = \mathcal{E}_0(t) + \mathcal{E}_1(t)$, with

$$\langle \mathcal{E}_0(t)\mathcal{E}_0(t')\rangle = 2T_0\gamma_0\delta(t-t'), \qquad \langle \mathcal{E}_1(t)\mathcal{E}_1(t')\rangle = T_1\gamma_1/\tau_1 \mathrm{e}^{-(t-t')/\tau_1}, \tag{9.87}$$

and $\langle \mathcal{E}_1(t)\mathcal{E}_0(t')\rangle = 0$. In the limit $\alpha \to 1$, the parameter T_1 is meant to tend to T_0 in order to fulfill the EFDR of the second kind: $\langle \mathcal{E}'(t)\mathcal{E}'(t')\rangle = T_0\Gamma(t-t')$. Within this model, the dilute case is recovered if $\gamma_1 \to 0$. In this limit, the parameters γ_0 and T_0 coincide with Γ_{E} and $T_{\mathrm{tr}}^{\mathrm{E}}$ of the Enskog theory [67]. The exponential form of the memory kernel can be justified within the mode-coupling approximation scheme (see Ref. [53] for details).

The model (9.85) predicts $C = f_C(t)$ and $R = f_R(t)$ with

$$f_{C(R)} = \mathrm{e}^{-gt}[\cos(\omega t) + a_{C(R)} \sin(\omega t)]. \tag{9.88}$$

The variables g, ω, a_C, and a_R are known algebraic functions of γ_0, T_0, γ_1, τ_1, and T_1. In particular, the ratio $a_C/a_R = [T_0 - \Omega(T_1 - T_0)]/[T_0 + \Omega(T_1 - T_0)]$, with $\Omega = \gamma_1/[(\gamma_0 + \gamma_1)(\gamma_0/M\tau_1 - 1)]$. Hence, in the elastic ($T_1 \to T_0$) as well as in the dilute limit ($\gamma_1 \to 0$), one gets $a_C = a_R$ and recovers the EFDR $C(t) = R(t)$. In Figure 9.4, the continuous lines show the result of the best fits obtained using Eq. (9.88) for the correlation function, at restitution coefficient $\alpha = 0.6$ and for different values of the packing fraction ϕ. The functional form fits very well the numerical data. A fit of measured C and R against Eq. (9.88), together with a measure of $\langle V^2\rangle$, yields five independent equations to determine the five parameters entering the model. We used the external parameters mentioned before, changing α or the box area A (to change ϕ) or the intruder's radius R, in order to change ϕ keeping or not keeping constant $\gamma_g \sim 1/l_0 \to 0$ (indeed different dilute limits can be obtained, where collisions matter or not). Fits from numerical simulations suggest the following identification for the parameters: $\gamma_0 \sim \Gamma_{\mathrm{E}}$, $T_0 \sim T_{\mathrm{tr}}$, and $T_1 \sim T_{\mathrm{b}}$. The coupling time τ_1 increases with the packing fraction and, weakly, with the inelasticity. In the most dense cases, it appears that $\gamma_1 \sim \gamma_{\mathrm{g}}^{\mathrm{E}} \propto \phi$: such an observation, however, does not hold in the dilute limit at constant collision rate, where $\gamma_1 \to 0 \ll \gamma_{\mathrm{g}}^{\mathrm{E}}$. It is also interesting to notice that at high density $T_{\mathrm{tr}} \sim T_{\mathrm{g}} \sim T_{\mathrm{g}}^{\mathrm{E}}$, which is probably due to the stronger correlations among particles. Finally, we notice that, at large ϕ, $T_{\mathrm{tr}} > T_{\mathrm{tr}}^{E}$, which is coherent with the idea that correlated collisions dissipate *less* energy.

Looking for an insight of the relevant physical mechanisms underlying such a phenomenology and in order to make clear the meaning of the parameters, it is useful to map Eq. (9.85) onto a Markovian equivalent model by introducing an auxiliary field (see Refs [51, 53] for details on how the mapping is achieved):

$$\begin{aligned} M\dot{V} &= -\Gamma_E(V-U) + \sqrt{2\Gamma_E T_g}\,\mathcal{E}_V, \\ M'\dot{U} &= -\Gamma' U - \Gamma_E V + \sqrt{2\Gamma' T_b}\,\mathcal{E}_U, \end{aligned} \tag{9.89}$$

where $\mathcal{E}_V$ and $\mathcal{E}_U$ are independent white noises of unitary variance and we have exploited the numerical observations discussed above (i.e., $T_0 \sim T_g$, $T_1 \sim T_b$, and $\gamma_0 \sim \Gamma_E$), while $\Gamma' = \gamma_0^2/\gamma_1$ and $M' = \gamma_0^2\tau_1/\gamma_1$. In the chosen form (9.89), the dynamics of the tracer is remarkably simple: indeed V follows a memoryless Langevin equation in a *Lagrangian frame* with respect to a local field U, which is the *local average velocity field* of the gas particles colliding with the tracer. Extrapolating such an identification to higher densities, we are able to understand the value for most of the parameters of the model: the self-drag coefficient of the intruder in principle is not affected by the change of reference to the Lagrangian frame, so that $\gamma_0 \sim \Gamma_E$; for the same reason, $T_0 \sim T_{tr}$ is roughly the temperature of the tracer; τ_1 is the main relaxation time of the average velocity field U around the Brownian particle; γ_1 is the intensity of coupling felt by the surrounding particles after collisions with the intruder; finally, T_1 is the "temperature" of the local field U, easily identified with the bath temperature $T_1 \sim T_b$: indeed, thanks to momentum conservation, inelasticity does not affect the average velocity of a group of particles that almost only collide with themselves.

9.6.3
Generalized FDR and Entropy Production

The model discussed above, Eq. (9.89), is able to reproduce the violations of EFDR, as shown in Figure 9.5, which depicts correlation and response functions in a

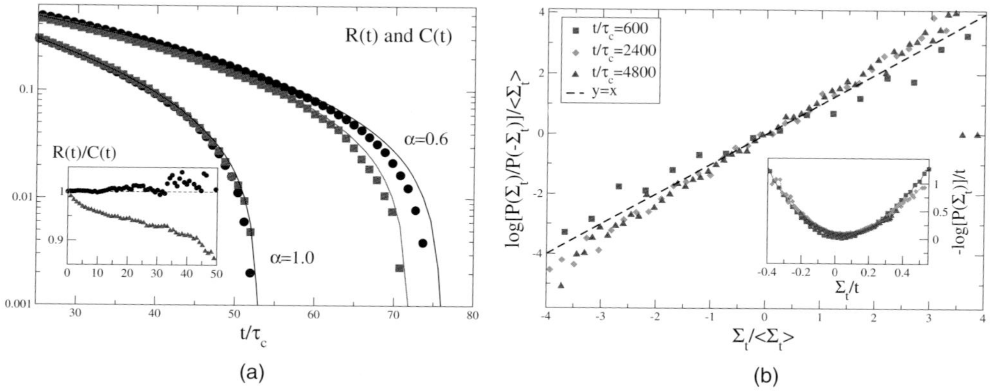

Figure 9.5 (a) Correlation function $C(t)$ (•) and response function $R(t)$ (■) for $\alpha = 1$ and $\alpha = 0.6$, at $\phi = 0.33$. Continuous lines show the best fits curves obtained with Eq. (9.88). *Inset*: The ratio $R(t)/C(t)$ is reported in the same cases. (b) Check of the fluctuation relation (9.92) in the system with $\alpha = 0.6$ and $\phi = 0.33$. *Inset*: Collapse of the rescaled probability distributions of Σ_t at large times onto the large deviation function.

dense case (elastic and inelastic). In the inelastic case, deviations from EFDR $R(t) = C(t)$ are observed. In the inset of Figure 9.5, the ratio $R(t)/C(t)$ is also reported. As shown in Section 9.2, a relation between the response and correlations measured in the unperturbed system still exists, but – in the nonequilibrium case – must take into account the contribution of the cross-correlation $\langle V(t)U(0)\rangle$, that is,

$$R(t) = aC(t) + b\langle V(t)U(0)\rangle, \tag{9.90}$$

with $a = [1 - \gamma_1/M(T_0 - T_1)\Omega_a]$ and $b = (T_0 - T_1)\Omega_b$, where Ω_a and Ω_b are known functions of the parameters. At equilibrium, where $T_0 = T_1$, the EFDR is recovered.

An important independent assessment of the effectiveness of model in Eq. (9.89) comes from the study of the fluctuating entropy production [72] that quantifies the deviation from detailed balance in a trajectory. Given the trajectory in the time interval $[0, t]$, $\{V(s)\}_0^t$, and its time reversal $\{\mathcal{I}V(s)\}_0^t \equiv \{-V(t-s)\}_0^t$, as discussed in Section 9.5.3, the entropy production for the model (9.89) takes the form [52]

$$\Sigma_t = \log \frac{P(\{V(s)\}_0^t)}{P(\{\mathcal{I}V(s)\}_0^t)} \approx \gamma_0 \left(\frac{1}{T_0} - \frac{1}{T_1}\right) \int_0^t \mathrm{d}s\, V(s)U(s). \tag{9.91}$$

This functional vanishes exactly in the elastic case, $\alpha = 1$, where equipartition holds, $T_1 = T_0$, and is zero on average in the dilute limit, where $\langle VU\rangle = 0$.

Formula (9.91) reveals that the leading source of entropy production is the energy transferred by the "force" $\gamma_0 U$ on the tracer, weighed by the difference between the inverse temperatures of the two "thermostats." Therefore, to measure entropy production, we need to measure the fluctuations of U: from the above discussion we have for U the interpretation of a spontaneous local velocity field interacting with the tracer. Therefore, it can be measured by performing a local average of particles' velocity in a circle of radius l centered on the tracer. Details on how to choose in a reliable way the proper l are given in Ref. [53]; for instance, following such procedure, in the case $\phi = 0.33$ and $\alpha = 0.6$, we estimate for the correlation length $l \sim 9r \sim 6l_0$. Then, measuring the entropy production from Eq. (9.91) along many trajectories of length t, we can compute the probability $P(\Sigma_t = x)$ and compare it to $P(\Sigma_t = -x)$, in order to verify the fluctuation relation

$$\log \frac{P(\Sigma_t = x)}{P(\Sigma_t = -x)} = x. \tag{9.92}$$

In Figure 9.5b, we report our numerical results. The main frame confirms that at large times the fluctuation relation (9.92) is well verified within the statistical errors. The inset shows the collapse, for large times, of $\log P(\Sigma_t)/t$ onto the Cramer function $S(y = \Sigma_t/t)$, introduced in Section 9.4. Notice also that formula (9.91) does not contain further parameters but the ones already determined by correlation and response measure; that is, the slope of the graph is not adjusted by further fits. Indeed, a wrong evaluation of the weighing factor $(1/T_0 - 1/T_1) \approx (1/T_{\mathrm{tr}} - 1/T_{\mathrm{b}})$ or of the "energy injection rate" $\gamma_0 U(t)V(t)$ in Eq. (9.91) could produce a completely different slope in Figure 9.5b.

To conclude this section, we stress that velocity correlations $\langle V(t)U(t')\rangle$ between the intruder and the surrounding velocity field are responsible for both the violations of the EFDR and the appearance of a nonzero entropy production, provided that the two fields are *at different temperatures.* We also mention that larger violations of EFDR can be observed using an intruder with a mass equal or similar to that of other particles [66], with the important difference that in such a case a simple "Langevin-like" model for the intruder's dynamics is not available.

9.7
Conclusions and Perspectives

We have reviewed a series of recent results on fluctuation–dissipation relations for out-of-equilibrium systems. The leitmotiv of our discussion is the importance of correlations for nonequilibrium response, much more relevant than in equilibrium systems: indeed the generalized fluctuation–dissipation relation discussed in Sections 9.3, 9.5, and 9.6 deviates from its equilibrium counterpart for the appearance of additional contributions coming from correlated degrees of freedom. This is the case, for instance, of the linear response in the diffusion model on a comb lattice, where – in the presence of a net drift – the linear response takes a non-negligible additive contribution (see Eq. (9.50)). The same occurs in the general two-variable Langevin model: additional contribution to the equilibrium linear response appears when the main field V is coupled to the second field U, which only appears out of equilibrium (see Eq. (9.70)). The granular intruder is a realistic many-body instance of such a coupling scenario, where we have seen the strong influence of correlated degrees of freedom to the linear response (see Eq. (9.89)).

Remarkably, in some cases, one may explicitly verify that the coupled field that contributes to the linear response in nonequilibrium setups is also involved in the violation of detailed balance: such a violation is measured by the fluctuating entropy production. The connection between nonequilibrium couplings and entropy production has been discussed for general coupled linear Langevin models (see Eq. (9.78)) and then verified for the entropy production of the granular intruder (Eq. (9.91)).

In conclusion, many results point in the same direction, suggesting a general framework for linear response in systems with nonzero entropy production. Even in out-of-equilibrium configurations, a clear connection between response and correlation in the unperturbed systems exists. A further step is looking for more accessible observables that could make easier the prediction of linear response: indeed, both proposed formulas, Eqs. (9.13) and (9.30), require the measurement of variables that, in general, depend on full phase space (microscopic) observables and can be strictly model dependent. Such a difficulty also explains why, in a particular class of slowly relaxing systems with several well-separated timescales, such as spin or structural glasses in the aging dynamics, that is, after a sudden quench below some dynamical transition temperature, approaches involving "effective temperatures" have been used in a more satisfactory way [85]. More recent interpretations of the additional nonequilibrium contributions have been proposed in Ref. [25], but its predictive power is not yet fully investigated and represents an interesting line of ongoing research.

Acknowledgments

We thank A. Baldassarri, F. Corberi, M. Falcioni, E. Lippiello, U. Marini Bettolo Marconi, L. Rondoni, and M. Zannetti for a long collaboration on the issues discussed here.

References

1 Einstein, A. (1905) On the movement of small particles suspended in a stationary liquid demanded by the molecular-kinetic theory of heat. *Ann. Phys.*, **17**, 549.

2 Onsager, L. (1931) Reciprocal relations in irreversible processes. I. *Phys. Rev.*, **37**, 405–426.

3 Onsager, L. (1931) Reciprocal relations in irreversible processes. II. *Phys. Rev.*, **38**, 2265–2279.

4 Kubo, R. (1966) The fluctuation–dissipation theorem. *Rep. Prog. Phys.*, **29**, 255.

5 Evans, D.J., Cohen, E.G.D., and Morriss, G.P. (1993) Probability of second law violations in shearing steady flows. *Phys. Rev. Lett.*, **71**, 2401.

6 Evans, D.J. and Searles, D.J. (1995) Steady states, invariant measures, and response theory. *Phys. Rev. E*, **52**, 5839.

7 Gallavotti, G. and Cohen, E.G.D. (1995) Dynamical ensembles in stationary states. *J. Stat. Phys.*, **80**, 931.

8 Crooks, G.E. (2000) Path ensemble averages in systems driven far from equilibrium. *Phys. Rev. E*, **61**, 2361.

9 Jarzynski, C. (1997) Nonequilibrium equality for free energy differences. *Phys. Rev. Lett.*, **78**, 2690.

10 Boltzmann, L. (1995) *Lectures on Gas Theory*, Dover (1896 first German edition).

11 Gibbs, J.W. (1902) *Elementary Principles in Statistical Mechanics*, Yale University Press.

12 Einstein, A. (1904) Zur allgemeinen molekularen theorie der wärme. *Ann. Phys.*, **14**, 354.

13 Perrin, J. (1913) *Les Atomes*, Alcan, Paris.

14 Langevin, P. (1908) Sur la theorie du mouvement brownien. *C. R. Acad. Sci. (Paris)*, **146**, 530; translated in *Am. J. Phys.* **65**, 1079 (1997)

15 Rose, R.H. and Sulem, P.L. (1978) Fully developed turbulence and statistical mechanics. *J. Phys. (Paris)*, **39**, 441.

16 Deker, U. and Haake, F. (1975) Fluctuation–dissipation theorems for classical processes. *Phys. Rev. A*, **11**, 2043.

17 Falcioni, M., Isola, S., and Vulpiani, A. (1990) Correlation functions and relaxation properties in chaotic dynamics and statistical mechanics. *Phys. Lett. A*, **144**, 341.

18 Boffetta, G., Lacorata, G., Musacchio, S., and Vulpiani, A. (2003) Relaxation of finite perturbations: beyond the fluctuation–response relation. *Chaos*, **13**, 806.

19 van Kampen, N.G. (1971) The case against linear response theory. *Phys. Norv.*, **5**, 279.

20 Kubo, R. (1986) Brownian motion and nonequilibrium statistical mechanics. *Science*, **32**, 2022.

21 Cessac, B. and Sepulchre, J.-A. (2007) Linear response, susceptibility and resonance in chaotic toy models. *Physica D*, **225**, 13.

22 Kraichnan, R.H. (1959) Classical fluctuation–relaxation theorem. *Phys. Rev.*, **113**, 118.

23 Kraichnan, R.H. (2000) Deviations from fluctuation–relaxation relations. *Physica A*, **279** 30.

24 Lippiello, E., Corberi, F., and Zannetti, M. (2005) Off-equilibrium generalization of

the fluctuation dissipation theorem for Ising spins and measurement of the linear response function. *Phys. Rev. E*, **71**, 036104.

25 Baiesi, M., Maes, C., and Wynants, B. (2009) Fluctuations and response of nonequilibrium states. *Phys. Rev. Lett.*, **103**, 010602.

26 Corberi, F., Lippiello, E., Sarracino, A., and Zannetti, M. (2010) Fluctuation–dissipation relations and field-free algorithms for the computation of response functions. *Phys. Rev. E*, **81**, 011124.

27 Gomez-Solano, J.R., Petrosyan, A., Ciliberto, S., Chetrite, R., and Gawedzki, K. (2009) Experimental verification of a modified fluctuation–dissipation relation for a micron-sized particle in a non-equilibrium steady state. *Phys. Rev. Lett.*, **103**, 040601.

28 Gomez-Solano, J.R., Petrosyan, A., Ciliberto, S., and Maes, C. (2011) Fluctuations and response in a non-equilibrium micron-sized system. *J. Stat. Mech.*, P01008.

29 Lebowitz, J.L. and Bergmann, P.G. (1957) Irreversible Gibbsian ensembles. *Ann. Phys.*, **1**, 1.

30 Cugliandolo, L.F., Kurchan, J., and Parisi, G. (1994) Off equilibrium dynamics and aging in unfrustrated systems. *J. Phys. I*, **4**, 1641.

31 Bouchaud, J.P. and Georges, A. (1990) Anomalous diffusion in disordered media: statistical mechanisms, models and physical applications. *Phys. Rep.*, **195**, 127.

32 Gu, Q., Schiff, E.A., Grebner, S., Wang, F., and Schwarz, R. (1996) Non-Gaussian transport measurements and the Einstein relation in amorphous silicon. *Phys. Rev. Lett.*, **76**, 3196.

33 Castiglione, P., Mazzino, A., Muratore-Ginanneschi, P., and Vulpiani, A. (1999) On strong anomalous diffusion. *Physica D*, **134**, 75.

34 Metzler, R. and Klafter, J. (2000) The random walk's guide to anomalous diffusion: a fractional dynamics approach. *Phys. Rep.*, **339**, 1.

35 Burioni, R. and Cassi, D. (2005) Random walks on graphs: ideas, techniques and results. *J. Phys. A: Math. Gen.*, **38**, R45.

36 Bénichou, O. and Oshanin, G. (2002) Ultraslow vacancy-mediated tracer diffusion in two dimensions: the Einstein relation verified. *Phys. Rev. E*, **66**, 031101.

37 Metzler, R., Barkai, E., and Klafter, J. (1999) Anomalous diffusion and relaxation close to thermal equilibrium: a fractional Fokker–Planck equation approach. *Phys. Rev. Lett*, **82**, 3563.

38 Barkai, E. and Fleurov, V.N. (1998) Generalized Einstein relation: a stochastic modeling approach. *Phys. Rev. E*, **58**, 1296.

39 Chechkin, A.V. and Klages, R. (2009) Fluctuation relations for anomalous dynamics. *J. Stat. Mech.*, L03002.

40 Lizana, L., Ambjörnsson, T., Taloni, A., Barkai, E., and Lomholt, M.A. (2010) Foundation of fractional Langevin equation: harmonization of a many-body problem. *Phys. Rev. E*, **81**, 51118.

41 Villamaina, D., Puglisi, A., and Vulpiani, A. (2008) The fluctuation–dissipation relation in sub-diffusive systems: the case of granular single-file diffusion. *J. Stat. Mech.*, L10001.

42 Hahn, K., Kärger, J., and Kukla, V. (1996) Single-file diffusion observation. *Phys. Rev. Lett.*, **76**, 2762.

43 Redner, S. (2001) *A Guide to First-Passages Processes*, Cambridge University Press.

44 Havlin, S. and Ben Avraham, D. (1987) Diffusion in disordered media. *Adv. Phys.*, **36**, 695.

45 Lippiello, E., Corberi, F., Sarracino, A., and Zannetti, M. (2008) Nonlinear susceptibilities and the measurement of a cooperative length. *Phys. Rev. B*, **77**, 212201.

46 Lippiello, E., Corberi, F., Sarracino, A., and Zannetti, M. (2008) Nonlinear response and fluctuation–dissipation relations. *Phys. Rev. E*, **78**, 041120.

47 Corberi, F., Lippiello, E., Sarracino, A., and Zannetti, M. (2010) Fluctuations of two-time quantities and non-linear response functions. *J. Stat. Mech.*, P04003.

48 Evans, D.J. and Searles, D.J. (2002) The fluctuation theorem. *Adv. Phys.*, **52**, 1529.

49 Andrieux, D. and Gaspard, P. (2007) Fluctuation theorem for currents and Schnakenberg network theory. *J. Stat. Phys.*, **127**, 107.

50 Lebowitz, J.L. and Spohn, H. (1999) A Gallavotti–Cohen-type symmetry in the large deviation functional for stochastic dynamics. *J. Stat. Phys.*, **95**, 333.

51 Villamaina, D., Baldassarri, A., Puglisi, A., and Vulpiani, A. (2009) Fluctuation–dissipation relation: how does one compare correlation functions and responses? *J. Stat. Mech.*, P07024.

52 Puglisi, A. and Villamaina, D. (2009) Irreversible effects of memory. *Europhys. Lett.*, **88**, 30004.

53 Sarracino, A., Villamaina, D., Gradenigo, G., and Puglisi, A. (2010) Irreversible dynamics of a massive intruder in dense granular fluids. *Europhys. Lett.*, **92**, 34001.

54 Cugliandolo, L.F., Kurchan, J., and Peliti, L. (1997) Energy flow, partial equilibration, and effective temperatures in systems with slow dynamics. *Phys. Rev. E*, **55**, 3898.

55 Onsager, L. and Machlup, S. (1953) Fluctuations and irreversible processes. *Phys. Rev.*, **91**, 1505.

56 van Zon, R. and Cohen, E.G.D. (2003) Extension of the fluctuation theorem. *Phys. Rev. Lett.*, **91**, 110601.

57 Puglisi, A., Rondoni, L., and Vulpiani, A. (2006) Relevance of initial and final conditions for the fluctuation relation in Markov processes. *J. Stat. Mech.*, P08010.

58 Bonetto, F., Gallavotti, G., Giuliani, A., and Zamponi, F. (2006) Chaotic hypothesis, fluctuation theorem and singularities. *J. Stat. Phys.*, **123**, 39.

59 Risken, H. (1989) *The Fokker–Planck Equation: Methods of Solution and Applications*, Springer, Berlin.

60 Jaeger, H.M., Nagel, S.R., and Behringer, R.P. (1996) Granular solids, liquids, and gases. *Rev. Mod. Phys.*, **68**, 1259.

61 Brilliantov, N.K. and Poschel, T. (2004) *Kinetic Theory of Granular Gases*, Oxford University Press.

62 Brey, J.J., Maynar, P., and Garcia de Soria, M.I. (2009) Fluctuating hydrodynamics for dilute granular gases. *Phys. Rev. E*, **79**, 051305.

63 Dufty, J.W. and Santos, A. (2006) Dynamics of a hard sphere granular impurity. *Phys. Rev. Lett.*, **97**, 058001.

64 Santos, A. and Dufty, J.W. (2001) Critical behavior of a heavy particle in a granular fluid. *Phys. Rev. Lett.*, **86**, 4823.

65 Garzó, V. (2004) On the Einstein relation in a heated granular gas. *Physica A*, **343**, 105.

66 Puglisi, A., Baldassarri, A., and Vulpiani, A. (2007) Violations of the Einstein relation in granular fluids: the role of correlations. *J. Stat. Mech.*, P08016.

67 Sarracino, A., Villamaina, D., Costantini, G., and Puglisi, A. (2010) Granular Brownian motion. *J. Stat. Mech.*, P04013.

68 Orpe, A.V. and Kudrolli, A. (2007) Velocity correlations in dense granular flows observed with internal imaging. *Phys. Rev. Lett.*, **98**, 238001.

69 Fiege, A., Aspelmeier, T., and Zippelius, A. (2009) Long-time tails and cage effect in driven granular fluids. *Phys. Rev. Lett.*, **102**, 098001.

70 Kubo, R., Toda, M., and Hashitsume, N. (1991) *Statistical Physics II: Nonequilibrium Statistical Mechanics*, Springer.

71 Marini Bettolo Marconi, U., Puglisi, A., Rondoni, L., and Vulpiani, A. (2008) Fluctuation–dissipation: response theory in statistical physics. *Phys. Rep.*, **461**, 111.

72 Seifert, U. (2005) Entropy production along a stochastic trajectory and an integral fluctuation theorem. *Phys. Rev. Lett.*, **95**, 040602.

73 Kurchan, J. (1998) Fluctuation theorem for stochastic dynamics. *J. Phys. A*, **31**, 3719.

74 Bonetto, F., Gallavotti, G., Giuliani, A., and Zamponi, F. (2006) Fluctuations relation and external thermostats: an application to granular materials. *J. Stat. Mech.*, P05009.

75 Feitosa, K. and Menon, N. (2004) Fluidized granular medium as an instance of the fluctuation theorem. *Phys. Rev. Lett.*, **92**, 164301.

76 Puglisi, A., Visco, P., Barrat, A., Trizac, E., and van Wijland, F. (2005) Fluctuations of internal energy flow in a vibrated granular gas. *Phys. Rev. Lett.*, **95**, 110202.

77 Williams, D.R.M. and MacKintosh, F.C. (1996) Driven granular media in one dimension: correlations and equation of state. *Phys. Rev. E*, **54**, R9.

78 van Noije, T.P.C., Ernst, M.H., Trizac, E., and Pagonabarraga, I. (1999) Randomly driven granular fluids: large-scale structure. *Phys. Rev. E*, **59**, 4326.

79 Puglisi, A., Loreto, V., Marconi, U.M.B., Petri, A., and Vulpiani, A. (1998) Clustering and non-Gaussian behavior in granular matter. *Phys. Rev. Lett.*, **81**, 3848.

80 Puglisi, A., Baldassarri, A., and Loreto, V. (2002) Fluctuation–dissipation relations in driven granular gases. *Phys. Rev. E*, **66**, 061305.

81 Puglisi, A., Visco, P., Trizac, E., and van Wijland, F. (2006) Dynamics of a tracer granular particle as a nonequilibrium Markov process. *Phys. Rev. E*, **73**, 021301.

82 Visco, P. (2006) Work fluctuations for a Brownian particle between two thermostats. *J. Stat. Mech.*, P06006.

83 Berne, B.J., Boon, J.P., and Rice, S.A. (1966) On the calculation of autocorrelation functions of dynamical variables. *J. Chem. Phys.*, **45**, 1086.

84 Zamponi, F., Bonetto, F., Cugliandolo, L. F., and Kurchan, J. (2005) A fluctuation theorem for non-equilibrium relaxational systems driven by external forces. *J. Stat. Mech.*, P09013.

85 Crisanti, A. and Ritort, F. (2003) Violation of the fluctuation–dissipation theorem in glassy systems: basic notions and the numerical evidence. *J. Phys. A*, **36**, R181.

10
Anomalous Thermal Transport in Nanostructures

Gang Zhang, Sha Liu, and Baowen Li

10.1
Introduction

Low-dimensional nanoscale systems such as nanowires and nanotubes are promising building blocks for future electronic, optoelectronic, and phononic/thermal devices. They have potential applications in novel power devices [1, 2], thermoelectric (TE) materials [3–6], and biological sensors [7, 8]. In addition to the electrical and optical properties, the thermal properties of nanoscale materials are also important. On the one hand, the study of thermal property of nanoscale materials is valuable to understand the underlying physics of phonon transport in low-dimensional systems. In fact, it is an outstanding and fundamental problem in nonequilibrium statistical physics to understand the macroscopic transport phenomena from microscopic dynamics [9]. In the past several decades, different models and theories have been proposed to describe the one-dimensional (1D) phonon transport (see Ref. [10] and references cited therein). Low-dimensional nanostructures such as nanowires and nanotubes provide a test bed for the conjectures and hypotheses. On the other hand, understanding of thermal property of nanoscale materials is very important for the development of future nanoelectronic devices. Typical integrated circuit chips consisting of billions of transistors can generate huge heat fluxes in a very small area, which is called hot spot [11]. The hot spot is becoming a bottleneck of the further development of future generations of electronic devices. One possible way to dissipate heat from hot spot is to use material with high thermal conductivity. The other way is to use thermoelectric materials to cool the hot spot. It has been shown that nanomaterials are promising candidates for both solutions. Therefore, the applications of nanomaterials will be obviously advanced by a systematic investigation of their thermal property.

Moreover, the studies of phonon transport in nonlinear lattice models have led to a novel field – phononics [12] – a science and technology of controlling and managing heat flow and processing information by phonons. Indeed, some theoretical models have been proposed for elementary phononic devices such as thermal diodes [13], thermal transistors [14], thermal logic gates [15], and thermal memory [16]. The nanostructured low-dimensional materials have been found to be ideal

Nonequilibrium Statistical Physics of Small Systems: Fluctuation Relations and Beyond, First Edition.
Edited by Rainer Klages, Wolfram Just, and Christopher Jarzynski.

candidates to realize these phononic functions. For example, the thermal rectification has been observed theoretically in a variety of nanostructures, such as nanotube molecular junctions [17], nanohorns [18], carbon nanocones [19], silicon–amorphous polyethylene interfaces [20], graphene nanoribbons (GNRs) [21], and Moebius graphene strips [22], and experimentally confirmed in asymmetric nanotubes [23].

In macroscopic systems, heat conduction is typically governed by Fourier's law as

$$J = -\kappa \nabla T, \tag{10.1}$$

where J is the local heat flux, ∇T is the temperature gradient, and κ is the thermal conductivity. For bulk material, κ is size independent; it depends only on the composite of material. This is based on the assumption that phonons transport diffusively in the bulk material. However, for the low-dimensional systems, in particular one-dimensional system, except for simple harmonic oscillator chains, we do not have any rigorous mathematical proof if the normal diffusion process can happen. Therefore, it is still an open and debatable question whether Fourier's law is valid in one-dimensional systems. Especially, the full set of sufficient and necessary conditions for Fourier's law is not clear yet.

There exists a large amount of literature and studies on different aspects of thermal property of nanomaterials, but due to the limit of space we address only the most fundamental one in low-dimensional nanomaterials, in particular the violation of Fourier's law. For the comprehensive review of nanoscale thermal conductivity, please refer to Refs [24, 25]. In this chapter, we focus on the anomalous behavior of thermal conductivity and its underlying mechanism – anomalous diffusion in low-dimensional nanostructures, including both numerical and experimental results in Sections 10.2 and 10.3. Then we will also discuss the underlying physical mechanism about the anomalous thermal conduction and diffusion in one-dimensional toy models in Section 10.4. These results are helpful in understanding the phonon transport in low-dimensional nanostructures.

10.2 Numerical Study on Thermal Conductivity and Heat Energy Diffusion in One-Dimensional Systems

For nanoscale materials, the fundamental question one may ask is whether Fourier's law is still valid. The question is not trivial, since nanoscale structure is of finite size (ranges from few nanometers to thousand nanometers), and the number of atoms in such a system is finite and far from the thermodynamic limit. Because there is no rigorous mathematical proof, first we need to investigate this issue by using simulation.

Carbon nanotubes (CNTs) [26] are one of the promising nanoscale materials discovered in the 1990s. They have many exceptional physical and chemical properties, which make them ranked among the best electronic devices so far [27]. It is measured experimentally that at room temperature the thermal conductivity of a single CNT is about $3000\,\mathrm{W\,m^{-1}\,K^{-1}}$ [28]. Theoretically, it is estimated that the thermal conductivity should be as high as $6600\,\mathrm{W\,m^{-1}\,K^{-1}}$ [29]. Recently, it has been demonstrated that the tube–tube interaction in a CNT rope has an important impact on the thermal conductivity due to the appearance of a narrow gap in the frequency spectrum of acoustic phonons [30, 31].

Yamamoto *et al.* have shown that even for the metallic nanotubes, the electrons give limited contribution to thermal conductivity of single-walled CNTs (SWNTs) at low temperature, and this part decreases quickly with the increase of temperature [32]. As the thermal conductivity of SWNT is dominated by phonons, it is ideally suited for molecular dynamics (MD) investigations of the validity of Fourier's law in low-dimensional systems. By using MD simulations, it is found that the thermal conductivity of SWNT diverges with the length of system as $\kappa \sim L^{\beta}$, where the exponent β depends on the temperature and SWNT diameters, and the value of β is between 0.12 and 0.4 [33] and between 0.11 and 0.32 [34]. This shows that the thermal conductivity of SWNT does not obey Fourier's law; that is, the thermal conductivity diverges with tube length.

To understand the physical mechanism for the length-dependent thermal conductivity in SWNT, we look at how the vibration energy is transported along SWNT [35]. The results are compared with the vibration energy transport in a 1D carbon atom chain. In our calculation, the Tersoff empirical potentials are used (more details can be found in Ref. [35]). We first thermalize the SWNT and 1D chain to a temperature T, and then excite a heat pulse (a packet of energy) in the middle of the tube/chain and study how it spreads along the tube/chain. Here free boundary condition along longitudinal direction is used; thus, there is reflection from the boundary. To suppress statistical fluctuations, an average over 10^3 realizations is performed. We calculate the heat pulse spreading at 2 and 300 K. Figure 10.1a and b shows two representatives of

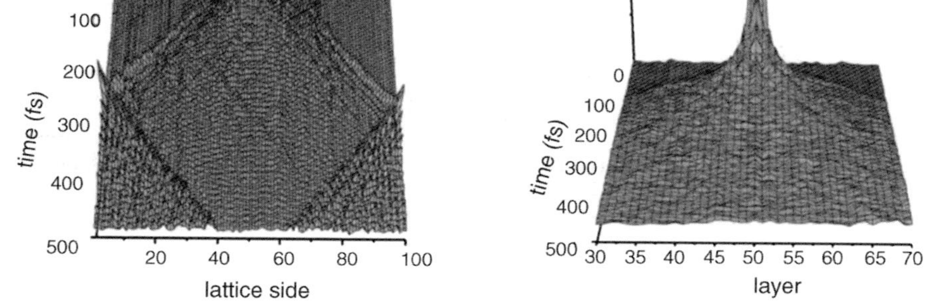

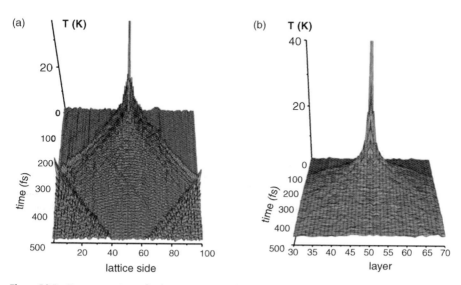

Figure 10.1 Representatives of pulse propagation along the (a) 1D lattice and (b) SWNT at 2 K. For further details, see Ref. [35].

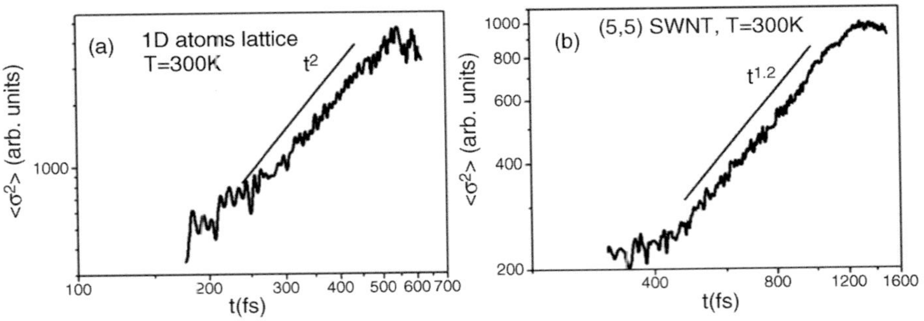

Figure 10.2 Energy diffusion in (a) 1D carbon chain and (b) SWNT at room temperature. For further details, see Ref. [35].

pulse propagation along the lattice and tube at 2 K. The pulse diffusion profiles at room temperature are very similar to those at low temperature. The differences between the SWNT and 1D atom chain are obvious. In the 1D lattice, a peak splits into two peaks and each peak spreads to the ends of the lattice. However, in the SWNT, one single peak expands with time, and no peak split is observed.

To describe the spread quantitatively, we calculate the width of the pulse by its second moment:

$$\sigma^2(t) = \frac{\sum_i [E_{i,t} - E_0](\vec{r}_{i,t} - \vec{r}_{i,0})^2}{\sum_i [E_{i,t} - E_0]}. \tag{10.2}$$

Here $E_{i,t}$ is the energy distribution of atom i at time t and $r_{i,0}$ is the position of energy pulse at $t = 0$. The averaged energy profile spreads as

$$\langle \sigma^2 \rangle = 2Dt^{\alpha}, \quad \text{with} \quad 0 < \alpha \leq 2, \tag{10.3}$$

where $\langle \cdot \rangle$ means the ensemble average over different realizations.

In Figure 10.2a and b, we show $\langle \sigma^2(t) \rangle$ versus time in double logarithmic scale, so that the slope of the curve before the turning point (due to boundary reflection) gives the value of α. More details of the numerical calculation can be found in Ref. [35]. For both SWNT and 1D lattice at 2 K (results not shown here), α is 2, which means that energy transports ballistically. It is clearly seen that for 1D lattice at a temperature of 300 K, the slope is still 2. This means that energy transports ballistically. However, at room temperature in SWNT, energy transports superdiffusively with $\alpha \approx 1.2$ (Figure 10.2b). This is slower than ballistic transport ($\alpha = 2$) but faster than normal diffusion ($\alpha = 1$). At low temperature, the vibration displacement in transverse direction is much smaller than the one along the tube axis and is negligible; that is, the phonons transport ballistically in both SWNT and 1D lattice. However, at room temperature, the situation changes dramatically. For 1D lattice, the increase of temperature does not change the harmonic character because of the strength of carbon–carbon interaction, while in SWNT at room temperature, the anharmonic term appears because of excitation of the transverse vibrational modes.

Based on billiard gas channel models, Li and Wang [36] have proposed a phenomenological relationship that connects anomalous heat conduction with anomalous diffusion. This connection indicates that only when the heat carrier undergoes normal diffusion like in the bulk material, the thermal conductivity is independent of system size; thus, heat conduction obeys Fourier's law. In superdiffusion case, $\alpha > 1$, the exponent $\beta > 0$, which means that thermal conductivity diverges with system size; this is the case for carbon nanotubes. Thus, superdiffusion is responsible for the divergent thermal conductivity of carbon nanotubes.

In addition to SWNTs, silicon nanowire (SiNW) is one of the promising nanomaterials that has drawn significant attention because of the ideal interface compatibility with the conventional Si-based technology. In the past few years, more and more theoretical efforts have been made to understand the thermal properties of SiNWs. The impacts of temperature, surface reconstruction, diameter, and isotopic doping have been reported by different simulation methods [37–41]. By using nonequilibrium molecular dynamics (NEMD) simulation with Stillinger–Weber (SW) potential, we have studied the length dependence of thermal conductivity of SiNWs [42]. The NW is a perfect one without rough surface; therefore, it is a momentum-conserved system. The dependence of thermal conductivity on the length of SiNW is shown in Figure 10.3. Nosé–Hoover and Langevin (please see references cited in Ref. [42]) heat baths are applied. Nosé–Hoover heat bath is an example of deterministic heat bath, and Langevin heat bath is an example of stochastic heat bath. Both types of heat baths give rise to the same results. It is obvious that the thermal conductivity increases with the length as $\kappa \propto L^{\beta}$, even when the wire length is as long as 1.1 μm. This demonstrates that the thermal transport in SiNW does not obey Fourier's law.

Traditionally, the phonon mean free path (λ) is a characteristic length scale beyond which phonons scatter and lose their phase coherence. By using the

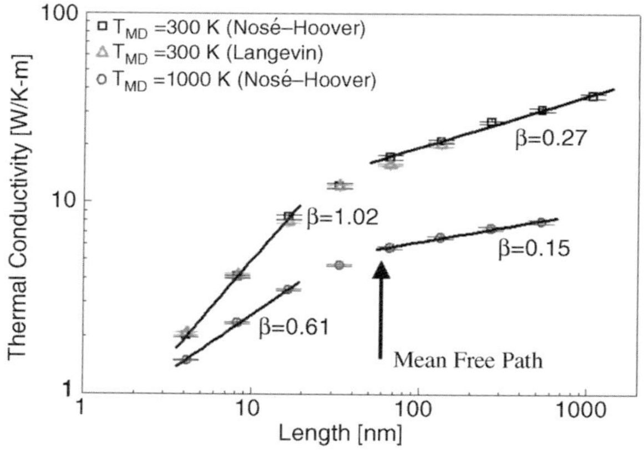

Figure 10.3 The dependence of thermal conductivity of SiNWs on the longitude length L_z. The results by the Nosé–Hoover method coincide with those by Langevin methods, indicating that the results are independent of the heat bath used. The solid curves are the power law fitting curves (linear in log–log scale). For more details, see Ref. [42].

phonon relaxation time (~10 ps) in SiNWs [37], and the group velocity of phonon as 6400 m s^{-1} [42], the mean free path λ is estimated as about 60 nm. The maximum SiNW length (1.1 μm) in our study is obviously much longer than the phonon mean free path of SiNW. It is well known that in three-dimensional systems, Fourier's law is obeyed when system scale $L \gg \lambda$. However, our NEMD results demonstrate that in SiNWs, Fourier's law is broken even when the length is obviously longer than the traditional mean free path.

More interestingly, it is found that the length dependence of thermal conductivity is different in different length regimes. At room temperature, when SiNW length is less than about 60 nm, the thermal conductivity increases with the length linearly ($\beta \approx 1$). For the longer wire ($L > 60$ nm), the diverged exponent β reduces to about 0.27. This critical length (60 nm) is in good agreement with the estimated value of mean free path. There is weak interaction among phonons when the length of SiNW is shorter than mean free path. So the phonons transport ballistically, like in the harmonic lattice. However, when the length of SiNW is longer than the mean free path, phonon–phonon scattering dominates the process of phonon transport and the phonon cannot flow ballistically. In addition, the diverged exponent β also depends on temperature. At 1000 K, β is only about 0.15 when $L > 60$ nm. The decrease of β can be understood from the temperature-dependent phonon coupling. In SiNWs at high temperature, the displacement of atoms increases, which induces more phonon–phonon interaction and reduction in β.

Now let us turn to the energy diffusion process in SiNWs. The details of the numerical calculation can be found in Ref. [42] and the parts about heat diffusion in CNT in this chapter. In Figure 10.4, we show $\langle \sigma^2(t) \rangle$ versus time in double logarithmic scale, so that the slope of the curve gives the value of α. For the NWs of 140 nm length, we obtain $\alpha > 1$ at both 300 and 1000 K, which corresponds to superdiffusion. Combining the results in carbon nanotube, these numerical results

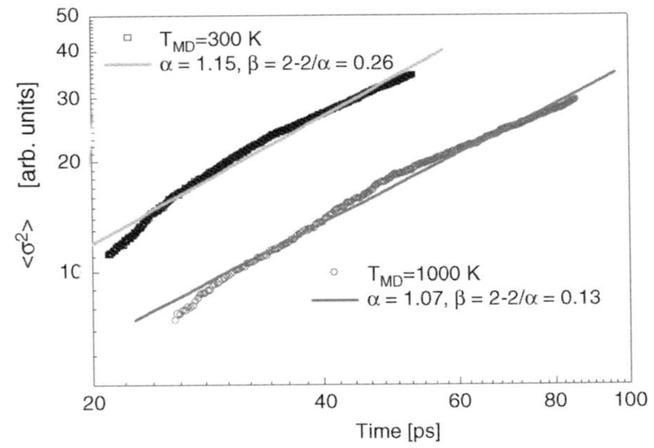

Figure 10.4 Energy diffusion in SiNW at room temperature and at 1000 K. The length of SiNW is 140 nm. For more details, see Ref. [42].

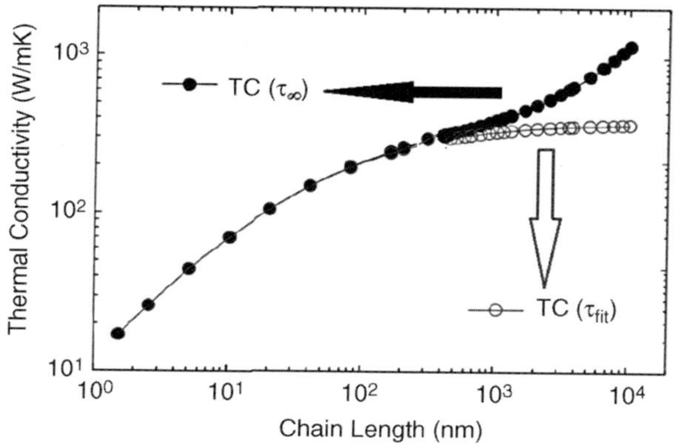

Figure 10.5 Thermal conductivity for a single chain. The results of the τ_∞ curve assume that the modes with wavelengths larger than 40 unit cells (about 10 nm) have infinite phonon–phonon relaxation times. The τ_{fit} curves use extrapolated values for the relaxation times of these modes. For more details, see Ref. [43].

demonstrate that the superdiffusion is responsible for the length-dependent thermal conductivity of one-dimensional nanostructures.

It is worth noting that the length-dependent thermal conductivity was also observed in polymer chain by Chen's group at MIT [43, 44]. Generally, a bulk polymer is a thermal insulator that has very low thermal conductivity of less than $0.1\,\text{W m}^{-1}\,\text{K}^{-1}$. Inspired by the study of heat conduction in 1D nonlinear lattice models [45], Henry and Chen performed computer simulation [46] for an individual chain of polyethylene employing both the Green–Kubo approach and a modal decomposition method. The phonon relaxation time is determined by using a modal analysis approach. As shown in Figure 10.5, they demonstrated a length-dependent thermal conductivity, which means that the thermal conductivity can be very high. This has been confirmed by experiment on ultradrawn polyethylene nanofibers with diameters of 50–500 nm and lengths up to tens of millimeters. The thermal conductivity of the nanofibers is as high as $\sim 104\,\text{W m}^{-1}\,\text{K}^{-1}$, which is about the order of magnitude of thermal conductivity of pure metals [44].

10.3
Breakdown of Fourier's Law: Experimental Evidence

The length-dependent thermal conductivity of carbon nanotube has been confirmed experimentally by Chang *et al.*, although the exponent differs from theoretical prediction [47]. Their experimental procedure is shown in Figure 10.6a; the inset shows the multiwalled nanotubes (MWNTs) attached to the test fixture via two contacts, which are formed from $(CH_3)_3(CH_3C_5H_4)Pt$. The right contact attaches the nanotube to the top of a preformed vertical "rib" on the right thermal

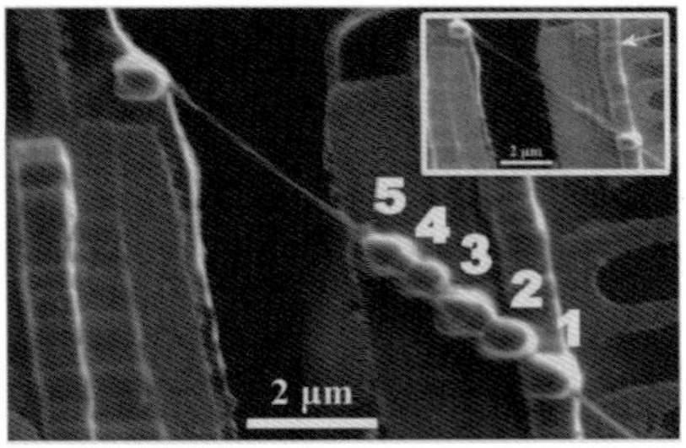

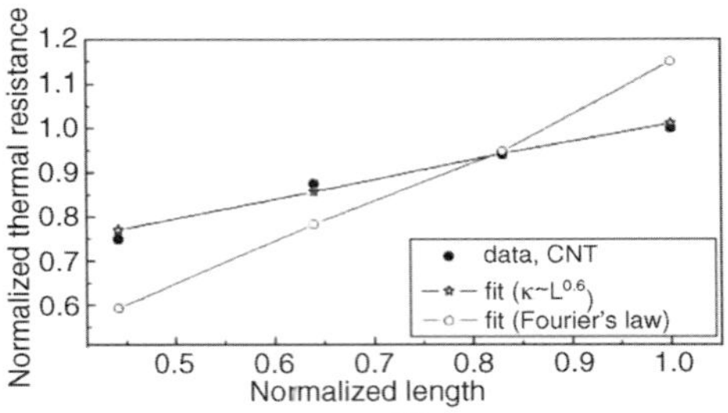

Figure 10.6 (a) A SEM image of a thermal conductivity test fixture with a nanotube after five sequences of $(CH_3)_3(CH_3C_5H_4)Pt$ deposition. The numbers denote the *n*th deposition. The inset shows the SEM image after the first $(CH_3)_3(CH_3C_5H_4)Pt$ deposition. (b) Normalized thermal resistance versus normalized sample length for CNT (solid circles), best fit assuming $\beta = 0.6$ (open stars), and best fit assuming Fourier's law (open circles). For more details, see Ref. [47].

pad, to ensure that the nanotube is suspended between the contacts. The thermal conductance of the nanotube is then measured. Next, an additional thermal contact post is deposited, again using $(CH_3)_3(CH_3C_5H_4)Pt$, inward of the original right-hand contact, which reduces the suspended length of the nanotube, thus reducing the "effective" length of nanotube. The thermal conductance is again measured. This process is repeated with a series of lengths. From the analysis of MWNT thermal conductance versus sample length, it is clear that thermal conductivity of nanotube does not follow Fourier's law, as shown in Figure 10.6b. Chang *et al.* found that the thermal conductivity diverges with tube length; the value of β ranges from 0.5 to 0.6. It is worth noting that the observed β on MWNTs differs from the theoretical predictions on SWNTs. The possible physical origin for this discrepancy might be the intershell phonon scattering in MWNTs.

In addition to nanotubes and nanowires that are based on one-dimensional or quasi-one-dimensional structure, graphene has garnered great interest due to its many remarkable physical properties. The room-temperature thermal conductivity of suspended single-layer graphene has been measured with a micro-Raman technique [48]. Very recently, it was found experimentally that the thermal conductivity of graphene nanoribbons also changes with size [49]. In their experiment, thermal conductivities of both suspended and supported few-layer graphene were measured from 77 to 350 K by using a suspended thermal bridge configuration in vacuum. Figure 10.7 shows scanning electron microscope (SEM) images of the microelectrothermal system. Three samples S1–S3 are supported by SiN_x with a length (L) of 5.0, 2.0, and 1.0 µm, respectively. S1 shows a room-temperature thermal conductivity of 1250 W m^{-1} K^{-1}, while the thermal conductivity of S2 is remarkably lower than that of S1 with a room-temperature value of 327 W m^{-1} K^{-1}, and S3 has an even lower thermal conductivity of only 150 W m^{-1} K^{-1}, as shown in Figure 10.7. As the thermal conductivity of S1 is way beyond scale, it is not shown in this figure.

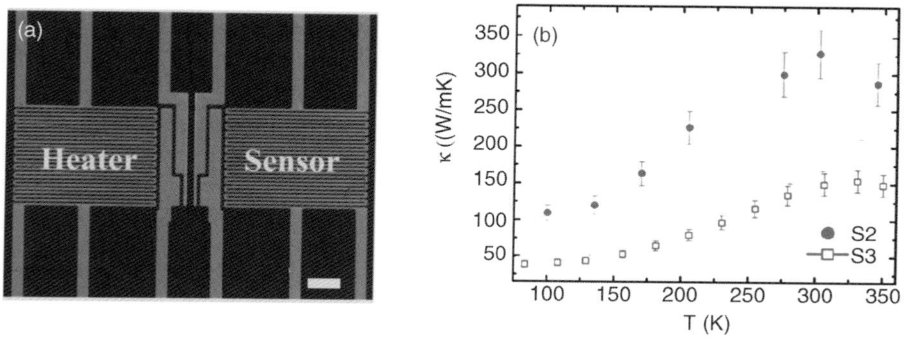

Figure 10.7 (a) SEM images of the METS for graphene samples; scale bar: 5 μm. (b) Measured thermal conductivity of different samples as a function of temperature. The lengths of samples S2 and S3 are 2.0 and 1.0 μm, respectively. For more details, see Ref. [49].

10.4 Theoretical Models

In this section, we discuss the underling physics of the anomalous thermal conduction in low-dimensional nanostructures.

Physicists and mathematicians have studied this problem for several decades to understand the necessary and sufficient condition, if any, for Fourier's law. As early as 1984, Casati *et al.* [50] suggested that the exponential instability might be the necessary condition, and this conjecture has been further supported by the study of heat conduction in a quasi-one-dimensional Lorenz gas channel [51]. However, the results from disordered Ehrenfest gas channel [52], triangle gas channel [53], polygonal gas channel [54], and an alternate mass hard-core potential chain [55] show that a system with zero Lyapunov exponent can also exhibit a normal diffusion and the heat conduction obeys Fourier's law. These studies have overturned the early conjecture that chaos is a necessary condition.

The numerical results from 1D nonlinear lattice, in particular from the Fermi–Pasta–Ulam (FPU) model [56–61], show that the heat conductivity diverges with system size L as L^{β} ($\beta > 0$), which means that heat conduction violates Fourier's law. It was conjectured early that the anomalous thermal conductivity is due to the solitary wave or soliton propagation along the chain [58]. But later it was found that it is not soliton but effective phonons [59]. It has been shown theoretically from the mode-coupling theory [62] and the Peierls equation [63] that the divergent exponent β should be 2/5, which is supported by the numerical results from different groups [60, 62]. For a generic 1D lattice, we now understand from the conclusion given by Prosen and Campbell [64] that momentum conservation will lead to a divergent thermal conductivity.

The 2/5 power law divergence for 1D nonlinear lattice seems like quite general. However, an analysis from the renormalization group argued that the divergent exponent should be 1/3 for a momentum-conserved one-dimensional system [65].

With mode-coupling theory, Wang and Li [66] have studied a quasi-1D system by allowing the lattice vibrations in both longitudinal and transverse directions. It is found that the exponent is 1/3 when the longitudinal modes are coupled with transverse modes, whereas it is 2/5 when the coupling is absent.

If the momentum conservation is broken by introducing an on-site potential like the Frenkel–Kontorova (FK) model [67] and the ϕ^4 model [60, 68], the heat conduction then obeys Fourier's law.

In addition to the anomalous thermal conductivity, another interesting topic is the connection between thermal conductivity and heat diffusion. As discussed in Section 10.2, it has been demonstrated that the superdiffusion in nanotube/nanowire is responsible for the breakdown of Fourier's law. However, the quantitative relationship is also a key point. Based on billiard gas channel models, in which heat carriers do not interact with each other, Li and Wang [36] proposed a phenomenological formula that connects anomalous heat conduction with the anomalous diffusion of heat carrier; that is, if the heat carrier undergoes a diffusion like $\langle \Delta x^2 \rangle = 2Dt^{\alpha}$, then the thermal conductivity is $\kappa \sim L^{\beta}$, with

$$\beta = 2 - 2/\alpha. \tag{10.4}$$

This connection indicates that only when the heat carrier undergoes normal diffusion like in the bulk material, the thermal conductivity is independent of system size; thus, heat conduction obeys Fourier's law. In superdiffusion case, $\alpha > 1$, the exponent $\beta > 0$, which means that thermal conductivity diverges with system size; this is the case for most nonlinear lattices studied so far. In subdiffusion case, $\alpha < 1$, the exponent $\beta < 0$, which means that the thermal conductivity vanishes when the system size goes to thermal dynamic limit.

This phenomenological derivation does not consider the interactions of the heat carriers. Recently, a theoretical approach has been developed to describe energy diffusion by Zhao [69], which can be applied to lattice systems. This approach uses the autocorrelation of energy fluctuation in *thermal equilibrium* to describe a nonequilibrium diffusion phenomenon, so it is a kind of linear response theory similar to the Green–Kubo formula of heat conductivity where the nonequilibrium heat flux is expressed in terms of equilibrium heat flux autocorrelation function. So rigorously speaking, this approach can only describe near-equilibrium energy diffusion. However, in most cases, it is of main interest. The virtue of the equilibrium description allows us to build stationary canonical correlation functions of quantities under investigation.

Using this method, Liu and Li have analytically derived a relation between (anomalous) diffusion and (anomalous) heat conduction [70]. Their starting point is to consider a 1D continuous system in thermal equilibrium. Let $e(x,t) = \varepsilon(x,t) - \langle \varepsilon(x,t) \rangle$ denote the energy density fluctuation, where $\varepsilon(x,t)$ is the energy density and $\langle \cdot \rangle$ denotes ensemble average. It follows the microscopic energy continuity equation

$$\frac{\partial e(x,t)}{\partial t} + \frac{\partial j(x,t)}{\partial x} = 0, \tag{10.5}$$

where $j(x,t)$ is heat current density. For homogeneous system at thermal equilibrium, the canonical correlations $\langle e(x,t)e(x',t')\rangle$, $\langle j(x,t)e(x',t')\rangle$, $\langle e(x,t)j(x',t')\rangle$, and $\langle j(x,t)j(x',t')\rangle$ depend only on $x-x'$ and $t-t'$. So we can define $C_{AB}(x-x',t-t') = \langle A(x,t)B(x',t')\rangle = C_{BA}(x'-x,t'-t)$. Then it is easy to verify that

$$\frac{\partial^2 C_{ee}(x,t)}{\partial t^2} = \frac{\partial^2 C_{jj}(x,t)}{\partial x^2}, \tag{10.6}$$

due to the even parity of $C_{ee}(x,t)$ and $C_{jj}(x,t)$ and the odd parity of $C_{je}(x,t)$ and $C_{ej}(x,t)$.

The heat current autocorrelation function in the Green–Kubo formula [71] is defined as

$$C(t) = \lim_{L\to\infty} \frac{1}{L}\langle J_L(t)J_L(0)\rangle, \tag{10.7}$$

where $J_L(t) = \int_{-L/2}^{L/2} j(x,t)\mathrm{d}x$ is the total heat current in 1D system with length L. Using periodic boundary conditions, it can be expressed as

$$C(t) = \lim_{L\to\infty} \frac{1}{L} \iint\limits_{-L/2<x,x'<L/2} \langle j(x,t)j(x',0)\rangle \mathrm{d}x\, \mathrm{d}x' = \int_{-\infty}^{\infty} C_{jj}(x,t)\mathrm{d}x. \tag{10.8}$$

According to Ref. [69], the energy density profile in energy diffusion can be described by the stationary energy correlation, that is,

$$\varrho(x,t) = \frac{1}{A}\langle e(x,t)e(0,0)\rangle = \frac{1}{A} C_{ee}(x,t), \tag{10.9}$$

where A is the normalization constant, which should be $\int_{-\infty}^{\infty} C_{ee}(x,0)\mathrm{d}x$. Similar to Eqs. (10.7) and (10.8), it can be expressed as $\lim_{L\to\infty}(1/L)\langle \Delta E_L(0)\Delta E_L(0)\rangle$. According to equilibrium statistical mechanics, $\langle \Delta E_L(0)\Delta E_L(0)\rangle = k_B T^2 C_V$ [72], where C_V is the heat capacity. So $A = k_B T^2 c_V$, with c_V the heat capacity per unit length. From above equations, we can get

$$\frac{\mathrm{d}^2\langle \sigma^2(t)\rangle}{\mathrm{d}t^2} = \frac{1}{A}\int_{-\infty}^{\infty} x^2 \frac{\partial^2 C_{jj}(x,t)}{\partial x^2}\mathrm{d}x = \frac{2}{A}\int_{-\infty}^{\infty} C_{jj}(x,t)\mathrm{d}x = \frac{2C(t)}{k_\mathrm{B} T^2 c_V}. \tag{10.10}$$

For normal heat conduction, the heat conductivity can be obtained from the Green–Kubo formula [71]:

$$\kappa = \frac{1}{k_\mathrm{B} T^2} \lim_{t\to\infty} \int_0^t C(t)\mathrm{d}t. \tag{10.11}$$

In the anomalous case, the size-dependent heat conductivity is usually obtained by putting a cutoff time $t_c \sim L/v_s$, with v_s the sound velocity and L the length, to the upper limit of the above formula [10, 73]. So

$$\kappa_L \sim \int_0^{L/v_s} \frac{C(t)}{k_\mathrm{B} T^2}\mathrm{d}t \sim \frac{c_V}{2}\int_0^{L/v_s} \frac{\mathrm{d}^2\langle \sigma^2(t)\rangle}{\mathrm{d}t^2}\mathrm{d}t \sim \frac{c_V}{2}\frac{\mathrm{d}\langle \sigma^2(t)\rangle}{\mathrm{d}t}\bigg|_{t\sim L/v_s}. \tag{10.12}$$

If energy spreads as $\langle\sigma^2(t)\rangle \sim t^\alpha$, it is easy to obtain $\kappa_L \sim L^{\alpha-1}$, which means

$$\beta = \alpha - 1. \tag{10.13}$$

It should be stressed that the relation of Eq. (10.13), which starts from the energy continuity equation, does not rely on any specific random walk model [74] such as Levy walk from which Denisov *et al.* also proposed the same formula [75]. Moreover, it can be noticed that although the above derivation in this work is only for 1D model, it can be generalized to any higher dimensional isotropic system straightforwardly. Furthermore, the generation to the discrete lattice model can be made by replacing the spatial derivative with difference or by simply regarding the lattice models as special cases of continuous models where energy and current density are δ-function. Therefore, Eq. (10.10) should be universal to any isotropic system with any dimension. So is Eq. (10.13), provided the cutoff time reasoning is valid, which is widely accepted [10, 59, 73, 76].

It is interesting that different formulas to connect thermal conductivity and heat diffusion have been derived, such as Eqs. (10.4) and (10.13). It is well accepted that the study of thermal property of nanoscale materials is valuable to understand the physics of phonon transport in low-dimensional systems. Thus, here we use the numerical results in Section 10.2 to compare the accuracy of these two equations. In a (5,5)-SWNT, from Figure 10.1, we have obtained $\alpha \approx 1.2$ from the analysis of diffusion; thus, $\beta = 2 - 2/\alpha = 0.33$ according to Eq. (10.4), which is very close to (with less than 3% discrepancy) the value $\beta = 0.32$ obtained independently by Maruyama in 2002 by the direct NEMD method [34]. If we use Eq. (10.13), $\beta = \alpha - 1$, then we obtain $\beta = 0.2$, which deviates 60% from $\beta = 0.32$ obtained by the NEMD simulation. Moreover, for the NWs of 140 nm length, from Figure 10.4, we obtain $\alpha = 1.15$ and 1.07 at 300 and 1000 K, respectively. With these values of diffusion exponents, one can obtain $\beta = 0.26$ (300 K) and 0.13 (1000 K) from Eq. (10.4), which are very close to those values (0.27 and 0.15, respectively) calculated directly from the NEMD simulation. It should be noted that if we use $\beta = \alpha - 1$, then we obtain $\beta = 0.15$ and 0.07, respectively, which are about 56 and 67% different from the values 0.27 and 0.15 obtained directly from the NEMD simulation. These numerical results from nanotube/nanowire agree very well with Eq. (10.4). Other existing numerical data ranging from billiard gas channel models to lattice models (see Refs [36, 45] for details) also support this formula.

However, contrary to the simulation result in nanostructures, the accurate study on the following 1D toy models agrees with Eq. (10.13) better than Eq. (10.4), for instance, the hard-point particles model studied by Cipriani *et al.* ($\alpha = 4/3$) [77] and by Grassberger *et al.* ($\beta = 1/3$) [78], which agrees with the renormalization group theory [65], and the FPU-β model by Zhao ($\alpha = 1.4$) [69] and by Wang ($\beta = 0.4$) [76], which agrees with the mode-coupling theory [62] and the Boltzmann equation approach [63].

In Figure 10.8, we show the α–β relationship based on Eqs. (10.4) and (10.13), and the simulation results. It is interesting to find that both Eqs. (10.4) and (10.13) have their separate numerical "supporters." The discrepancy between Eq. (10.13) and the simulation results in nanostructures may be attributed to the following reasons. First, the quasi-1D nanowires and nanotubes are not perfectly

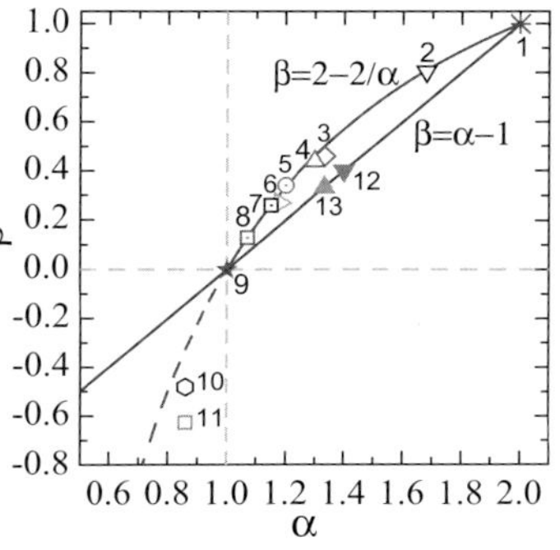

Figure 10.8 The relationship between α and β based on Eqs. (10.4) and (10.13), and different numerical results. Point 1 is harmonic lattice, Toda lattice; point 2 is for 1D Ehrenfest gas channel from Ref. [36]; point 3 is for triangle–square channel from Ref. [45]; point 4 is for polygonal billiard channel with one angle $(\sqrt{5}-1)\pi/4$ and one angle $\pi/3$ from Refs [45, 54]; point 5 is for (5,5)-SWNT from Refs [34, 35]; point 6 is for rational triangle from Ref. [45]; point 7 is for SiNW at 300 K from Ref. [42]; point 8 is for SiNW at 1000 K from Ref. [42]; point 9 is for normal diffusion; point 10 is for polygonal gas channel with one angle $(\sqrt{5}-1)\pi/4$ and one angle $\pi/4$, length $40 \leq L \leq 80$ from Refs [45, 54]; point 11 is for polygonal gas channel with one angle $(\sqrt{5}-1)\pi/4$ and one angle $\pi/4$, length $1 \leq L \leq 40$ from Refs [45, 54]; point 12 is for the FPU model from Refs [69, 78]; and point 13 is for hard-point gases from Refs [75, 77].

isotropic. Second, Eq. (10.13) is a linear response result, which is valid only when the energy excitation is small enough. However, in simulation in "real" nanomaterials, the initial heat pulse should be sufficiently large, so it will not immerse into the environment quickly. Each of these reasons may quantitatively affect the calculated value of α and β. We should point out that a better, more fundamental theory, which is valid for both equilibrium and nonequilibrium conditions, however, is still missing, and needs further study. For the problems of Fourier's law and different exponents discussed here, we also refer the interested readers to the excellent treatise by Klages [79] together with the original literature cited therein.

10.5 Conclusions

In this chapter, we give a brief overview of the recent developments in experimental and theoretical studies of heat transport in low-dimensional nanomaterials, in particular the anomalous thermal conductivity and anomalous diffusion. Thermal property in nanostructures differs significantly from that in bulk materials because phonons transport superdiffusively – a process faster than random walk but slower than ballistic motion, which leads to a length-dependent thermal conductivity. These nanostructured materials are promising platforms to testify fundamental phonon transport theories. For instance, we have demonstrated that heat conduction in individual nanotubes and nanowires does not obey Fourier's law even though the system length is much longer than the phonon mean free path. Moreover, the study of thermal property of nanomaterials is also important for potential applications in phononic devices; in particular, the length-dependent thermal conductivity might be very useful in designing functional thermal materials to control and manage heat flow. Given that thermal management devices generally can be considered for applications that require power ranging from milliwatts up to several thousand watts, the general range of applications where they are indispensable is stupendous. These technologies will serve as major players in global sustainable energy sources. Nevertheless, much work remains to be done in this area.

Acknowledgments

This work was supported in part by a grant from the Asian Office of Aerospace R&D of the US Air Force (AOARD-114018) and a grant from SERC of A*STAR (R-144-000-280-305), Singapore. G.Z. was supported by the Ministry of Science and Technology of China (Grant No. 2011CB933001) and Ministry of Education, China (Grant No. 20110001120133).

References

1 Hu, L. and Chen, G. (2007) *Nano Lett.*, **7**, 3249.
2 Li, J.S., Yu, H.Y., Wong, S.M., Zhang, G., Sun, X.W., Lo, L.Q., and Kwong, D.L. (2009) *Appl. Phys. Lett.*, **95**, 033102.
3 Hochbaum, A.I., Chen, R., Delgado, R.D., Liang, W., Garnett, E.C., Najarian, M., Majumdar, A., and Yang, P. (2008) *Nature (London)*, **451**, 163.
4 Boukai, A.I., Bunimovich, Y., Kheli, J.T., Yu, J.-K., Goddard, W.A., III, and Heath, J. R. (2008) *Nature*, **451**, 168.
5 Zhang, G., Zhang, Q., Bui, C.-T., Lo, G.-Q., and Li, B. (2009) *Appl. Phys. Lett.*, **94**, 213108.
6 Zhang, G., Zhang, Q.-X., Kavitha, D., and Lo, G.-Q. (2009) *Appl. Phys. Lett.*, **95**, 243104.
7 Cui, Y., Wei, Q., Park, H., and Lieber, C.M. (2001) *Science*, **293**, 1289.
8 Zhang, G.-J., Zhang, G., Chua, H.J., Chee, R.-E., Wong, E.H., Agarwal, A., Buddharaju, K.D., Singh, N., Gao, Z., and Balasubramanian, N. (2008) *Nano Lett.*, **8**, 1066.
9 Bonetto, F. *et al.* (2000) in *Mathematical Physics 2000* (eds A. Fokas *et al.*), Imperial College Press, London, pp. 128–150, and references cited therein.
10 Dhar, A. (2008) *Adv. Phys.*, **57**, 457.
11 Rowe, D.M. (ed.) (2006) *Thermoelectrics Handbook: Macro To Nano*, Taylor & Francis Group, Boca Raton, FL.
12 (a) Wang, L. and Li, B. (2008) *Phys. World*, **23**, 27. (b) Li, N., Ren, J., Wang, L., Zhang, G., Hänggi, P. and Li, B. (2012) *Rev. Mod. Phys.*, **84**, 1045.
13 (a) Li, B., Wang, L., and Casati, G. (2004) *Phys. Rev. Lett.*, **93**, 184301. (b)Segal, D., and Nitzan, A. (2005) *Phys. Rev. Lett.*, **94**, 034301. (c)Hu, B., Yang, L., and Zhang, Y. (2006) *Phys. Rev. Lett.*, **97**, 124302. (d)Yang, N., Li, N., Wang, L., and Li, B. (2007) *Phys. Rev. B*, **76**, 020301.
14 (a) Li, B., Wang, L., and Casati, G. (2006) *Appl. Phys. Lett.*, **88**, 143501. (b)Lo, W.-C., Wang, L., and Li, B. (2008) *J. Phys. Soc. Jpn.*, **77**, 054402.
15 Wang, L. and Li, B. (2007) *Phys. Rev. Lett.*, **99**, 177208.
16 Wang, L. and Li, B. (2008) *Phys. Rev. Lett.*, **101**, 267203.
17 Wu, G. and Li, B. (2007) *Phys. Rev. B*, **76**, 085424.
18 Wu, G. and Li, B. (2008) *J. Phys.: Condens. Matter*, **20**, 175211.
19 Yang, N., Zhang, G., and Li, B. (2008) *Appl. Phys. Lett.*, **93**, 243111.

20 Hu, M., Keblinski, P., and Li, B. (2008) *Appl. Phys. Lett.*, **92**, 211908.
21 (a) Hu, J., Ruan, X., and Chen, Y.P. (2009) *Nano Lett.*, **9**, 2730. (b)Yang, N., Zhang, G., and Li, B. (2009) *Appl. Phys. Lett.*, **95**, 033107.
22 Jiang, J.-W., Wang, J.-S., and Li, B. (2010) *Europhys. Lett.*, **89**, 46005.
23 Chang, C.W., Okawa, D., Majumdar, A., and Zettl, A. (2006) *Science*, **314**, 1121.
24 Cahill, D.G., Ford, W.K., Goodson, K.E., Mahan, G.D., Majumder, A., Maris, H.J., Merlin, R., and Phillpot, S.R. (2003) *J. Appl. Phys.*, **93**, 793.
25 Zhang, G. and Li, B. (2010) *Nanoscale*, **2**, 1058.
26 Iijima, S. (1991) *Nature (London)*, **354**, 56.
27 Javey, A. and Kong, J. (2009) *Carbon Nanotube Electronics*, Springer, Berlin.
28 Kim, P., Shi, L., Majumdar, A., and McEuen, P.L. (2001) *Phys. Rev. Lett.*, **87**, 215502.
29 Berber, S., Kwon, Y.-K., and Tománek, D. (2000) *Phys. Rev. Lett.*, **84**, 4613.
30 Donadio, D. and Galli, G. (2007) *Phys. Rev. Lett.*, **99**, 255502.
31 Savin, A.V., Kivshar, Y.S., and Hu, B. (2009) *Europhys. Lett.*, **88**, 26004.
32 Yamamoto, T., Watanabe, S., and Watanabe, K. (2004) *Phys. Rev. Lett.*, **92**, 075502.
33 Zhang, G. and Li, B. (2005) *J. Chem. Phys.*, **123**, 114714.
34 Maruyama, S. (2002) *Physica B*, **323**, 193.
35 Zhang, G. and Li, B. (2005) *J. Chem. Phys.*, **123**, 014705.
36 Li, B. and Wang, J. (2003) *Phys. Rev. Lett.*, **91**, 044301.
37 Volz, S.G. and Chen, G. (1999) *Appl. Phys. Lett.*, **75**, 2056.
38 Schelling, P.K., Phillpot, S.R., and Keblinski, P. (2002) *Phys. Rev. B*, **65**, 144306.
39 Chen, Y., Li, D., Lukes, J., and Majumdar, A. (2005) *J. Heat Transfer*, **127**, 1129.
40 Bodapati, A., Schelling, P.K., Phillpot, S. R., and Keblinski, P. (2006) *Phys. Rev. B*, **74**, 245207.
41 Yang, N., Zhang, G., and Li, B. (2008) *Nano Lett.*, **8**, 276.
42 Yang, N., Zhang, G., and Li, B. (2010) *Nano Today*, **5**, 85.
43 Henry, A. and Chen, G. (2009) *Phys. Rev. B*, **79**, 144305.
44 Shen, S., Henry, A., Tong, J., Zheng, R., and Chen, G. (2010) *Nat. Nanotechnol.*, **5**, 251.
45 Li, B., Wang, J., Wang, L., and Zhang, G. (2005) *Chaos*, **15**, 015121.
46 Henry, A. and Chen, G. (2008) *Phys. Rev. Lett.*, **101**, 235502.
47 Chang, C.W., Okawa, D., Garcia, H., Majumdar, A., and Zettl, A. (2008) *Phys. Rev. Lett.*, **101**, 075903.
48 Balandin, A.A., Ghosh, S., Bao, W., Calizo, I., Teweldebrhan, D., Miao, F., and Lau, C. N. (2008) *Nano Lett.*, **8**, 902.
49 Wang, Z.-Q., Xie, R.-G., Bui, C.-T., Liu, D., Ni, X.-X., Li, B., and Thong, J.T.L. (2011) *Nano Lett.*, **11**, 113.
50 Casati, G., Ford, J., Vivaldi, F., and Visscher, W.M. (1984) *Phys. Rev. Lett.*, **52**, 1861.
51 Alonso, D., Artuso, R., Casati, G., and Guarneri, I. (1999) *Phys. Rev. Lett.*, **82**, 1859.
52 Li, B., Wang, L., and Hu, B. (2002) *Phys. Rev. Lett.*, **88**, 223901.
53 Li, B., Casati, G., and Wang, J. (2003) *Phys. Rev. E*, **67**, 021204.
54 (a) Alonso, D., Ruiz, A., and de Vega, I. (2002) *Phys. Rev. E*, **66**, 066131. (b)Alonso, D., Ruiz, A., and de Vega, I. (2004) *Physica D*, **187**, 184.
55 Li, B., Casati, G., Wang, J., and Prosen, T. (2004) *Phys. Rev. Lett.*, **92**, 254301.
56 Kaburaki, H. and Machida, M. (1993) *Phys. Lett. A*, **181**, 85.
57 Lepri, S., Livi, R., and Politi, A. (1997) *Phys. Rev. Lett.*, **78**, 1896.
58 (a) Fillipov, A., Hu, B., Li, B., and Zeltser, A. (1998) *J. Phys. A*, **31**, 7719. (b)Zhao, H. (2006) *Phys. Rev. Lett.*, **96**, 140602.
59 Li, N.-B., Li, B., and Flach, S. (2010) *Phys. Rev. Lett.*, **105**, 054102.
60 Hu, B., Li, B., and Zhao, H. (2000) *Phys. Rev. E*, **61**, 3828.
61 Aoki, K. and Kusnezov, D. (2001) *Phys. Rev. Lett.*, **86**, 4029.
62 Lepri, S. (1998) *Phys. Rev. E*, **58**, 7165.
63 Pereverzev, A. (2003) *Phys. Rev. E*, **68**, 056124.
64 Prosen, T. and Campbell, D.K. (2000) *Phys. Rev. Lett.*, **84**, 2857.
65 Narayan, O. and Ramaswamy, S. (2002) *Phys. Rev. Lett.*, **89**, 200601.

66 (a) Wang, J.-S. and Li, B. (2004) *Phys. Rev. Lett.*, **92**, 074302. (b)Wang, J.-S., and Li, B. (2004) *Phys. Rev. E*, **70**, 021204.
67 Hu, B., Li, B., and Zhao, H. (1998) *Phys. Rev. E*, **57**, 2992.
68 Aoki, K. and Kusnezov, D. (2000) *Phys. Lett. A*, **265**, 250.
69 Zhao, H. (2006) *Phys. Rev. Lett.*, **96**, 14.
70 Liu, S., Xu, X., Xie, R., Zhang, G. and Li, B. (2012), *European Physical Journal B*, **85**, 337. (b) Liu, S., Li, N., Ren J. and Li B. (2012) arxiv:1103.2835.
71 (a) Green, M.S. (1954) *J. Chem. Phys.*, **22**, 398. (b)Kubo, R., Yokota, M., and Nakajima, S. (1957) *J. Phys. Soc. Jpn.*, **12**, 1203.
72 Huang, K. (1987) *Statistical Mechanics*, 2nd edn, John Wiley & Sons, Inc., New York, p. 146.
73 Lepri, S., Livi, R., and Politi, A. (1998) *Europhys. Lett.*, **43**, 271.
74 Li, B. and Wang, J. (2004) *Phys. Rev. Lett.*, **92**, 089402.
75 Denisov, S., Klafter, J., and Urbakh, M. (2003) *Phys. Rev. Lett.*, **91**, 194301.
76 Wang, L. and Wang, T. (2011) *Europhys. Lett.*, **93**, 54002.
77 Cipriani, P., Denisov, S., and Politi, A. (2005) *Phys. Rev. Lett.*, **94**, 244301.
78 Grassberger, P., Nadler, W., and Yang, L. (2002) *Phys. Rev. Lett.*, **89**, 180601.
79 Klages, R. (2007) *Microscopic Chaos, Fractals and Transport in Nonequilibrium Statistical Mechanics, Advanced Series in Nonlinear Dynamics*, World Scientific, Singapore.

11
Large Deviation Approach to Nonequilibrium Systems

Hugo Touchette and Rosemary J. Harris

11.1
Introduction

Nonequilibrium systems are being increasingly studied using methods borrowed from the mathematical theory of large deviations, as developed in the 1960s and 1970s by Donsker and Varadhan, and Freidlin and Wentzell (see Refs [1–3] for historical references). Indeed, the central concepts and quantities of this theory – for example, the large deviation principle, rate functions, generating functions, and so on – have now entered the standard jargon of driven nonequilibrium systems modeled as discrete- or continuous-time Markov processes (see, for example, Refs [3–5]).

With hindsight, one can argue that this evolution, although relatively recent, was to be expected: large deviation theory has been used successfully in equilibrium statistical mechanics for well over 30 years [3, 6–8], and so it is not surprising that this success finds its way into nonequilibrium statistical mechanics. However, there is more, in that the two scenarios – equilibrium and nonequilibrium – share many ideas, concepts, and even a theoretical structure that happen to find a clear and precise expression in the language of large deviations. It is natural, therefore, to see this language being used for both theories.

By viewing equilibrium statistical mechanics from the point of view of large deviation theory, one gets a clear sense, for example, of why there is a Legendre transform in thermodynamics connecting the entropy and the free energy, when this Legendre transform is valid, how equilibrium states relate to the notions of concentration and typicality, and how these states arise out of variational principles such as the maximum entropy principle or the minimum free energy principle. Similar ideas and results of nonequilibrium statistical mechanics are also made clear by viewing them through the prism of large deviation theory. In addition, one sees, as mentioned, essential similarities between equilibrium and nonequilibrium.

Our goal in this chapter is to explain these points and to illustrate them with simple examples, mainly involving continuous-time Markov processes. We start in the next section by recalling the basis and essential concepts of equilibrium statistical mechanics, and by discussing their analogues in nonequilibrium statistical mechanics. Among these, we mention the notions of statistical ensemble,

Nonequilibrium Statistical Physics of Small Systems: Fluctuation Relations and Beyond, First Edition.
Edited by Rainer Klages, Wolfram Just, and Christopher Jarzynski.

stationarity, typicality, fluctuations, and scaling limit (e.g., thermodynamic limit, hydrodynamic or macroscopic limit, and long-time limit). In Section 11.3, we re-express these concepts in the language of large deviations to define them in a precise, mathematical way and to emphasize the theoretical structure that underlies both equilibrium and nonequilibrium statistical mechanics. We illustrate this structure with a variety of applications in Section 11.4 and then end in Section 11.5 with some concluding remarks and open problems.

Much of the large deviation content explored in the present contribution can be found with more details in Ref. [3]. Here we focus on discussing the common goals, concepts, and results of equilibrium and nonequilibrium statistical mechanics, rather than providing a complete review of either subject, and on proposing a clear approach to studying nonequilibrium systems which parallels that used for studying equilibrium systems. We draw inspiration in doing so from the works of Oono [9–11], Eyink [12–14], and Maes *et al.* [15–18], among others, which show the emergence of similar ideas and views as early as the late 1980s.

11.2 From Equilibrium to Nonequilibrium Systems

Before we discuss how large deviation concepts enter in equilibrium and nonequilibrium statistical mechanics, we recall in this section the basis of each theory and emphasize some concepts shared by both. Of these, the most important to keep in mind is the concept of *typicality*, connected mathematically to the law of large numbers and the concentration of probability distributions.

11.2.1 Equilibrium Systems

The goal of *equilibrium* statistical mechanics, as is well known, is to explain and predict the emergence of macroscopic equilibrium states of systems composed of many particles by treating their microscopic states in a probabilistic way. The main properties of equilibrium states are that they are stationary in time, stable against small perturbations, and described by only a few variables; that is, they are low-dimensional *macro*states compared to the high-dimensional *micro*states used for describing many-particle systems at the microscopic level.

Equilibrium states are also defined with respect to a given macrostate. Physicists often say that a system is *at equilibrium* or is *in a state of equilibrium*, but what is meant, to be more precise, is that the system has, say, an equilibrium energy or an equilibrium magnetization. Thus, an equilibrium state is a particular state or a value of a macrostate or a collection of macrostates. This is different from saying that a system *is an equilibrium system*. We shall give a definition of the latter concept later, after describing the analogue of an equilibrium state for nonequilibrium systems.

For now, let us recall how equilibrium states are modeled in statistical mechanics. The basic ingredients are well known. Consider a system of N particles, which for simplicity we take to be a classical system, and let the sequence $\omega = (\omega_1, \omega_2, \ldots, \omega_N)$ denote the microscopic configuration or *microstate* of the system, where ω_i is the state of the ith particle. To study the statistical properties of this system, we consider a *prior* probability distribution $P(\omega)$, interpreted as the stationary distribution of the microscopic dynamics. The state space of one particle is denoted by Λ, so that $P(\omega)$ is a probability distribution over the N-particle space $\Lambda_N = \Lambda^N$.

Next, we consider a *macrostate* M_N, corresponding mathematically to a function $M_N(\omega)$ of the microstates, and proceed to compute the probability distribution of this random variable obtained under $P(\omega)$ using

$$P(M_N = m) = \int_{\Lambda_N} \delta(M_N(\omega) - m) P(\omega) \mathrm{d}\omega. \tag{11.1}$$

If $P(\omega)$ is a valid model of an equilibrium system and the macrostate M_N is chosen properly, then what should be observed is that $P(M_N)$ is concentrated around certain highly probable values and that this concentration gets more pronounced as N gets larger. It is these *most probable* or *typical* values (points or states) of M_N that we call *equilibrium states* of M_N.

Mathematically, the concentration of $P(M_N)$ is akin to a law of large numbers, in the sense that there exist sets B of values of M_N such that

$$\lim_{N\to\infty} P(M_N \in B) = 1 \quad \text{and} \quad \lim_{N\to\infty} P(M_N \notin B) = 0. \tag{11.2}$$

The smallest set B having this property corresponds to the set of equilibrium values of M_N or, more loosely, the set of equilibrium states of the system (as defined with respect to M_N).

We shall see in the next section that an essential property of the concentration of $P(M_N)$ on B is that it is *exponential* as a function of N (or, more generally, the volume of the system), which means that the probability that M_N deviates from one of its equilibrium values is exponentially small with the system size, N. Physically, these deviations are termed *fluctuations*, and so we say that the probability of fluctuations from equilibrium states is exponentially small with N.

The exponential concentration of $P(M_N)$ explains why large deviation theory is used in equilibrium statistical mechanics. Physically, it is also the reason why equilibrium states correspond to *typical* values of M_N and not, as often claimed, to *average* values of M_N. The fundamental property of equilibrium states is indeed that they do not fluctuate, or at least appear not to fluctuate at the macroscopic level, so the fact that M_N has a well-defined average value is obviously not enough: M_N must converge in probability to some typical values.

The reason why average values are used in statistical mechanics is arguably that they are conceptually simpler than typical (or concentration) values and that, if M_N has a unique typical value, then its average is the same as its typical value (we say in this case that M_N is *self-averaging*). However, the use of averages is somewhat misleading as it detracts us from the essential property of equilibrium states, which is again that these states arise probabilistically from the concentration of a probability distribution. Equilibrium states are first and foremost *typical* states arising from the scaling limit that is the *thermodynamic limit* [19, 20].

This is an important point that leads us to the discussion of extensivity versus intensivity. Physically, it should be clear that the total energy H_N of an N-body system with short-range interactions does not concentrate in the thermodynamic limit, simply because the energy is *extensive* for such a system, and so diverges with N. As a result, one cannot formally say that the system has an *equilibrium energy* in the thermodynamic limit. Rather, the correct macrostate having an equilibrium value, that is, the one that concentrates in the thermodynamic limit, is the energy per particle or mean energy $h_N = H_N/N$, which is an *intensive* quantity.

To explain this point more clearly, let us consider a sum

$$S_N = \sum_{i=1}^{N} X_i \tag{11.3}$$

of N independent and identically distributed random variables $X_1, \ldots, X_N$. If the mean $\langle X_1 \rangle = \mu$ of these random variables is finite, then as $N \to \infty$ the distribution of S_N/N concentrates to the mean in such a way that

$$\lim_{N\to\infty} P(|S_N/N - \mu| > \varepsilon) = 0 \tag{11.4}$$

for all $\varepsilon > 0$, in accordance with the law of large numbers. The point to note about this result, which is equivalent to the second limit shown in (11.2), is that it holds for the mean sum S_N/N and not for the sum S_N: the distribution of $P(S_N)$ does not concentrate; in fact, it flattens as $N \to \infty$. Similarly, it is easy to show that the distribution of S_N/N^α flattens for all $\alpha \in (0, 1)$ and concentrates in a trivial way at zero for all $\alpha > 1$. Hence, the only normalization of a sum S_N of independent (or near-independent) random variables with finite mean that yields a nontrivial concentration point is S_N/N.

The same observation applies to equilibrium states. The reason again why equilibrium states appear stable at the macroscopic level is that the fluctuations around these states are very improbable and become the more unlikely the bigger the system gets. Mathematically, the way to make sense of this observation is to consider random variables that have the property of concentrating in the thermodynamic limit $N \to \infty$. From this point of view, the total energy H_N is not a "good" macrostate to consider because it does not concentrate in this limit. The same goes, similarly to S_N, for the macrostates $H_N/\sqrt{N}$ or H_N/N^2: the distribution of the former flattens, whereas the distribution of the latter concentrates trivially to zero. We get a nontrivial concentration only for H_N/N.

This at least is generally true for short-range interacting systems. For long-range interacting systems, such as gravitational systems or mean-field systems, the "good" energy macrostate to consider might be H_N/N^2 or, more generally, H_N/N^α with $\alpha \geq 2$ [21]. The choice of α will depend on the system considered, but the requirement again is that the distribution of the macrostate that is studied should concentrate when $N \to \infty$. From this point of view, one might have to consider different thermodynamic (or scaling) limits and macrostates in order to correctly describe the equilibrium states of a system.

11.2.2
Nonequilibrium Systems

The study of nonequilibrium systems is conceptually more difficult than that of equilibrium systems because one is interested in describing not only the stationary fluctuations, as is done for equilibrium systems, but also the fluctuating dynamics of the system arising in time and the fluctuations of macrostates or observables integrated over time. Thus, in addition to considering the number of particles present in a system (or its volume), one also needs to consider the evolution of that system in time. This implies that different scaling limits may be taken depending on the system and macrostate or observable studied.

For definiteness, we consider here nonequilibrium systems modeled by Markovian processes. To simplify the presentation of these models, we assume for now that the stochastic evolution takes place in discrete time (although continuous-time models will also be discussed in the following sections). In this case, the *microstate* ω that represented before the configuration of an equilibrium system at a fixed (yet unspecified) instant of time is now a complete trajectory $\omega = \{\omega_i\}_{i=1}^n$ consisting of n time steps. The assumption that the process is Markovian then amounts to assuming that the prior distribution $P(\omega)$ can be decomposed according to a Markov chain

$$P(\omega) = P(\omega_1)P(\omega_2|\omega_1)\cdots P(\omega_n|\omega_{n-1}), \tag{11.5}$$

with initial distribution $P(\omega_1)$ and transition matrix elements $P(\omega_i|\omega_{i-1})$, which, in most cases, are assumed to be time homogeneous (i.e., time independent). This form of prior is, from a pragmatic point of view, our stochastic model for ω from which all distributions of macrostates or observables are computed. Thus, so far, the formalism is abstractly the same as for equilibrium systems: a system is described by its microstate ω and a probability distribution $P(\omega)$ on the space of microstates. What changes for nonequilibrium systems is the interpretation of ω as a time trajectory of a system of one or more particles.

This difference allows us to consider many types of macrostates or observables. For example, one can consider a *fixed-time* or *static* observable $M(\omega_i)$ which is a function of the state of the system at a specific time step i. One can also consider *dynamic* observables of the form

$$M_n(\omega) = \frac{1}{n}\sum_{i=1}^{n} f(\omega_i), \tag{11.6}$$

referred to mathematically as *additive observables* or *additive functionals* [3], which involve states at different times. Another type of *dynamic* observable, which arises in the context of particle currents, is

$$M_n(\omega) = \frac{1}{n}\sum_{i=1}^{n-1} f(\omega_i, \omega_{i+1}). \tag{11.7}$$

Whatever the observable chosen, the goal when studying nonequilibrium systems is to compute the probability distribution $P(M = m)$ of a given observable M

starting from the prior distribution $P(\omega)$ defining our model of that system, and to see whether this distribution concentrates, in some scaling limit, over specific values of M. The scaling limit that needs to be considered depends on the system and observable chosen: it can be the infinite-volume limit $N \to \infty$ for a fixed-time observable $M(\omega_i)$, the infinite-time limit $n \to \infty$ for additive or current-like observables, or a combination of these two limits if the latter observables involve many particles. Other limits can also be conceived, for example, the small-noise limit of dynamical systems perturbed by noise, the continuous-time limit of discrete-time systems, or the continuous-space limit of discrete-space systems [3].

Some of these limits will be explored in the following sections. The essential point to note is that these *scaling* or *hydrodynamic limits* are expected to give rise to a concentration of the probability distribution $P(M)$ similar to the one discussed for equilibrium systems, and thus to the emergence of typical states for the system or observable studied.

11.2.3
Equilibrium Versus Nonequilibrium Systems

So far we have not attempted to distinguish equilibrium from nonequilibrium systems in any precise way other than to hint that the distribution $P(\omega)$ describes a "static" random variable in the case of equilibrium systems and a "dynamic" random variable, that is, a *stochastic process*, in the case of nonequilibrium systems. But what makes a system an equilibrium or a nonequilibrium system?

To answer this question in a mathematical way, we need to consider the stochastic time evolution of a system and study how the prior distribution $P(\omega)$ of its complete trajectory ω behaves when the time ordering of ω is reversed. To be more precise, consider the discrete-time trajectory $\omega = (\omega_1, \omega_2, \ldots, \omega_{n-1}, \omega_n)$ containing n time steps, and define the *time-reversed* trajectory ω^{R} associated with ω as the trajectory obtained by reordering the states of ω in reverse order, that is, $\omega^{\mathrm{R}} = (\omega_n, \omega_{n-1}, \ldots, \omega_2, \omega_1)$. Then we say that the system modeled by $P(\omega)$ is an *equilibrium system* if $P(\omega) = P(\omega^{\mathrm{R}})$ for all ω [4].[1] If this condition is not satisfied, then we say that the system is a *nonequilibrium system*. Mathematically, this condition is equivalent to the notion of *reversibility* or *detailed balance*, stated here at the level of complete trajectories rather than the more usual level of transition rates. Thus, the stochastic dynamics of an equilibrium system satisfy detailed balance, whereas those of a nonequilibrium system do not.

The rationale behind this definition is that equilibrium prior distributions, such as the microcanonical and canonical distributions, are stationary distributions of stochastic dynamics verifying detailed balance, and that the very notion of detailed balance captures the physical observation that equilibrium systems are systems in which fluctuations arise with no "preferred" direction in time. Nonequilibrium systems, by contrast, have a stochastic dynamics that is not symmetric under time reversal. This does not mean that nonequilibrium systems do not have stationary

1) For simplicity, we assume that the ω's themselves have even parity under time reversal.

distributions; they often do, but the form of these distributions is generally much more complicated than equilibrium distributions.

11.3
Elements of Large Deviation Theory

We show in this section how the mathematical theory of large deviations makes precise the observation that probability distributions of macrostates concentrate *exponentially* with some scaling parameter (e.g., number of particles, volume, integration time, noise power, etc.). This exponential concentration is the source, for equilibrium systems, of the Legendre transform connecting the entropy and the free energy, and, therefore, of the Legendre structure of thermodynamics. For nonequilibrium systems, it also gives rise to a Legendre transform between quantities that are the nonequilibrium analogues of the entropy and the free energy.

11.3.1
General Results

To explain the central ideas and results of large deviation theory, we first consider a general random variable or macrostate A_n indexed by the parameter n which can, for example, be the number of particles or the number of time steps.

The starting point of large deviation theory is the observation that the probability distribution $P(A_n)$ of A_n is, for many random variables of interest, decaying to zero exponentially fast with n. The exponential decay is in general not exact; rather, what often happens is that the dominant term in the expression of $P(A_n)$ is a decaying exponential with n, so that we can write

$$P(A_n = a) \approx \mathrm{e}^{-nI(a)}, \tag{11.8}$$

with $I(a)$ the rate of decay. When $P(A_n)$ has this form, we say that $P(A_n)$ or A_n satisfies a *large deviation principle* (LDP). To be more precise, we say that $P(A_n)$ or A_n satisfies an LDP if the limit

$$\lim_{n\to\infty} -\frac{1}{n}\ln P(A_n = a) = I(a) \tag{11.9}$$

exists. The decay function $I(a)$ defined by this limit is called the *rate function*. The factor n in the exponential is called the *speed* of the LDP [2, 3, 8].

The interest in large deviations arises because many random variables and stochastic processes satisfy such a principle (although not all do; see, for example, Ref. [3]). The goal of large deviation theory, in this context, is to provide methods for proving that a given random variable or process satisfies an LDP and for obtaining the rate function controlling the rate of decay of the LDP.

Among these methods, let us mention two that are especially useful. The first is known as the *Gärtner–Ellis theorem* [8] and proceeds by calculating the following function:

$$\lambda(k) = \lim_{n\to\infty} \frac{1}{n} \ln\langle e^{nkA_n}\rangle, \tag{11.10}$$

known as the *scaled cumulant generating function* (SCGF). The statement of the Gärtner–Ellis theorem, in its simplified form, is that, if $\lambda(k)$ is differentiable for all $k \in \mathbb{R}$, then A_n satisfies an LDP with rate function given by the *Legendre–Fenchel transform* of $\lambda(k)$:

$$I(a) = \max_{k\in\mathbb{R}} \{ka - \lambda(k)\}. \tag{11.11}$$

In many physical applications, $\lambda(k)$ is actually differentiable *and* strictly convex, and in this case the Legendre–Fenchel transform reduces to the better-known *Legendre transform*, written as

$$I(a) = k(a)a - \lambda(k(a)), \tag{11.12}$$

with $k(a)$ the unique solution of $\lambda'(k) = a$.

The Gärtner–Ellis theorem is useful in practice because it bypasses the direct calculation of $P(A_n)$. By calculating the SCGF of A_n and by checking that this function is differentiable, we instantly prove that $P(A_n)$ satisfies an LDP and obtain the rate function controlling the concentration of $P(A_n)$ in the limit $n \to \infty$.

For certain random variables, $\lambda(k)$ can be calculated but is not differentiable; see Ref. [3] for examples. In this case, it can be proved that the Legendre–Fenchel transform of $\lambda(k)$ yields only the convex envelope of $I(a)$ [3]. To obtain the full rate function, one may use other methods, such as the *contraction principle* [2, 3, 8, 22]. The basis of this method is to express A_n as a function $f(B_n)$ of some random variable B_n satisfying an LDP with rate function $J(b)$, that is,

$$P(B_n = b) \approx e^{-nJ(b)}. \tag{11.13}$$

If such a random variable and function can be found, then the contraction principle states that A_n also satisfies an LDP with rate function given by

$$I(a) = \min_{b:f(b)=a} J(b). \tag{11.14}$$

The minimization appearing in this result is a natural consequence of the approximation method known as Laplace's principle [2, 3, 8] as applied to the integral

$$P(A_n = a) = \int_{b:f(b)=a} P(B_n = b)\mathrm{d}b. \tag{11.15}$$

Assuming that the probability distribution $P(B_n)$ decays exponentially with n, this integral is dominated by the largest exponential term such that $f(b) = a$, which means that we can write

$$P(A_n = a) \approx \exp\left(-n \min_{b:f(b)=a} J(b)\right) \tag{11.16}$$

with subexponential correction factors in n. Thus, we see that $P(A_n)$ satisfies an LDP with rate function given by Eq. (11.14).

The name "contraction" arises in the context of this result from the fact that the function f can in general be a many-to-one function, in which case we are "contracting" the fluctuations of B_n down to the fluctuations of A_n in such a way that the probability of the fluctuation $A_n = a$ is the probability of the most probable (yet exponentially improbable) fluctuation $B_n = b$ leading to $A_n = a$.

This interplay between the appearance of exponentially small terms in integrals and the possibility to approximate these integrals by their largest term using Laplace's principle also explains the appearance of the "max" in the Gärtner–Ellis theorem and the Legendre transform connecting rate functions and SCGFs [3]. In this sense, large deviation theory can be thought of as a "calculus" of exponentially decaying probability distributions, connecting the properties of integrals such as $\langle e^{nkA_n}\rangle$, which are exponential in n, with the exponential properties of $P(A_n)$ itself.

11.3.2
Equilibrium Large Deviations

The application of the results stated above to equilibrium systems is straightforward. For definiteness, we consider a general macrostate $M_N(\omega)$ involving N particles and study its probability distribution $P_\beta(M_N)$ in the canonical ensemble defined by the prior probability distribution

$$P_\beta(\omega) = \frac{e^{-\beta H_N(\omega)}}{Z_N(\beta)}, \qquad Z_N(\beta) = \int_{\Lambda_N} e^{-\beta H_N(\omega)}\, d\omega, \tag{11.17}$$

where H_N is the Hamiltonian of the system considered.

If $P_\beta(M_N)$ satisfies an LDP, then the limit

$$\lim_{N\to\infty} -\frac{1}{N}\ln P(M_N = m) = I_\beta(m) \tag{11.18}$$

exists and defines the rate function $I_\beta(m)$ of M_N in the canonical ensemble at fixed inverse temperature β. The SCGF associated with this rate function is

$$\lambda_\beta(k) = \lim_{N\to\infty} \frac{1}{N}\ln\langle e^{NkM_N}\rangle_\beta, \tag{11.19}$$

where

$$\langle e^{NkM_N}\rangle_\beta = \int_{\Lambda_N} e^{NkM_N(\omega)}\, P_\beta(\omega) d\omega. \tag{11.20}$$

If $\lambda_\beta(k)$ is differentiable in k, then by the Gärtner–Ellis theorem we have that $I_\beta(m)$ is the Legendre–Fenchel transform of $\lambda_\beta(k)$.

The connection between the LDP of M_N and its equilibrium values comes directly by writing the LDP informally in the form

$$P_\beta(M_N = m) \approx e^{-NI_\beta(m)}. \tag{11.21}$$

Since rate functions are always positive, this result shows that $P_\beta(M_N)$ decays exponentially fast with N except at points where $I_\beta(m)$ vanishes. As noted before, these points must correspond to the points where $P_\beta(M_N)$ concentrates in the limit $N \to \infty$, and so to the equilibrium values of M_N. Mathematically, we therefore define the set $\mathcal{E}_\beta$ of equilibrium values of M_N in the canonical ensemble as the set of global minima and zeros of the rate function $I_\beta(m)$:

$$\mathcal{E}_\beta = \{m : I_\beta(m) = 0\}. \tag{11.22}$$

A similar definition can be given for the equilibrium values of M_N in the microcanonical ensemble or any other ensemble by replacing $P_\beta(\omega)$ with the prior probability distribution defining these ensembles (see Section 5.3 of Ref. [3]).

The rate function describes of course not only the equilibrium states but also the fluctuations around these states. In particular, if the rate function $I_\beta(m)$ admits a Taylor expansion of the form

$$I_\beta(m) = a(m - m^*)^2 + O(|m - m^*|^3) \tag{11.23}$$

around a given equilibrium value m^*, then the *small* fluctuations of M_N around m^* are Gaussian distributed. The rate function, however, is rarely an exact parabola, which means that the *larger* fluctuations of M_N away from m^* are in general not Gaussian distributed. Their distribution is determined by the rate function $I_\beta(m)$ and depends on the system studied.

This explains the word "large" in large deviation: contrary to the central limit theorem, which gives only information about the distribution of random variables around their mean, large deviation theory gives information about this distribution not only near the mean but also away from the mean – that is, it gives information about both the *small* and the *large* fluctuations or deviations of random variables. From this point of view, large deviation theory can be thought of as generalizing both the law of large numbers and the central limit theorem.

These observations are valid for any random variable. A more specific connection with equilibrium systems can be established by studying the large deviations of the mean energy $h_N = H_N/N$ with respect to the uniform distribution $P(\omega) = 1/|\Lambda_N|$. In this case, the integral

$$P(h_N = u) = \int_{\Lambda_N} \delta(h_N(\omega) - u) P(\omega) \mathrm{d}\omega \tag{11.24}$$

is, up to a multiplicative constant, the density of states giving the number of microstates having a given mean energy $h_N(\omega) = u$. For most (if not all) equilibrium systems, the density of states is known to grow exponentially with N (or, more generally, the volume) and this directly implies an LDP for $P(h_N = u)$, which we write as

$$P(h_N = u) \approx \mathrm{e}^{Ns(u)}. \tag{11.25}$$

In this form, it is clear that the function $s(u)$ obtained with the limit

$$s(u) = \lim_{N\to\infty} \frac{1}{N} \ln P(h_N = u) \tag{11.26}$$

is the thermodynamic entropy associated with h_N. To be more precise, it is the entropy of h_N, as usually calculated from the density of states, minus an unimportant additive constant (see Section 5.2 of Ref. [3]).

To complete the connection with thermodynamics, note that the generating function

$$\langle e^{Nkh_N} \rangle = \int_{\Lambda_N} e^{Nkh_N(\omega)} P(\omega) d\omega \tag{11.27}$$

can be interpreted as the canonical partition function, whereas the SCGF of h_N, defined as

$$\lambda(k) = \lim_{N\to\infty} \frac{1}{N} \ln\langle e^{Nkh_N} \rangle, \tag{11.28}$$

can be seen as the analogue of the free energy function of the canonical ensemble. To be more precise, redefine the partition function $Z_N(\beta)$ by including the prior uniform distribution $P(\omega)$ in the integral over Λ_N:

$$Z_N(\beta) = \int_{\Lambda_N} e^{-\beta H_N(\omega)} P(\omega) d\omega \tag{11.29}$$

and define the free energy[2)] by

$$\varphi(\beta) = \lim_{N\to\infty} -\frac{1}{N} \ln Z_N(\beta). \tag{11.30}$$

Then, it is easy to see that $\varphi(\beta) = -\lambda(k)$ with $k = -\beta$. From this connection, it is also easy to see that the Gärtner–Ellis theorem implies that, if $\varphi(\beta)$ is differentiable, then

$$s(u) = \min_{\beta\in R} \{\beta u - \varphi(\beta)\}. \tag{11.31}$$

This Legendre–Fenchel transform, with its inverse transform expressing $\varphi(\beta)$ in terms of $s(u)$ (see Section 5.2 of Ref. [3]), is the formal expression of the Legendre transform appearing in thermodynamics. The large deviation derivation of this transform makes it clear that it is valid under specific mathematical conditions (namely, the differentiability of φ) and that it arises because of the exponential nature of both $P(h_N)$ and $Z_N(\beta)$ and the Laplace's principle linking these two functions in the thermodynamic limit. In this sense, the Legendre transform of thermodynamics does not arise out of any physical requirement – it is a consequence of the large deviation structure of statistical mechanics, which appears in the thermodynamic limit.

11.3.3
Nonequilibrium Large Deviations

Let us now consider a nonequilibrium macrostate or observable $M_n(\omega)$ involving n time steps of a Markov process $\omega = (\omega_1, \omega_2, \ldots, \omega_n)$ described by the transition matrix elements $P(\omega_i|\omega_{i-1})$. The large deviation properties of M_n can be studied

2) In thermodynamics, the free energy is more commonly defined with an additional factor $1/\beta$ in front of the logarithm.

similarly as done for equilibrium macrostates by calculating the SCGF $\lambda(k)$ associated with M_n and by obtaining the rate function of M_n as the Legendre–Fenchel transform of $\lambda(k)$ provided that $\lambda(k)$ is differentiable.

The Markov structure of the process underlying M_n can be used here to obtain more explicit expressions for $\lambda(k)$. In the case of the additive observable shown in Eq. (11.6), for example, we have

$$\lambda(k) = \ln \zeta(\mathbf{P}_k), \tag{11.32}$$

$\zeta(\mathbf{P}_k)$ being the largest eigenvalue of the so-called *tilted* transition matrix $\mathbf{P}_k$ with elements

$$P_k(\omega_i|\omega_{i-1}) = P(\omega_i|\omega_{i-1})\mathrm{e}^{kf(\omega_i)}. \tag{11.33}$$

For the current-like observable M_n shown in Eq. (11.7), we have the same result but with $\mathbf{P}_k$ now given by

$$P_k(\omega_i|\omega_{i-1}) = P(\omega_i|\omega_{i-1})\mathrm{e}^{kf(\omega_i,\omega_{i-1})}. \tag{11.34}$$

These results are valid if the state space of the Markov chain is bounded. For unbounded state spaces, $\lambda(k)$ is not necessarily given by the logarithm of the dominant eigenvalue of $\mathbf{P}_k$. We shall see a related example in Section 11.4.4.

For Markov processes evolving continuously in time, the above statements translate into the following ones. For an additive functional of the form

$$M_T(\omega) = \frac{1}{T}\int_0^T f(\omega_t)\mathrm{d}t, \tag{11.35}$$

the SCGF $\lambda(k)$, calculated in the limit $T \to \infty$, is given by the largest eigenvalue of the *tilted generator*,

$$G_k(\omega',\omega) = G(\omega',\omega) + kf(\omega)\delta_{\omega',\omega}, \tag{11.36}$$

with $G(\omega',\omega)$ the elements of the generator of the original process. Note the absence of the logarithm here, as we are dealing with the generator, not the transition matrix. For current-like observables having the form

$$M_T(\omega) = \frac{1}{T}\sum_{i=0}^{N(T)-1} f(\omega_{t_i},\omega_{t_{i+1}}), \tag{11.37}$$

where the sum is over the random transitions between states occurring at times $\{t_0, t_1, \ldots, t_{N(T)-1}\}$, $\lambda(k)$ is given instead by the largest eigenvalue of a tilted generator with elements

$$G_k(\omega',\omega) = G(\omega',\omega)\mathrm{e}^{kf(\omega,\omega')}. \tag{11.38}$$

Observables involving other scaling limits, in addition to the time or particle number limits, can be treated in a similar way, as discussed, for example, in Section 11.4.5. In all cases, the LDPs that we obtain give us information about the fluctuations of the observable of interest similar to the information obtained from equilibrium LDPs. In particular, the global minima and zeros of the rate function determine the typical values of that observable, which are physically interpreted as typical steady states or hydrodynamic states, depending on the observable studied. Then, as for random variables in general, the shape of the rate function around its minima determines the behavior of the small and large fluctuations of the observable around its typical values.

With the knowledge of the rate function of an observable, it is possible, for example, to determine whether this observable satisfies the so-called fluctuation relation symmetry. Consider, to be specific, an observable M_T integrated over the time T and assume that M_T satisfies an LDP with rate function $I(m)$. We say that M_T satisfies a *Gallavotti–Cohen-type fluctuation relation* if

$$\frac{P(M_T = m)}{P(M_T = -m)} \approx \mathrm{e}^{Tcm}, \tag{11.39}$$

with c a positive constant. This means that the positive fluctuations of M_T are exponentially more probable than negative fluctuations of equal magnitude. In large deviation terms, it is easy to see that a sufficient condition for having this result is that $I(m)$ satisfy the following symmetry relation:

$$I(-m) - I(m) = cm. \tag{11.40}$$

In terms of the SCGF, we have equivalently

$$\lambda(k) = \lambda(-k - c). \tag{11.41}$$

The next section includes an example of a very simple Markov process having this fluctuation symmetry. For other more complicated examples, see, for example, Refs [3, 23].

11.4
Applications to Nonequilibrium Systems

In this section, we aim to illustrate, more concretely, how the large deviation formalism of the preceding section can be applied to nonequilibrium systems as introduced in Section 11.2.2. We concentrate on Markov processes in continuous time, providing a detailed pedagogical treatment for toy models of random walkers and indicating how the same techniques can be applied to more complicated (and hence more interesting) models of interacting particles. Among other sources, we draw here on the comprehensive review of Derrida [4] in which further details of many-particle applications can be found.

11.4.1
Random Walkers in Discrete and Continuous Time

To fix ideas, let us start by analyzing perhaps the simplest possible model – a random walker in discrete space. Specifically, we consider a particle moving on a one-dimensional lattice of L sites with, for now, periodic boundary conditions. The microstate of the model is the particle's position $\omega \in \{1, 2, \ldots, L\}$ and, in discrete time, the probabilities $P(\omega_i|\omega_{i-1})$ to move between those positions are contained in the transition matrix $\mathbf{P}$. We assume that the particle has probability p per time step, with $0 < p < 1$, to move in the clockwise direction and probability q per time step, with $0 < q \leq 1 - p$, to move in the anticlockwise direction. Note that, if $p + q < 1$, the particle also has a finite probability to remain stationary in a given time step. The position of the particle on the state space $\{1, 2, \ldots, L\}$ is thus a Markov chain with transition matrix

$$\mathbf{P} = \begin{pmatrix} 1-p-q & p & 0 & \cdots & q \\ q & 1-p-q & p & \cdots & 0 \\ 0 & q & 1-p-q & \cdots & 0 \\ \vdots & \vdots & \vdots & \ddots & \vdots \\ p & 0 & 0 & \cdots & 1-p-q \end{pmatrix}. \tag{11.42}$$

It is a simple exercise to show that this Markov chain has a limiting (stationary) distribution with probability $1/L$ for the particle to be found on any given site – an intuitively obvious result! Slightly more interesting, from the nonequilibrium point of view, is the particle current $J_n(\omega)$ that we define as the net number of clockwise jumps made by the particle *per time step*. This is a function of the form (11.7) with

$$f(\omega_i, \omega_{i+1}) = \delta_{\omega_i+1,\omega_{i+1}} - \delta_{\omega_i-1,\omega_{i+1}}. \tag{11.43}$$

A straightforward calculation shows that the mean stationary current is given by $\langle J_n \rangle = p - q$. We shall see in Section 11.4.3 that this corresponds to the concentration point of the probability distribution $P(J_n = j)$ which satisfies an LDP. For now, note that there is an obvious qualitative difference between the case of $p = q$ (zero mean current) and the case of $p \neq q$ (nonzero mean current). Mathematically, this difference is just the distinction between a reversible and a nonreversible Markov chain, in the sense of detailed balance. As explained in Section 11.2.3, we identify the former case as an equilibrium system and the latter as an nonequilibrium system.

The continuous-time version of this random walk can be understood by associating a physical time increment Δt with each discrete time step (where above we implicitly assumed $\Delta t = 1$), setting the hopping probabilities per time step to $p\Delta t$ and $q\Delta t$ and then taking the limit $\Delta t \to 0$. Formally, our particle then remains at a given site for an exponentially distributed waiting time, with mean $1/(p + q)$, before moving clockwise, with probability $p/(p + q)$, or anticlockwise, with probability $q/(p + q)$. Note, in particular, that p and q are now interpreted as rates rather than probabilities and can each be greater than unity. The infinitesimal generator corresponding to this process is

$$\mathbf{G} = \begin{pmatrix} -p-q & p & 0 & \cdots & q \\ q & -p-q & p & \cdots & 0 \\ 0 & q & -p-q & \cdots & 0 \\ \vdots & \vdots & \vdots & \ddots & \vdots \\ p & 0 & 0 & \cdots & -p-q \end{pmatrix}. \tag{11.44}$$

Unsurprisingly, a picture similar to the discrete-time case emerges: the process has a stationary state that has mean density $1/L$ on each site and mean current $p - q$. Following the previous sections, we shall be interested next in deriving such mean values as concentration points of LDPs. To be specific, we shall illustrate the general discussion below with explicit calculations related to the continuous-time random walk model and various modifications of it. We note that, as mentioned in Section 11.2.2, the appropriate scaling limit for which an LDP holds depends on the observable we wish to consider.

11.4.2
Large Deviation Principle for Density Profiles

A general interacting particle system has a state space consisting of all possible particle configurations and, on a coarse-grained scale, one is often interested in the probability of observing a particular fixed-time density profile in space. This leads to the concept of a density function LDP which can be straightforwardly extended from equilibrium to nonequilibrium (see, for example, Ref. [4]).

The appropriate scaling limit expressing such an LDP is the infinite-volume (and infinite particle number) limit. To be precise, for a system defined on a lattice of linear size L in d dimensions, one considers taking the thermodynamic limit $L \to \infty$ while rescaling the coordinates $\mathbf{r}$ to $\mathbf{x} = \mathbf{r}/L$. The probability of seeing a given density profile $\varrho(\mathbf{x})$ is then expected to obey

$$P[\varrho(\mathbf{x})] \approx \exp\{-L^d \mathcal{F}[\varrho(\mathbf{x})]\} \tag{11.45}$$

as $L \to \infty$. This is a *functional LDP*, as P and $\mathcal{F}$ are both functionals of $\varrho(\mathbf{x})$.[3)] The use of the square brackets emphasizes this point.

For equilibrium systems with short-range interactions, the form of the large deviation rate functional $\mathcal{F}$ is obtained from the knowledge of $f(\varrho)$, the free energy per site, as

$$\mathcal{F}[\varrho(\mathbf{x})] = \int_0^1 [f(\varrho(\mathbf{x})) - f(\varrho^*) - (\varrho(\mathbf{x}) - \varrho^*)f'(\varrho^*)]\mathrm{d}\mathbf{x}, \tag{11.46}$$

where ϱ^* is just the mean number of particles per site [4]. We see here that $\mathcal{F} = 0$ for the uniform density profile $\varrho(x) = \varrho^*$. In other words, as expected, ϱ^* is the typical density about which the probability distribution concentrates in the large volume limit.

To illustrate these results, let us consider a collection of independent random walkers in one dimension, discussed as a special case in Ref. [24]. To facilitate later generalizations, we now choose to work with open boundaries rather than periodic boundary conditions. Specifically, we modify our setup to consider L sites coupled to left and right boundary reservoirs with densities ϱ_L and ϱ_R, respectively, so that

3) $\mathcal{F}$ is the large deviation analogue of the Ginzburg–Landau free energy expressed as a function of the particle density in the grand canonical ensemble.

particles are input from the left reservoir with rate $p\varrho_L$ and from the right reservoir with rate $q\varrho_R$. In the bulk, and for exiting the system, each particle independently has the dynamics of the single random walker defined in Section 11.4.1.

For equal reservoir densities, $\varrho_L = \varrho_R = \varrho^*$, it is a relatively simple exercise to show that for any p, q, the number of particles on each site is a Poisson distribution with mean ϱ^*. The corresponding free energy (see, for example, Ref. [24]) leads to a large deviation functional of the form

$$\mathcal{F}[\varrho(\mathbf{x})] = \int_0^1 \left[\varrho^* - \varrho(\mathbf{x}) + \varrho(\mathbf{x})\ln\frac{\varrho(x)}{\varrho^*}\right]\mathrm{d}\mathbf{x}. \tag{11.47}$$

A Taylor expansion of the integrand readily demonstrates Gaussian fluctuations about $\varrho(x) = \varrho^*$. Note that, in this special case of equal reservoir densities, the form of $\mathcal{F}$ is the same for $p = q$ and $p \neq q$: it is a local, convex, functional as is typically the case for equilibrium. Furthermore, an ensemble equivalence argument suggests that in the $L \to \infty$ limit, the density large deviation functional would be the same for periodic boundary conditions with a fixed mean density ϱ^*.

For interacting particle systems with *nonequal* reservoir densities (i.e., boundary driving), the situation is much more interesting. In particular, the density large deviation functional is generically expected to have a nonlocal structure reflecting the long-range spatial correlations characteristic of nonequilibrium. This is seen, for example, in the case of the symmetric simple exclusion process [25, 26] which can be treated analytically by using the well-known matrix product ansatz [27] and an associated additivity property. In the corresponding asymmetric simple exclusion process, $\mathcal{F}$ is nonconvex for some parameters indicating a phase transition [28, 29]. The form of $\mathcal{F}$ in certain models can be obtained by utilizing the macroscopic fluctuation theory of Bertini *et al.* [5, 30] to which we shall return in Section 11.4.5.

11.4.3
Large Deviation Principle for Current Fluctuations

The presence of nonzero currents is a generic feature of nonequilibrium stationary states. For Markov processes, one generically finds that a given current J_T (e.g., the net number of particles hopping between two lattice sites) time averaged over the interval $[0, T]$ obeys a large deviation principle with speed T, that is,

$$P(J_T = j) \approx \mathrm{e}^{-TI(j)}. \tag{11.48}$$

The relevant scaling limit here is the long-time, $T \to \infty$, limit. Although this may be combined with an infinite-volume limit (as in Section 11.4.5), there is particular interest in current fluctuations in *small systems*, for example, trapped colloidal particles (see Chapter 2) or single-molecule biological experiments (see Chapter 5). In this spirit, we use this section to illustrate the calculation and properties of $I(j)$ for a single random walker. For a treatment of general Markov diffusions, see Refs [16, 31].

As follows from the general discussion in Section 11.3.3, the rate function $I(j)$ can be obtained as the Legendre–Fenchel transform of the SCGF $\lambda(k)$ of J_T, which, for a continuous-time process with finite state space, is given by the principal eigenvalue $\zeta(k)$ of a tilted generator. For example, for the single-particle random walker on a ring with a current J_T defined, as in Section 11.4.1, to be the net number of clockwise jumps the particle makes per unit time, then we need the principal eigenvalue of the matrix

$$\mathbf{G}(k) = \begin{pmatrix} -p-q & pe^{k} & 0 & \dots & qe^{-k} \\ qe^{-k} & -p-q & pe^{k} & \dots & 0 \\ 0 & qe^{-k} & -p-q & \dots & 0 \\ \vdots & \vdots & \vdots & \ddots & \vdots \\ pe^{k} & 0 & 0 & \dots & -p-q \end{pmatrix}. \tag{11.49}$$

It is easy to show that the normalized vector $(1/L, 1/L, \dots, 1/L)$ is a left eigenvector of this matrix with eigenvalue

$$-p(1-e^{k}) - q(1-e^{-k}) \tag{11.50}$$

and an appeal to Perron–Frobenius theory [32] supports the assertion that this is the desired principal eigenvalue $\zeta(k)$ which is equal to the SCGF $\lambda(k)$. From here it is a straightforward, albeit tedious, exercise to calculate $I(j)$:

$$I(j) = \max_{k} \{kj - \lambda(k)\} = p + q - \sqrt{j^2 + 4pq} + j \ln \frac{j + \sqrt{j^2 + 4pq}}{2p}. \tag{11.51}$$

Note that this result can also be obtained by arguing that the clockwise jumps form a Poisson process with rate p, whereas the anticlockwise jumps form a Poisson process with rate q. Considering the long-time limit of the Poisson process, we hence have separate large deviation functions for the clockwise current J_+ and the anticlockwise current J_- with respective rate functions

$$I(j_+) = p - j_+ + j_+ \ln\frac{j_+}{p}, \qquad I(j_-) = q - j_- + j_- \ln\frac{j_-}{q}. \tag{11.52}$$

The rate function for the net current $J_T = J_+ - J_-$ can then be obtained by the method of contraction discussed in Section 11.3.1.

Notice that the rate function for the current of the random walker obeys the fluctuation relation symmetry

$$I(-j) - I(j) = cj, \tag{11.53}$$

with $c = \ln(p/q)$, so that

$$\frac{P(J_T = j)}{P(J_T = -j)} = e^{Tcj}. \tag{11.54}$$

This result is a simple example of a fluctuation relation of the Gallavotti–Cohen type [33–35]. As mentioned in Section 11.3.3, this result can also be expressed as the SCGF property $\lambda(k) = \lambda(-k - c)$ which derives ultimately from a straightforwardly verified symmetry of the tilted generator:

$$\mathbf{G}(k)^T = \mathbf{G}(-k - c), \tag{11.55}$$

where T denotes here transpose (not time). Note that for other currents (e.g., counting the jumps across just a single bond), $\mathbf{G}(k)^T$ is no longer equal to $\mathbf{G}(-k - c)$ but, under quite general conditions, is related to it by a similarity transform so that the two matrices have identical eigenvalues and the symmetry (11.54) still holds. The diagonal change-of-basis matrix in the similarity transform is related to current boundary terms which, for finite state space, are irrelevant in the long-time limit. The plethora of different finite-time fluctuation relations can be associated with different choices for these boundary terms (see, for example, Refs [23, 36] and elsewhere in this book).

We shall see with an example below that, for infinite state space, the boundary terms may become relevant. In passing, we also note that while the fluctuation theorem can be elegantly expressed as a property of the large deviation rate function, the existence of a large deviation principle is not, as sometimes believed, a *necessary* prerequisite for the existence of a fluctuation relation of form (11.54). A simple counterexample is provided by a random walker with right and left hopping rates increasing in time as $p \times t$ and $q \times t$, respectively. It is easy to show that such a system has no stationary state (the mean current increases indefinitely), but since the ratio of rates of right and left steps is constant, a relation of the form (11.54) still holds.

11.4.4
Interacting Particle Systems: Features and Subtleties

Thus far, the explicit examples of this section have been concerned with single-particle random walks or noninteracting collections thereof. The same general formalism applies for interacting particle systems [37], although analytically tractable models are the exception rather than the rule. Paradigmatic examples include the symmetric and asymmetric simple exclusion processes, mentioned already in Section 11.4.2, and the zero-range process (ZRP) [38]. Among other results, the current large deviations in the open-boundary asymmetric exclusion process have recently been calculated [39, 40]. Here we concentrate on the ZRP with open-boundary conditions [41] (connected to so-called Jackson networks of queueing theory [42]) in order to exemplify subtleties arising in systems with unbounded state space.

For our purposes, it suffices to consider the ZRP on a one-dimensional open lattice, although related issues have also been examined for queuing models on more complicated geometries [43]. In one dimension, each site $l \in \{1, 2, \dots, L\}$ contains an integer number of particles n_l that can hop to the nearest neighbor sites according to a continuous-time dynamics. Specifically, in the bulk the topmost particle on each site hops to the right (or to the left) with rate pw_n (respectively, qw_n), where w_n is a function of the number of particles n on the departure site. Particles are injected onto site 1 (or L) with rate α (respectively, δ) and extracted with rate γw_n (respectively, βw_n).

The properties of the model depend crucially on the function w_n. The choice $w_n \propto n$ corresponds to noninteracting particles, such as the random walkers considered above, whereas other forms represent an effective on-site attraction or repulsion. In particular, if w_n is bounded as $n \to \infty$, that is,

$$\lim_{n\to\infty} w_n = a < \infty, \tag{11.56}$$

then the model exhibits a condensation transition where, for some choices of boundary rates, particles "pile up" indefinitely at one or more sites and there is no stationary state. For boundary rates outside this regime, the stationary state of the model is a product measure characterized by a site-dependent fugacity. The mean current across each bond (i.e., between each pair of sites) depends on the rates p, q, α, β, γ, and δ, but not explicitly on w_n. However, the form of w_n determines the relationship between the fugacity and the particle density and also the relationship between α and δ and effective reservoir densities at the boundaries.

To calculate the fluctuations around the stationary-state current, we can try to look, as above, for the principal eigenvalue of a tilted generator (which can be represented in terms of tensor products of matrices encoding the particle dynamics on each site). The form of this tilted generator will depend on the bond(s) across which we choose to measure the current. In the case where w_n is unbounded, in the sense that $w_n \to \infty$ as $n \to \infty$, then the spectrum of the tilted generator is always gapped and the Legendre transform of the principal eigenvalue, which can be explicitly calculated in terms of the transition rates, gives the large deviation rate function for all values of current. Furthermore, as might be expected, the principal eigenvalues, and hence the current fluctuations, are the same for currents across all bonds.

On the other hand, for w_n bounded, the spectrum of the tilted generator becomes gapless for some values of k and certain boundary terms can also diverge. Mathematically, this means that $\lambda(k)$ is no longer simply given by the principal eigenvalue. Physically, this possibility is related to the fact that, over long but finite timescales, an arbitrarily large number of particles can accumulate on each site. This manifests in the following properties of the current large deviation function:

- It is bond inhomogeneous so that the probability of seeing extreme current fluctuations depends on where the current is measured.
- It depends on the initial probability distribution of the system.
- It does not obey the Gallavotti–Cohen fluctuation relation for large current fluctuations.

All of these properties are seen even in the single-site zero-range process, for which the complete spectrum of the tilted generator and the form of $\lambda(k)$ for all k can be explicitly obtained [36, 44]. A phase diagram in $x - k$ space, where x is the fugacity characterizing the initial state, reveals that there are two types of "phase transitions" in $\lambda(k)$ analogous to first-order and continuous transitions in equilibrium. At the former, $\lambda(k)$ has a nondifferentiable point, while at the latter it remains differentiable. An attentive reader may question how we can then obtain $I(j)$, since the Gärtner–Ellis theorem of Section 11.3.1 requires differentiability of

$\lambda(k)$ for all k. In fact, the Legendre–Fenchel transform yields the convex envelope of $I(j)$ which contains linear sections corresponding to the nondifferentiable points of $\lambda(k)$. For the ZRP, one can argue on physical grounds that this *is* the correct form of $I(j)$ because the most probable way to realize an average current j in the linear regime involves a phase separation in time with the system spending part of its history in a state with one average current and part in a state with another average current. This argument relies on the system having only short-range correlations in time (just as the analogous Maxwell construction in equilibrium requires short-range correlations in space), so it might be expected to fail in, for example, non-Markovian systems.

11.4.5
Macroscopic Fluctuation Theory

Underlying the so-called *macroscopic fluctuation theory* is the concept of the hydrodynamic limit which describes the emergence of a deterministic coarse-grained description from stochastic microscopic rules [45]. Recall from Sections 11.2 and 11.3 that such a nonfluctuating macroscopic state, corresponding to the concentration point of some probability distribution, is expected to be given by the zero of a large deviation rate function. In this subsection, we sketch the approach of Bertini *et al.* [5, 30] for calculating this rate function.

We are interested here in systems with particle conservation in the bulk and a key ingredient is the functional form of the dependence of the instantaneous local current on the density. The correct scaling required so that the *joint* distribution of current and density profiles concentrates in the limit $L \to \infty$ depends on the form of this current–density relationship. Specifically, we focus our attention here on *diffusive* processes for which the relevant macroscopic coordinates are $\mathbf{x} = \mathbf{r}/L$ and $\tau = t/L^2$. This class of systems includes symmetric and weakly asymmetric versions of both the exclusion process and the zero-range process, but not their asymmetric counterparts for which *Euler scaling* $\tau = t/L$ is needed. To illustrate loosely the procedure involved in taking the hydrodynamic limit, we now return to our favorite example of random walkers.

Consider a collection of noninteracting particles on a one-dimensional lattice with boundary reservoirs ϱ_L and ϱ_R, as in Section 11.4.2, and a weak asymmetry in the bulk hopping dynamics, namely, $p = 1/2 + E/2L$ and $q = 1/2 - E/2L$. The starting point for the hydrodynamic description is to consider the mean current between two neighboring lattice sites, say, l and $l+1$. In terms of the site densities (mean occupation numbers) ϱ_l and ϱ_{l+1}, the mean current is

$$\langle J_{l,l+1} \rangle = p\varrho_l - q\varrho_{l+1}, \tag{11.57}$$

which yields a lattice continuity equation of the form

$$\frac{\partial \varrho_l(t)}{\partial t} = \langle J_{l-1,l} \rangle - \langle J_{l,l+1} \rangle = [p\varrho_{l-1}(t) - q\varrho_l(t)] - [p\varrho_l(t) - q\varrho_{l+1}(t)]. \tag{11.58}$$

Now writing $t = L^2\tau$ and $l = xL$, we have

$$\frac{1}{L^2}\frac{\partial \varrho(x,\tau)}{\partial \tau} = \left[p\varrho\left(x - \frac{1}{L}, \tau\right) - q\varrho(x,\tau)\right] - \left[p\varrho(x,\tau) - q\varrho\left(x + \frac{1}{L}, \tau\right)\right]. \tag{11.59}$$

Then assuming *local stationarity* and carrying out a Taylor expansion to second order yields

$$\frac{\partial \varrho(x,\tau)}{\partial \tau} = -\frac{\partial}{\partial x}\left[E\varrho - \frac{1}{2}\frac{\partial \varrho}{\partial x}\right] \equiv -\frac{\partial}{\partial x}\hat{J}(x,\tau), \tag{11.60}$$

with $\hat{J}(x,\tau)$ a rescaled current.

Similarly, for general d-dimensional diffusive systems obeying Fick's law at the macroscopic level, we expect

$$\hat{\mathbf{J}}(\mathbf{x},t) = \sigma[\varrho(\mathbf{x},t)]\mathbf{E} - D[\varrho(\mathbf{x},t)]\nabla\varrho(\mathbf{x},t), \tag{11.61}$$

where $D[\varrho(\mathbf{x},t)]$ is the diffusivity associated with the density profile $\varrho(\mathbf{x},t)$ and $\sigma[\varrho(\mathbf{x},t)]$ is the corresponding mobility. The hydrodynamic equation, describing the deterministic or macroscopic limit as $L \to \infty$, is

$$\frac{\partial \varrho(\mathbf{x},\tau)}{\partial \tau} = -\nabla \cdot \hat{\mathbf{J}}(\mathbf{x},\tau). \tag{11.62}$$

Together, Eqs. (11.61) and (11.62) only represent the macroscopic or typical behavior obtained in the hydrodynamic limit. To describe the fluctuations around this limit, let us now find the joint rate function of the density and current. Specializing to the one-dimensional case with $E = 0$, one observes (see, for example, Ref. [4]) that for boundary reservoirs with equal density ϱ^* the fluctuations of the microscopic current across each bond can be characterized by

$$\lim_{t\to\infty}\frac{\langle J^2\rangle}{t} = \frac{\sigma(\varrho^*)}{L}. \tag{11.63}$$

At the macroscopic level, in d-dimensions, this motivates adding to $\hat{\mathbf{J}}(\mathbf{x},\tau)$ a term representing Gaussian white noise with variance $\sigma[\varrho(x,t)]$, which leads to an LDP for the joint probability of seeing a particular density and current profile having the form

$$P\left[\varrho(\mathbf{x},\tau), \hat{\mathbf{J}}(\mathbf{x},\tau)\right] \sim \exp\left\{-L^d\int_0^T\int_0^1 \frac{(\hat{\mathbf{J}}(\mathbf{x},\tau) + D[\varrho(\mathbf{x},\tau)]\nabla\varrho(\mathbf{x},\tau))^2}{2\sigma[\varrho(\mathbf{x},\tau)]}\mathrm{d}\mathbf{x}\,\mathrm{d}\tau\right\}, \tag{11.64}$$

with $\hat{\mathbf{J}}$ and ϱ linked by the continuity equation (11.62).

In principle, from here one can use the tools of variational calculus to look for the optimal density profile $\varrho(\mathbf{x},\tau)$ leading to a given final density $\varrho(\mathbf{x})$. This contraction leads to an implicit expression for the density rate function $\mathcal{F}[\varrho(\mathbf{x})]$ defined in Section 11.4.2. Finding explicit solutions is a difficult task, since in general the optimal profile is time dependent. However, successful treatments along these lines include the one-dimensional zero-range and Kipnis–Marchioro–Presutti models [30, 46]; in

Ref. [30], it is also shown that the approach is consistent with the independent results of Derrida *et al.* [25, 26] for the symmetric simple exclusion process.

It turns out to be easier to calculate the current large deviation function [47], in particular, if one assumes that the optimal profile leading to a particular current fluctuation is independent of time and that the optimal current is constant in space. In one dimension, the first assumption can be shown [48] to be equivalent to the additivity principle of Bodineau and Derrida [49], which is known to break down in systems with dynamical phase transitions; the second assumption is important for higher dimensional systems [50]. Under these conditions, the rate function for the (rescaled) current is then given by

$$I(\hat{\mathbf{J}}) = \min_{\varrho(\mathbf{x})} \int_0^1 \frac{(\hat{\mathbf{J}} + D[\varrho(\mathbf{x})]\nabla\varrho(\mathbf{x}))^2}{2\sigma[\varrho(\mathbf{x})]}\,\mathrm{d}\mathbf{x}. \tag{11.65}$$

For any particular model, this integral must be minimized with $\varrho(\mathbf{x})$ matched to the reservoir densities at the boundaries. This generically yields a current distribution with non-Gaussian tails even though the fluctuations themselves are locally Gaussian.

Various scenarios for the form of the macroscopic current large deviation function (including those indicating dynamical phase transitions) are discussed in Ref. [47]. The example models treated there include the one-dimensional ZRP with w_n unbounded, the special case $w_n = n$ corresponding again to noninteracting random walkers. It has recently been pointed out that, if Eq. (11.65) holds, the optimal density profile is the same for all currents with the same magnitude $|\hat{\mathbf{J}}|$ leading to what has been dubbed an *isometric fluctuation relation* [51]. Macroscopic fluctuation theory has also been generalized to treat models with dissipated energy [52].

11.5
Final Remarks

In this chapter, we have merely skimmed the surface of the way in which large deviation theory can provide a framework for understanding existing results in the theory of nonequilibrium systems and probing for new ones. We conclude here with some pointers to other relevant work and ideas for future research directions.

First, we note that the study of fluctuation theorems and relations, as briefly touched on in Sections 11.3.3 and 11.4.3, is a vast subject which percolates through many of the contributions in this volume; for overviews, see Chapters 1, 2, 3 and 7. In particular, we have not discussed here the important concept of entropy production and its distribution, which is central for many fluctuation relation statements and can often be expressed in a large deviation form.

A second topic worth mentioning is the extension of the concept of statistical ensemble, discussed for equilibrium systems in Section 11.3.2, to nonequilibrium systems. Just as one distinguishes equilibrium systems with fixed energy (microcanonical ensemble) from systems with the energy fixed on average via a conjugate Lagrange parameter (canonical ensemble), one can construct a microcanonical ensemble of trajectories for a nonequilibrium system that is constrained to realize a particular observable value (e.g., a particular current) or a canonical ensemble of trajectories which realize that constraint on average. An example of the former ensemble, obtained for the ASEP on a ring conditioned on enhanced flux, has recently been analyzed in Ref. [53]. A study of canonical-type nonequilibrium ensembles, which are also known as *biased ensembles*, can be found in the work of Sollich and Jack [54]. In the context of large deviation theory, these ensembles can be understood in terms of conditional LDPs and the so-called Gibbs conditioning principle [2].

Related to the topic of nonequilibrium ensembles is the issue of determining configurations or states giving rise to fluctuations. We can already get information about the most probable (typical) way to realize a given current fluctuation from the eigenvector corresponding to the principal eigenvalue of the tilted generator (11.38). More work is still needed to understand the properties and correlations of such current-carrying states and constrained states in general. The associated inverse problem of determining microscopic rates that are most likely to yield given macroscopic properties has been studied by Evans [55, 56] and Monthus [57].

Throughout this chapter, we have assumed that nonequilibrium systems of interest are modeled by Markov processes. However, the memoryless property may be an inappropriate approximation for the description of many systems where long-range temporal correlations are known to be important (see, for example, Ref. [58] and references cited therein). Recent work to characterize the large deviation properties of certain classes of history-dependent models can be found in Refs [59, 60]. In particular, it is shown in Ref. [59] that modifying a continuous-time random walker (as introduced in Section 11.4.1) so that the hopping rates at time t depend on the average current up to time t can lead to an altered "speed" (i.e., power of time T) in the LDP for current. Fluctuation relations with the right-hand side of (11.39) replaced by $e^{T^{\alpha} cm}$ have also appeared in the context of anomalous dynamics (see Chapter 8). There is much scope for future work investigating many-particle non-Markovian processes and establishing a common framework for the results. In this regard, and in a more general way, we expect the large deviation formalism to continue playing an important role in quantifying nonequilibrium fluctuations in small systems.

Acknowledgments

H.T. is grateful for the hospitality of the École Normale Supérieure of Lyon and the National Institute for Theoretical Physics at Stellenbosch University, where parts of this chapter were written.

References

1 Deuschel, J.D. and Stroock, D.W. (1989) *Large Deviations*, Academic Press, Boston, MA.

2 Dembo, A. and Zeitouni, O. (1998) *Large Deviations Techniques and Applications*, 2nd edn, Springer, New York.

3 Touchette, H. (2009) The large deviation approach to statistical mechanics. *Phys. Rep.*, **478** (1–3), 1–69.
4 Derrida, B. (2007) Non-equilibrium steady states: fluctuations and large deviations of the density and of the current. *J. Stat. Mech.*, P07023.
5 Bertini, L., Sole, A.D., Gabrielli, D., Jona-Lasinio, G., and Landim, C. (2007) Stochastic interacting particle systems out of equilibrium. *J. Stat. Mech.*, P07014.
6 Ruelle, D. (1969) *Statistical Mechanics: Rigorous Results*, W.A. Benjamin, Amsterdam.
7 Lanford, O.E. (1973) Entropy and equilibrium states in classical statistical mechanics, in *Statistical Mechanics and Mathematical Problems*, Lecture Notes in Physics, vol. 20 (ed. A. Lenard), Springer, Berlin, pp. 1–113.
8 Ellis, R.S. (1985) *Entropy, Large Deviations, and Statistical Mechanics*, Springer, New York.
9 Oono, Y. (1989) Large deviation and statistical physics. *Prog. Theor. Phys. Suppl.*, **99**, 165–205.
10 Paniconi, M. and Oono, Y. (1997) Phenomenological framework for fluctuations around steady state. *Phys. Rev. E*, **55** (1), 176–188.
11 Oono, Y. and Paniconi, M. (1998) Steady state thermodynamics. *Prog. Theor. Phys. Suppl.*, **130**, 29–44.
12 Eyink, G.L. (1990) Dissipation and large thermodynamic fluctuations. *J. Stat. Phys.*, **61** (3), 533–572.
13 Eyink, G.L. (1996) Action principle in nonequilibrium statistical dynamics. *Phys. Rev. E*, **54**, 3419–3435.
14 Eyink, G.L. (1998) Action principle in statistical dynamics. *Prog. Theor. Phys. Suppl.*, **130**, 77–86.
15 Maes, C., Netočný, K., and Shergelashvili, B. (2006) A selection of nonequilibrium issues, in *Methods of Contemporary Mathematical Statistical Physics*, Lecture Notes in Mathematics, vol. **1970** (eds. M. Biskup, A. Bovier, F. den Hollander, D. Ioffe, F. Martinelli, K. Netočný, and C. Toninelli), Springer, Berlin, pp. 1–60.
16 Maes, C. and Netočný, K. (2008) Canonical structure of dynamical fluctuations in mesoscopic nonequilibrium steady states. *Europhys. Lett.*, **82** (3), 30003.
17 Maes, C. and Netočný, K. (2007) Minimum entropy production principle from a dynamical fluctuation law. *J. Math. Phys.*, **48**, 053306.
18 Maes, C. and Netočný, K. (2003) Time-reversal and entropy. *J. Stat. Phys.*, **110** (1), 269–310.
19 Goldstein, S. (2001) Boltzmann's approach to statistical mechanics, in *Chance in Physics: Foundations and Perspectives*, Lecture Notes in Physics, vol. **574** (eds J. Bricmont, D. Dürr, M.C. Galavotti, G.C. Ghirardi, F. Petruccione, and N. Zanghi), Springer, Berlin, pp. 39–54.
20 Goldstein, S. and Lebowitz, J.L. (2004) On the (Boltzmann) entropy of non-equilibrium systems. *Physica D*, **193** (1–4), 53–66.
21 Campa, A., Dauxois, T., and Ruffo, S. (2009) Statistical mechanics and dynamics of solvable models with long-range interactions. *Phys. Rep.*, **480** (3–6), 57–159.
22 Touchette, H. (2010) Methods for calculating nonconcave entropies. *J. Stat. Mech.*, P05008.
23 Harris, R.J. and Schütz, G.M. (2007) Fluctuation theorems for stochastic dynamics. *J. Stat. Mech.*, P07020.
24 Derrida, B. and Gerschenfeld, A. (2009) Current fluctuations in one dimensional diffusive systems with a step initial density profile. *J. Stat. Phys.*, **137** (5–6), 978–1000.
25 Derrida, B., Lebowitz, J.L., and Speer, E.R. (2001) Free energy functional for nonequilibrium systems: an exactly solvable case. *Phys. Rev. Lett.*, **87** (15), 150601.
26 Derrida, B., Lebowitz, J.L., and Speer, E.R. (2002) Large deviation of the density profile in the steady state of the open symmetric simple exclusion process. *J. Stat. Phys.*, **107** (3–4), 599–634.
27 Derrida, B., Evans, M.R., Hakim, V., and Pasquier, V. (1993) Exact solution of a 1D asymmetric exclusion model using a matrix formulation. *J. Phys. A: Math. Gen.*, **26** (7), 1493–1517.
28 Derrida, B., Lebowitz, J.L., and Speer, E.R. (2002) Exact free energy functional for a

driven diffusive open stationary nonequilibrium system. *Phys. Rev. Lett.*, **89** (3), 030601.

29 Derrida, B., Lebowitz, J.L., and Speer, E.R. (2003) Exact large deviation functional of a stationary open driven diffusive system: the asymmetric exclusion process. *J. Stat. Phys.*, **110** (3–6), 775–810.

30 Bertini, L., Sole, A.D., Gabrielli, D., Jona-Lasinio, G., and Landim, C. (2002) Macroscopic fluctuation theory for stationary non-equilibrium states. *J. Stat. Phys.*, **107** (3–4), 635–675.

31 Maes, C., Netočný, K., and Wynants, B. (2008) Steady state statistics of driven diffusions. *Physica A*, **387** (12), 2675–2689.

32 Minc, H. (1988) *Nonnegative Matrices*, John Wiley & Sons, Inc., New York.

33 Evans, D.J. and Searles, D.J. (1994) Equilibrium microstates which generate second law violating steady states. *Phys. Rev. E*, **50** (2), 1645–1648.

34 Gallavotti, G. and Cohen, E.G.D. (1995) Dynamical ensembles in nonequilibrium statistical mechanics. *Phys. Rev. Lett.*, **74** (14), 2694–2697.

35 Lebowitz, J.L. and Spohn, H. (1999) A Gallavotti–Cohen-type symmetry in the large deviation functional for stochastic dynamics. *J. Stat. Phys.*, **95** (1–2), 333–365.

36 Rákos, A. and Harris, R.J. (2008) On the range of validity of the fluctuation theorem for stochastic Markovian dynamics. *J. Stat. Mech.*, P05005.

37 Liggett, T.M. (1985) *Interacting Particle Systems*, Springer, New York.

38 Evans, M.R. and Hanney, T. (2005) Nonequilibrium statistical mechanics of the zero-range process and related models. *J. Phys. A: Math. Gen.*, **38** (19), R195–R240.

39 Lazarescu, A. and Mallick, K. (2011) An exact formula for the statistics of the current in the TASEP with open boundaries. *J. Phys. A: Math. Theor.*, **44** (31), 315001.

40 de Gier, J. and Essler, F.H.L. (2011) Large deviation function for the current in the open asymmetric simple exclusion process. *Phys. Rev. Lett.*, **107**, 010602.

41 Levine, E., Mukamel, D., and Schütz, G. M. (2005) Zero-range process with open boundaries. *J. Stat. Phys.*, **120** (5–6), 759–778.

42 Jackson, J.R. (1957) Networks of waiting lines. *Oper. Res.*, **5** (4), 518–521.

43 Chernyak, V.Y., Chertkov, M., Goldberg, D.A., and Turitsyn, K. (2010) Non-equilibrium statistical physics of currents in queuing networks. *J. Stat. Phys.*, **140** (5), 819–845.

44 Harris, R.J., Rákos, A., and Schütz, G.M. (2006) Breakdown of Gallavotti–Cohen fluctuation theorem for stochastic dynamics. *Europhys. Lett.*, **75** (2), 227–233.

45 Kipnis, C. and Landim, C. (1999) *Scaling Limits of Interacting Particle Systems*, Grundlehren der Mathematischen Wissenschaften, vol. **320**, Springer, Berlin.

46 Bertini, L., Gabrielli, D., and Lebowitz, J.L. (2005) Large deviations for a stochastic model of heat flow. *J. Stat. Phys.*, **121** (5–6), 843–885.

47 Bertini, L., Sole, A.D., Gabrielli, D., Jona-Lasinio, G., and Landim, C. (2006) Non equilibrium current fluctuations in stochastic lattice gases. *J. Stat. Phys.*, **123** (2), 237–276.

48 Bertini, L., Sole, A.D., Gabrielli, D., Jona-Lasinio, G., and Landim, C. (2005) Current fluctuations in stochastic lattice gases. *Phys. Rev. Lett.*, **94** (3), 030601.

49 Bodineau, T. and Derrida, B. (2004) Current fluctuations in non-equilibrium diffusive systems: an additivity principle. *Phys. Rev. Lett.*, **92** (18), 180601.

50 Pérez-Espigares, C., del Pozo, J.J., Garrido, P.L., and Hurtado, P.I. (2011) Large deviations of the current in a two-dimensional diffusive system, in *Nonequilibrium Statistical Physics Today: Proceedings of the 11th Granada Seminar on Computational and Statistical Physics*, AIP Conference Proceedings, vol. **1332**, American Institute of Physics, pp. 204–213.

51 Hurtado, P.I., Pérez-Espigares, C., del Pozo, J.J., and Garrido, P.L. (2011) Symmetries in fluctuations far from equilibrium. *Proc. Natl. Acad. Sci. USA*, **108** (19), 7704–7709.

52 Prados, A., Lasanta, A., and Hurtado, P.I. (2011) Large fluctuations in driven dissipative media. *Phys. Rev. Lett.*, **107**, 140601.

53 Popkov, V., Schütz, G.M., and Simon, D. (2010) ASEP on a ring conditioned on enhanced flux. *J. Stat. Mech.*, P10007.

54 Jack, R.L. and Sollich, P. (2010) Large deviations and ensembles of trajectories in stochastic models. *Prog. Theor. Phys. Suppl.*, **184**, 304–317.

55 Evans, R.M.L. (2004) Rules for transition rates in nonequilibrium steady states. *Phys. Rev. Lett.*, **92** (15), 150601.

56 Evans, R.M.L. (2005) Detailed balance has a counterpart in non-equilibrium steady states. *J. Phys. A: Math. Gen.*, **38** (2), 293–313.

57 Monthus, C. (2011) Non-equilibrium steady states: maximization of the Shannon entropy associated with the distribution of dynamical trajectories in the presence of constraints. *J. Stat. Mech.*, P03008.

58 Rangarajan, G. and Ding, M. (eds) (2003) *Processes with Long-Range Correlations: Theory and Applications,* Lecture Notes in Physics, vol. **621**, Springer, Berlin.

59 Harris, R.J. and Touchette, H. (2009) Current fluctuations in stochastic systems with long-range memory. *J. Phys. A: Math. Theor.*, **42** (34), 342001.

60 Maes, C., Netočný, K., and Wynants, B. (2009) Dynamical fluctuations for semi-Markov processes. *J. Phys. A: Math. Theor.*, **42** (36), 365002.

12
Lyapunov Modes in Extended Systems

Hong-Liu Yang and Günter Radons

12.1
Introduction

The modern theory of nonlinear dynamics is of great importance for the understanding of the fundamentals of statistical mechanics. This idea already dates back to the pioneering work of Krylov [1], where he pointed out the connection between the exponential instability of the system dynamics and the mixing properties needed by statistical mechanics. The latter was studied adopting the mathematical results for the instability of geodesic flows on compact manifolds with negative curvatures. Following such a geometric viewpoint, the ergodic theory of smooth dynamical systems has been developed and is making ongoing progress mainly for abstract dynamical systems [2]. Complementary to this, the geometric theory of Hamiltonian chaos for many-body systems of concrete physical relevance has been developed recently taking advantage of the progress in Riemannian geometry [3]. In the meantime, relations between quantities characterizing the microscopic dynamics of systems, for instance, the Kolmogorov–Sinai entropy and the Lyapunov exponents, and quantities used in the macroscopic description of the same dynamics, for instance, the transport coefficients, have been worked out [4].

In view of these developments, the recent discovery of hydrodynamic Lyapunov modes (HLMs) by Posch and Hirschl [5] is of potential importance, since it provides a new possibility to connect the reduced description of a many-body system to the microscopic information of its detailed dynamics. HLMs are special Lyapunov vectors (LVs) associated with near-zero Lyapunov exponents exhibiting long-wavelength structures and slow oscillations. They were originally reported in many-particle systems with hard-core interactions. Since then HLMs were suspected for a long time to exist only in hard-core systems because no clear global structures were identified in other systems [6, 7]. This numerical fact is in contrast to intuitive expectations and to results of theoretical attempts, indicating that continuous symmetries and conserved quantities are essential ingredients for HLMs. So far, by using the spectral analysis of LV correlation functions [8, 9], the existence of HLMs has been demonstrated to be a common feature of a large class of systems with continuous symmetries and conserved quantities, including many-particle

Nonequilibrium Statistical Physics of Small Systems: Fluctuation Relations and Beyond, First Edition.
Edited by Rainer Klages, Wolfram Just, and Christopher Jarzynski.

systems with hard-core [5, 6, 10–12] or soft potential interaction [9, 13], coupled map lattices (CMLs) [14–16], Hamiltonian lattice models [14, 17], and the Kuramoto–Sivashinsky (KS) equation [14].

Another important concept related to Lyapunov vectors is that of effective degrees of freedom (DOF) of partial differential equations (PDEs). In ubiquitous natural and laboratory situations, dissipative nonlinear PDEs are used to model pattern formation, space–time chaos and turbulence, and so on [18]. Despite their infinite-dimensional nature, most dissipative nonlinear PDEs are known to have a finite-dimensional attractor owing to the strong dissipation [18, 19]. It was further conjectured that albeit some trivial transient decaying process, the relevant dynamics of these PDEs occurs on a finite-dimensional manifold, named the *inertial manifold* (IM) [19, 20]. The concept of IM thus opens the possibility to model an infinite-dimensional PDE system by a finite-dimensional one. The existence of IM has been proved for a growing list of systems, including the Kuramoto–Sivashinsky equation, complex Ginzburg–Landau equation, and some reaction–diffusion equations [21, 22]. Although conceptually important, the merit of IM is largely unexplored, limited by the complexity of the necessary studying tools [20].

Motivated by the intuitive expectation that infinitesimal perturbations to an IM should behave somehow similar to the finite-size ones, the Lyapunov analysis method was applied recently to the study of IM [23]. By using covariant Lyapunov vectors (CLVs) [24], a hyperbolic separation between two sets of Lyapunov vectors was found in dissipative nonlinear PDEs, that is, the angle between the two subspaces spanned respectively by the two sets of vectors being bounded from zero [23]. The set of mutually entangled finite number of Lyapunov vectors associated with positive and weakly negative Lyapunov exponents, named "physical modes," was conjectured to represent the physically relevant dynamics of the original PDE systems. The linear space spanned by physical modes may serve as the local approximation of the IM. The remaining set of vectors corresponding to strongly negative Lyapunov exponents is believed to represent the trivial decaying process to the IM [23]. Note that similar results were also obtained for discrete system, for instance, coupled map lattices. Along this line of study, direct support to this conjecture is obtained via the analysis of the projection of the difference vectors between recurrent states and the reference state to the subspaces spanned by Lyapunov vectors.

In this chapter, we briefly review our recent results on the Lyapunov instabilities of extended dynamical systems. In Section 12.2, the numerical method used to calculate Lyapunov exponents and vectors is explained and the correlation functions used to characterize the spatiotemporal structures of LVs are defined. In Section 12.3, we show the existence of HLMs in a list of Hamiltonian and dissipative systems and show that Hamiltonian and dissipative systems belong to two different universality classes with respect to the nature of HLMs. In Section 12.4, we show in model systems of coupled map lattices that the hyperbolicity of the system, especially the angle between the Lyapunov vectors, crucially determines the significance of Lyapunov modes. In Section 12.5, we show that in a diatomic coupled map lattices system, the Lyapunov spectrum and the corresponding Lyapunov vectors split into two branches similar to the splitting of phonons in a diatomic crystal lattice. In Section 12.6, we study the relation between covariant and orthogonal

Lyapunov vectors with respect to Hydrodynamic Lypunov modes. In Section 12.7, we show that the information of the hyperbolicity of a partial differential equation can be used to infer the number of effective degrees of freedom involved in such an infinite-dimensional system. In Section 12.8, we go beyond the Lyapunov analysis to probe the geometric structures of the inertial manifold via the projection method in order to obtain a direct relation between the identified physical modes and the inertial manifold of PDEs. Details of these studies and our previous results on Lennard-Jones system can be found in Refs [23, 25–28, 48].

12.2 Numerical Algorithms and LV Correlations

Consider a dynamical system, which can be either time continuous or discrete, with the solution $X(t) = \phi(t, X(0))$ for the initial condition $X(0)$ where $\phi(t, X)$ is the evolution function of the dynamical system. In his seminal work about the multiplicative ergodic theorem, Oseledec [29] proved that the limit $\Xi = \lim_{t\to+\infty} [M^T(t,0) \cdot M(t,0)]^{1/2t}$ exists for almost every initial point of a nonlinear dynamical system, where $M(t,0)$ is the fundamental matrix governing the time evolution of perturbations $\delta X(t)$ to the reference trajectory $X(t)$ as $\delta X(t) = M(t,0) \cdot \delta X(0)$. The set of Lyapunov exponents are defined as $\lambda^{(\alpha)} = \ln \mu^{(\alpha)}$, where $\mu^{(\alpha)}$ are the eigenvalues of the so-called Oseledec matrix Ξ, that is, $\Xi \cdot g^{(\alpha)} = \mu^{(\alpha)} g^{(\alpha)}$.

In practice, the standard method [30] is used to calculate the orthogonal Lyapunov vectors (OLVs) and the corresponding Lyapunov exponents. For this purpose, an ensemble of linear equations governing the evolution of the offset vectors is integrated simultaneously with the original nonlinear evolution equations. The offset vectors have to be reorthogonalized periodically, either by means of Gram–Schmidt orthogonalization or QR decomposition. The time-averaged values of the logarithms of the renormalization factors are the Lyapunov exponents and the set of offset vectors $f^{(\alpha)}$ right after the reorthonormalization are the orthogonal Lyapunov vectors. The relation between the Oseledec eigenvectors $g^{(\alpha)}$ and the Lyapunov vectors obtained via the standard method is subtle. It is proved that the Lyapunov vectors $f^{(\alpha)}$ obtained via the standard method converge exponentially to the Oseledec eigenvectors for the inverse-time dynamics of the original system [31, 32]. In other words, as $t \to +\infty$, there is $f^{(\alpha)} \sim \bar{g}^{(\alpha)}$, where $\bar{g}^{(\alpha)}$ are eigenvectors of the matrix $\bar{\Xi} = \lim_{t\to+\infty} [\overline{M(t,0)}\,T \cdot \overline{M(t,0)}]^{1/2t}$ as well as its inverse $\bar{\Xi}-1 = \lim_{t\to+\infty} [M(t,0) \cdot M(t,0)^T]^{1/2t}$ and $\overline{M(t,0)} \equiv [M(t,0)]^{-1} = M(0,t)$ is the fundamental matrix of the inverse-time dynamics. See Refs [31, 32] for details.

Exactly in the same work [29], Oseledec also proved that for almost all initial conditions x, there is a splitting of the tangent space $TM(x)$:

$$TM(x) = E^1(x) \oplus E^2(x) \oplus \cdots \oplus E^s(x), \tag{12.1}$$

and there exist real numbers $\lambda_1(x) > \lambda_2(x) > \cdots > \lambda_s(x)$ such that

$$\lim_{n\to\pm\infty} \frac{1}{n} \ln \left\| Df^n|_{E^i} \right\| = \lambda_i(x), \tag{12.2}$$

where Df is the derivative governing the tangent space dynamics. The set of numbers $\lambda_i(x)$ with degeneracy $m_i = \dim E^i(x)$ composes the Lyapunov spectrum and the decomposition stated in Eq. (12.1) is called the Oseledec splitting. The spanning vectors $e^{(\alpha)}$ of the Oseledec subspace $E^i(x)$ are the CLVs, which are not orthogonal in general. In the simplest case of nondegenerate Lyapunov exponents, the $E^i(x)$ are one dimensional and the CLVs $e^{(i)}(x)$ are the directions in which the expansion/contraction rate is exactly $\lambda_i(x)$ in forward and backward directions, respectively.

In contrast to the popularity of OLVs, the use of CLVs was made feasible only recently owing to recently proposed efficient algorithms [24, 33]. The main idea of the algorithm by Ginelli *et al.* [24] is based on the fact that almost any randomly selected vector in a subspace spanned by the first n orthogonal Lyapunov vectors will evolve backward in time asymptotically to the nth covariant Lyapunov vector. The essential ingredient of this algorithm is the effective linearized dynamics restricted to the mentioned subspace. A representation of this effective dynamics in the coordinate space of OLVs is given by the R-matrix produced as a by-product from the Gram–Schmidt orthogonalization or QR decomposition performed in the standard method [30]. Based on the information obtained via the standard method, one additional integration of the linearized inverse-time dynamics is performed to get CLVs and the corresponding local expansion/contraction rates (finite-time Lyapunov exponents). Detailed formulas can be found in Ref. [24].

In the spirit of molecular hydrodynamics, we introduced in Refs [8, 9] a dynamical variable called *LV fluctuation density*,

$$u^{(\alpha)}(r,t) = \sum_{j=1}^{N} \delta x_j^{(\alpha)}(t) \cdot \delta(r - r_j(t)), \tag{12.3}$$

where $\delta(z)$ is Dirac's delta function, $r_j(t)$ is the position coordinate of the jth particle, and $\{\delta x_j^{(\alpha)}(t)\}$ is the coordinate part of the αth Lyapunov vector (OLV or CLV) at time t. For spatially discrete systems, such as the coupled map lattices studied in the following sections, $r_j(t) = j \cdot a$, where a is the lattice constant. The spatial structure of LVs is characterized by the *static LV structure factor* defined as

$$S_u^{(\alpha)}(k) = \int \langle u^{(\alpha)}(r,0)u^{(\alpha)}(0,0)\rangle e^{-jk\cdot r} dr, \tag{12.4}$$

which is simply the spatial power spectrum of the LV fluctuation density. In Eq. (12.4), $\langle \cdots \rangle$ denotes an average over time or ensemble. One can also define the *instantaneous static LV structure factor* as

$$s_u^{(\alpha)}(k,t) = \int u^{(\alpha)}(r,t)u^{(\alpha)}(0,t) e^{-jk\cdot r} dr. \tag{12.5}$$

Information on the dynamics of LVs can be extracted via the *dynamical LV structure factor*, which is defined as

$$S_u^{(\alpha)}(k,\omega) = \int\int \langle u^{(\alpha)}(r,t)u^{(\alpha)}(0,0)\rangle e^{-jk\cdot r} e^{j\omega t} dr\, dt. \tag{12.6}$$

12.3
Universality Classes of Hydrodynamic Lyapunov Modes

It is commonly believed that continuous symmetries are responsible for the appearance of hydrodynamic Lyapunov modes, long-wavelength structures in Lyapunov vectors associated with near-zero Lyapunov exponents. HLMs are therefore expected to appear in a large class of spatially extended systems with continuous symmetries. In this section, numerical and analytical results are provided to show that HLMs do exist in lattices of coupled Hamiltonian and dissipative maps and to show that Hamiltonian and dissipative systems belong to two universality classes with respect to the nature of the HLMs.

The two classes of coupled map lattices under investigation are of the form

$$v_{t+1}^l = (1-\gamma)v_t^l + \varepsilon^l \left[f\left(u_t^{l+1} - u_t^l\right) - f\left(u_t^l - u_t^{l-1}\right) \right], \tag{12.7a}$$

$$u_{t+1}^l = u_t^l + v_{t+1}^l, \tag{12.7b}$$

and

$$u_{t+1}^l = u_t^l + \varepsilon^l \left[f\left(u_t^{l+1} - u_t^l\right) - f\left(u_t^l - u_t^{l-1}\right) \right], \tag{12.8}$$

where $f(z)$ is a nonlinear map, t is the discrete time index, $l = \{1, 2, \ldots, L\}$ is the index of the lattice sites, and L is the system size. Unless it is explicitly stated, we set $f(z) = (1/2\pi)\sin(2\pi z)$, $\varepsilon^l = \varepsilon$, $\gamma = 0$ and use periodic boundary conditions $\{u_t^0 = u_t^L, u_t^{L+1} = u_t^1\}$ in the numerical simulations below.

It is quite obvious that both systems are invariant under an arbitrary translation in u-direction. It is expected that such a symmetry is responsible for the appearance of HLMs here. An essential difference between the two systems of coupled standard maps (SMs) of Eq. (12.7) and coupled circle maps (CMs) of Eq. (12.8) is that for $\gamma = 0$, the former is a Hamiltonian system, implying, for example, that the phase space volume is conserved by the time evolution, while the latter is dissipative. Equation (12.7) with $0 < \gamma < 1$ is also dissipative and interpolates between these extreme cases.

In Figure 12.1, the Lyapunov spectra and the static LV structure factors for the two systems (Eqs. (12.7) and (12.8)) are shown. The contour plots demonstrate that the LVs with $\lambda \approx 0$ are dominated by certain long-wavelength components, which indicates the existence of HLMs in both systems. This discovery shows that the existence of HLMs, which was previously demonstrated for many-particle systems [5, 6, 9–13], is not an exclusive feature of many-particle systems. It is instead a common property of a large class of systems with continuous symmetries. Furthermore, the finding of HLMs in coupled circle maps demonstrates that a Hamiltonian structure is not a necessary condition for the existence of HLMs.

An important characteristic feature for the spatial structure of HLMs is the $\lambda - k$ dispersion relation. As shown in Figure 12.1, for each Lyapunov vector, the spectrum $S_u^{(\alpha)}(k)$ has a dominant peak at $k_{\max}^{(\alpha)}$. In Figure 12.2, values of $\lambda^{(\alpha)}$ versus $k_{\max}^{(\alpha)}$ are presented for various systems. The difference between the two classes is quite obvious. Fitting data in the regime $\lambda \approx 0$ and $k_{\max} \approx 0$ to a power-law $\lambda \sim k_{\max}^{\beta}$

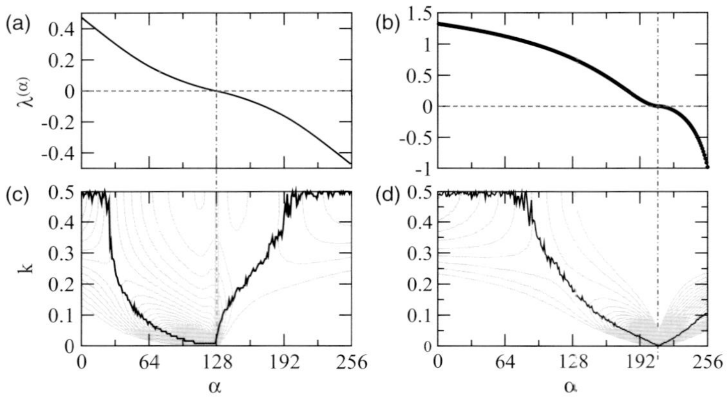

Figure 12.1 (a) Lyapunov spectrum and (b) contour plot of the static LV structure factors for coupled standard maps (Eq. (12.7)) with $\varepsilon = 0.6$. In (b), the graph of $k_{\text{max}}^{(\alpha)}$ is overlaid on the contour plot, where $k_{\text{max}}^{(\alpha)}$ is the wave number of the dominant peak in each spectrum $S_u^{(\alpha)}(k)$. (c) and (d) are similar to (a) and (b), but for coupled circle maps (Eq. (12.8)) with $\varepsilon = 2.3$.

yields $\beta \simeq 1.0$ for coupled standard maps, whereas $\beta \simeq 2.0$ for coupled circle maps. Varying the coupling strength ε and/or adopting other nonlinear functions $f(z)$, for instance, the logistic or the tent map, yields results similar to those shown above. The robustness of the discrimination according to the forms of the dispersion relations indicates the different nature of the two systems, which leads to the conjecture that they belong to two universality classes.

Due to the mathematical complexity of the problem, we cannot provide a general proof of the above conjecture. There are, however, special cases that can be treated analytically to show that the two systems stated in Eqs. (12.7) and (12.8) do have different $\lambda - k$ dispersion relations. If the nonlinear function $f(z) = (1/2\pi)\sin(2\pi z)$ in Eq. (12.8) is replaced by the Bernoulli shift map $f(z) = 2z(\text{mod } 1)$, one can easily obtain the Lyapunov exponents as $\lambda(k) \equiv \ln|1 - 4\varepsilon(1 - \cos k)|$. The associated Lyapunov vectors are of the form $\vec{e}^{(\alpha)} \equiv \{\delta u_l^{(\alpha)}\}$ with $\delta u_l^{(\alpha)} = \mathrm{d}_1 \cos(k \cdot l\alpha) + \mathrm{d}_2 \sin(k \cdot l\alpha)$, where l is the lattice site index, $k = (2\pi\alpha)/L$, with $\alpha = \{0, 1, 2, \ldots, L-1\}$, is the discrete wave number, and d_1, d_2 are constants. As $k \to 0$, the Lyapunov exponents are approximated by

$$\lambda(k) \simeq 2|\varepsilon|k^2. \tag{12.9}$$

This leads to the asymptotic dispersion relation $\lambda \sim k^2$ for the dissipative system (Eq. (12.8)) with $f(z) = 2z(\text{mod } 1)$ (BM). Similarly, Lyapunov exponents for the Hamiltonian system (Eq. (12.7)) with $f(z) = 2z(\text{mod } 1)$ are obtained as

$$\lambda(k) = \ln\left|\frac{\eta(k) + 2 \pm \sqrt{\eta^2(k) + 4\eta(k)}}{2}\right|, \tag{12.10}$$

with $\eta(k) = -4\varepsilon(1 - \cos k)$. The Lyapunov vectors are of the form $\{\vec{e}^{(\alpha)}; c(k)\vec{e}^{(\alpha)}\}$ for $1 \leq \alpha \leq L$ and $\{c(k)\vec{e}^{(\alpha)}; -\vec{e}^{(\alpha)}\}$ for $L + 1 \leq \alpha \leq 2L$, where the two parts of a Lyapunov vector correspond to perturbations of the coordinate and momentum parts of the state vector, respectively. The vectors $\vec{e}^{(\alpha)}$ are the same as given above and $c(k) = (-\eta + \sqrt{\eta^2 + 4\eta})/2$. For $k \to 0$, ($\varepsilon < 0$ in Eq. (12.7)), the Lyapunov exponents are approximated by

$$\lambda(k) \simeq \sqrt{2|\varepsilon|}k. \tag{12.11}$$

Thus, the dispersion relation for small k for the Hamiltonian system (Eq. (12.7)) with $f(z) = 2z(\text{mod } 1)$ is $\lambda \sim k$ (HBM). So far, we have shown analytically that for $f(z) = 2z(\text{mod } 1)$, the dissipative system (Eq. (12.8)) has a quadratic dispersion relation $\lambda \sim k^2$ for small k, while the Hamiltonian system (Eq. (12.7)) is characterized by a linear dispersion relation $\lambda \sim k$. Results of numerical simulations are presented in Figure 12.2. As expected, they are in agreement with analytical calculations.

To further check the validity of our conjecture, we also did simulations for the dynamical XY model with Hamiltonian $H = \sum_i \dot{\theta}_i + \varepsilon \sum_{ij} [1 - \cos(\theta_j - \theta_i)]$ (XY) [34] and the Kuramoto–Sivashinsky equation $h_t = -h_{xx} - h_{xxxx} - h_x^2$ [35]. As shown in Figure 12.2, all data in the regime $\lambda \approx 0$ and $k_{\max} \approx 0$ collapse on two master curves characterized by $\lambda \sim k_{\max}^2$ for dissipative systems and by $\lambda \sim k_{\max}$ for Hamiltonian cases. These results further confirm the existence of two universality classes of HLMs.

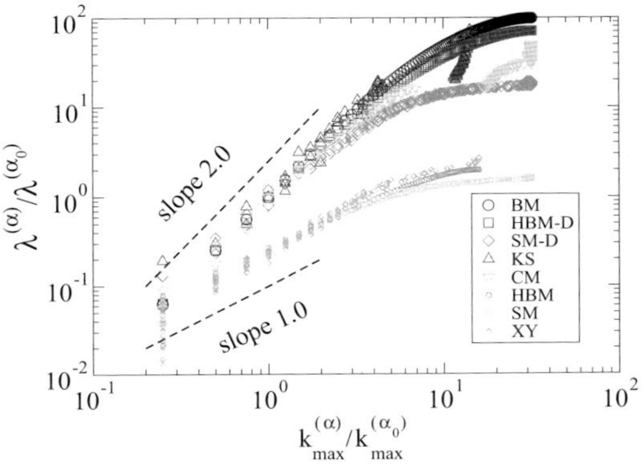

Figure 12.2 The $\lambda - k$ dispersion relations for various extended systems with continuous symmetries. Normalized data for different systems collapse on two master curves. These results strongly support our conjecture that there exist two universality classes with $\lambda \sim k$ and $\lambda \sim k^2$, respectively. Here, systems in the former group include Eq. (12.7) with $f(z) = (1/2\pi)\sin(2\pi z)$ (SM), Eq. (12.7) with $f(z) = 2z(\text{mod } 1)$ (HBM), and the 1D XY model (XY) [34]. Systems belonging to the class $\lambda \sim k^2$ are Eq. (12.8) with $f(z) = (1/2\pi)\sin(2\pi z)$ (CM), Eq. (12.8) with $f(z) = 2z(\text{mod } 1)$ (BM), Eq. (12.7) with $f(z) = (1/2\pi)\sin(2\pi z)$ and $\gamma = 0.7$ (SM-D), and Eq. (12.7) with $f(z) = 2z(\text{mod } 1)$ and $\gamma = 0.7$ (HBM-D) and the 1D Kuramoto–Sivashinsky (KS) equation.

Actually, the difference in the $\lambda - k$ dispersion relation, that is, in the spatial structure of the HLMs, is not the only discriminating feature of the two classes of systems. As shown in Figure 12.3, the dynamical LV structure factors (Eq. (12.6)) of coupled standard maps (Eq. (12.7)) have two sharp side peaks. Moreover, the side peaks shift linearly with k as $\omega_u \simeq c_u k$ for $k \simeq 2\pi/L$, which implies that the HLMs in these systems are propagating waves. In contrast, the spectra of coupled circle maps possess only a single central peak. They can be fitted quite well by a Lorentzian curve $S(k,\omega) = a(k)/(\omega^2 + a^2(k))$, which is typical of diffusive processes.

To summarize this section, we have shown that HLMs exist in a large class of systems with continuous symmetries. We provided numerical and analytical evidence that these systems belong to two universality classes with respect to the structure of the HLMs. Hamiltonian systems have a linear dispersion relation $\lambda \sim k$ and the HLMs are propagating; in the dissipative cases, a quadratic dispersion relation $\lambda \sim k^2$ is observed and the HLMs move only diffusively. Note that numerical simulations of equilibrium many-particle systems with soft-core or hard-core interactions yield linear $\lambda - k$ dispersion relations [5, 9]. Although up to now there are no reports about the nonequilibrium case, it would be interesting to see whether a similar classification also holds there. Note that results presented in this section were obtained for OLVs. We found that the use of CLVs instead of OLVs will not change these results regarding the universality of HLMs. Details about a comparison between CLVs and OLVs will be given in Section 12.6.

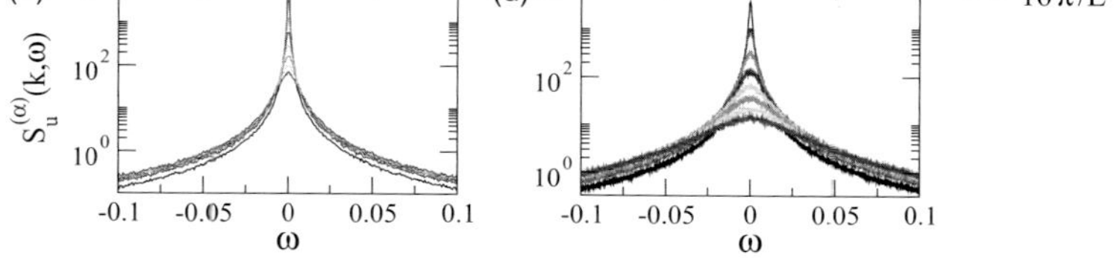

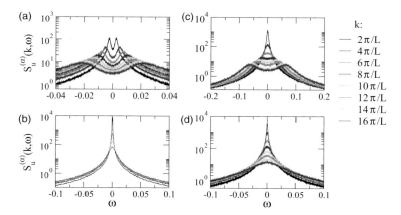

Figure 12.3 Dynamical LV structure factors $S_u^{(\alpha)}(k,\omega)$ for (a) coupled standard maps, Eq. (12.7) with $\varepsilon = 1.3$; (b) coupled circle maps, Eq. (12.8) with $\varepsilon = 1.3$; (c) Eq. (12.7) with $\varepsilon = 1.3$, $\gamma = 0.2$; and (d) Eq. (12.7) with $\varepsilon = 1.3$, $\gamma = 0.7$. Equal-distantly sampled k-values are used for panels. In (a), each spectrum has two sharp symmetric side peaks located at $\pm\omega_u$. Furthermore, $\omega_u \simeq \pm c_u k$ for $k \geq 2\pi/L$.

12.4 Hyperbolicity and the Significance of Lyapunov Modes

Hydrodynamic Lyapunov modes were originally reported in many-particle systems with hard-core interactions [5]. Even though the controversy over the existence of HLMs in systems with soft potential interactions has been successfully resolved [9, 13] by using the spectral method invented in Refs [8, 9], there are still some important questions left open. For instance, the long-wavelength global structures in LVs of hard-core systems can be easily identified by the naked eye and they can be fitted well to a sinusoidal function directly [5, 6, 10, 11], whereas such collective structures in soft potential systems can only be detected indirectly via spectral analysis [9, 13]. One then asks naturally: What specific nature of a system determines the significance of HLMs?

In the past, the hyperbolicity of hard-core systems was considered as being relevant for the appearance of significant HLMs [10]. The resulting (good) separation between stable and unstable manifolds, however, is a too coarse criterion and obviously not sufficient since Lyapunov vectors provide a splitting of the tangent space on a much fine scale. Previous numerical investigations indicate that large fluctuations in finite-time Lyapunov exponents may cause entanglement among unstable subspaces corresponding to different Lyapunov exponents, which damages HLMs. In order to have good HLMs, these unstable subspaces with different Lyapunov exponents should separate uniformly in the same manner as unstable and stable manifolds of hyperbolic sets. A useful concept is the domination of the Oseledec splitting (DOS), saying that each Oseledec subspace is uniformly more/less expanding than the next subspace corresponding to a larger/smaller Lyapunov exponent [36]. Our hypothesis is that partial DOS with respect to subspaces associated with near-zero Lyapunov exponents is crucial for observing good HLMs. Results for a simple system of coupled map lattices are presented to support our proposal.

We use the coupled map lattice model (Eq. (12.7)) with the local map $f(z)$ having the properties of the skewed tent (T) or Bernoulli shift map (B):

$$f_{\mathrm{B,T}}(z) = \begin{cases} z'/r, & \text{for} \quad 0 < z' \leq r, \\ (z'-r)/(1-r), & \text{for} \quad r < z' < 1\,(\mathrm{B}), \\ (1-z')/(1-r), & \text{for} \quad r < z' < 1\,(\mathrm{T}), \end{cases} \tag{12.12}$$

with $z' = z(\mathrm{mod}\,1)$. In this section, we concentrate on a homogeneous case, that is, Eq. (12.7), with $\varepsilon^l = \varepsilon$ and $\gamma = 0$. A splitting parameter p_{split} was defined in Ref. [26] to characterize the fluctuations in the *instantaneous* Lyapunov exponents $\lambda_i(t)$ corresponding to the time evolution of the ith Oseledec subspace. It reads

$$p^i_{\mathrm{split}} \equiv \max_{t,j\neq i}\{p_{ij}(t)\}, \tag{12.13}$$

with $p_{ij}(t)$ defined as $\lambda_j(t)/\lambda_i(t)$ or $\lambda_i(t)/\lambda_j(t)$ for j larger or smaller than i, respectively. The case $p_{\mathrm{split}} < 1$ indicates the satisfaction of the domination of the Oseledec splitting, a situation where the hyperbolic splitting among Oseledec subspaces (Lyapunov vectors) enables the appearance of good Lyapunov modes. In addition, one can define the fraction of the DOS violation time as $\nu^i = \langle \Theta(\max_{j \neq i}\{p_{ij}(t)\})\rangle_t$, where $\Theta(t)$ is the Heaviside step function. In case of degeneracy, the correspondence between the index i of the Oseledec subspaces and the Lyapunov index α is not one-to-one. Here, we concentrate only on the quantities with odd Lyapunov indices $\alpha = 2i$ and estimate the corresponding splitting parameters as $p^i = \max_{x_t}\{\lambda^{(2i+1)}(x_t)/\lambda^{(2i)}(x_t)\}$ and $\nu(\lambda_{i+1} > \lambda_i) \equiv \langle \Theta(\lambda^{(2i+1)}(x_t) - \lambda^{(2i)}(x_t))\rangle$, irrespective of the actual degeneracy.

Figure 12.4 depicts numerical results for a skewed tent map case $f_T(z)$. Strong fluctuations in instantaneous Lyapunov exponents $\lambda^{(\alpha)}(t)$ indicated by the standard deviations of $\lambda^{(\alpha)}(t)$ show that the chaoticity of the dynamics of this system is not uniform. The splitting parameter p_{split} increases gradually as the Lyapunov exponents $\lambda^{(\alpha)}$ increase from zero and it crosses the line $p_{\text{split}} = 1$ at $i_c/L \simeq 0.4$. In accordance with this, the fraction of DOS violation time ν^i is constantly 0 in the regime $i > i_c$.

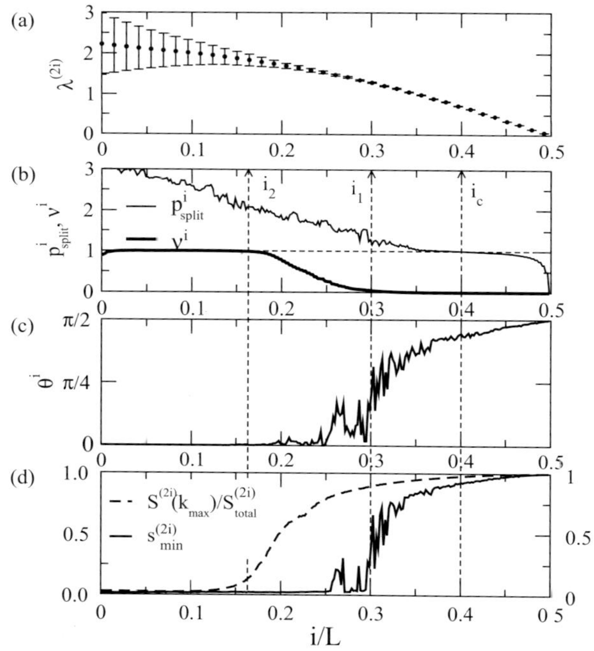

Figure 12.4 (a) Lyapunov spectrum $\lambda^{(\alpha)}$ with bars indicating the standard deviations of instantaneous Lyapunov exponents $\lambda^{(\alpha)}(t)$ along a trajectory, (b) the splitting parameters p_{split} and ν, (c) the minimal angle between Oseledec subspaces θ, and (d) $S^{(\alpha)}(k_{\max})/S^{(\alpha)}_{\text{total}}$ and $s^{(\alpha)}_{\min}$ for Eq. (12.7) with the skewed tent map $f_T(z)$ in Eq. (12.12), $\gamma = 0$, $\varepsilon' = 1.3$, $r = 0.02$, and $L = 256$. In case of degeneracy, the correspondence between the index i of the Oseledec subspaces and the Lyapunov index α is not one-to-one. We plot here only the quantities with odd Lyapunov indices $\alpha = 2i$ irrespective of the actual degeneracy.

Let us see how the dynamics and the significance of HLMs change with the splitting parameters p_{split} and ν. To characterize the entanglement among Lyapunov modes, we use the quantity θ^i that is defined as the smallest angle between the ith Lyapunov mode (Oseledec subspace) and any other with different Lyapunov exponents. As can be seen from Figure 12.4c, in the regime $p_{\text{split}} < 1$, that is, $i > i_c$, the angle θ^i stays close to $\pi/2$ and it decreases with increasing p_{split}. An interesting point is that the value of θ^i fluctuates strongly in the regime $i_2 < i < i_c$ and experiences a sudden fall to near-zero values at i_1, the boundary for the regime of the steep increase of the violation of DOS. Taking into account that $\theta^i = 0$ implies the presence of tangencies or entanglements among Oseledec subspaces, the observation that θ^i is bounded away from zero in the regime $i > i_c$ confirms that the DOS prevents the presence of tangencies. Surprisingly, the minimal angle θ^i for decreasing i becomes zero in a regime near i_1, far below the threshold value i_c for the DOS. The reason for the delayed response of θ^i to the violation of DOS is discussed in Ref. [25]. To measure the significance of HLMs, we use a quantity $S^{(\alpha)}(k_{\max})/S^{(\alpha)}_{\text{total}}$, where $S^{(\alpha)}(k)$ is the static LV structure factor (Eq. (12.4)), $k_{\max}$ is the wave number of the dominant peak in $S^{(\alpha)}(k)$, and $S^{(\alpha)}_{\text{total}} \equiv \sum_k S^{(\alpha)}(k)$. The significance measure attains a value 1 if the HLM is a pure Fourier mode. As can be seen from Figure 12.4d, in the regime, where $p_{\text{split}} < 1$, the measure $S^{(\alpha)}(k_{\max})/S^{(\alpha)}_{\text{total}}$ stays always very close to 1. It decreases monotonically with increasing p_{split} in the regime $i_2 < i < i_c$ and drops drastically to zero in a regime around i_2, the boundary value for nearly permanent violation of DOS. To further quantify the change in the significance of HLMs, we use $s^{(\alpha)}_{\min} \equiv \min_t \{s^{(\alpha)}(k_{\max}, t)/s^{(\alpha)}_{\text{total}}(t)\}$, the minimal value of the instantaneous LV structure factor $s^{(\alpha)}(k_{\max}, t)$ defined in Eq. (12.5). We find that the time evolution of $s^{(\alpha)}(k_{\max}, t)$ is also strongly fluctuating for $i_2 < i < i_c$. Moreover, it is shown in Figure 12.4d that corresponding to the sudden fall of the variable θ^i, a sudden jump in $s^{(\alpha)}_{\min}$ occurs at i_1. The strong correlation in the evolution of the two quantities implies that entanglement among Oseledec subspaces as indicated by the minimal angle θ^i is the origin for the steep decrease of the significance of HLMs as demonstrated by the variation of $S^{(\alpha)}(k_{\max})/S^{(\alpha)}_{\text{total}}$ and $s^{(\alpha)}_{\min}$. Note that even the fluctuations of the minimal angle θ^i are reflected in the instantaneous spatial structure of the Lyapunov vectors as indicated by $s^{(\alpha)}_{\min}$ (Figure 12.4c and d). These findings show that p_{split} and ν faithfully indicate the tendency of change in the significance of HLMs. Moreover, the significance of HLMs is quite sensitive to the change in the values of splitting parameters, especially in the regime $i_2 < i < i_c$. Our results thus show that DOS can be used to explain the variation in the significance of HLMs as the Lyapunov exponents increase from zero.

We also studied how the significance of HLMs changes with the splitting parameter p_{split} in the regime $p_{\text{split}} > 1$. As shown in Figure 12.5, for all cases considered, the significance measure $S^{(\alpha)}(k_{\max})/S^{(\alpha)}_{\text{total}}$ decreases monotonically with increasing ν, the fraction of time for which the DOS is violated. The monotonic decrease of $S^{(\alpha)}(k_{\max})/S^{(\alpha)}_{\text{total}}$ with increasing ν shows that HLMs are more significant if less DOS is violated.

These numerical results demonstrate clearly the existence of an essential relation between the hyperbolicity of the system dynamics and the significance of coherent

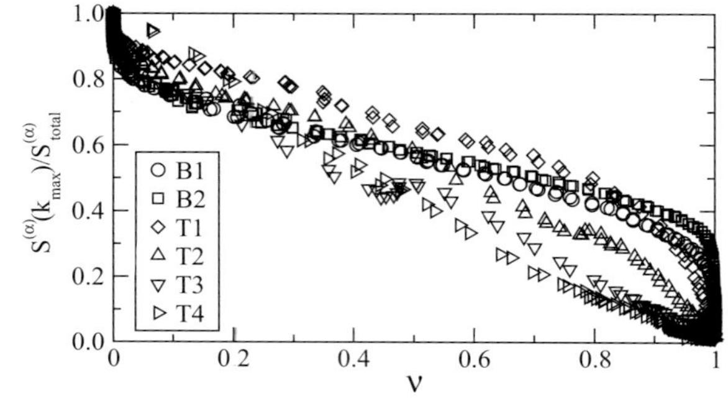

Figure 12.5 The significance measure of HLMs $S^{(\alpha)}(k_{\max})/S^{(\alpha)}_{\text{total}}$ as function of the fraction of time with DOS violation for several cases with different parameters. The cases presented include $f_B(z)$ with $\varepsilon = -1.3$, $r = 0.1$ (B1) and 0.2 (B2); $f_T(z)$ with $\varepsilon = 1.3$, $r = 0.02$ (T1), 0.05 (T2), 0.1 (T3), and 0.2 (T4).

structures in Lyapunov vectors, that is, the relevance of the domination of the Oseledec splitting for observing good HLMs.

12.5
Lyapunov Spectral Gap and Branch Splitting of Lyapunov Modes in a "Diatomic" System

The concept of collective excitations is very important in modern physics. For instance, vibrational normal modes in a crystal lattice, phonons, are known to play an essential role for many physical properties of solids. In the past, encouraged by the success of this concept in solids, there have already been some attempts to extend the concept of phonons and to find their counterpart in fluids. Such an idea may date back to Maxwell, who suggested that the dynamics of liquids at short times is similar to that of solids. One recent contribution to this line of research consists in the concept of the so-called *instantaneous normal modes* (INMs) [37], which are eigenvectors of the Hessian matrix evaluated from instantaneous states of liquids.

On the other hand, questions of the connection of the recently found hydrodynamic Lyapunov modes to other physical quantities were posed right after their discovery. It has already been noticed that the appearance of HLMs relies on the same mechanisms as phonons and INMs, that is, the spontaneous breaking of certain symmetries of the system Hamiltonians [6, 38]. Moreover, all three sets of modes, phonons, INMs, and HLMs, are related to the Hessian matrix. According to the geometric theory of Hamiltonian chaos, both phonons and HLMs represent certain eigendirections characterizing stabilities of geodesics of certain manifolds with suitable metrics [3, 17, 39, 40]. Similar to INMs, the calculation of HLMs relies on Hessian matrices evaluated from instantaneous states of the system. HLMs, however, encode additional information on the time correlations among these instantaneous states. In view of these facts, it is natural to ask whether there are connections between these modes and whether HLMs are able to serve as the counterpart of phonons in systems with strong anharmonic dynamics. As a step toward the understanding of such problems, we compare the dynamics of HLMs in a "diatomic" system with that of phonons. It is known that in diatomic crystal lattices, the frequency spectrum of phonons has a gap and phonons split into two branches, acoustic and optical ones, respectively. Our current investigation demonstrates that, similar to the phonon case, mass imparity may induce a gap in the Lyapunov spectrum, and the two corresponding branches of Lyapunov modes behave acoustic-like and optical-like, respectively. A major difference between LMs and phonons, however, is that a large enough mass difference beyond a certain threshold value is necessary for the appearance of the gap in the Lyapunov spectrum and the splitting of the modes.

The coupled map lattices (Eq. (12.7)) with $f(z)$ being the skewed tent map is investigated in this section. The quantity $1/\varepsilon^l$ plays the same role as mass in mechanical systems and ε^l takes the values ε_1 and ε_2 for the odd and even lattice sites, respectively. In the following simulations, $\varepsilon_2 = 1$ is fixed and the value ε_1 is tuned to study the influence of mass differences. The Lyapunov exponents and orthogonal Lyapunov vectors are obtained via the so-called standard method [30].

We show in Figure 12.6 the variation of the Lyapunov spectra with the mass ratio $\kappa \equiv \varepsilon_2/\varepsilon_1$. For the Hamiltonian system considered, the Lyapunov spectrum has the symmetry $\lambda^{(2L-1-\alpha)} = -\lambda^{(\alpha)}$. As can be seen, gaps appear in the plot in the middle of each half of the spectrum if the mass imparity is large. With increasing κ toward 1, the spectral gap shrinks and disappears eventually. These facts imply that Lyapunov exponents play a similar role for Lyapunov modes as do the frequencies for

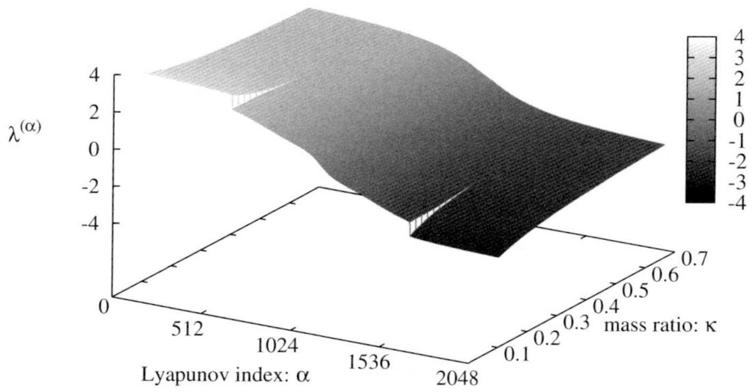

Figure 12.6 Lyapunov exponents as function of the Lyapunov index α and the mass ratio $\kappa \equiv \varepsilon_2/\varepsilon_1$. Here, $L = 1024$ and $r = 0.2$. Note that the gaps in the Lyapunov spectra disappear as κ increases beyond a certain threshold value κ_c.

phonons and the mass difference between neighboring sites does induce gaps in the Lyapunov spectrum.

The Lyapunov spectrum of an extended system is proven to be a continuous curve in the thermodynamic limit for many cases. In the numerical simulation of a system of size L, the increments between neighboring Lyapunov exponents are not zero, but on the order of $1/L$. Thus, one may suspect that the observed disappearance of the spectral gap in Figure 12.6 is only a numerical artifact, that is, the spectral gap just becomes too small to be detected in the simulation using a finite L. In order to clarify this point, we present in Figure 12.7 the system size dependence of the spectral gap $\delta\lambda \equiv \lambda^{(L/2-1)} - \lambda^{(L/2)}$. Obviously, in the two regimes on each side of the threshold value κ_c, the spectral gap size $\delta\lambda$ behaves differently. For $\kappa < \kappa_c$, $\delta\lambda$ is independent of the system size L, while $\delta\lambda$ vanishes for large L as L^{-1} for $\kappa > \kappa_c$. As to the κ-dependence, $\delta\lambda$ decreases with increasing κ in the regime $\kappa < \kappa_c$, while it is nearly constant in the regime $\kappa > \kappa_c$. Thus, one expects that in the thermodynamic limit, the spectral gap size $\delta\lambda$ decreases gradually to zero as κ approaches κ_c from the side $\kappa < \kappa_c$, while it stays to be zero in the regime $\kappa > \kappa_c$. Figure 12.7a shows that data of $\delta\lambda$ from simulations with increasing system size L have the tendency to approach such a master curve. This indicates that the spectral gap does disappear at the threshold value κ_c and excludes the possibility of numerical artifacts.

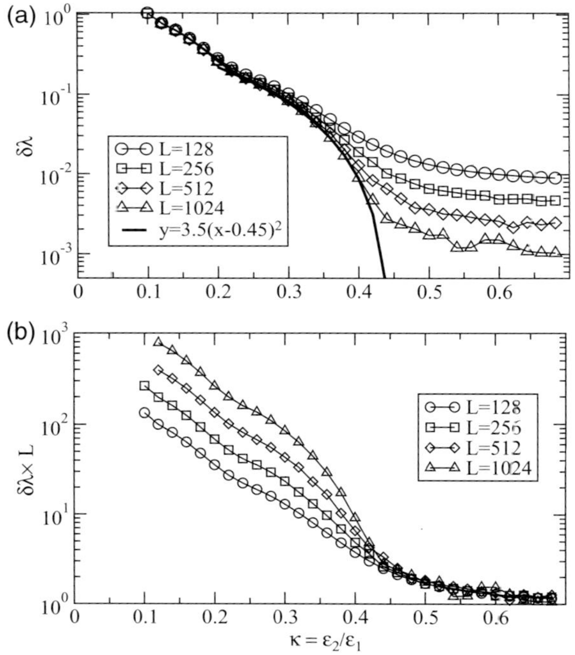

Figure 12.7 The Lyapunov spectral gap size $\delta\lambda \equiv \lambda^{(L/2-1)} - \lambda^{(L/2)}$ versus the mass ratio κ with $r = 0.2$. Obviously, $\delta\lambda$ has different system size dependencies as κ is below/beyond the threshold value $\kappa_c \approx 0.45$.

Now we turn to the study of the influence of mass imparity on Lyapunov modes. Inspired by the scenario of changes in phonons, we also expect to observe two types of Lyapunov modes, acoustic and optical ones, respectively. To check this, we consider the tangent space dynamics of neighboring sites for two Lyapunov modes belonging to the two branches, respectively. Results for an example with $\kappa = 0.125 < \kappa_c$ are shown in Figure 12.8. Here, $\delta u_1^{(\alpha)}$ denotes the coordinate component of the αth Lyapunov mode for the first lattice site, and $\delta u_2^{(\alpha)}$ denotes the second site correspondingly. The plot of $\delta u_2^{(\alpha)}$ versus $\delta u_1^{(\alpha)}$ in Figure 12.8 shows that the distribution of phase points is highly anisotropic and tends to align along different directions for the two Lyapunov modes. In a phonon context, one would associate the two modes presented with the edge of the first Brillouin zone in a diatomic harmonic chain, one in the acoustic branch and one in the optical branch. Mass differences in diatomic systems induce differences in oscillating amplitudes of the two sorts of particles. The dynamics of these zone-boundary phonons then becomes quite simple, that is, one or the other of the two sublattices is at rest. We expect that the anisotropy observed in Figure 12.8 has an origin similar to the corresponding zone-boundary phonon dynamics. The anisotropy of the tangent space dynamics may be further made evident by the sharp peak in the probability distributions of $\theta(t) \equiv (1/\pi) \arctan \left(\delta u_2^{(\alpha)}/\delta u_1^{(\alpha)}\right)$. Simulations for other Lyapunov vectors show that despite the chaotic nature of the system, the evolution also exhibits qualitatively in many respects a behavior similar to that of phonons in diatomic systems.

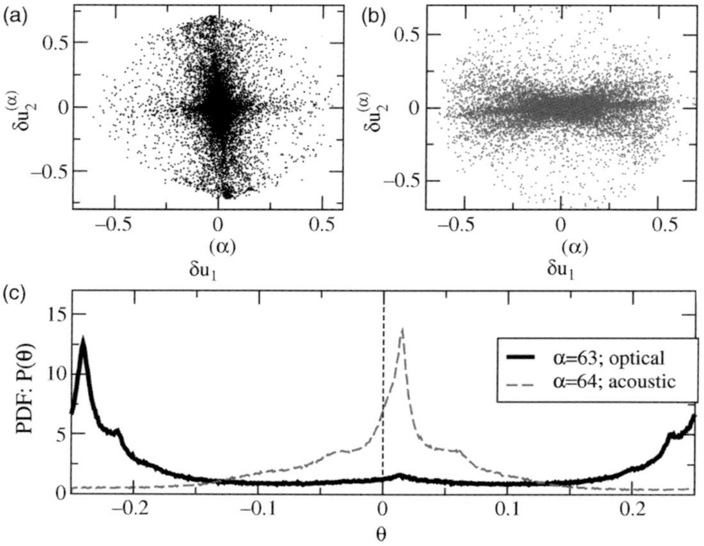

Figure 12.8 $\delta u_1^{(\alpha)}$ versus $\delta u_2^{(\alpha)}$ for Lyapunov modes with (a) $\alpha = L/2 - 1$ and (b) $\alpha = L/2$, which belong to the acoustic and optical branches, respectively. Here, $L = 128$, $\kappa = 0.125$, and $r = 0.3$. (c) The probability distributions of $\theta(t) = 1/\pi \arctan \left(\delta u_2^{(\alpha)}/\delta u_1^{(\alpha)}\right)$.

Our main finding in this section is that the mass difference induces the appearance of gaps in the Lyapunov spectrum and the splitting of Lyapunov modes into acoustic and optical branches. Such a similarity in response to mass differences has its root in the similarity of the mathematical form of the tangent space dynamics of our system and that of the lattice dynamics of harmonic crystals. It suggests and partially confirms the existence of a certain correspondence between Lyapunov modes and phonons even in the strongly anharmonic chaotic regime. This finding on one hand is important for understanding the physical relevance of Lyapunov modes in relation to normal modes such as phonons. On the other hand, it suggests a potential relevance of Lyapunov modes for understanding strongly anharmonic dynamics such as conformational transformations of proteins.

12.6
Comparison of Covariant and Orthogonal HLMs

So far, the orthogonal Lyapunov vectors obtained via the standard method were used to study HLMs except in Section 12.4. Recently, the application of another set of vectors called covariant Lyapunov vectors was made feasible via an efficient algorithm proposed by Ginelli *et al.* [24]. CLVs have been shown being suitable for the characterization of hyperbolicity of high dimensional systems since they are expected to span the local stable and unstable subspaces of the investigated systems. In view of the obvious difference, it becomes necessary to study the relation between CLVs and the previously used orthogonal Lyapunov vectors. Details of this study can be found in Ref. [28].

We use again the coupled map lattice models (Eqs. (12.7) and (12.8)). Two options of the local map are used, the sinusoidal map $f(z) = (1/2\pi)\sin(2\pi z)$ and the skewed tent map (Eq. (12.12)). With the parameter r being close to zero, the system from Eqs. (12.7) and (12.8) with the skewed tent map is highly hyperbolic, whereas Eqs. (12.7) and (12.8) with the sinusoidal map describe a nonhyperbolic system. Especially the Hamiltonian systems (Eq. (12.7)) with the two options of the local map are similar to the well-studied cases of hard-core systems and soft potential systems, respectively. Alternatively, tuning the parameter r of Eq. (12.12) from 0 to 0.5 leads to a smooth variation of the dynamics of Eq. (12.7) from hard-core-like to soft potential-like.

We show in Figure 12.9 the contour plot of the static CLV structure factors $S_u^{(\alpha)}(k)$ (Eq. (12.4)). For both dissipative and Hamiltonian systems, either with the skewed tent map or the sinusoidal map, a clear ridge structure can be seen in the regime $(k, \lambda) \sim (0, 0)$, which indicates the existence of long-wavelength structures in CLVs associated with near-zero Lyapunov exponents. These numerical results demonstrate that HLMs formerly detected via OLVs survive if CLVs are used instead.

In Section 12.3 we showed that dissipative systems have a quadratic $\lambda - k$ dispersion, while Hamiltonian systems have a linear one. We show that it is the same if CLVs are used instead.

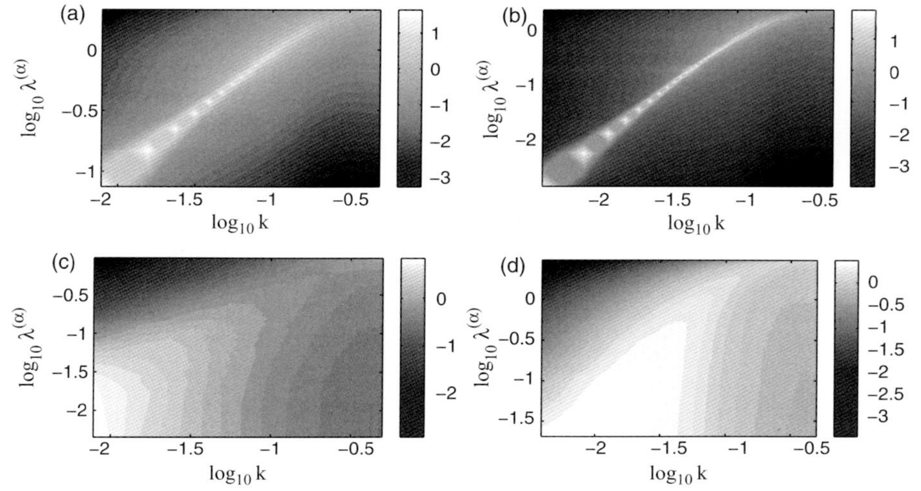

Figure 12.9 Contour plot of the static CLV structure factors for (a and c) Hamiltonian and (b and d) dissipative coupled map lattices. The local dynamics used is the skewed tent map (a and b) and the sinusoidal map (c and d), respectively. Other parameters are $\varepsilon = 1.3$ and $r = 0.15$. A ridge structure can be clearly seen in the small (λ, k) regime, which indicates the existence of HLMs.

The cases with the skewed tent map as local dynamics are shown in Figure 12.10. For the used parameter setting, the dispersion curves for CLVs agree very well with those of OLVs. Cases with the sinusoidal map as local dynamics are shown in Figure 12.11. For both dissipative and Hamiltonian systems, CLV dispersions are converging to the expected asymptotic forms, even better than OLVs. Note also that, as shown in Figure 12.11b, for Hamiltonian systems, CLV dispersions for positive and negative Lyapunov exponents follow the same curve, while OLV dispersions behave differently in the positive and negative Lyapunov exponent regimes. For the used system size, only the positive Lyapunov exponent branch of OLV dispersion is close

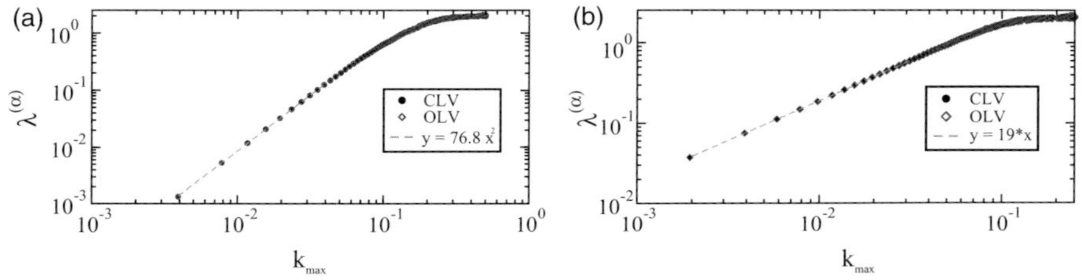

Figure 12.10 Dispersion relations $\lambda - k_{max}$ obtained from CLVs and OLVs for (a) dissipative and (b) Hamiltonian systems, respectively. The local map is the skewed tent map with $\varepsilon = 1.3$ and $r = 0.15$. Note the perfect agreement between data from CLVs and OLVs for the highly hyperbolic cases shown here.

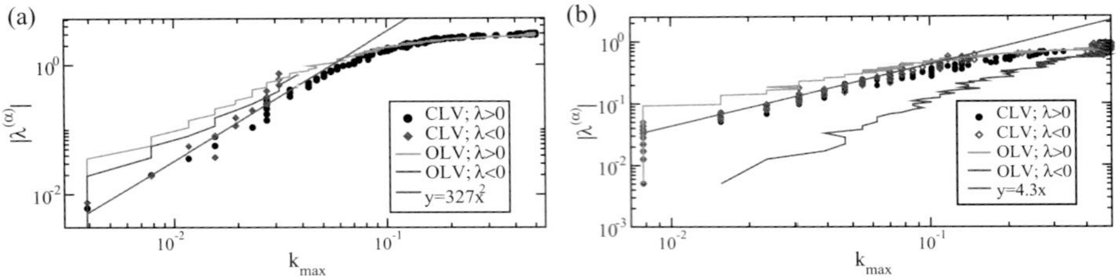

Figure 12.11 Similar to Figure 12.10, but the local map is the sinusoidal map and $\varepsilon = 1.3$. Note that for the nonhyperbolic cases shown in this figure, CLVs and OLVs still have the same asymptotic behavior.

to the asymptotic form. Further discussion regarding these differences will be given in the following sections. Nevertheless, for the two representative cases, the investigated CLV dispersions follow well the previously reported classification of the universality classes of HLMs (see Section 12.3).

By definition, CLVs are in general not mutually orthogonal in contrast to OLVs. This difference has some interesting consequences in Hamiltonian systems. As reported in Ref. [15], a conjugate pair of OLVs with $\lambda^{(\alpha)} = -\lambda^{(2L-1-\alpha)}$ has the symmetry $\delta u^{(\alpha)} = \pm\delta v^{(2L-1-\alpha)}$ and $\delta v^{(\alpha)} = \mp\delta u^{(2L-1-\alpha)}$. Here, δu and δv denote the coordinate and momentum parts of LVs, respectively. The physical origin of this symmetry lies in the symplectic structure of Hamiltonian system. OLVs as the eigenvectors of the matrix $\Xi = \lim_{t\to+\infty}\left[\overline{M(t,0)}\,T \cdot \overline{M(t,0)}\right]^{1/2t}$ (Section 12.2) are thus forced to have the observed symmetry. Examples of conjugate pairs of OLVs are shown in Figure 12.12.

As can be seen from the same plot, CLVs behave differently. The relations $\delta u^{(\alpha)} = \pm\delta u^{(2L-1-\alpha)}$ and $\delta v^{(\alpha)} = \mp\delta v^{(2L-1-\alpha)}$ seem to work well instead. Note also that for CLVs, the amplitude of the wave structure in the momentum part is much smaller than that in the corresponding coordinate part.

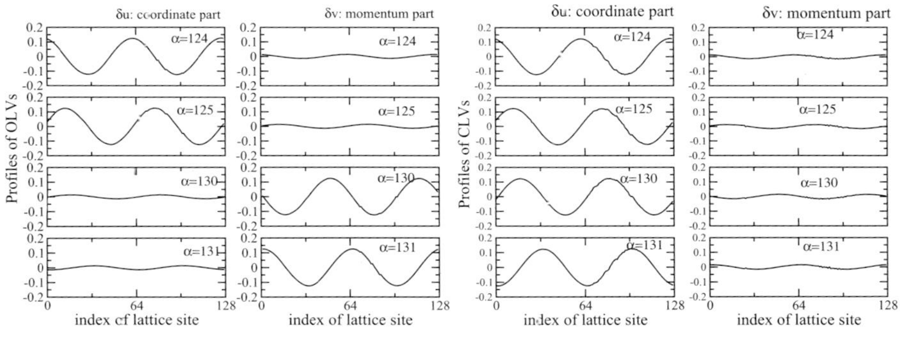

Figure 12.12 Instantaneous profiles of two conjugate pairs of CLVs and OLVs for the Hamiltonian case. The local map is the skewed tent map with $\varepsilon = 1.3$ and $r = 0.15$. The system size used is $N = 128$. Note that OLVs and CLVs have different symmetry relations for the conjugate pair.

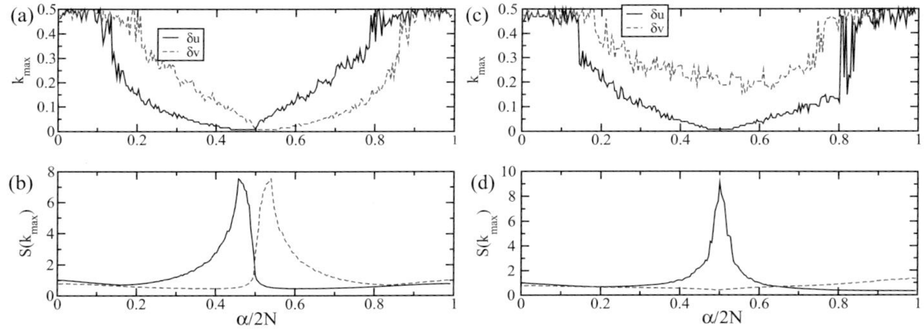

Figure 12.13 Comparison of the dominant wave numbers k_{max} and $S^{(\alpha)}(k_{max})$ obtained from the coordinate and momentum parts of OLVs (a and b) and CLVs (c and d), respectively. The local map is the sinusoidal map and $\varepsilon = 1.3$.

As going to the nonhyperbolic cases with the sinusoidal map as the local dynamics, the mentioned simple relations between δu and δv of instantaneous Lyapunov vectors are no longer valid. However, as can be seen from Figure 12.13, the conjugate pairs of CLVs have nearly identical dominant wave numbers k_{max} and $S^{(\alpha)}(k_{max})$, that is, they are statistically indistinguishable. This interesting feature of CLVs is believed to stem from the microreversibility of Hamiltonian system [28].

Besides the mentioned symmetry resulting from the general Hamiltonian property, the conjugate pair of CLVs in systems with continuous symmetries such as Eq. (12.7) has some unexpected interesting features. The angle θ between a pair of CLVs is used to characterize their relation, with $\cos(\theta) \equiv |e^{(\alpha)} \cdot e^{(\beta)}|$. For the highly hyperbolic cases, for instance, Eq. (12.7), with the special skewed tent map as the local dynamics, CLVs corresponding to near-zero Lyapunov exponents of the *same sign* are nearly orthogonal to each other as expected. The near orthogonal nature of these CLVs explains the observed similarity between CLVs and OLVs for the current parameter setting. In contrast, a conjugate pair of CLVs tends to the same orientation as approaching the zero Lyapunov exponents. With changing the local dynamics to the sinusoidal map, the near orthogonal regime disappears completely, whereas the qualitative behavior of the angle between conjugate pairs is barely influenced [28].

In this section, we have explored similarities and differences between CLVs and OLVs, especially with respect to hydrodynamic Lyapunov modes. With the replacement of OLVs by CLVs, the formerly detected long-wavelength structure in Lyapunov vectors can also be seen. Moreover, the CLV $\lambda - k$ dispersion relation is linear for Hamiltonian system, while quadratic for dissipative system as found for OLVs. We should also mention that for CLVs, the significance of HLMs is more pronounced than using OLVs, especially for highly nonhyperbolic systems. For Hamiltonian systems, the symmetry between the two halves of the modes corresponding to positive and negative Lyapunov exponents, respectively, is not the same for CLVs and OLVs. It is known that a dynamical system has two sets of OLVs, backward and forward ones. Only the backward OLVs, which can be calculated numerically via the standard method, are discussed here. We found that for Hamiltonian systems, they are similar to the set of CLVs with indices $0 \leq \alpha < N$. Similar results are expected for the forward OLVs except that for Hamiltonian systems, they would bear a similarity with the other half of the set of CLVs with indices $N \leq \alpha < 2N$.

12.7
Hyperbolicity and Effective Degrees of Freedom of Partial Differential Equations

It is well known that despite an underlying discreteness, many phenomena in nature and in laboratory are well described by partial differential equations. Examples include the Navier–Stokes equation for fluid dynamics, the complex Ginzburg–Landau equation, and the Kuramoto–Sivashinsky equation for pattern formation and space–time chaos.

An immediate difficulty for studying the dynamics of PDE systems is that the number of involved degrees of freedom is infinite. For instance, numerical simulations of such systems, which are usually the only available way to obtain information of their complex behavior, often start from discretizing space and time. The infinite-dimensional nature of PDE systems seems to demand an infinitesimal discretization step, which is obviously impossible for any computer. Fortunately, owing to the strong dissipative nature, such infinite-dimensional systems often possess a finite-dimensional global attractor. Thus, a finite-dimensional smooth manifold, the inertial manifold, is assumed to exist for many interesting examples of PDE systems [41]. All essential dynamics of these systems are believed to take place just inside the inertial manifold, which thus opens the possibility to model an infinite-dimensional PDE system by a finite-dimensional one. Although conceptually important for understanding many issues of PDE systems, the application of the inertial manifold for the investigation of nonlinear systems is yet very limited, partially due to the absence of a practical method to determine it accurately.

On the other hand, in the study of space–time chaos in extensive dynamical systems, it was recognized that not every information about the dynamics of each DOF is relevant for macroscopic features of such systems [18]. Instead, the extensive nature of space–time chaotic systems suggests that distant parts of such systems are uncorrelated and that the whole system could further on be viewed as consisting of many such uncorrelated components. Therefore, a probabilistic description in the spirit of statistical mechanics or a reduced description based on coarse-grained models would be a better choice. To this aim, seeking the effective DOFs entering such descriptions becomes an essential step.

In a pioneering work, Ruelle conjectured the existence of a large volume limit of the Lyapunov spectrum of extensive chaotic systems [42]. It implies the existence of an intensive quantity, that is, the Lyapunov dimension density $\delta \equiv D/V$, where D is the Kaplan–Yorke dimension of the attractor and V is the system volume. A length scale can be defined correspondingly as $\xi_\delta \equiv \delta^{-1/d}$, where d is the dimensionality of the considered system. The physical significance of this length scale is that regions of size ξ_δ contain on average one nontrivial degree of freedom.

In this section, the Lyapunov instability of a one-dimensional PDE system is investigated. In contrast to most works in the literature, we attempt to see how the spatial resolution of simulations influences the obtained Lyapunov characteristics. Our simulations show that in tangent space of a PDE system, a finite number of covariant Lyapunov vectors are hyperbolically separated from the rest set of infinite number of CLVs. This is demonstrated by the fact that the angle between two subspaces spanned by the two sets of CLVs is bounded from zero. The finite number of CLVs associated with positive and weakly negative Lyapunov exponents, named physical modes, is thought to represent the nontrivial active DOFs for the underlying system. Moreover, the dimension of this manifold shows a stepwise increase with system size. These features are expected to be common for a large class of PDEs.

Our results were obtained from simulations of the one-dimensional Kuramoto–Sivashinsky equation in the form

$$\partial_t u = -\partial_x^2 u - \partial_x^4 u - (\partial_x u)^2, \quad x \in [0, L], \tag{12.14}$$

and by applying periodic boundary conditions. It is known that the system size L is the only control parameter of this system and the system becomes mostly spatio-temporally chaotic for $L \geq 50$. The algorithm proposed in Ref. [24] is used to calculate covariant Lyapunov vectors $\vec{e}^{\,(\alpha)}$ and the corresponding local Lyapunov exponents $\lambda_\tau^{(\alpha)}$, which is defined as the average value of the instantaneous Lyapunov exponents along a trajectory segment of length τ.

Figure 12.14 shows how the spatial resolution of simulations influences the Lyapunov spectrum. Several cases with the same L but with different space resolutions $\delta x \equiv L/N$ are presented in the main panel, where N is the number of space grid

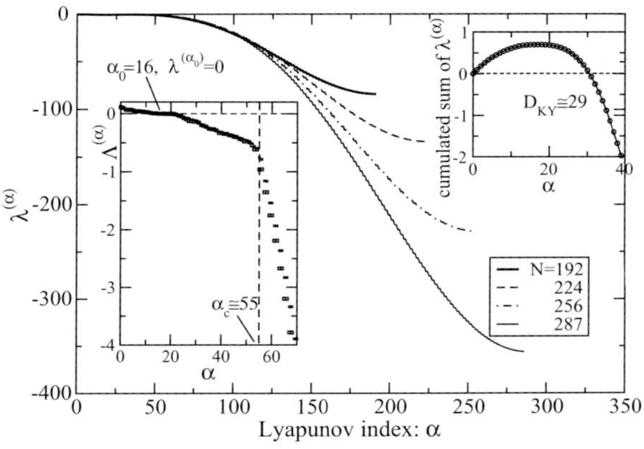

Figure 12.14 Lyapunov spectra for varying space resolutions $\delta x \equiv L/N$ with the system size fixed as $L = 133.12$. The enlargement in the left inset shows the robust splitting of the spectrum into two parts: a smoothly varying part independent of space resolution and the other part with size two-step structures changing with the spatial resolution.

points used. Improving space resolution means more DOFs entering the simulated dynamics and therefore more Lyapunov exponents are obtained. As can be seen clearly from the enlargement of spectra in the left inset, the Lyapunov spectrum of this system splits naturally into a smoothly varying part for $\alpha < \alpha_c$ and a part consisting of size two-step structures. Note that the threshold value α_c lies deeply in the regime of negative Lyapunov exponents. Moreover, the smooth part of the Lyapunov spectrum and the threshold value α_c of the splitting are not influenced by the variation of the spatial resolution as demonstrated by the data collapse in this regime. In contrast, the number of Lyapunov exponents in the other part and the value of these Lyapunov exponents do change by varying the spatial resolution. The pairing nature of Lyapunov exponents, however, is not influenced by the variation of space resolution. These numerical results suggest that the dynamics associated with the smooth part of the Lyapunov spectrum does not depend on the simulation detail and thus would be more interesting than that corresponding to the discrete part. For simulations performed via the pseudo-spectral method, Lyapunov spectrum shows a similar splitting between a smooth part and a stepwise part. Owing to the higher order accuracy of that method, one can indeed see that the stepwise part of the spectra for different resolutions also collapse on a master curve [23].

To characterize the detailed hyperbolicity, we use the quantities p_{split} and $\theta_{\min}$ [23]. The quantity p_{split} measures the fluctuations of *finite-time* Lyapunov exponents $\lambda_\tau^{(\alpha)}(t)$, which is the local expansion/contract rate in the time interval $[t, t+\tau]$, and the latter $\theta_{\min}$ denotes the minimal angle between a selected Oseledec subspace (or the corresponding CLVs) and all the other Oseledec subspaces (the corresponding CLVs). We define $p_{\text{split}}^{(\alpha)}(\tau) = \max_{\beta \neq \alpha t}\{p_\tau^{\alpha\beta}(t)\}$ and $p_\tau^{\alpha\beta}$ is given by $p_\tau^{\alpha\beta}(t) = \lambda_\tau^{(\alpha)}(t) - \lambda_\tau^{(\beta)}(t)$ for $\beta > \alpha$ and $p^{\beta\alpha} = p^{\alpha\beta}$. Note that the definition of p_{split} used here is slightly different from that in Section 12.4 and instead negative values of p_{split} and corresponding nonzero values of $\theta_{\min}$ indicate the fulfillment of the domination of the Oseledec splitting. Figure 12.15 shows the variation of the hyperbolicity measures p_{split} and $\cos(\theta_{\min})$ with Lyapunov index α for several cases with the same L but with different resolutions δx and times τ. Look first at the case $\delta x = 0.52$ and $\tau = 0.4$, which is denoted by filled circles in the plot. Obviously, the behavior of both quantities p_{split} and $\cos(\theta_{\min})$ is clearly distinct in the two regimes separated by a threshold value α_c, corresponding to the splitting of the Lyapunov spectrum shown in Figure 12.14. One can easily see that p_{split} always attains positive values in the regime $\alpha \leq \alpha_c$, while for $\alpha > \alpha_c$ it is always negative and its value decreases with increasing α. Correspondingly, $\cos(\theta_{\min})$ stays close to 1 for $\alpha \leq \alpha_c$ and it departs from 1 gradually with increasing α beyond α_c. This indicates the fulfillment of the (partial) domination of the Oseledec splitting with respect to the Lyapunov vectors with $\alpha \geq \alpha_c$. The linear subspace spanned by CLVs with $\alpha \leq \alpha_c$ is hyperbolically isolated from that spanned by CLVs with $\alpha > \alpha_c$.

Now let us specify more clearly the meaning of this isolation and/or the implication of the claimed strong hyperbolicity. The quantity $\theta_{\min}$ being nonzero for $\alpha > \alpha_c$ means that during the time evolution, any one of the CLVs with $\alpha > \alpha_c$ has no chance to become tangent to any other CLV associated with a different Lyapunov exponent. Combined with their covariant nature, one can thus easily see that each of these CLVs evolves just freely, that is, having no interaction with other CLVs.

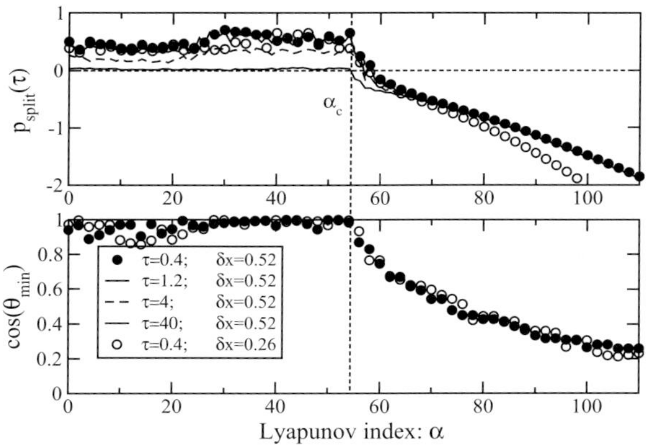

Figure 12.15 Hyperbolicity measures (a) $p_{\text{split}}(\tau)$ and (b) $\cos\theta_{\min}$ for cases either with $\delta x = 0.52$ and $\tau = 0.4$, 1.2, 4, and 40, respectively (filled circles and lines), or with $\delta x = 0.26$ and $\tau = 0.4$ (open circles). The system size is fixed to $L = 133.12$. Note that the definition of p_{split} is slightly different from the one in Figure 12.4.

Imagine adding to the system dynamics a small perturbation along some CLV with $\alpha > \alpha_c$. The amplitude of this perturbation will decay exponentially to zero as indicated by the negative value of $\lambda^{(\alpha)}$. Moreover, perturbations along this specified direction would not induce perturbations along other directions with different $\lambda^{(\alpha)}$ and vice versa. In contrast, zero values of $\theta_{\min}$ for $\alpha \leq \alpha_c$ indicate the appearance of (near) tangencies between CLVs in that regime. Therefore, perturbations initially introduced along some CLV with $\alpha \leq \alpha_c$ may eventually induce activity along all other directions with $\alpha \leq \alpha_c$. Especially, perturbations introduced along CLVs with negative Lyapunov exponents in the regime $\alpha \leq \alpha_c$ are able to induce motions along directions with positive Lyapunov exponents and eventually lead to a drastic change of the dynamical evolution. In this sense, the dynamics along CLVs with $\alpha \leq \alpha_c$ is highly entangled, while decoupled from the decaying dynamics along directions with $\alpha > \alpha_c$ and therefore the manifold spanned by CLVs with $\alpha \leq \alpha_c$ is called dynamically isolated.

Other cases with the same δx and different τ and one case with smaller δx shown in Figure 12.15 demonstrate that the existence of an isolated manifold is an intrinsic property of the underlying PDE system and simulation details have no or ignorable influence on it. The isolated manifold identified here is expected to be right the so-called inertial manifold proposed previously [20].

As can be seen in Figure 12.14, the dimension of the isolated manifold D_{IM} is much larger than the Kaplan–Yorke dimension of the attractor. Thus, a new characteristic length different from the dimension length ξ_δ may be defined. For this purpose, one needs to study the system size dependence of the isolated manifold and to check its extensivity. Figure 12.16 shows a stepwise increase of D_{IM} with system size L as expected. Moreover, the value of D_{IM} jumps by two between nearest neighboring steps, which is related to the double degeneration of Lyapunov exponents and is assumed to result from the spatial translational invariance of the system. Remarkably, the variation of D_{IM} follows perfectly a function $D_{\text{IM}} = 1 +$

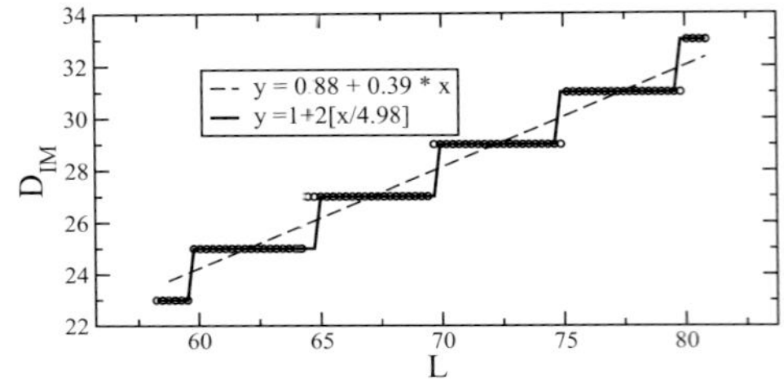

Figure 12.16 Stepwise increasing of the dimension D_{IM} of the isolated manifold with system size L. Numerical fitting yields $D_{IM} = 0.39L + 0.88$. Moreover, numerical data agree well with a function $D_{IM} = 1 + 2[L/\xi_{IM}]$ with $\xi_{IM} \approx 4.98$ and $[x]$ denoting the integer part of x.

$2[L/\xi_{IM}]$ with $[x]$ denoting the integral part of x. The extension of the steps $\xi_{IM} \approx 4.98$ may be defined as a new length scale for characterizing space–time chaos.

We conclude the section with some comments. First, our proposition of Lyapunov instability analysis provides a practical method to probe information on the number of the effective degrees of freedom (the dimension of the inertial manifold) of PDEs. Second, although the numerical results presented here were obtained for the Kuramoto–Sivashinsky equation, similar results were found for the complex Ginzburg–Landau equation [23]. We expect our discovery to hold for a large class of dissipative PDE systems. Finally, we would like to emphasize that besides the conceptual importance for understanding features of PDE systems, our method for an accurate estimation of D_{IM} may also provide helpful hints to continuum system related to practical applications.

12.8
Probing the Local Geometric Structure of Inertial Manifolds via a Projection Method

The concept of IM allows the reduction of an infinite-dimensional PDE system to a finite-dimensional one. But due to the complexity of the necessary studying tools, the use of IM in practical situations is rather rare. We proposed to use the Lyapunov analysis in Section 12.7 as an alternative to address the issue based on the intuitive expectation that infinitesimal perturbations to an IM should behave somehow similar to the finite-size ones [23]. We showed in Section 12.7 that there exists a hyperbolic separation between a set of mutually entangled finite number of physical modes and the remaining infinite number of spurious modes. The set of finite number of physical modes associated with positive and weakly negative Lyapunov exponents was conjectured to represent the physically relevant dynamics on the inertial manifold of original PDE systems. The linear space spanned by physical modes may serve as the local approximation of IM. The remaining set of vectors corresponding to strongly negative Lyapunov exponents is believed to represent the trivial decaying process to the IM [23].

In this section, direct support to this conjecture is obtained from the fully nonlinear dynamics by the following idea: we consider a state at some point in the time evolution of the system, a reference state, and consider recurrences of the system trajectory into a neighborhood of this reference state. This defines the so-called recurrent states with distance r to the reference state. The difference vector between recurrent and reference states is then projected onto subspaces spanned by some Lyapunov vectors at the reference state. The deviation of the difference vector from its projection defines a projection error. We find below that two competing factors can contribute to the projection error depending on the completeness of the spanned subspace compared to the tangent space of the IM. A sharp transition is observed as the number of spanning Lyapunov modes is increased beyond the IM dimension estimated via Lyapunov analysis [35]. This finding provides the first direct evidence for the relation between physical modes and the state space geometry of the IM. It confirms the interpretation that the set of physical modes span a linear space approximating the IM locally. Furthermore, the specific variation behavior of the projection error reflects the local quadratic curvature of the IM in general nonlinear systems. It should be emphasized that differences between recurrent states and the reference state used here are of finite size in contrast to the infinitesimal perturbations considered in a Lyapunov analysis [35]. Note that our projection method can be accomplished also with the standard, orthogonal Lyapunov vectors, for which sophisticated methods were developed for obtaining them from experimental data [43, 44]. One can hope that the existence of IMs, their dimensions, and their curvature properties can also be identified from experimental data.

To demonstrate our finding, we use as an example the one-dimensional Kuramoto–Sivashinsky equation (Eq. (12.14)) [18, 45]. The system is numerically integrated by using the pseudo-spectral scheme and the exponential time differencing method [46] with periodic boundary conditions and a Fourier basis of size 128. The size L is the only control parameter of the system and we use the value $L = 22$ as in Ref. [45], which is sufficiently large to have a structurally stable chaotic attractor and also allows a detailed exploration of the state space geometry of the IM. The standard method of Benettin *et al.* [30] is used to calculate the orthogonal Lyapunov vectors $f^{(\alpha)}$ and the algorithm of Ginelli *et al.* [24] is adopted to get supplementary information from covariant Lyapunov vectors.

We assume that the spectrum of Lyapunov exponents is always arranged in descending order and the associated Lyapunov vectors are ordered correspondingly. The values of the Lyapunov exponents shown in Figure 12.17 are in good agreement with Refs [45].[1] The angle between two subspaces spanned by the first n CLVs (i.e., $\alpha \leq n$) and the remaining ones, respectively, is a fluctuating quantity and is denoted as the manifold angle θ in the following. It is calculated as in Ref. [35, 47]. Distributions $p(\theta)$ of the manifold angle for several values of n are

1) See the more recent result at http://chaosbook.org/version13/Maribor11.shtml, which is in complete agreement with our Figure 12.17 that one more (near) zero Lyapunov exponent appears compared to that in Ref. [45], possibly due to the elimination of the Galilean invariance in Ref. [45].

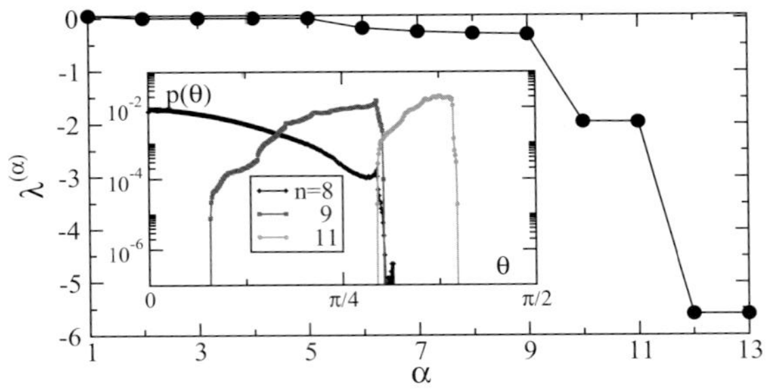

Figure 12.17 The largest Lyapunov exponents (of 128) arranged in descending order for the Kuramoto–Sivashinsky equation with $L = 22$. The largest exponent has the value $\lambda^{(1)} = 0.048$.[1] The inset shows the distribution $p(\theta)$ of the manifold angle for several cutting dimensions n.

presented in the inset of Figure 12.4. It can be seen that the distribution is bounded from zero for $n = 9$ and 11 in contrast to $n = 8$ (and lower values), which indicates the (lowest) IM dimension to be $M = 9$ for $L = 22$ as in Refs [45].[1]

By definition, the complete set of Lyapunov vectors span the tangent space of the dynamical system at a given reference state. Analogously, assuming the existence of an IM of dimension M, the first $n = M$ Lyapunov vectors span the tangent space of the IM at the same reference state, a local linear approximation of the IM. To quantify the accuracy of the approximation, one can define the deviation of the difference vector of reference and recurrent states on the IM from its projection on the spanned linear space as the projection error. In general, the IM is a nonlinear manifold and in the neighborhood of a reference state, the quadratic nonlinearity dominates. The projection error for the case with $n = M$ is expected solely due to the nonlinearity/curvature of the IM and would thus decrease quadratically as the reference state is approached. In contrast, a linear space spanned by fewer, say the set of the first $n < M$, Lyapunov vectors cannot approximate the IM as well due to its incompleteness compared to the tangent space of the IM. Consequently, the nonzero projections to the missing Lyapunov vectors provide a new source of the projection error and the variation of the projection error with the distance to the reference state will behave differently. Such differences in the projection error by varying the number of spanning Lyapunov vectors can be used to detect the existence of an IM and to estimate its dimension as well.

To be more specific, to probe the local geometry of the IM at a given reference state u_0, we use the collection of recurrent states u_r with the Euclidean distance to the reference state being $r = \|u_r - u_0\|$. The projection of the state difference vector $\delta u_r = u_r - u_0$ to the linear subspace spanned by the first n Lyapunov vectors can be simply calculated as the sum of projections to each Lyapunov vector individually if OLVs are used.[2] Denoting the length of the projection of the state difference vector δu_r on the αth normalized OLV $f^{(\alpha)}$ as $p(\alpha, r) = |\delta u_r \cdot f^{(\alpha)}|$, the projection error of δu_r to the subspace spanned by the first n OLVs can be calculated as

2) In case CLVs are used, the projection error can be calculated via a method similar to that used in Ref. [47] for the angle between subspaces.

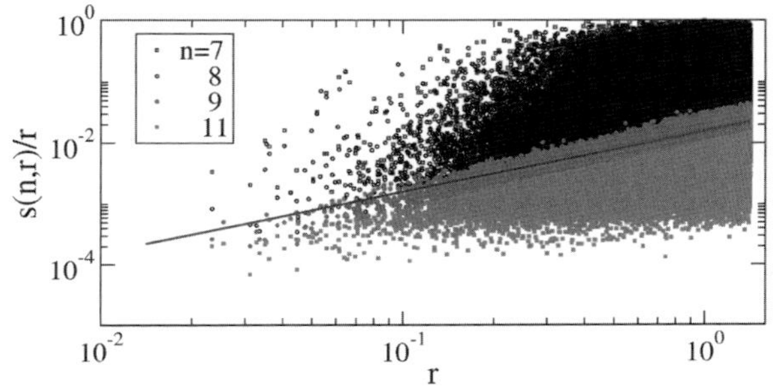

Figure 12.18 Variation of the normalized projection error $s(n,r)/r$ with the distance to the reference state r with $n = 7$, 8, 9, and 11, respectively. To improve the statistics, data from 200 reference states are presented together. A line with slope 1 is shown to guide the eyes. (For a color version of this figure, please see the color plate at the end of this book.)

$s(n,r) = \sqrt{r^2 - \sum_{\alpha=1}^{n} p^2(\alpha,r)}$. Equivalently, the projection error can also be obtained as the Euclidean norm of the sum of projections to all OLVs with indices larger than n. We study in the following the dependence of the projection error $s(n,r)$, respectively, the normalized error $\hat{s}(n,r) = s(n,r)/r$, on the distance r to the reference state and the cutting dimension n. $\hat{s}(n,r)$ is simply the projection error obtained from normalized difference vectors $\delta u_{\text{r}}/\|\delta u_{\text{r}}\|$ and obeys $0 \le \hat{s}(n,r) \le 1$.

The variation of the normalized projection error $\hat{s}(n,r)$ with r is shown in Figure 12.18 for several values of n. To improve the statistics data from 200, randomly selected and uncorrelated reference states, each with over 10^4 recurrent states, are plotted together. A clear difference between the cases with $n < M = 9$ and those with $n \ge M$ can be seen in the variation of the upper bound of $\hat{s}(n,r)$ with r.[3)] For $n \ge M$, the upper bound of the normalized projection error $\hat{s}$ changes linearly with r, which indicates a quadratic dependence of the upper bound of $s(n,r)$ on r. This is consistent with the expectation that the projection error for $n \ge M$ is solely due to the nonlinearity of the IM and the quadratic behavior of the upper bound indicates the local quadratic behavior of the IM. For $n < M$, the upper bound of $\hat{s}(n,r)$ is simply the value of 1 independent of the distance r to the reference state, which means that the projection error s can take any value up to size r. This is simply due to the nonzero projection of the state difference vector δu_{r} on the subspace spanned by the missing Lyapunov vectors with indices from $n+1$ to M. The value 1 of $\hat{s}(n,r)$ corresponds to the extreme situation that the state difference vector δu_{r} is completely contained in the subspace spanned by the missing Lyapunov vectors.

We consider now in Figure 12.19 the averaged normalized projection error $\hat{S}(n,r) \equiv \langle \hat{s}(n,r) \rangle$, where $\langle \ldots \rangle$ denotes an average over both the recurrent states and reference states. With decreasing r, the mean error $\hat{S}(n,r)$ for $n < M$ shows a clear tendency of saturation to a nonzero value. In contrast, for $n \ge M$, the mean error $\hat{S}(n,r)$ decays continuously in the whole range of r considered. Fitting data to

3) Actually the upper bound value has always a tiny deviate from 1, on the order of $O(r^2)$, due to the nonlinearity of the IM.

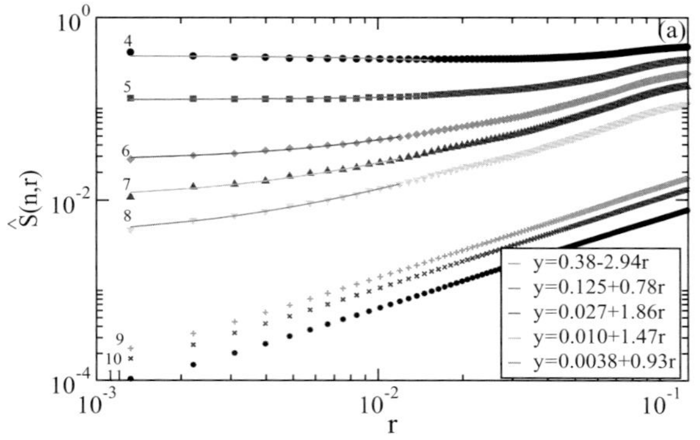

Figure 12.19 The normalized average projection error $S(n, r)/r$ versus the distance to the reference state r with n from 4 to 11. Linear fittings of data for $n < M$ confirms the saturation of $S(n, r)/r$ to a nonzero constant value. (For a color version of this figure, please see the color plate at the end of this book.)

a power law shows that its exponent is almost 1, which indicates a linear decrease of $\hat{S}(n, r)$ with r for $n \geq M$. For $n < M$, the variation of $\hat{S}(n, r)$ with r in the small r regime can also be fitted well with a linear function $y = a_0 + a_1 r$, but now with $a_0 \neq 0$. The nonzero values of a_0 confirm the saturation of $\hat{S}(n, r)$ with decreasing r. Note that here also the linear behavior has the same origin as the linear decay of $\hat{S}(n, r)$ for $n \geq M$, that is, the local quadratic nonlinearity of the IM. The nonzero saturation value a_0 for $n < M$ has a different origin and results from the nonzero projections of state difference vectors to the missing set of OLVs e_m with indices $n < m \leq M$. The value $a_0 = \lim_{r \to 0} \hat{S}(n, r)$ becoming zero for $n \geq M$ indicates the completeness of the set of the first $n = M$ Lyapunov vectors as a basis of the tangent space of the IM. It can thus be used to detect the existence of an IM and to estimate its dimension M.

12.9 Summary

Two main topics were addressed in this chapter, HLMs reflecting collective behavior in Lyapunov vectors and the relation between hyperbolicity and the effective degrees of freedom of PDEs. For the collective structures in Lyapunov vectors, we showed that the existence of HLMs is generic for a large class of extended system with continuous symmetries, irrespective of a Hamiltonian structure. Numerical and analytical evidence was provided to show that Hamiltonian and dissipative systems belong to two different universality classes with respect to the nature of the HLMs. Hyperbolicity was found to be essential for the significance of HLMs. To be precise, domination of the Oseledec splitting is a necessary condition for the appearance of good/clean HLMs. Extensive numerical simulations showed that the use of CLVs instead does not change the general conclusions drawn from the previous studies based on OLVs. Certainly one can see some quantitative changes in the significance of HLMs especially for highly nonhyperbolic systems. For Hamiltonian systems, the two halves of modes corresponding to positive and negative Lyapunov exponents bear a different symmetry for CLVs and OLVs.

We found for nonlinear dissipative PDEs a hyperbolic separation between a finite number of mutually entangled physical modes and the remaining set containing an infinite number of spurious modes. The set of physical modes are conjectured to represent physically relevant dynamics in the inertial manifold of PDEs. A projection method provides direct evidence for the connection between the physical modes and the inertial manifold. The local geometry of inertial manifolds can be probed by studying the projection error of the difference vectors between recurrent states and the corresponding reference state to the linear subspaces spanned by the first n Lyapunov vectors. The dependence of the projection error on the distance r between the recurrent state and the reference state shows a sharp transition as the number of spanning vector n is increased beyond the IM dimension M. These changes in projection errors at $n = M$ indicate unambiguously the completeness/incompleteness of the linear subspaces spanned by the first n Lyapunov vectors as the tangent space of IM. It supports on one hand the geometric interpretation that the set of physical modes span the tangent space of the IM at the given reference state. The finite number of the physical modes indicates on the other hand the finite dimension of the IM. The projection method can serve as an alternative fast method to detect IMs and to estimate their dimensions, even from time series of experimental observations.

Acknowledgments

We acknowledge discussions with Hugues Chaté, Francesco Ginelli, Wolfram Just, Roberto Livi, Arkady Pikovsky, Antonio Politi, Harald Posch, and Kazumasa Takeuchi; the financial support from the Deutsche Forschungsgemeinschaft; computing time from NIC Jülich; and algorithmic support from G. Rünger and M. Schwind. Special thanks go to Hugues Chaté, Francesco Ginelli, and Kazumasa Takeuchi for the collaboration on PDE systems.

References

1 Krylov, N.S. (1979) *Works on the Foundations of Statistical Mechanics*, Princeton University Press, Princeton.
2 Sinai, Ya.G. (2000) *Dynamical Systems, Ergodic Theory and Applications*, Springer, Berlin.
3 Pettini, M., Casetti, L., Cerruti-Sola, M., Franzosi, R., and Cohen, E.G.D. (2005) Weak and strong chaos in Fermi–Pasta–Ulam models and beyond. *Chaos*, **15**, 015106.
4 Dorfman, J.P. (1999) *An Introduction to Chaos in Nonequilibrium Statistical Mechanics*, Cambridge University Press, Cambridge.
5 Posch, H.A. and Hirschl, R. (2000) Simulation of billiards and of hard body fluids, in *Hard Ball Systems and the*

Lorentz Gas (ed. D. Szàsz), Springer, Berlin.

6 Forster, C., Hirschl, R., Posch, H.A., and Hoover, W.G. (2004) Perturbed phase-space dynamics of hard-disk fluids. *Physica D*, **187**, 294–310.

7 Hoover, W.G., Posch, H.A., Forster, C., Dellago, C., and Zhou, M. (2002) Lyapunov modes of two-dimensional many-body systems; soft disks, hard disks, and rotors. *J. Stat. Phys.*, **109**, 765–775.

8 Radons, G. and Yang, H.L. (2004) Static and dynamic correlations in many-particle Lyapunov vectors. *arXiv*:nlin.CD/0404028.

9 Yang, H.L. and Radons, G. (2005) Lyapunov instability of Lennard-Jones fluids. *Phys. Rev. E*, **71**, 036211.

10 Eckmann, J.-P., Forster, C., Posch, H.A., and Zabey, E. (2005) Lyapunov modes in hard-disk systems. *J. Stat. Phys.*, **118**, 813–847.

11 Taniguchi, T. and Morriss, G.P. (2003) Boundary effects in the stepwise structure of the Lyapunov spectra for quasi-one-dimensional systems. *Phys. Rev. E*, **68**, 026218.

12 Taniguchi, T. and Morriss, G.P. (2005) Time-oscillating Lyapunov modes and the momentum autocorrelation function. *Phys. Rev. Lett.*, **94**, 154101.

13 Forster, C. and Posch, H.A. (2005) Lyapunov modes in soft-disk fluids. *New J. Phys.*, **7**, 32.

14 Yang, H.L. and Radons, G. (2006) Universal features of hydrodynamic Lyapunov modes in extended systems with continuous symmetries. *Phys. Rev. Lett.*, **96**, 074101.

15 Yang, H.L. and Radons, G. (2006) Hydrodynamic Lyapunov modes in coupled map lattices. *Phys. Rev. E*, **73**, 016202.

16 Yang, H.L. and Radons, G. (2006) Dynamical behavior of hydrodynamic Lyapunov modes in coupled map lattices. *Phys. Rev E*, **73**, 016208.

17 Yang, H.L. and Radons, G. (2006) Hydrodynamic Lyapunov modes and strong stochasticity threshold in Fermi–Pasta–Ulam models. *Phys. Rev. E*, **73**, 016201.

18 Cross, M.C. and Hohenberg, P.C. (1993) Pattern formation outside of equilibrium. *Rev. Mod. Phys.*, **65**, 851–1112.

19 Témam, R. (1988) *Infinite Dimensional Dynamical Systems in Mechanics and Physics*, Springer, New York.

20 Foias, C., Sell, G.R., and Témam, R. (1988) Inertial manifolds for nonlinear evolutionary equations. *J. Diff. Equ.*, **73**, 309.

21 (a) Foias, C. *et al.* (1988) Inertial manifolds for the Kuramoto–Sivashinsky equation and an estimate of their lowest dimension. *J. Math. Pures. Appl.*, **67**, 197 (b) Jolly, M.S., Rosa, R., and Témam, R. (2001) Accurate computations on inertial manifolds. *SIAM J. Sci. Comput.*, **22**, 2216.

22 Mallet-Paret, J. and Sell, G.R. (1988) Inertial manifolds for reaction diffusion equations in higher space dimensions. *J. Am. Math. Soc.*, **1**, 805.

23 (a) Yang, H-.L., Takeuchi, K.A., Ginelli, F., Chaté, H., and Radons, G. (2009) Hyperbolicity and the effective dimension of spatially extended dissipative systems. *Phys. Rev. Lett.*, **102**, 074102 (b) Takeuchi, K.A., Yang, H.-L., Ginelli, F., Radons, G., and Chaté, H. (2011) Hyperbolic decoupling of tangent space and effective dimension of dissipative systems. *Phys. Rev. E*, **84**, 046214.

24 Ginelli, F., Poggi, P., Turchi, A., Chaté, H., Livi, R., and Politi, A. (2007) Characterizing dynamics with covariant Lyapunov vectors. *Phys. Rev. Lett.*, **99**, 130601.

25 Yang, H.L. and Radons, G. (2007) Lyapunov spectral gap and branch splitting of Lyapunov modes in a diatomic system. *Phys. Rev. Lett.*, **99**, 164101.

26 Yang, H.L. and Radons, G. (2008) When can one observe good hydrodynamic Lyapunov modes? *Phys. Rev. Lett.*, **100**, 024101.

27 Yang, H.L. and Radons, G. (2008) in *Lyapunov Instabilities of Extended Systems*, vol. **39** (eds G. Münster, D. Wolf, and M. Kremer), NIC Series, pp. 349–358.

28 Yang, H.-L. and Radons, G. (2010) Comparison between covariant and orthogonal Lyapunov vectors. *Phys. Rev. E*, **82**, 046204.

29 Oseledec, V.I. (1968) A multiplicative ergodic theorem and Lyapunov characteristic numbers for dynamical

systems. *Trans. Moscow Math. Soc.*, **19**, 197.

30 Benettin, G., Galgani, L., and Strelcyn, J. M. (1976) Kolmogorov entropy and numerical experiments. *Phys. Rev. A*, **14**, 2338–2345.

31 Goldhirsch, I., Sulem, P.L., and Orszag, S.A. (1987) Stability and Lyapunov stability of dynamical systems: a differential approach and a numerical method. *Physica D*, **27**, 331.

32 Ershov, S.V. and Potapov, A.B. (1998) On the concept of stationary Lyapunov basis. *Physica D*, **118**, 167.

33 (a) Wolfe, C.L. and Samelson, R.M. (2007) An efficient method for recovering Lyapunov vectors from singular vectors. *Tellus*, **59**, 355 (b) This method was introduced to the physics community by Szendro, I.G., Pazó, D., Rodrǵuez, M.A., and López, J.M. (2007) Spatiotemporal structure of Lyapunov vectors in chaotic coupled-map lattices. *Phys. Rev. E*, **76**, 025202.

34 (a) Escande, D., Kantz, H., Livi, R., and Ruffo, S. (1994) Self-consistent check of the validity of Gibbs calculus using dynamical variables. *J. Stat. Phys.*, **76**, 605 (b) Antoni, M. and Ruffo, S. (1995) Clustering and relaxation in Hamiltonian long-range dynamics. *Phys. Rev. E*, **52**, 2361.

35 (a) Kuramoto, Y. and Tsuzuki, T. (1976) Persistent propagation of concentration waves in dissipative media far from thermal equilibrium. *Prog. Theor. Phys.*, **55**, 356 (b) Sivashinsky, G.I. (1977) Nonlinear analysis of hydrodynamic instability in laminar flames. *Acta Astronaut.*, **4**, 1177.

36 Bochi, J. and Viana, M. (2005) The Lyapunov exponents of generic volume-preserving and symplectic maps. *Ann. Math.*, **161**, 1423–1485.

37 Keyes, T. (1997) Instantaneous normal mode approach to liquid state dynamics. *J. Phys. Chem. A*, **101**, 2921–2930.

38 de Wijn, A.S. and van Beijeren, H. (2004) Goldstone modes in Lyapunov spectra of hard sphere systems. *Phys. Rev. E*, **70**, 016207.

39 Pettini, M., Casetti, L., Cerruti-Sola, M., Franzosi, R., and Cohen, E.G.D. (2005) Weak and strong chaos in Fermi-Pasta-Ulam models and beyond. *Chaos*, **15**, 015106.

40 Yang, H.L. and Radons, G. (2008) Hydrodynamic Lyapunov modes and strong stochasticity threshold in the dynamic XY model: an alternative scenario. *Phys. Rev. E*, **77**, 016203.

41 Foias, C., Sell, G.R., and Témam, R. (1988) Inertial manifolds for nonlinear evolutionary equations. *J. Diff. Equ.*, **73**, 309–353.

42 Ruelle, D. (1982) Large volume limit of the distribution of characteristic exponents in turbulence. *Commun. Math. Phys.*, **87**, 287–302.

43 (a) Wolf, A., Swift, J.B., Swinney, H.L., and Vastano, J.A. (1985) Determining Lyapunov exponents from a time series. *Physica D*, **16**, 285 (b) Sano, M. and Sawada, Y. (1985) Measurement of the Lyapunov spectrum from a chaotic time series. *Phys. Rev. Lett.*, **55**, 1082 (c) Eckmann, J.P. *et al.* (1986) Liapunov exponents from time series. *Phys. Rev. A*, **34**, 4971.

44 Kantz, H. and Schreiber, T. (2003) *Nonlinear Time Series Analysis*, Cambridge University Press, Cambridge.

45 Cvitanović, P., Davidchack, R.L., and Siminos, E. (2010) On the state space geometry of the Kuramoto–Sivashinsky flow in a periodic domain. *SIAM J. Appl. Dyn. Syst.*, **9**, 1.

46 Cox, S.M. and Matthews, P.C. (2002) Exponential time differencing for stiff systems. *J. Comput. Phys.*, **176**, 430.

47 Kuptsov, P.V. and Kuznetsov, S.P. (2009) Violation of hyperbolicity in a diffusive medium with local hyperbolic attractor. *Phys. Rev. E*, **80**, 016205.

48 Yang, H.L. and Radons, G. (2012) Geometry of Inertial Manifolds Probed via a Lyapunov Projection Method, *Phys. Rev. Lett.* **108**, 154101.

13
Study of Single-Molecule Dynamics in Mesoporous Systems, Glasses, and Living Cells

Stephan Mackowiak and Christoph Bräuchle

13.1
Introduction

Since the first single-molecule detection experiments of pentacene in *p*-terphenyl crystals at liquid helium temperature by Moerner and Kador [1] and Orrit and Bernard [2] more than two decades ago, single-molecule microscopy (SMM) has developed significantly, thanks to improvements in technologies and dyes. Nowadays, SMM is an essential tool in many different fields in biology, chemistry, and physics.

SMM offers the opportunity to observe single molecules in a wide variety of systems and can thereby provide information about static and dynamic properties of these single entities. Furthermore, the distribution of the properties of single molecules can be measured. The type of the distribution as well as the moments of the distribution such as the mean, variance, skewness, and kurtosis can provide important insights into the characteristics of the studied system. In particular, information about protein conformational changes [3, 4], individual reaction rates of single molecules [5, 6], and dynamic heterogeneity in supercooled liquids can be obtained [7, 8]. Furthermore, studies in biological systems (for a review see Ref. [9]), such as the pathway of virus entry into the cell [10] and the movement of single motor proteins on actin filaments, are feasible [11]. In addition, the tracking of single molecules offers the possibility to investigate many open questions, for example, in the field of confined diffusion in physical systems, such as mesopores [12], and in biological systems, such as living cells [13]. These are only some of the applications where SMM has been used with great success.

13.1.1
Experimental Method

SMM experiments can be performed using different types of microscopes such as a confocal, a total internal reflection fluorescence (TIRF), or a wide-field epifluorescence microscope. In the following, studies will be presented that employ epifluorescence wide-field microscopy. Thus, we will limit the methods discussion to this setup. For a review about the other techniques, see Refs [14, 15].

Nonequilibrium Statistical Physics of Small Systems: Fluctuation Relations and Beyond, First Edition.
Edited by Rainer Klages, Wolfram Just, and Christopher Jarzynski.

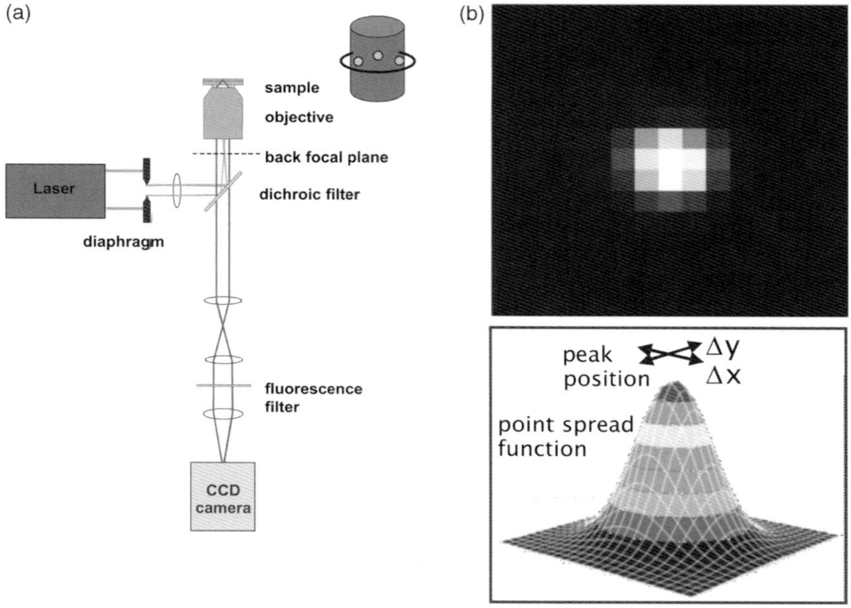

Figure 13.1 (a) Setup of wide-field epifluorescence microscope. The cylinder shows the excitation volume of a wide-field microscope. (b) *Top*: Fluorescence intensity distribution of a single molecule. *Bottom*: Two-dimensional Gaussian curve that describes the fluorescence intensity pattern of a single molecule.

In an epifluorescence setup (cf. Figure 13.1a), the sample is excited with a laser beam through an objective and the fluorescence is collected by the same objective. For wide-field excitation, the sample is excited with a collimated laser beam, resulting in a cylindrical excitation volume (cf. Figure 13.1a). The objectives usually have a high numerical aperture, such as oil objectives with a numerical aperture of 1.4. The numerical aperture (NA) is related to the half collection angle α of the objective and the refractive index n of the medium by

$$\mathrm{NA} = n \sin \alpha. \tag{13.1}$$

A high numerical aperture is usually required for SMM because the numerical aperture is proportional to the number of collected photons. The emitted fluorescence from the sample is separated from the excitation beam by a dichroic mirror. Subsequently, the fluorescence is detected by a high-sensitivity electron multiplying CCD (EMCCD). Employing a wide-field fluorescence microscope provides the opportunity to detect many single molecules in parallel and allows data acquisition with frame rates up to 100 Hz. Compared to a confocal microscope, the wide-field microscope is more prone to out-of-focus fluorescence. This problem can be reduced by using thin samples with thicknesses of only a few hundred nanometers or less. From the fluorescence image of the single molecules, their position can be determined. Thus, the intensity distribution of a single molecule can be described as a two-dimensional Gaussian distribution (cf. Figure 13.1b) according to the following relation:

$$f(x, y, A_0, w) = A_0\, e^{-(x-x_0)^2/2w^2}\, e^{-(y-y_0)^2/2w^2}. \tag{13.2}$$

Here, $f(x, y, A_0, w)$ is the recorded intensity, A_0 is the amplitude of the signal, w is width of the Gaussian curve, and x_0, y_0 are the coordinates of the center of the peak. The accuracy of the position, as determined by the standard error of the mean σ, depends on the photon noise (s_i^2/N), the error due to the finite pixel size of the detector $((a^2/12)/N)$, and the background $(8\pi s_i^2 b^2/a^2 N^2)$, which are related to each other by the following relation [11, 16]:

$$\sigma_{x_0, y_0, i} = \sqrt{\frac{s_i^2}{N} + \frac{a^2/12}{N} + \frac{8\pi s_i^2 b^2}{a^2 N^2}}. \tag{13.3}$$

Here, s_i is the standard deviation of the distribution in the x- or y-direction, a is the pixel size of the detector, and b is the standard deviation of the background. The number of detected photons N appears in all three terms and consequently is the main factor that determines the positioning accuracy. Considering that the position accuracy scales with the square root of the number of detected photons, the dyes used in SMM are crucial and they often set the limits in SMM experiments. Therefore, the dyes used should have a long lifetime before photobleaching, that is, being able to emit a high number of photons. In the following studies, derivatives of terrylene diimide (TDI) are mainly used because they have an excellent photostability and emit more than 10^8 photons [17] before they photodegrade.

It is important to distinguish between localization accuracy and resolution. The resolution of the microscope describes the minimum distance at which two objects can still be distinguished. The wide-field microscope obeys Abbe's law, and therefore the resolution is about half the excitation wavelength in the lateral plane and about the excitation wavelength in the axial plane. For excitation with visible light, the lateral resolution is about 250 nm and the axial resolution is approximately 500 nm. To circumvent the resolution problem, the samples are usually prepared with dye concentrations of about $10^{-10} - 10^{-11}$ M. These concentrations ensure that most molecules have a distance to each other much larger than the resolution limit. In contrast, the localization accuracy is a measure of how well a single molecule can be localized by fitting its intensity distribution with a two-dimensional Gaussian curve according to Eq. (13.2). In the following studies, the position accuracy will be crucial because it will determine what kind of features can be detected in the SM tracking experiments.

13.1.2
Analysis of the Single-Molecule Trajectories

After the position of a single molecule has been determined in each frame, the positions are connected frame by frame to give a SM trajectory. Using the SM trajectory, the mean square displacement (MSD) is calculated for all possible time lags according to

$$\langle r(\tau)^2 \rangle = \frac{1}{N}\sum_{i=1}^{N} (r(t+\tau) - r(t))^2. \tag{13.4}$$

Here, r indicates the position of the molecule in the focal plane, that is, $r = [x, y]$, assuming z indicates the out-of-plane position. The z-coordinate is not determined in the experiments because a wide-field microscope has a low sensitivity toward out-of-plane movement. The specific time origin is indicated by t and N is the number of time origins over which the MSD is averaged for each specific time lag τ. The time averaging is indicated by the angle brackets. Thus, we assume an equilibrium system, where all time points in the trajectory are equal. Subsequently, the MSD $\langle r(\tau)^2 \rangle$ is fitted with the Einstein–Smoluchowski equation:

$$\langle r(\tau)^2 \rangle = 2dDt. \tag{13.5}$$

This equation describes normal diffusion in $d = 1$, 2, or 3 dimensions. The number of dimensions is determined by the system being studied. In the case of nanopores, where the SM can only travel along the nanopore, it will be one-dimensional. Mesoporous films will be treated as two-dimensional systems because the molecules can move in the plane of the film. For three-dimensional system, the wide-field microscope produces a two-dimensional projection of the three-dimensional diffusion. Therefore, the number of dimension will be 2 here as well. The diffusion constant D will be obtained from the MSD versus time plot by fitting the data with a line. Here, the slope will indicate the diffusion constant D. The specific cases will be discussed in more detail in the following sections.

13.2
Investigation of the Structure of Mesoporous Silica Employing Single-Molecule Microscopy

13.2.1
Mesoporous Silica

Mesoporous silica is characterized by a high pore volume of up to $1\,cm^3\,g^{-1}$, a large surface area exceeding $1000\,m^2\,g^{-1}$ [18], and can show various morphologies. These properties make mesoporous silica an interesting material for many different applications such as molecular sieves [19, 20], catalysis [21], and drug delivery [22], to mention only some of them. It is synthesized using a precursor solution consisting of tetraethyl orthosilicate (TEOS) (Figure 13.2a) and a structure directing agent such as cetyltrimethylammonium bromide (CTAB), Brij56, or Pluronic F127 (Figure 13.2b).

For functionalization of the mesopores, with an amino, a thiol, or a cyano group, a co-condensation approach is employed, in which functionalized silicate (RTES) is added to the precursor solution [18, 24].

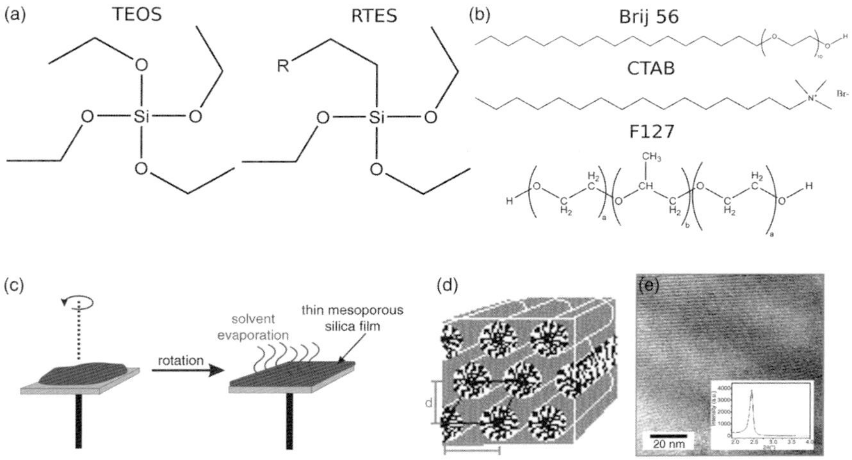

Figure 13.2 (a) Structure of TEOS and functionalized silicate (RTES). R = amino, thiol, or cyano group. (b) Structure directing agents Brij56, CTAB, and Pluronic F127. (c) Schematic of EISA. (d) Structure of a hexagonal mesoporous silica. (e) TEM image of linear and parallel aligned pores of hexagonal mesoporous silica system. Inset shows the XRD pattern. (Images (d) and (e) are taken from Ref. [23]).

The silica films are obtained by evaporation-induced self-assembly (EISA) according to Brinker *et al.* [25]. Here, the precursor solution is spin coated, during which solvent evaporation occurs and the silica molecules self-assemble (Figure 13.2c). The resulting film can exhibit different morphologies such as cubic, lamellar, hexagonal, or a mixture thereof, depending on the concentration of the structure directing agent [25]. Figure 13.2d shows a schematic of a hexagonal mesoporous film. The alignment and shape of the pores is usually determined by transmission electron microscopy (TEM) and the morphology is inferred from the X-ray diffraction (XRD) pattern and the TEM image. A TEM image and the corresponding XRD pattern of a hexagonal mesoporous film with linear, parallel aligned pores are shown in Figure 13.2e and inset.

Besides mesoporous silica films, mesoporous silica nanoparticles can be synthesized as well. Here, the inside of the nanoparticle can have a different functionalization from the surface of the nanoparticle [18, 24]. This bifunctionalization of the nanoparticles opens up the possibility to covalently bind a cargo inside the nanopores and a targeting ligand on the surface of the nanoparticle. Such mesoporous silica nanoparticles are used for drug delivery, a very competitive new field in nanomedicine [26].

In summary, mesoporous silica is a versatile type of material, and we anticipate that it will be utilized in a wide range of applications, thanks to its large pore volume, its large surface area, and its functionalization with various chemical groups.

13.2.2
Combining TEM and SMM for Structure Determination of Mesoporous Silica

To investigate the structure of mesoporous materials, transmission electron microscopy is often employed because its high resolution allows to determine the pore structure of the mesoporous system. However, TEM faces some drawbacks such as invasive sample preparation, a rather small field of view, and access to static information only. Furthermore, it cannot give any information on how a possible guest molecule in the mesoporous system would interact with and diffuse through the structure. In contrast, SMM can provide information about the structure of the nanopores as well as linking the pore structure with the guest molecule's dynamics by observing the movement of fluorescent single molecules inside the structure. Zürner *et al.* have studied mesoporous silica using both TEM and SMM [27].

They synthesized a 100 nm thick hexagonal mesoporous silica film with a pore-to-pore distance of ≈7 nm, in which a fluorescent terrylene diimide derivative and 280 nm polystyrene beads were incorporated. The polystyrene beads were used as markers and allowed the overlay of the TEM images with the SMM image (Figure 13.3a). The TEM image was fast Fourier transformed (FFT) to show the orientation of the nanopores more clearly. The FFT directors (black bars in Figure 13.3a) indicate the orientation and ordering of the structure. For the SMM experiments, the position of the fluorescent molecules was determined in each frame by fitting the single-molecule intensity distribution to a two-dimensional

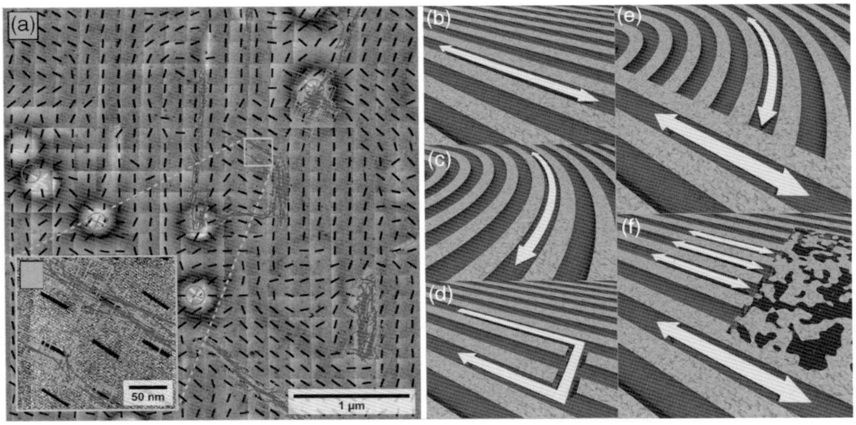

Figure 13.3 (a) Overlay of TEM image (gray) and FFT directors (black bars) with single-molecule trajectories (dark blue). The polystyrene beads, used for the overlay, are indicated by the yellow (TEM) and red (SMM) crosses. The light blue boxes show the positioning error of the SM trajectories. (b–f) Possible movement patterns of a single molecule in various structural features found in the hexagonal mesoporous films. Images are taken from Ref. [27]. (For a color version of this figure, please see the color plate at the end of this book.)

Gaussian curve (cf. Section 13.1.1). Subsequently, the SM trajectory is obtained by connecting the SM position frame by frame (dark blue traces in Figure 13.3a). From the SM trajectories, the pore structure could be inferred. The positioning error is indicated by the light blue boxes in the inset of Figure 13.3a. In these experiments, the SM could be located to an ensemble of 5–10 channels as can be determined from the positioning accuracy. By comparing the FFT directors as obtained from the TEM image with the SM trajectories, it can be concluded that the single molecules diffuse along the pores. This confirms the concept that single molecules can be used as molecular beacons to obtain information about the structure of the mesoporous silica equivalent to information obtained by TEM. In addition, the SM trajectories show that in areas where the structure is less ordered, the single molecule moves in a random way as well. From the SM trajectories, it can be concluded that channels are linear and curved (cf. Figure 13.3b and c) as well as have dead ends and defects in the channel walls (cf. Figure 13.3d–f). The proof of principle that SMM can elucidate structural features of the mesoporous silica film, equivalent to TEM, offers new possibilities for employing SMM. Due to the possibility to obtain a structural characterization of the mesopores relatively easy and in a noninvasive way with SMM, it can be used as an analytical tool to optimize the synthesis of mesoporous silica, as will be applied in the following section.

13.2.3 Applications of SMM to Improve the Synthesis of Mesoporous Systems

Mesoporous films should ideally consist of linear and parallel aligned nanopores in domains, extending over tens of microns. Moreover, it is crucial that molecules can travel from one side of the mesoporous material to the other side. Such a well-ordered mesoporous material would be useful in many different applications, for example, membranes for molecular sieving, electrophoresis, catalysis, and others. However, the current synthesis approaches can only produce mesoporous systems with domain sizes of a few hundred nanometers. In order to improve the mesopore synthesis, it is necessary to get fast feedback about the mesoporous structure as made possible by SMM. Recently, Ruehle *et al.* reported that they improved the synthesis of the mesoporous silica in such a way that they were able to synthesize linear and parallel aligned nanopores in domains, which are hundreds of microns large [28]. Instead of using the EISA approach as described in Section 13.2.1, a capillary flow approach using a polydimethylsiloxane (PDMS) stamp with microgrooves was employed (Figure 13.4a). For this approach, the PDMS stamp was placed on a glass cover slip and the precursor solution, consisting of TEOS:F127:ethanol:HCl:water, was placed on both open sides of the PDMS stamp. Subsequently, the precursor solution was pulled into the microgrooves of the stamp by capillary forces. After 2 days of drying, the solution was solidified and the PDMS stamp was removed. The structure of the dried film was investigated with SMM. Here, the fluorescent single molecules (TDI), which were incorporated into the mesoporous film during the synthesis, are located in each individual frame by fitting their intensity distribution with a two-dimensional Gaussian curve (cf. Section 13.1.1). Subsequently, the SM

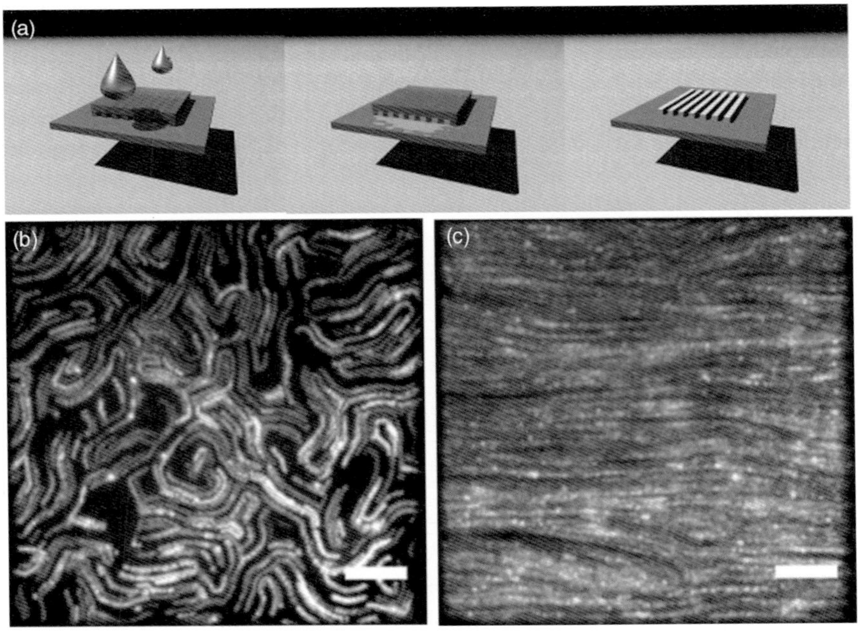

Figure 13.4 (a) Schematic presentation of capillary flow approach for the synthesis of linear and parallel aligned nanopores. Precursor solution is placed on a glass cover slip close to both open sides of PDMS stamp and pulled into microgrooves of PDMS stamp by capillary forces. (b) Trajectories of single molecules, shown in a maximum projection technique, for a mesoporous material with nonaligned nanopores. (c) Maximum projection of single molecules moving through highly oriented mesopores. The scale bar is 5 μm. (Images are adapted from Ref. [28]).

movement from frame to frame is followed to obtain a SM trajectory. To elucidate the mesoporous structure, the SM trajectory is usually time averaged. However, this method overemphasizes nonmoving molecules over moving molecules. In order to remove this bias toward static molecules, the trajectories can be mapped with a maximum projection technique instead. Here, for each SM position only the maximum intensity in the entire trajectory is retained. Figure 13.4b and c shows the maximum projections of a mesoporous material synthesized with the EISA approach and with the capillary flow approach, respectively. While the EISA approach gives nonlinear randomly oriented mesopores (cf. Figure 13.4b), mesoporous silica films synthesized via the capillary flow approach are characterized by linear and parallel aligned channels in micron size domains (cf. Figure 13.4c).

The maximum projection can give important insights about the mesoporous structure, but it is less well suited to allow conclusions about the abundance of defects in the walls of the nanopores as well as dead ends of the nanopores. To obtain this information, it is necessary to track the guest molecules throughout the entire movie and analyze the trajectories in detail.

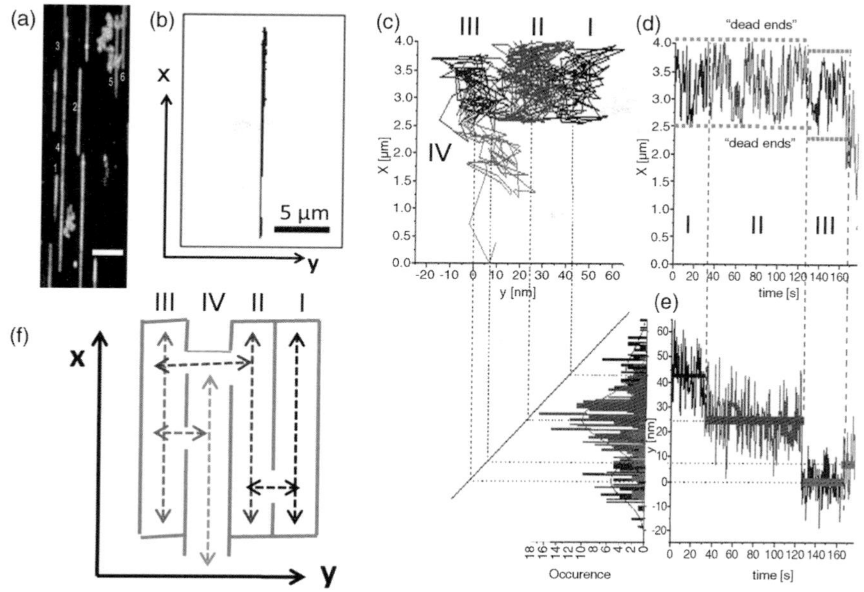

Figure 13.5 (a) Maximum projections of 10 molecules moving through highly oriented mesopores. (b) Trajectory of molecule 1 from (a). (c) Projection of x- and y-coordinate of molecule 1 moving through multiple pores. Phases I, II, III, and IV indicate different time windows. The time progresses from phase I to phase IV. The trajectories are zoomed into the y-axis. The y-axis has a nanometer resolution while the x-axis has a micron resolution. The molecule was followed for 475 frames with a frame rate of 2.5 Hz. (d) The time dependence of the x-coordinates of the single molecule, indicating dead ends in the nanopores. (e) The time dependence of the y-coordinate shows the movement of the single molecule through four different channels by using defects in the walls of the nanopores. (f) Channel structure as mapped by molecule 1. (Images are adapted from Ref. [28]).

Figure 13.5a shows the maximum projections of molecules moving inside a mesoporous system, which has been synthesized with the capillary flow approach. The scale bar is 5 μm in this figure and it emphasizes the large domain size that was achieved by employing this novel synthesis approach. The SM trajectories permit a more in-depth analysis of the movement of the single molecules in the channels and can provide further information about the mesoporous structure. As an example, the movement of molecule 1 from Figure 13.5a will be discussed in more detail in the following. The trajectory of molecule 1 is shown in Figure 13.5b. This trajectory is enlarged in Figure 13.5c and extends up to 4 μm in the x-direction, which is chosen as the channel direction. The y-axis is enlarged into the nanometer range. It clearly indicates four different distinguishable regions in the y-direction, labeled I–IV. From this x–y plot of the trajectory, two separate graphs $x(t)$ and $y(t)$ can be obtained (cf. Figure 13.5d and e). The time dependence of the y-coordinate in Figure 13.5e demonstrates that the molecule jumps in the y-direction in four consecutive steps. The jumps can be correlated to the molecule switching between neighboring channels. The corresponding channel structure is shown in Figure 13.5f. With this figure the jumps of the molecule from region I to region IV can be clearly interpreted as switches between neighboring channels. So the molecule starts in channel I and then jumps after 31.6 s through a defect into channel II, where it then moves for further 93.9 s. Then it switches again to another adjacent channel III and stays there for another 39.2 s before finally moving into channel IV. Additional information can be obtained from the time dependence of the x-coordinate, which is oriented along the channels. This plot shows that the molecule oscillates between two dead ends for 165 s over a range of 13.5 μm. Interestingly, the dead ends for channel II and III are almost at the same position as for channel I. In the case of channel IV, the dead ends are located at different positions and the molecule oscillates in a range of 2.5 μm. Thus, the molecule reaches the full extension of the trajectory in the x-direction not in one channel, but by switching into neighboring channels through defects in the pore walls.

In conclusion, a molecule in such a system mainly moves up and down in a linear channel. Occasionally, the channels might be blocked by dead ends. To circumvent these dead ends, the molecule can switch into an adjacent channel and might then move further through the mesoporous system. Hence, the defects in the walls of the nanopores are crucial to enable the molecules to move through the entire mesoporous structure and are therefore beneficial.

13.3 Investigation of the Diffusion of Guest Molecules in Mesoporous Systems

13.3.1 A More Detailed Look into the Diffusional Dynamics of Guest Molecules in Nanopores

Besides the apparent structural features of the nanopores, measuring the diffusional dynamics of a single molecule over a longer timescale can give important insights into the interactions of the guest molecule with the nanopore walls and the structure of the nanopores as well. Kirstein *et al.* observed the diffusion of a single TDI molecule in mesoporous silica for 755 frames with a frame rate of 10 Hz [12]. In the course of the measurement, the molecule moved through three different domains in the mesoporous film (cf. Figure 13.6a).

Here, the average channel orientations of domains A and B are perpendicular to each other and the channels of domains B and C are connected by a turn. Analyzing the channel systems 1, 2, and 3 in domain B in more detail provides some interesting insights into the diffusional dynamics of the TDI molecule. It elucidates a more detailed picture about the nanopore structure, besides the orientation of the pores and the presence of defects in the walls of the nanopore as done in the previous sections. For this analysis, the squared displacement of the single molecule $r^2(t) = (\mathbf{x}(\tau + t) - \mathbf{x}(\tau))^2$ is calculated. It is assumed that every point in the trajectory is equal and $\mathbf{x}(\tau)$ gives the molecule's x-, y-, and z-coordinate at time origin τ

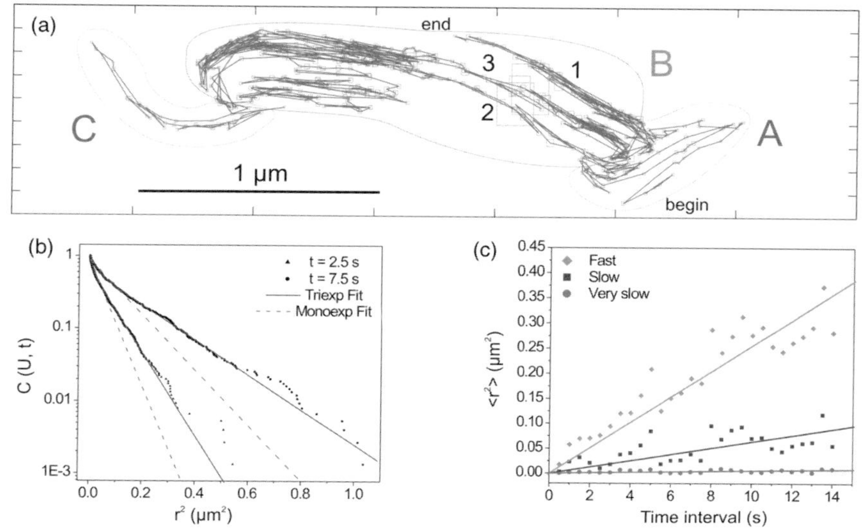

Figure 13.6 (a) Trajectory of a TDI molecule diffusing through three different domains, labeled A, B, and C. Domain B contains the three channel systems 1, 2, and 3. (b) Cumulative probability $C(U, t)$ for the mean squared displacement for two time lags. The dashed lines are single exponential fits and the solid lines are triexponential fits. (c) Plot of characteristic mean squared displacement $\langle r(t)^2 \rangle$ against time lag t. The solid lines show the fit according to $\langle r(t)^2 \rangle = 2Dt$. (Images are taken from Ref. [12] and the movie to (a) can be found as movie 8 in the supporting information of Ref. [12]).

and t is the time lag. Moreover, we assume isotropic diffusion, which allows us to characterize three-dimensional diffusion by measuring the diffusion only in one plane. For a specific time lag t, the squared displacements $r^2(t)$ are ranked in a descending order with a rank j. Dividing the rank j of each displacement $r^2(t)$ by the total number of squared displacements for this specific time lag, denoted N, that is, j/N, gives the cumulative probability $C(U, t)$ for the specific time lag t. The cumulative probability of the squared displacement $C(U, t)$ is a measure for the probability that a squared displacement $r^2(t)$ is smaller than a squared displacement U for a time lag t. Figure 13.6b shows $C(U, t)$ for two different time lags, $t = 2.5$ s and $t = 7.5$ s, calculated from the trajectories in the channel systems 1, 2, and 3. For random diffusion, $C(U, t)$ can be described by a single exponential function according to Eq. (13.6) with $n = 1$:

$$C(U, t) = \sum_{i=1}^{n} A_i \, e^{U/\langle r_i(t)^2 \rangle}. \tag{13.6}$$

Here, $\langle r(t)^2 \rangle$ describes a characteristic mean squared displacement for the time lag t. The dashed lines in Figure 13.6b indicate a single exponential fit. It appears that this fit does not describe $C(U, t)$ in a satisfying way. In contrast, fitting $C(U, t)$ with a triexponential fit (Eq. (13.6), $n = 3$) produced a satisfactory fit (solid lines in Figure 13.6b). The necessity of a triexponential fit indicates that the diffusion of the molecule is described by at least three different diffusive behaviors. By calculating $C(U, t)$ for all possible time lags t and fitting it with a triexponential function, three characteristic mean squared displacements $\langle r(t)^2 \rangle$ for each time lag t can be obtained (cf. Figure 13.6c). Fitting the mean squared displacement to the one-dimensional Einstein–Smoluchowski equation, $\langle r(t)^2 \rangle = 2Dt$, gives three different diffusion constants $D = 1.3 \times 10^{-2}$, 3.2×10^{-3}, and $2.8 \times 10^{-4}\ \mu m^2\ s^{-1}$. Further analysis reveals that the three different diffusion regimes are not correlated to a spatial region. Instead, the environment within one channel changes along the pathway of the single molecule. These structural heterogeneities in a single channel manifest themselves in a continuous change of the single molecule's mode of motion between at least three different diffusional constants. In principle, it is possible to obtain the diffusion constants from the velocity autocorrelation function as well. However, it appears that for an accurate calculation of the velocity autocorrelation function more data points are required than for the calculation of the MSD. Hence, for most single-molecule experiments, the calculation of the MSD is preferred over the velocity autocorrelation function to determine the diffusion constant.

In summary, these SMM experiments demonstrate how a detailed analysis of single-molecule trajectories and single-molecule dynamics can shed light on the structural features of the mesopores, which cannot be obtained by standard methods such as TEM and XRD.

13.3.2
Modification of the Flow Medium in the Nanopores and Its Influence on Probe Diffusion

As discussed in the previous sections, SMM is extremely useful to investigate the dynamics of the probe molecule inside the mesoporous system. In particular, the single-molecule dynamics are influenced by the interactions between the guest molecule and the host. To investigate these interactions, Jung *et al.* followed the diffusion of TDI molecules in mesoporous silica [23]. The interactions depend on the properties of the probe and the nanopores as well as the flow medium inside the nanopores. In the following experiments, the flow medium is altered from water to chloroform by changing the atmosphere, surrounding the mesoporous system, from an air atmosphere with 40% relative humidity (RH) to a chloroform atmosphere. In case of TDI in silica nanopores with a CTAB template and water as a lubricant, it was observed that the TDI molecule exhibits barely any diffusional movement. The time-averaged image of TDI molecules, which were tracked for 1000 frames, shows that the TDI molecules did not move significantly (cf. Figure 13.7a).

This lack of dynamics can be explained by strong interactions between the electron lone pairs of the carbonyl groups of the TDI and the positively charged CTAB head groups (cf. Figure 13.7b). These interactions result in sticking of the TDI molecules onto the nanopore wall. Changing the atmosphere from 40% RH air to chloroform significantly alters the diffusive dynamics of the TDI [23]. Figure 13.7c

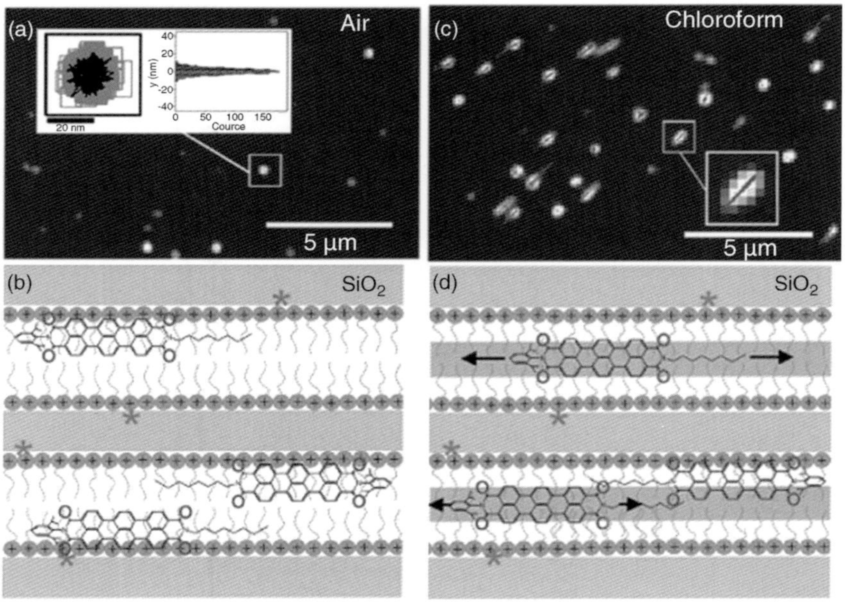

Figure 13.7 (a) Time-averaged image of 1000 frames of TDI in nanopores under an air atmosphere. The fluorescence appears circular, indicating no movement of TDI on this timescale. *Inset*: Zoom in on one molecule, confirming only very small positional changes of the molecule, which can be assigned to the localization accuracy. (b) Schematic of interactions between the TDI and the nanopore in an air atmosphere. The TDI is immobilized. (c) Time-averaged image of single molecules in nanopores under a chloroform atmosphere. A linear movement of the molecules is apparent. (d) The solvation of the TDI by chloroform alters the interactions between the TDI and the nanopore wall and thereby makes the TDI mobile.

shows the time-averaged image of TDI molecules, which has been tracked for 1000 frames. The circular fluorescence pattern with water as flow medium has changed to a linear pattern with chloroform as flow medium, indicating movement of TDI. In a chloroform atmosphere, the TDI is surrounded by a layer of chloroform, which blocks the interactions between the electron lone pairs on the TDI carbonyl groups and the positively charged head groups of the CTAB molecules (Figure 13.7d). Hence, the TDI is not adhered to the nanopore wall anymore and can move through the nanopore.

In conclusion, changing the flow medium in the nanopores can have significant effects on the observed dynamics of the probe. This is due to the fact that the probe interacts with the walls of the nanopores, and the lubricant can influence the interactions between the probe and the nanopore. These studies demonstrate that SMM can contribute significant insights into the diffusive behavior of probes in confined spaces, which other techniques such as TEM and XRD cannot offer because they do not provide any information about the guest molecule and its dynamics. Therefore, SMM can help to tune the interactions between guest molecules and host system, which is important in order to tune the guest molecule's diffusion constant as well as for making it possible to load various cargoes into the nanopores as will be discussed in the next section.

13.3.3
Loading of Cargoes into Mesopores: A Step toward Drug Delivery Applications

The tuning of the interactions between the guest molecule and the nanopores is essential for future applications of these mesoporous materials, for example, in drug delivery applications [22]. Due to their mesoporous structure, they exhibit a large free volume, which should make it possible to fill them with a high load of cargo. In order to introduce larger molecules than fluorophores, such as DNA, into the mesopores, it is necessary to remove the template from the mesopores (Figure 13.8a).

In addition, the interactions between the possible cargo and the host have to be tuned so that they favor uptake of the cargo by the mesopores. In order to achieve favorable interactions, the functionalization of the pore wall can be modified and the flow medium inside the pore can be changed as has been discussed in the previous section. Feil *et al.* have synthesized mesopores with a diameter of 4 nm and

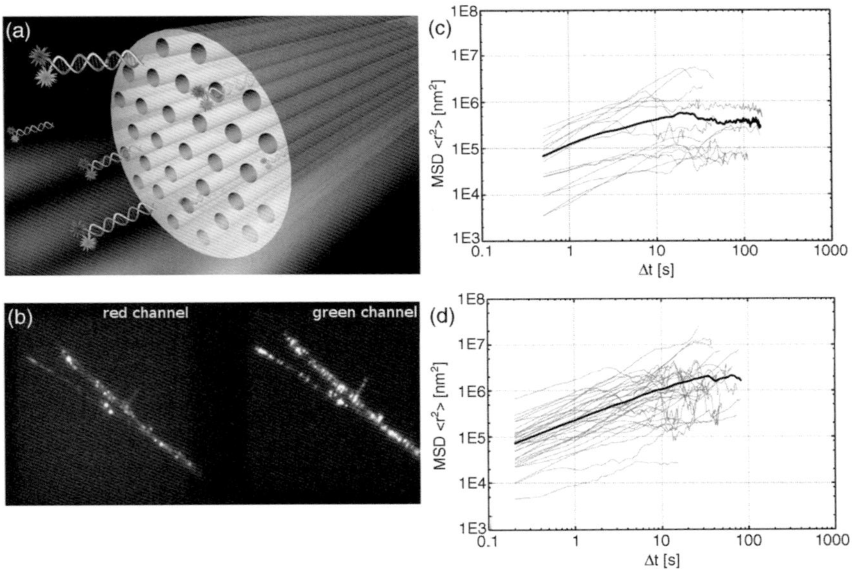

Figure 13.8 (a) Template-free functionalized silica nanopores loaded with 15 bp dsDNA. The one end of the DNA is labeled with cy3 and cy5 (light and dark stars) to allow FRET measurements. (b) FRET signal of dsDNA in the red and green channels, proving the integrity of dsDNA inside the nanopores. (c) MSD of dsDNA inside the mesopores, water serving as the flow medium. (d) MSD of TDI inside the mesopores with a chloroform/water mixture serving as the flow medium.

pore walls, which are bifunctionalized with amino and nitrile-propyl groups [29]. Loading of double-stranded DNA (dsDNA) into the bifunctionalized mesopores under a saturated water atmosphere at room temperature took about a week. The importance of the wall functionalization becomes obvious, considering that loading of dsDNA into unfunctionalized or monofunctionalized mesopores with nitrile-propyl groups was not achieved on this timescale. Furthermore, only under a high-humidity atmosphere movement of dsDNA inside the mesopores was detectable. Under low-humidity conditions, movement of dsDNA in the mesopores ceased. Besides the loading of the dsDNA into the pores, it is important that the dsDNA stays intact inside the mesopores. The integrity of the dsDNA was checked by measuring the Förster resonance energy transfer (FRET) signal between the cy3 and cy5 dyes, which are both attached to one end of the dsDNA (cf. Figure 13.8a). The presence of the FRET signal proves the integrity of the dsDNA inside the mesoporous system (cf. Figure 13.8b). For investigation of the dynamics inside the pores, the dsDNA was tracked and their mean square displacements were obtained from the trajectories (Figure 13.8c). Similarly, monofunctionalized nitro-propyl pores were loaded with TDI and the diffusion inside the pores with a chloroform/water mixture as lubricant was investigated (Figure 13.8d). In both cases, the top of the film was coated with a gold layer to quench any molecule outside the mesoporous system. Thereby, it was ensured that only dsDNA and TDI molecules were observed, which are inside the mesopores. Both MSD plots show a linear regime in the early time regime and a leveling off of the MSD in the later time regime. The diffusion constant for dsDNA and TDI is obtained by fitting the linear part of the MSD with the one-dimensional form of the Einstein–Smoluchowski equation according to $\mathrm{MSD} = 2Dt$. It appears that the TDI moves about twice as fast as dsDNA in the mesopores, giving a diffusion constant for TDI of $D = 1.1 \times 10^5\ \mathrm{nm^2\,s^{-1}}$ compared to a diffusion constant of $D = 5.9 \times 10^4\ \mathrm{nm^2\,s^{-1}}$ for dsDNA. This difference in diffusional dynamics can be attributed to the bigger size of dsDNA compared to TDI as well as the different interactions between the guest molecule and the host originating from the different flow medium in the nanopores and the different functionalization of the nanopore walls in the experiments.

In summary, this work supports the concept that tuning the interactions between the cargo and the nanopores, by functionalization of the pore walls and choosing a particular flow medium inside the pores, is crucial for loading cargo into the mesoporous system. This is, however, an important step toward the application of mesoporous silica for, for example, drug delivery purposes.

13.4
A Test of the Ergodic Theorem by Employing Single-Molecule Microscopy

So far single-molecule data have only been used to infer properties of the host system or to obtain information about the single molecule itself. In the following, we will show how single-molecule techniques can help to answer open questions in physics, which cannot be answered by ensemble methods alone. In particular, we will demonstrate how SMM was useful to verify experimentally the ergodic theorem [30]. The ergodic theorem states that the time average of a property of a single molecule is equivalent to the ensemble average of this property over many molecules. We will study the diffusion coefficient as an example for a property, where the ensemble value and the time-averaged value should correspond to each other. An often used method to measure ensemble diffusion constants is pulsed field gradient (PFG) NMR diffusometry. For the determination of the time-averaged single-molecule diffusion constants, we will employ wide-field epifluorescence SMM. To confirm the ergodic theorem through experiments, it is essential to study the same system under the same conditions with both an ensemble and a single-molecule method. Fulfilling this condition required to overcome two major obstacles. First, for NMR experiments a high concentration of the investigated molecule is required while for SMM experiments a very low concentration is required. Therefore, the concentration of the studied molecules must be varied for both experiments in such a way that the used concentrations are comparable. Second, for NMR spectroscopy the diffusion constants of the studied molecules should be high while SMM requires low diffusion constants. Feil *et al.* were able to bring these contradictory requirements into agreement by choosing the fluorophore Atto532 as the studied species and observing it in a nanoporous glass [30]. The tuning of the fluorophore's diffusion constant, to a value detectable by both methods, was achieved by choosing a nanoporous silica glass with a mean pore diameter of 3 nm and methanol as the flow medium. In the case of the PFG NMR experiments, the diffusion of Atto532 inside and outside the nanoporous glass was measured for five different concentrations. Figure 13.9a shows the normalized spin-echo attenuation and the biexponential fit, from which two diffusion constants were obtained. The faster diffusion constant originates from the molecules moving outside the nanoporous glass while the slower diffusion constant is caused by the molecules moving inside. For SMM experiments, molecules moving inside the nanoporous glass were localized frame by frame and SM trajectories were calculated by connecting the SM positions. From the SM trajectories, the MSD was calculated and the diffusion constant was obtained by fitting the MSD with the two-dimensional form of the Einstein–Smoluchowski equation according to $\mathrm{MSD} = 4Dt$, as described in Section 13.1.2. In total, the movement of 170 molecules was observed for three different dye concentrations. Figure 13.9b and c shows the MSD of 70 dye molecules and the cumulative probability of the diffusion constants, calculated analogous to the method described in Section 13.3.1, for a dye concentration of $3.2 \times 10^{-11}\,\mathrm{mol\,l^{-1}}$. The cumulative probability is fitted with a log-normal distribution (solid line in Figure 13.9c), indicating heterogeneous dynamics.

Comparing the mean diffusion constant from the PFG NMR experiments with the mean diffusion constant from the SMM experiments shows that both techniques deliver the same results within the range of experimental accuracy (cf. Figure 13.9d). The agreement of the diffusion constants obtained by ensemble measurements and single-molecule measurements demonstrates the validity of the ergodic theorem.

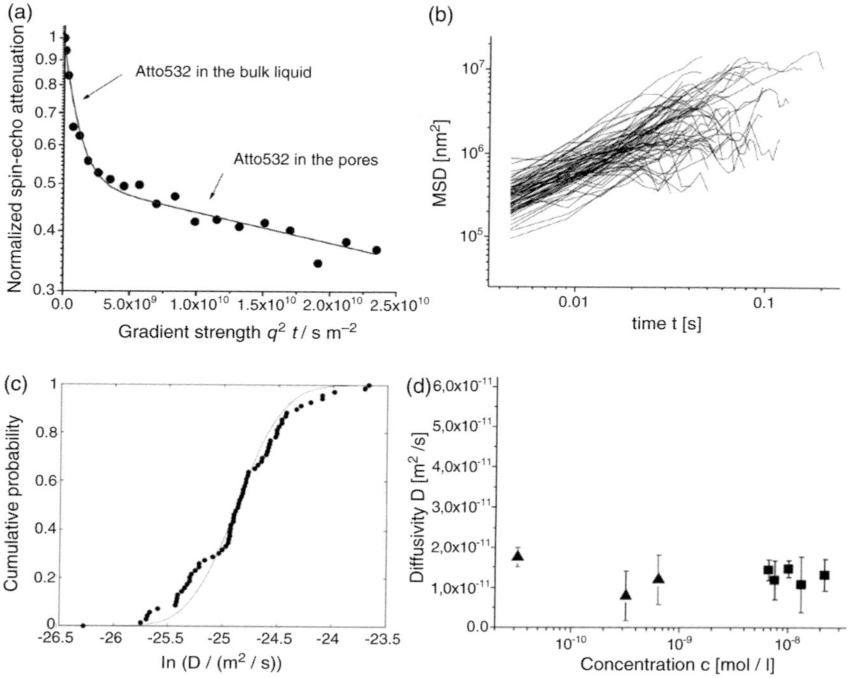

Figure 13.9 (a) PFG NMR spin-echo diffusion attenuation by dye ensemble. The data were fitted with a biexponential function. (b) MSD of 70 Atto532 molecules, measured by SMM. (c) Cumulative probability of logarithm of diffusion constants of the 70 single molecules. The solid line is a fit to a log-normal distribution. (d) Concentration dependence of mean diffusion constants from ensemble measurements via PFG NMR spectroscopy (squares) and single-molecule microscopy (triangles).

To sum up, by measuring the diffusion constant of Atto532 in a nanoporous glass in the same system under the same conditions using PFG NMR spectroscopy and SMM for the ensemble and single-molecule data, respectively, it was possible to verify experimentally the ergodic theorem.

13.5 Single-Particle Tracking in Biological Systems

SMM has provided many new insights into the structure of mesopores and the dynamics of confined diffusion. The useful properties of SMM such as measuring single-molecule dynamics and to avoid ensemble averaging as well as elucidating the diffusion of single molecules in confined and crowded spaces can help to investigate and solve many problems in the area of biophysics as well. In this way, SMM experiments have produced many important contributions in areas such as the movement of myosin V along actin filaments [11], the infection pathway of viruses [10], the protein movement in bacteria [31], and single-molecule gene expression in bacteria [32]. For a recent review, see Ref. [9]. In the field of drug delivery, single-particle microscopy can provide important insights regarding the question of nanoparticle uptake by the cell. In particular, it is interesting to investigate questions such as how do nanoparticles attach to the cell, how does their uptake occur, what happens with the nanoparticles after their uptake by the cell, and are these mechanisms different, for example, for nanoparticles with and without a receptor ligand. To study nanoparticle uptake by cells, flow cytometry is usually employed. However, flow cytometry is an ensemble method and can only answer the question how many particles have been uptaken by the cell. It cannot provide information about the mechanism of the uptake and the dynamics of the nanoparticle in the cell. de Bruin *et al.* have studied the uptake of nanoparticles by cells employing single-particle wide-field epifluorescence microscopy [13]. The uptake of nanoparticles with and without targeting ligands was studied. The nanoparticles were made of linear polyethylenimine (LPEI), so-called polyplexes. They have attracted significant attention because they can be fused with DNA and have shown good transfection efficacy. As a targeting ligand, the epidermal growth factor (EGF) was used, which is overexpressed on many types of tumor cells. The polyplexes were labeled with cy5 to detect them by their fluorescence. In these experiments, HuH7 cells were used because they express a high number of EGF receptors on their cell surface. Figure 13.10a shows the trajectory of a polyplex with EGF starting with the attachment of the polyplex at the cell.

In the beginning of the trajectory (labeled phase I in the inset of Figure 13.10a), the polyplex moves in a very limited range. After about 160 s, the dynamics of the polyplex seems to increase (labeled phase II in the inset of Figure 13.10a) and after 240 s there appears directed motion of the polyplex (cf. Figure 13.10a, phase III) until the end of the measurement after 270 s. The motion in these three phases can be characterized as slow directed motion, normal and anomalous diffusion, and finally fast directed motion along the microtubule, respectively. The length of the different phases differs from particle to particle. However, the existence of the three phases as well as their characteristics is similar for all particles investigated. Plots of the MSD for all three phases for 10 nanoparticles with EGF are shown in Figure 13.10c–e. Comparing these to the MSD of normal, directed, anomalous, and confined diffusion (cf. Figure 13.10b) provides some interesting insights into the characteristics of nanoparticle uptake by cells. Normal two-dimensional diffusion is characterized by a linear behavior of the MSD versus time, that is, $\alpha = 1$ when $\text{MSD} = 4Dt^{\alpha}$. Directed diffusion manifests itself by a quadratic dependence of the MSD with time, and anomalous as well as confined diffusion exhibits a time dependence with $\alpha < 1$. Comparing the MSD of phase I (cf. Figure 13.10c) with the four types of diffusion (cf. Figure 13.10b), it can be concluded that phase I exhibits directed motion and consequently the single-particle MSD can be fitted with $\text{MSD} = v^2t^2 + 4Dt$, where v indicates the velocity, D the diffusion constant, and t the time. From the 10 trajectories shown in Figure 13.10c, the average velocity

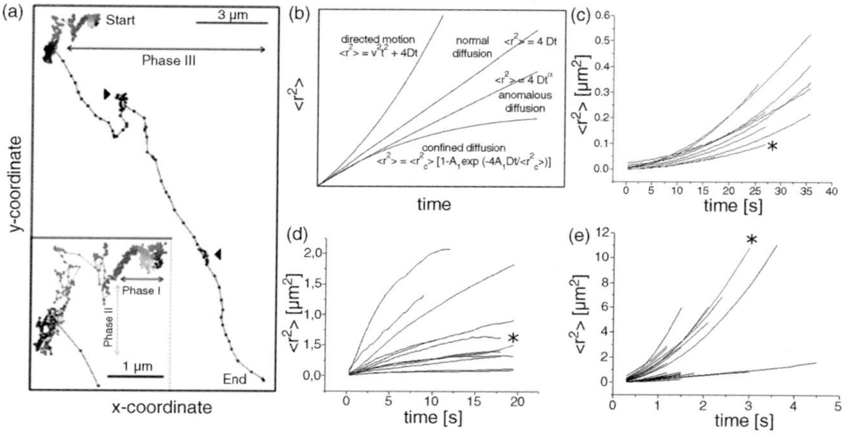

Figure 13.10 (a) Trajectory of polyplex with EGF, attached to HuH7 cells. The inset shows the three phases of movement. The first phase ranges from the time origin to 160 s, the second until 240 s, followed by the third phase. (b) Simulated MSD plots for directed, normal, anomalous, and confined diffusion. The relation between the MSD and the time is given by the equations. (c) MSD plots for 10 typical polyplexes exhibiting phase I dynamics. The plot labeled with an asterisk is the polyplex from (a). (d) MSD plots for polyplexes in phase II indicating anomalous and confined diffusion. (e) MSD plots for particles in phase III. Similar to (c), the MSD plots show a quadratic time dependence, indicating directed diffusion. But the values of the MSDs are significantly higher than those in phase I. (Images are adapted from Ref. [13]).

$v_I = 0.015 \pm 0.003\,\mu m\,s^{-1}$ and the average diffusion constant $D_I = 4 \times 10^{-4} \pm 4 \times 10^{-4}\,\mu m^2\,s^{-1}$ are obtained. The large standard deviation of the diffusion constant originates from the significantly different single-particle MSDs. The large standard deviation emphasizes that the overall characteristics in terms of the three phases are particle independent, while the specific characteristics of the particle dynamics such as MSD, velocity, and length of the various phases are particle dependent. The MSDs of phase II (cf. Figure 13.10d) shows normal diffusion in the short time regime and anomalous (subdiffusive) as well as confined diffusion in the late time regime. This indicates that the molecule diffuses freely in the beginning of phase II, then senses obstacles, and finally becomes corralled at MSD values between 0.3 and 2.0 μm. Only after the molecule explores the environment for a long time, the MSD changes from normal diffusion to anomalous and confined diffusion. Because the confinement is only experienced at longer times, molecules having shorter phase II durations do not show confined diffusion (cf. Figure 13.10d). The MSDs of phase III (cf. Figure 13.10e) show directed motion similar to phase I. However, both phases differ significantly in terms of the magnitude of the MSDs. The mean velocity in phase III is $v_{III} = 0.7 \pm 0.4\,\mu m\,s^{-1}$ and the mean diffusion constant is $D_{III} = 0.1 \pm 0.1\,\mu m^2\,s^{-1}$ as obtained from the 10 trajectories shown in Figure 13.10e. Thus, it can be concluded that the mechanism for the polyplex' motion in phases I and III must differ. To investigate the underlying reason for the different dynamics in the three phases, various experiments were performed. In the first step, a membrane-impermeable fluorescence quencher, trypan blue, was added to the cell medium in order to distinguish nanoparticles, which are inside the cell and outside the cell. The fluorescence of nanoparticles inside the cell will not be quenched, whereas nanoparticles outside the cell will be quenched by the presence of trypan blue. Tracking the polyplexes revealed that most particles exhibiting phase I dynamics were quenched, while no particles exhibiting phase II and phase III were quenched. This leads to the conclusion that particles in phases II and III are internalized while particles in phase I are predominantly not internalized. In addition, the quenching experiments show that internalized particles are not recycled back to the surface of the cell or exocytosed because no particle, which exhibits phase II or III behavior, was quenched any time during the course of the experiment. To elucidate the underlying mechanism of phases II and III, the cells were treated with nocodazole. Nocodazole is a microtubule-disrupting drug. It is assumed that the directed motion of phase III is due to the transport of the particles by motor proteins along microtubule. As a confirmation of this hypothesis, the movement of particles, exhibiting phase III dynamics, ceased upon addition of nocodazole.

While the results presented above were for nanoparticles with EGF, studies using particles without EGF showed mainly the same results. Independent of whether the particles have the EGF ligand or not, the appearance of all three phases was detected. Here, the length of phases II and III is about equal. The particles with and without EGF attached differ, however, only in the length of phase I, that is, the time at which the particles are, for example, on the membrane surface and not internalized by the cell. Phase I is typically only a few minutes for particles with EGF and typically more than 30 min for particles without EGF. This is in line with expectations, considering that the EGF ligand targets the EGF receptor on the cell surface. Consequently, the presence of the EGF ligand on the nanoparticle should result in a faster attachment and receptor-mediated endocytosis of the nanoparticle. This will result in a faster uptake of nanoparticles with EGF compared to nanoparticles without EGF.

In conclusion, single-particle tracking is a useful tool in the field of drug delivery because it helps to elucidate the dynamics and mechanisms of nanoparticle uptake. SMM revealed three dynamic phases exhibiting directed, normal, anomalous, and confined diffusion, during nanoparticle uptake by the cell.

13.6 Conclusion and Outlook

Single-molecule microscopy has proven its usefulness in many different fields, by providing results that cannot be obtained by ensemble methods. This has led to many new insights into the structure of mesoporous silica, the improvement of the synthesis of the mesopores itself, and the loading of various cargoes into the mesopores. Furthermore, it has shed light on the dynamics of single molecules/particles in confined systems and demonstrated the presence of normal, anomalous (subdiffusive), and directional diffusion in both physical and biological systems. In addition, with the help of SMM it was possible to verify the ergodic theorem, experimentally.

In the future, we expect that due to further improvements of technologies as well as fluorophore's properties, SMM will reveal many new insights into various fields of biology, chemistry, and physics.

Acknowledgments

The authors are very grateful to all coworkers and collaborators named in their publications. Special thanks go to the collaborators Thomas Bein and Ernst Wagner both at LMU Munich, and Jörg Kärger at University of Leipzig. These studies were supported by the Excellence Clusters Nanosystems Initiative Munich (NIM), Center for Integrated Protein Science Munich (CIPSM), and the SPP 1313 grant of the DFG.

References

1 Moerner, W.E. and Kador, L. (1989) Optical detection and spectroscopy of single molecules in a solid. *Phys. Rev. Lett.*, **62**, 2535–2538.

2 Orrit, M. and Bernard, J. (1990) Single pentacene molecules detected by fluorescence excitation in a *p*-terphenyl crystal. *Phys. Rev. Lett.*, **65**, 2716–2719.

3 Lewis, R., Dürr, H., Hopfner, K.P., and Michaelis, J. (2008) Conformational changes of a Swi2/Snf2 ATPase during its mechano-chemical cycle. *Nucleic Acids Res.*, **36** (6), 1881–1890.

4 Mapa, K., Sikor, M., Kudryavtsev, V., Waegemann, K., Kalinin, S., Seidel, C.A., Neupert, W., Lamb, D.C., and Mokranjac, D. (2010) The conformational dynamics of the mitochondrial Hsp70 chaperone. *Mol. Cell*, **38** (1), 89–100.

5 Lu, H.P., Xun, L., and Xie, X.S. (1998) Single-molecule enzymatic dynamics. *Science*, **282** (5395), 1877–1882.

6 Roeffaers, M.B.J., Sels, B.F., Uji-i, H., De Schryver, F.C., Jacobs, P.A., De Vos, D.E., and Hofkens, J. (2006) Spatially resolved observation of crystal-face-dependent catalysis by single turnover counting. *Nature*, **439** (7076), 572–575.

7 Zondervan, R., Kulzer, F., Berkhout, G.C. G., and Orrit, M. (2007) Local viscosity of supercooled glycerol near T_g probed by rotational diffusion of ensembles and single dye molecules. *Proc. Natl. Acad. Sci. USA*, **104** (31), 12628–12633.

8 Mackowiak, S.A., Leone, L.M., and Kaufman, L.J. (2011) Probe dependence of spatially heterogeneous dynamics in supercooled glycerol as revealed by single molecule microscopy. *Phys. Chem. Chem. Phys.*, **13** (5), 1786–1799.

9 Lord, S.J., and Lee, H.L.D., and Moerner, W.E. (2010) Single-molecule spectroscopy and imaging of biomolecules in living cells. *Anal. Chem.*, **82** (6), 2192–2203.

10 Seisenberger, G., Ried, M.U., Endreß, T., Büning, H., Hallek, M., and Bräuchle, C. (2001) Real-time single-molecule imaging of the infection pathway of an adeno-associated virus. *Science*, **294** (5548), 1929–1932.

11 Yildiz, A., Forkey, J.N., McKinney, S.A., Ha, T., Goldman, Y.E., and Selvin, P.R. (2003) Myosin V walks hand-over-hand: single fluorophore imaging with 1.5-nm localization. *Science*, **300** (5628), 2061–2065.

12 Kirstein, J., Platschek, B., Jung, C., Brown, R., Bein, T., and Bräuchle, C. (2007) Exploration of nanostructured channel systems with single-molecule probes. *Nat. Mater.*, **6** (4), 303–310.

13 de Bruin, K., Ruthardt, N., von Gersdorff, K., Bausinger, R., Wagner, E., Ogris, M., and Bräuchle, C. (2007) Cellular dynamics of EGF receptor-targeted synthetic viruses. *Mol. Ther.*, **15** (7), 1297–1305.

14 Moerner, W.E. and Fromm, D.P. (2003) Methods of single-molecule fluorescence spectroscopy and microscopy. *Rev. Sci. Instrum.*, **74** (8), 3597–3619.

15 Kulzer, F. and Orrit, M. (2004) Single-molecule optics. *Annu. Rev. Phys. Chem.*, **55** (1), 585–611.

16 Thompson, R.E., Larson, D.R., and Webb, W.W. (2002) Precise nanometer localization analysis for individual fluorescent probes. *Biophys. J.*, **82** (5), 2775–2783

17 Jung, C., Müller, B.K., Lamb, D.C., Nolde, F., Müllen K., and Bräuchle, C. (2006) A new photostable terrylene diimide dye for applications in single molecule studies and membrane labeling. *J. Am. Chem. Soc.*, **128** (15), 5283–5291.

18 Cauda, V., Schlossbauer, A., Kecht, J., Zürner, A., and Bein, T. (2009) Multiple core-shell functionalized colloidal mesoporous silica nanoparticles. *J. Am. Chem. Soc.*, **131** (32), 11361–11370.

19 Tanev, P.T. and Pinnavaia, T.J. (1996) Mesoporous silica molecular sieves prepared by ionic and neutral surfactant templating: a comparison of physical properties. *Chem. Mater.*, **8** (8), 2068–2079.

20 de Clippel, F., Harkiolakis, A., Ke, X., Vosch, T., Van Tendeloo, G., Baron, G.V., Jacobs, P. A., Denayer, J.F.M., and Sels, B.F. (2010) Molecular sieve properties of mesoporous silica with intraporous nanocarbon. *Chem. Commun.*, **46** (6), 928–930.

21 Taguchi, A. and Schüth, F. (2005) Ordered mesoporous materials in catalysis. *Microporous Mesoporous Mater.*, **77** (1), 1–45.

22 Slowing, I., Trewyn, B., Giri, S., and Lin, V.Y. (2007) Mesoporous silica nanoparticles for drug delivery and biosensing applications. *Adv. Funct. Mater.*, **17** (8), 1225–1236.

23 Jung, C., Kirstein, J., Platschek, B., Bein, T., Budde, M., Frank, I., Müllen, K., Michaelis, J , and Bräuchle, C. (2008) Diffusion of oriented single molecules with switchable mobility in networks of long unidimensional nanochannels. *J. Am. Chem. Soc.*, **130** (5), 1638–1648.

24 Kecht, J., Schlossbauer, A., and Bein, T. (2008) Selective functionalization of the outer and inner surfaces in mesoporous silica nanoparticles. *Chem. Mater.*, **20** (23), 7207–7214.

25 Brinker, C.J., Lu, Y., Sellinger, A., and Fan, H. (1999) Evaporation-induced self-assembly nanostructures made easy. *Adv. Mater.*, **11** (7), 579–585.

26 Ambrogio, M.W., Thomas, C.R., Zhao, Y. L., Zink, J.I., and Stoddart, J.F. (2011) Mechanized silica nanoparticles: a new frontier in theranostic nanomedicine. *Acc. Chem. Res.*, **44** (10), 903–913.

27 Zürner, A., Kirstein, J., Doblinger, M., Bräuchle, C., and Bein, T. (2007) Visualizing single-molecule diffusion in mesoporous materials. *Nature*, **450** (7170), 705–708.

28 Ruehle, B., Davies, M., Lebold, T., Bräuchle, C., and Bein, T. (2012) Highly oriented mesoporous silica channels synthesized in microgrooves and visualized with single molecule diffusion. *ACS Nano*, **6** (3), 1948–1960.

29 Feil, F., Cauda, V., Bein, T., and Bräuchle, C. (2012) Direct visualization of dye and oligonucleotide diffusion in silica filaments with collinear mesopores. *Nano Lett.* **12** (3), 1354–1361.

30 Feil, F., Naumov, S., Michaelis, J., Valiullin, R., Enke, D., Kärger, J., and Bräuchle, C. (2012) Single-particle and ensemble diffusivities – test of ergodicity. *Angew. Chem., Int. Ed.*, **51** (5), 1152–1155.

31 Kim, S.Y., Gitai, Z., Kinkhabwala, A., Shapiro, L., and Moerner, W.E. (2006) Single molecules of the bacterial actin MreB undergo directed treadmilling motion in *Caulobacter crescentus*. *Proc. Natl. Acad. Sci. USA*, **103** (29), 10929–10934.

32 Yu, J., Xiao, J., Ren, X., Lao, K., and Xie, X. S. (2006) Probing gene expression in live cells, one protein molecule at a time. *Science*, **311** (5767), 1600–1603.

Index

Nonequilibrium Statistical Physics of Small Systems: Fluctuation Relations and Beyond, First Edition.
Edited by Rainer Klages, Wolfram Just, and Christopher Jarzynski.